Advancing Sustainable Science and Technology for a Resilient Future

About the Conference

"Advancing Sustainable Science and Technology for a Resilient Future" is the proceedings book from the First International Conference on Sustainable Societies, Science and Technology (ICSST 2023). This pioneering virtual event took place on July 29th and 30th, 2023, gathering researchers, scientists, engineers, technologists, and industry experts worldwide. With a focus on Science and Technology, the conference aimed to exchange knowledge, present cutting-edge research, and explore solutions for sustainability.

ICSST 2023 addressed the theme of "Advancing Sustainable Science and Technology for a Resilient Future," highlighting the role of scientific and technological advancements in driving sustainable development. Tracks covered renewable energy, clean technologies, environmental science and engineering, sustainable materials and manufacturing, green chemistry, sustainable processes, water and sanitation technology, and smart systems, cities, and urban technology.

The conference fostered in-depth discussions, collaborations, and knowledge sharing among experts. It aimed to accelerate the implementation of sustainable Science and Technology solutions, promoting resilient practices. The book serves as a comprehensive resource, capturing the knowledge presented at ICSST 2023. It provides valuable insights into developments, breakthroughs, and solutions in sustainability. Whether you're a researcher, scientist, engineer, technologist, or interested in sustainable Science and Technology, this book is an essential reference for understanding advancements shaping a resilient future.

Advancing Sustainable Science and Technology for a Resilient Future

Editors

Sai Kiran Oruganti
Dimitrios A Karras
Srinesh Singh Thakur

First edition published 2024
by CRC Press
4 Park Square, Milton Park, Abingdon, Oxon, OX14 4RN

and by CRC Press
2385 NW Executive Center Drive, Suite 320, Boca Raton FL 33431

British Library Cataloguing-in-Publication Data
A catalogue record for this book is available from the British Library

ISBN: 978-1-032-79020-6 (pbk)
ISBN: 978-1-003-49021-0 (ebk)

DOI: 10.1201/9781003490210

Typeset in Times LT Std
by Aditiinfosystems

Advancing Sustainable Science and Technology for a Resilient Future – Sai Kiran Oruganti et al. (eds)
© 2024 Taylor & Francis Group, London, ISBN 978-1-032-79020-6

Contents

Advancing Sustainable Science and Technology for a Resilient Future – Sai Kiran Oruganti et al. (eds)
© 2024 Taylor & Francis Group, London, ISBN 978-1-032-79020-6

List of Figures

Advancing Sustainable Science and Technology for a Resilient Future – Sai Kiran Oruganti et al. (eds)
© 2024 Taylor & Francis Group, London, ISBN 978-1-032-79020-6

List of Tables

Advancing Sustainable Science and Technology for a Resilient Future – Sai Kiran Oruganti et al. (eds)
© 2024 Taylor & Francis Group, London, ISBN 978-1-032-79020-6

About the Editors

Sai Kiran Oruganti

Indian Institute of Technology Patna India.

ORCID: https://orcid.org/0000-0003-4601-2907

Areas: Wireless Power Transfer, Wireless technologies, IoT, Radio Science, Electromagnetics & applications.

Profile Summary

Prof. Dr. Sai Kiran Oruganti is with the School of Electrical and Automation Engineering, Jiangxi University of Science and Technology, Ganzhou, People's Republic of China as a full Professor since October 2019. He is responsible for establishing an advanced wireless power transfer technology laboratory as a part of the international specialists team for the Center for Advanced Wirless Technologies. Between 2018-2019, he served as a senior researcher/Research Professor at Ulsan National Institute of Science and Technology. Previously, his PhD thesis at Ulsan National Institute of Science and Technology, South korea, led to the launch of an University incubated enterprise, for which he served as a Principal Engineer and Chief Designer in 2017-2018. After his PhD in 2016, he served Indian Institute of Technology, Tirupati in the capacity of Assistant Professor (Electrical Engineering) between 2016-2017.

Research

Prof. Dr. Oruganti, prime research focus is in the development of Wireless Power Transfer(WPT) for applications- Internet of Things (IoT) device charging, Agriculture, Electric Vehicle Charging, Biomedical device charging, Electromagnetically induced transparancy techniques for military and defence applications, Secured shipping containers, Nano Energy Generators.

Achievements

Prof. Dr. Oruganti has more than 21 patents pending on his credit and with several of those patent applications passing the NoC stage. As of 2021, 16 of 21 patents have been granted. He is credited with the pioneering work in the field of Zenneck Waves based Wireless Power Transfer system. Most notably, he has been regarded as one of the only few researchers in the field of WPT to be able to conduct power and signal transmission across partial Faraday shields. His recent paper accepted by Nature Scientific Reports has generated a lot of interest and excitement in the field. International Union of Radio Science(URSI) recognized his research efforts and awarded him Young Scientist Award in 2016. He is also recipient of IEEE sensors council letters of appreciation.

Dimitrios A. Karras

University of Athens (NKUA), Greece

https://orcid.org/0000-0002-2759-8482

Areas: Intelligent Systems, Computational intelligence, machine learning and deep learning, Adaptive systems and distributed systems, Multi agent Systems, Signal and Image processing, Communication Systems

Dimitrios A. Karras received his Diploma and M.Sc. Degree in Electrical and Electronic Engineering from the National Technical University of Athens (NTUA), Greece in 1985 and the Ph. Degree in Electrical Engineering, from the NTUA, Greece in 1995, with honours. From 1990 and up to 2004 he collaborated as visiting professor and researcher with several universities and research institutes in Greece. Since 2004 he has been with the Sterea Hellas Institute of Technology, Automation Dept., Greece as associate professor in Digital Systems and Signal Processing, till 12/2018, as well as with the Hellenic Open University, Dept. Informatics as a visiting professor in Communication Systems (the latter since 2002 and up to 2010). Since 1/2019 is Associate Prof. in Digital Systems and Intelligent Systems, Signal Processing , in National & Kapodistrian University of Athens,

Greece, School of Science, Dept. General. He is, also, adjunct professor with GLA University. Mathura, India and BIHER, BHARATH univ. India as well as with EPOKA and CIT universities Tirana. Moreover, he is with AICO EDV-Beratung GmbH as senior researcher as well as Director of Research and Documentation at ADIafrica N. G. O. He has published more than 80 research refereed journal papers in various areas of pattern recognition, image/signal processing and neural networks as well as in bioinformatics and more than 185 research papers in International refereed scientific Conferences. His research interests span the fields of pattern recognition and neural networks, image and signal processing, image and signal systems, biomedical systems, communications, networking and security. He has served as program committee member in many international conferences, as well as program chair and general chair in several international workshops and conferences in the fields of signal, image, communication and automation systems. He is, also, former editor in chief (2008-2016) of the International Journal in Signal and Imaging Systems Engineering (IJSISE), academic editor in the TWSJ, ISRN Communications and the Applied Mathematics Hindawi journals as well as associate editor in various scientific journals. He has been cited in more than 2560 research papers, his H/G-indices are 20/52 (Google Scholar) and his Erdos number is 5. His RG score is 32.78

Srinesh Thakur

Anvita Electronics PVT LTD India

Srinesh Thakur is an accomplished hardware engineer and successful entrepreneur with a remarkable career spanning over a decade. He has made significant contributions to the field of hardware engineering, working with prestigious institutions like Technische Hochschule Nürnberg and Fraunhofer Research Institute in Germany. As a co-founder of Anvita Electronics in Hyderabad, India, and Atya Technologies, Srinesh has demonstrated exceptional leadership and vision in establishing these ventures. His expertise lies in highly complex 32-layer PCB design, a skill that has set him apart in the industry. Throughout his career, Srinesh has primarily focused on the development of cutting-edge test equipment for the automotive testing industry. His passion for innovation and commitment to delivering high-end solutions have earned him a stellar reputation among his peers. With a deep understanding of hardware engineering principles and an entrepreneurial spirit, Srinesh Thakur continues to drive technological advancements in the field, revolutionizing automotive testing and setting new standards for the industry.

Advancing Sustainable Science and Technology for a Resilient Future – Sai Kiran Oruganti et al. (eds)
© 2024 Taylor & Francis Group, London, ISBN 978-1-032-79020-6

Internet of Things and Green Production: Implications for Market Research and Industrialization

1

Kannan Nova*
Data Scientist, Microsoft, 14239 NW 18th MNR,
Pembroke Pines, FL 33028, USA,

Vartika Kulshrestha[1]
Assistant Professor, Department of computer Science and
Engineering,
Alliance University, Bangalore

Chetan V. Hiremath[2]
Associate Professor, Department of Operations and Analytics,
Kirloskar Institute of Management, Yantrapur, Harihar,
Davangere, Karnataka, India

Karan R. Jagdale[3]
UG Scholar, Department of Computer Science and Engineering,
Alliance University, Bangalore

Abstract: The Industrial Internet of Things (IIoT) has become an effective tool with significant implications for industrialisation and Market Research (MR), especially in the field of green production. Green IIoT (GRIIoT) can be used to implement Green Production (GP) goals for the environment. The purpose of this study is to examine the drivers behind the adoption of GIIoT, MR, and industrialization decision-making, as well as the effects these drivers have on industrialization performance (IP). A structured questionnaire was used to gather information in order to evaluate the suggested study paradigm. The results indicate that institutional isomorphism influences the acceptance of GRIIoT in a favorable way. Furthermore, Green innovation (GI) activities that result in IP are favorably correlated with GIIoT. The potential effects of the various institutional isomorphisms discussed in this study can aid organizations in better understanding the responsibilities to protect and satisfying stakeholders, particularly as the adopt GIIoT to handle production problems and possible accordance pressures in the process.

Keywords: Green production (GP), Industrial internet of things (IIoT), Green IIoT (GRIIoT), Industrial performance, Market research

1. Introduction

IoT makes it possible for devices, systems, and equipment to connect and communicate with one another invisibly, enabling the gathering and analysis of data in real-time. The industrialization of IoT additionally changes the organizations run but also provides the path for automated solutions that enhance productivity, decrease downtime, and boost profitability (Suma, et al., 2019). The goal of GP, often referred to as sustainable development, is to reduce the harmful effects of industrial processes on the environment while optimizing resource efficiency (Saetta, et al., 2022). Gathering, evaluating, and interpreting data on a particular market or sector are all important steps in the process of doing MR. It gives organizations useful information about customer

preferences, market trends, and competitive environments to ensure that may make wise choices and create winning strategies (Nunan, et al., 2019). The IIoT intends to increase operational effectiveness, advance intelligent decision-making, and optimize industrial processes. The IIoT provides preventive maintenance, better resource allocation, and predictive analytics, leading to decreased downtime and greater productivity by collecting and analyzing data from numerous sources, including machines, equipment, and production lines (Lai, et al., 2020).Industrial behaviour and behavioural science have discussed to make assured management judgments, including that, like, and the appropriate way to make Industrial activities (Rajagopal, et al., 2022).

*Corresponding author: novakannan@gmail.com
[1]vartikakul@gmail.com, [2]hirechetan@gmail.com, [3]karanjagdale42@gmail.com

DOI: 10.1201/9781003490210-1

2. Related Works

Tabaa, et al., 2020 provide a general review of the Internet of Things (IoT) influence on the development of applications in connection to various revolutions, a list of several industrial revolutions, a general description of the GIIoT, as well as specific applications and perspectives. Wójcicki, et al., 2022 discussed the IoT, IIoT, and Industrial ideas. It highlights the potential outcomes, dangers, and challenges related to carrying out something. Radanliev, et al., 2019 provide significant advancements in this field in connection to the incorporation of cutting-edge IoT and cyber-physical systems into the digital economy in order to better comprehend cyber hazards, economic value, and risk effect. Its objective is to outline the present progress of the digital economy and the cyber threats it entails, as well as to examine potential future advancements in the IoT and Industry 4.0. Gupta, et al., 2022 recommended a new threat detection solution for IoT technology based on machine learning. The methodology produced exact results. The study reports that this technique may be used to predict and find malware in IoT-based devices.The purpose of this research is to develop a novel quantitative forecasting approach for anticipating HR demand using recurrent neural networks (RNNs) with grey wolf's optimization (GWO) (Rajagopal, et al., 2022).

Hypothesis development: H1a - GRIIoT and Green Product Innovation (GPdI) have a productive interaction. H1b-Positive correlation exists between GRIIoT and Green Process Innovation (GPcI). H1c - GRIIoT and Green Management Innovation (GMI) have a productive interaction. H2a - IP and GPdI are positively correlated. H2b - The GPcI and IP have a strong correlation. H2c - The association between GMI and IP is positive.

3. Methodology

Measures: The research modified the measurement elements from earlier studies in a simpler environment to keep the content valid. The scale of GRIIoT, GPdI, GPcI,GMI, and Industrial Performance (IP) was adopted. The cited demographic data of participants was also gathered since the research took gender, education, and experience into account as control factors.

Size of the sample, the gathering of data, and the method of analysis: Employees of industrial organizations in Harbin, China, made up the study's sample. China's industrial boom has made it very difficult for the country to maintain the equilibrium among ecological and economic needs. Additionally, the questionnaire was adapted into Chinese and then back into English by three bilingual researchers in order to confirm the authenticity of its content. Additional

adjustments were made to improve comprehension and suitability of Chinese methods of production. In November 2021, questionnaires were sent through an internet survey link to the staff members of manufacturing companies that had been randomly chosen. The research made sure, base the fiction, that a sample size of more than 200 is thought to be appropriate for using Structural Equation Modelling (SEM). Obtained 330 completed questionnaires in February 2022; 12 of them were returned unfinished. 318 valid replies were taken into account in the study's final analysis. The Smart Partial least squares (PLS) is utilized to do the SEM since it is among the top tools for running the SEM because the research used SEM to investigate the associations between variables.

Demographical Information: The information obtained from the test's participants is shown in Table 1. The findings show that 67% of respondents have degrees, 74% have more than a year of work experience, and 60.64 % of responders are males. As a result, the survey includes both female and male respondents who have education and experience.

Table 1.1 Demographical information

Features	No (%)
Gender	
Male	195 (60.64)
Female	123 (39.36)
Experience	
1-4 (years)	88 (26.99)
5-8 (years)	149 (47.23)
9-11 (years)	53 (15.98)
>11 (years)	29 (9.81)
Education	
UG	41 (12.2)
PG	140 (44.7)
Diploma	78 (23.8)
Other	59 (19.2)

4. Results

Measurement model: The average variance extracted (AVE) and Cronbach's alpha was tested to determine the composite reliability (CR). The convergent validity analysis results for Table 1.2 indicate that all values are larger than the restriction levels. As several Cronbach's alphas, CR, and AVE measurements are within the allowed range, convergent strength was not found to have a significant problem. The AVE values that exceed the value of the component for all variables are shown in Table 1.3.

Table 1.2   Results of converging validity

	Cronbach'sAlpha	rho_A	CR	AVE
GRIIoT	0.906	0.910	0.927	0.680
GMI	0.811	0.811	0.888	0.726
GPcI	0.920	0.921	0.944	0.807
GPdI	0.875	0.893	0.914	0.727
IP	0.908	0.910	0.931	0.731

Table 1.3   Component association and AVE square

	GRIIoT	GMI	GPcI	GPdI	IP
GRIIoT	0.824				
GMI	0.405	0.852			
GPcI	0.373	0.407	0.899		
GPdI	0.357	0.480	0.542	0.853	
OP	0.355	0.419	0.434	0.395	0.855

Table 1.4   Cross loadings

	GIIoT	GMI	GPcI	GPdI	IP
GRIIoT1	0.811	0.357	0.381	0.377	0.348
GMI1	0.327	0.854	0.351	0.420	0.346
GPcI1	0.355	0.412	0.892	0.508	0.381
GPdI1	0.330	0.418	0.478	0.879	0.357
OP1	0.366	0.368	0.430	0.374	0.820

The values are greater and cross-loadings of respective parallel variables are also higher then those of their explanatory variable, which is shown in Table 1.4, demonstrating the lack of a severe problem with discriminate validity.

The upper limit of the variables that value should be less than one to distinguish the two components is the heterotrait-monotrait ratio (HTMT). Table 1.5 displays the HTMT valuethat is beyond the permitted range while demonstrating that predictive relevance is impacted.

Modelling using structural equations Table 6 displays the path coefficient, p-value, and t-value. The findings show that there is a variation in GRIIoT of 23%, GPdI of 57%, GPcI of 35%, GMI of 58%, and Industrial performance of 27%.

Acceptance of GRIIoT, GPdI, and GPcI also has a substantial link to GPdI and GMI, which serve as the foundation for H1a, H1b, and H1c acceptance (Fig. 1). Similar to how H2a, H2b, and H2c are accepted, GPdI, GPcI, and GMI have considerable Industrial innovation. The suggested research paradigm is accepted by the study based on all outcomes.

5. Discussion

MR may reveal sustainable product markets, customer willingness to pay, and corporate strategy changes. MR can illuminate the competitive landscape and assist industries find new IIoT and GRIIoT potential. MR is essential to discover target markets, understand customer needs, and establish viable industries that follow GRIIoT principles while industrialization embraces them as needed.

Table 1.5   HTMT ratio standard

	GRIIoT	GMI	GPcI	GPdI			
GRIIoT	0.361						
GMI	0.425	0.469					
GPcI	0.419	0.404	0.470				
GPdI	0.414	0.390	0.568	0.602			
ORP	0.406	0.384	0.486	0.471	0.432	0.430	0.337

Table 1.6   SEM results

Hypotheses	Original Sample(O)	Sample Mean(M)	STDEV	T statistics (IO/STDEVI)	P values
H1a=GIIoT–GPdI	0.357	0.356	0.073	4.911	0.000
H1b=GIIoT–GPcI	0.373	0.372	0.071	5.279	0.000
H1c=GIIoT–GMI	0.405	0.407	0.059	6.909	0.000
H2a=GPdI–IP	0.135	0.135	0.065	2.076	0.038
H2b=GPcI–IP	0.260	0.258	0.058	4.502	0.000
H2c=GMI–IP	0.248	0.248	0.061	4.080	0.000

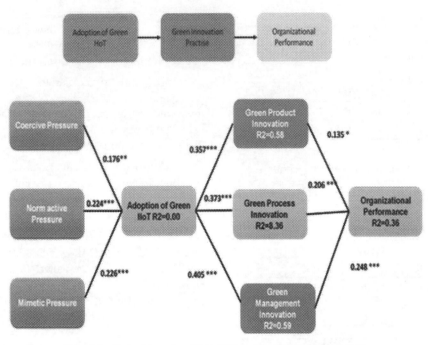

Fig. 1.1 Data from the SEM: ***P 0.001, **P 0.005, *P 0.01

6. Conclusion

The implementation of GRIIoT was associated with greater, according to the findings. These demands, such as requiring Industrialization to embrace GRIIoT, make them permanently dedicated to achieving their environmental requirements. Results also influence the acceptance of GRIIoT as a basis for implementing GI practices into an approach that will ensure the Industries' internal and external environments. GI expansion of the Industry's corporate performance by lowering costs, conserving resources, handling waste, and lowering carbon and hazardous emissions. According to research, GI techniques supported by GRIIoT have a significant influence on Industrialization performance by reducing production expenses, and waste disposal, and promoting effective decision-making. According to estimates, the GRIIoT will significantly alter our industry and result in a greener environment. GRIIoT lessens both the environmental effects of the IoT itself as well as the greenhouse effect in other industries. The management has so committed to using GRIIoT to decrease environmental risks, conserve energy, and emit less carbon dioxide. As a result, GRIIoT adoption by industries may enhance GI practices. Future research should thus be conducted in wealthy nations to have a better understanding. Future studies may build on this model by integrating these internal as well as external components to estimate the adoption of GRIIoT.

References

1. Suma, V., 2019. Towards sustainable industrialization using big data and the internet of things. Journal of ISMAC, 1(01), pp. 24–37.
2. Saetta, S., & Caldarelli, V. (2020). Lean production as a tool for green production: the Green Foundry case study. Procedia manufacturing, 42, 498–502.
3. Nunan, D. and Di Domenico, M., 2019. Rethinking the market research curriculum. International Journal of Market Research, 61(1), pp. 22–32.
4. Lai, X., Hu, Q., Wang, W., Fei, L. and Huang, Y., 2020. Adaptive resource allocation method based on deep Q network for industrial Internet of Things. IEEE Access, 8, pp. 27426–27434.
5. Rajagopal, N.K., Qureshi, N.I., Durga, S., Ramirez Asis, E.H., Huerta Soto, R.M., Gupta, S.K. and Deepak, S., 2022. Future of business culture: an artificial intelligence-driven digital framework for the organization decision-making process. Complexity, 2022.
6. Tabaa, M., Monteiro, F., Bensag, H. and Dandache, A., 2020. Green Industrial Internet of Things from a smart industry perspectives. Energy Reports, 6, pp. 430–446.
7. Leong, Y.R., Tajudeen, F.P. and Yeong, W.C., 2021. Bibliometric and content analysis of the internet of things research: a social science perspective. Online Information Review, 45(6), pp. 1148–1166.
8. Wójcicki, K., Biegańska, M., Paliwoda, B. and Górna, J., 2022. Internet of Things in Industry: Research Profiling,

Application, Challenges and Opportunities—A Review. Energies, 15(5), p. 1806.

9. Radanliev, P., De Roure, D., Nurse, J.R., Nicolescu, R., Huth, M., Cannady, S. and Montalvo, R.M., 2019. New developments in Cyber Physical Systems, the Internet of Things and the Digital Economy–discussion on future developments in the Industrial Internet of Things and Industry 4.0.

10. Gupta, S.K., Pattnaik, B., Agrawal, V., Boddu, R.S.K., Srivastava, A. and Hazela, B., 2022, September. Malware Detection Using Genetic Cascaded Support Vector Machine Classifier in the Internet of Things. In 2022 Second International Conference on Computer Science, Engineering and Applications (ICCSEA) (pp. 1–6). IEEE.

11. Rajagopal, N.K., Saini, M., Huerta-Soto, R., Vílchez-Vásquez, R., Kumar, J.N.V.R., Gupta, S.K. and Perumal, S., 2022. Human resource demand prediction and configuration model based on grey wolf optimization and recurrent neural network. Computational Intelligence and Neuroscience, 2022.

Note: All the figures and tables in this chapter were made by the authors.

Advancing Sustainable Science and Technology for a Resilient Future – Sai Kiran Oruganti et al. (eds)
© 2024 Taylor & Francis Group, London, ISBN 978-1-032-79020-6

Analysis of Solutions of a Fractional Differential Equation Coupled with a Boundary Value Issue

2

D. Saravanan*

Assistant Professor, Department of CSE, Sathyabama Institute of Science and Technology (Deemed to be University), Chennai, India

S. Vasundhara[1]

Assiatant Professor, Department of Mathematics, G. Narayanamma Institute of Rechnology and Science, Shaikpet, Hyderabad, India

K. Dhilipkumar[2]

Assistant Professor of Mathematics, KPR College of Arts Science and Research, Coimbatore

Pravin P.[3]

Department of Mechanical Engineering, Graphic Era Deemed to be University, Uttarakhand, India

Abstract: As a consequence of difficulties in extracting the features of financial data, the current approaches for predicting financial crises are lacking. Thus, an intelligent financial crisis prediction method for human resources (HR) is suggested using enhanced K-means clustering and fitness scaling improved spider monkey optimization (K-MC+FSISMO). The K-MC+FSISMO technique is used to mine the attributes of financial data using financial index data from publicly traded companies. Following the discoveries made via data feature mining, this study creates an index system for predicting financial crises. CCR measures the effectiveness of publicly listed companies' decision-making units, with a focus on those having the greatest impact on inputs and outputs. Experimental findings show that this technique may significantly shorten the time it takes to foresee financial crises and that its forecast accuracy is comparable to the real status of enterprises, as compared to traditional prediction methods.

Keywords: Human resource (HR), Financial crisis, Enhanced K-means clustering, Fitness scaling improved spider monkey optimization, CCR model

1. Introduction

A differential equation that incorporates fractional derivatives of a function and boundary conditions at the domain's ends is known as a fractional differential equation paired with boundary value issues (Qarout, et al., 2016). The addition of boundary conditions and derivatives of non-integer order makes it a generalization of ordinary differential equations. When compared to solutions of ordinary differential equations, the solutions of FDEs coupled with BVI frequently display non-local and memory effects, which may lead to strange or unexpected behavior (Rani, et al., 2018).

Investigating characteristics like stability, existence, uniqueness, and regularity may be necessary while analyzing the solutions of FDEs in combination with BVI (Härkönen, et al., 2022). The majority of the findings indicated that by utilizing nonlinear analytic methods, equations of fractional differentials had at least one and perhaps numerous positive solutions (Bhrawy and Zaky 2016). Regarding the following an adversely affected fractional differential equation, for instance, the authors examined the possibility of several positive solutions (Chhabra, et al., 2013).

$$\begin{cases} -T_{0+}^{b} y(d) = b(d)l(d,y)(d)) - q(d), d \in (0,1), \\ y(0) = y'(0) = 0, y(1) = 0, \end{cases} \quad (1)$$

where T_{0+}^{b} represents the usual Riemann-Liouville derivative, $2 < b \le 3$ is a valid integer, $o: (0,1) \to [0, +\infty)$ is integrable

*Corresponding author: saravanan.d.cse@sathyabama.ac.in
[1]vasucall123@gmail.com, [2]dhilipkumarmaths@gmail.com, [3]pravinp667@yahoo.com

DOI: 10.1201/9781003490210-2

in the Lebesgue sense and does not disappear uniquely across any interval of $(0,1)$. Krasnosel'skii's fixed point theorem was used to a cone to prove the existence of findings. Boundary value issues for FDE have limited findings in the literature about their uniqueness of solutions.

$$
\begin{cases}
T_d^\alpha x(d) = l\left(d, x(d), T_d^\beta x(d)\right), d \in (0,1), \\
x(0) = 0, T_d^\beta x(d) - \sum_{j=1}^{n-2} \zeta_j T_d^\beta x\left(\xi_j\right) = x_0,
\end{cases} \quad (2)
$$

Here $1 < \alpha \le 2$, $0 < \beta < 1$, $0 < \xi_j < 1(i\ 1, 2, ..., m - 2)$, $\zeta_j \ge 0$ with $\sum_{j=1}^{n-2} \zeta_j \xi_j^{\alpha-\beta-1} < 1$ and T_d^α illustrates Riemann-Liouville's Derivative of Fractions used in common applications. Utilizing the Banach fixed point theorem, they established that solutions exist and are unique. In light of the aforementioned research, we analyze the solutions to the following BVI for FDE.

$$
\begin{cases}
T^b y(d) + b(d)l(d, y(d)) + o(d) = 0, d \in (0,1) \\
y(0) = y'(0) = 0, y(1) = 0,
\end{cases} \quad (3)
$$

Where $2 < p \le 3$ is a valid integer. On the premise that $l(d, y)$ is a Lipschitz constant operator, by utilization of u0-positive function, for a fractional differential equation, we investigate whether or not a unique solution exists. The initial eigenvalues of the relevant operators are connected to the Lipschitz constant, which is an intriguing fact. $(Z_1)b: (0, 1) \to [0, +\infty)$ is continuous and not disappearing uniformly at any subinterval of $(0, 1)$ such that

$$
0 < \int_0^1 b(g)tg < +\infty \quad (4)
$$

$(Z_2)l:[0, 1] \times \mathbb{K} \to \mathbb{K}$ is constant.

$(Z_3)o:[0, 1] \to \mathbb{K}$ is constant and Lebesgue integrable. The study is set up such that Part II offers relevant work, Part III describes the analysis, Part IV shows the results and discussion, and Part V concludes with recommendations for future research.

2. Related Works

Wang, et al., 2016 provided a summary of our key findings about IFDE. They provide a variety of these equations with different boundary value constraints. Atangana and Owolabi., 2018 explore some of the key computing concerns, including the proficient handling of the permanent retention truncation and the resolution of nonlinear systems related to implicit methods, and they analyze two of the most successful groups of numerical techniques for solving problems with fractional variables. Garrappa, R. (2018) focused on the approximation

of FDEs via a spectral theory. Ahmad et al, 2017 focused on the recent progress in differential and integral equations, inclusions, and inequalities using the integral and Hadamard derivative. Kumar,et al., 2018 employed a variant of the Kudryashov technique to develop fresh precise solutions to a set of conformable fractional differential equations.

3. Proposed Methodology

Here, we provide some essential definitions from the field of fractional equations for the benefit of the reader. The latest monograph has these characteristics and definitions.

The Definition 1: The order-fractional integral Riemann-Liouville $b > 0$ of a function: $(0, \infty) \to \mathbb{K}$ is given by

$$
J^b l(d) = \frac{1}{\Gamma(b)} \int_0^d (d-g)^{b-1} l(g)tg, \quad (5)
$$

The right side is point-wise specified on .

The Definition 2: The derivative of the order fraction in the Riemann-Liouville system $b > 0$ of a continuous function $l: (0, \infty) \to \mathbb{K}$ is given by

$$
T^b l(d) = \frac{1}{\Gamma(m-b)} \left(\frac{t}{td}\right)^m \int_0^d \frac{l(g)}{(d-g)^{b-1+1}} tg \quad (6)
$$

where $m - 1 \le \alpha < m$, Assuming the right side has a point-wise definition, on $(0,\infty)$. In Banach space $E = V[0, 1]$ that the standard is established by $\|x\| = \max_{d \in [0,1]} |x(t)|$, we set $B = \{y \in V[0, 1] \mid y(d) \ge 0, \forall d \in [0, 1]\}$. P is a cone of positivity in $V[0, 1]$. B always provides incomplete sorting in this article.

The Definition 3: In this work, we define what is known as a bounded linear operator. $D : A \to A$ is u0-optimistic on the cone B if there exists $w_0 \in B\backslash\{\theta\}$ like that for each $y \in B\backslash\{\theta\}$ certain natural numbers exist m and positive constants $\alpha(y), \beta(y)$ such that

$$
\alpha(y) w_0 \le D^m y \le \beta(y) \quad (7)
$$

$\varphi^* \in A$ be is an eigenvalue of the linear operator that takes positive values D if $\phi^* \in B\backslash\{\theta\}$ and there exists $\lambda > 0$ like that $\lambda T \varphi^* = \varphi^*$. For our purposes, we need to know whether or not a solution to a linear boundary value issue exists, and if it does, whether or not that solution is unique.

Lemma 1: Let $e \in \mathbb{K}$, $\sigma \in V(0, 1) \cap F(0, 1)$ and $2 < b \le 3$, then the solution of

$$
\begin{cases}
T^b y(d) + \sigma(d) = 0, d \in (0,1), \\
y(0) = y'(0) = 0, y(1) = e,
\end{cases} \quad (8)
$$

is given by

$$
y(d) = ed^{b-1} + \int_0^1 S(d,g)\sigma(g)tg, \quad (9)
$$

Where is Green's function is presented by

$$S(d,g) = \begin{cases} \dfrac{(1-g)^{b-1}d^{b-1} - (d-g)^{b-1}}{\Gamma(b)}, 0 \le g \le d \le 1, \\[2mm] \dfrac{(1-g)^{b-1}d^{b-1}}{\Gamma(b)}, 0 \le d \le g \le 1 \end{cases} \quad (10)$$

Lemma 2: The operator $S(d, g)$ defined by (10) fulfills all of the criteria below:

$$d^{b-1}(1-d)g(1-g)^{b-1} \le \Gamma(b)S(d, g)$$
$$\le (b-1)g(1-g)^{b-1}, g \in (0,1), \quad (11)$$

$$d^{b-1}(1-d)g(1-g)^{b-1} \le \Gamma(b)S(d, g)$$
$$\le (b-1)g^{b-1}(1-g), d, g \in (0, 1), \quad (12)$$

Let the operators D and E be identified as

$$(Dy)(d) = \int_0^1 S(d,g)b(g)y(g)ts, d \in [0,1], y \in V[0,1] \quad (13)$$

$$(Ey)(d)$$
$$= \int_0^1 S(d,g)\big[o(g)+b(g)l(g,y(g))\big]tg, d \in [0,1], y \in V[0,1]$$
$$\quad (14)$$

Respectively. It is simple to confirm that $D : A \to A$ is an entirely continuous linear and $D(B) \subset B$.

Lemma 3: T is w_0-positive operator with $w_0(d) = d^{b-1}(1-d)$.

$$(Dy)(d) = \int_0^1 S(d,g)b(g)y(g)ts \le \frac{b-1}{\Gamma(b)}$$
$$\int_0^1 b(g)y(g)ts.d^{b-1}(1-d) \quad (15)$$

On the other hand, by Lemma 2 again, we have

$$(Dy)(d) = \int_0^1 S(d,g)b(g)y(g)ts \ge \frac{1}{\Gamma(b)}$$
$$\int_0^1 g(1-g)^{b-1}b(g)y(g)ts.d^{b-1}(1-d) \quad (16)$$

The resulting inequalities imply that D is the w_o-positive operator with $w_o(d) = d^{b-1}(1 - d)$. This concludes the argument.

Lemma 4: Consider this T: E $\to$ E is an entirely uninterrupted Linear Process and T (P) $\subset$ P. If Presuming existence $\psi \in$ E\ (−P) and a continuous c > 0 like that cT $\psi \ge \psi$, followed by the Radius of the Spectrum r(T)/= 0 and Since the initial eigenvalue of T is thus positive, the eigen function of T is also positive. $\lambda 1 = (r(T))−1$, i.e. $\phi = \lambda 1T \phi$. The spectral radius is determined by Lemma 3 and Lemma 4, as shown below.

T and $k(D) \ne 0$ have an affirmative eigenvalue $\varphi^*(t)$ concerning its initial eigenvalue $\lambda_1 = (k(D))^{-1}$.

Remark: Let φ^* be the affirmative eigenfunction of D equivalent to λ_1, thus $\lambda_1 D\varphi^* = \varphi^*$ after that, using Definition 3 and Lemma 3, there exist $r_1(\varphi^*)$, $r_2(\varphi^*) > 0$ like that

$$r_1\big(\varphi^*\big)w_0 \le D\varphi^* = \frac{1}{\lambda_1}\varphi^* \le r_2\big(\varphi^*\big)w_0 \quad (17)$$

Consequently, we found out D is φ^* - affirmative operator.

4. Result and Discussion

Theorem 1: Consider the possibility that like that

$$|l(d, c) - l(d, w)| \le r\lambda_1 |w - c|, \forall d\epsilon[0, 1], w, c\epsilon\mathbb{K} \quad (18)$$

where λ_1 is the initial eigenvalue of T. Then (3) had a solution y^* in A, and for any $y_0 \in A$, the continuous process $x_m = Ey_{0-1}$ ($m = 1, 2, ...$) merges to y^*.

Proof: It is simple to demonstrate that $E: b \to B$ is fully constant, therefore there is a solution.

For any given $x_0 \in A$, let $y_m = Ey_{0-1}$ ($m = 1, 2, ...$). By Remark and Lemma 3, there exists $\beta = \beta(|y_1 - y_0|) > 0$ like that

$$(D|y_1 - y_0|)(d) \le \beta\varphi^*(d), d \in [0, 1] \quad (19)$$

Notice for $n \in M$ that

$$|y_n(d) - y_{n+1}(d)| = |(Ey_n)(d) - (Ey_{n-1})(d)| \quad (20)$$

$$= |\int_0^1 S(d,g)b(g)l(g,y_n(g))tg - \int_0^1 S(d,g)b(g)l(g,y_{n-1}(g))tg|$$

$$\le \int_0^1 S(d,g)b(g) |l(g,y_n(g))tg - l(g,y_{n-1}(g))| tg$$

$$\le r\lambda_1 D\big(|y_n - y_{n-1}|\big)(d) \le ... \le r^n\lambda_1^n D^n\big(|y_1 - y_0|\big)(d) \quad (21)$$

$$\le r^n\lambda_1^n D^{n-1}\big(\beta\varphi^*(d)\big) = r^n\lambda_1^n\beta D^{n-1}\big(\varphi^*(d)\big) = r^n\beta\lambda_1\varphi^*(d)$$
$$\quad (22)$$

Thus for

$$|y_{m+n+1}(d) - y_m(d)| \big||y_{m+n}(d) - y_{m+n+1}(d)... + y_{m+1}(d) - y_m(d)\big|$$

$$\le |y_{m+n+1}(d) - y_{m+n}(d) + ... + y_{m+1}(d) - y_m(d)|$$

$$\le \beta\lambda_1\big[r^{n+n} + ... + r^n\big]\varphi^*(d) = \beta\lambda_1 \frac{r^n\big(1-r^{n+1}\big)}{1-r}\varphi^*(d)$$
$$\quad (23)$$

Therefore,

$$y_{m+n+1} - y_m \le \beta\lambda_1 \frac{r^n\big(1-r^{n+1}\big)}{1-r}\varphi^* \to 0, \text{ as } m, n \to \infty \quad (24)$$

By the completeness of A, there exists $ex^* \in A$ such that $\lim_{m \to \infty} y_m = y^*$ reaching the limit and entering $y_{n+1} = Ey_m$ and utilizing the fact that E is constant, it follows that y^* is a fixed point of E in A.

We then prove that there is no more than one fixed point of E in A. Let's pretend there are two substances. $y, x \in A$ with $y = Ey$ and $x = Ex$. By Lemma 2.3, there exists $\beta = \beta(|y - x|) > 0$ such that

$$(D(|y - x|))(d) \leq \beta \varphi^*(d), d \in [0, 1] \quad (25)$$

Then for all $m \in M$, so, we possess

$$|y(d) - x(d)| = |(E^m y)(d) - (E^m x)(d)| \leq r^m \beta \lambda_1 \varphi^*(d) \quad (26)$$

This is only possible if $y = x$. This suggests that E has only one fixed point at most. As a result, the only fixed point of E in A is y^*. With this, the case is closed.

Theorem 2: Suppose that there exist $y_0 \in A$, $r \in [0,1)$ satisfies the following conditions:

$$T^b y_0 (d) + o(d) + b(d)l(d, y_0 (d))$$
$$\leq 0, d \in [0,1), y_0 (0) = 0, y_0{}' (0), y_0(1) >= 0 \quad (27)$$

$$0 \leq l(d, w(d)) - l(d, c(d))$$
$$\leq r\lambda_1(w(d) - c(d)), w(d) >= c(d), \forall d \in [0,1], w, c \in \Omega \quad (28)$$

in which $\Omega = \{y \in A \mid y \geq y_0\}$ Then (3) has a solution y^* in Ω.

Proof: Lemma 1 implies that E is developing on T and $y_0 \leq Ey_0$, so we obtain $E(T) \subset T$. Let $y_m = Ey_{m-1} - 1 (m = 1, 2,...)$, then we have

$$y_0 \leq y_1 \leq \cdots. \leq y_m \leq \cdots. \quad (29)$$

By Lemma 3, there exists $\beta > 0$ such that

$$D(y_1 - y_0) \leq \beta \varphi^* \quad (30)$$

Then for $\forall m \in M$ and $d \in [0,1]$, we posses

$$0 \leq y_{m+1} (d) - y_n(d) = Ey_m(d) - Ey_{m-1} (d) \leq r\lambda_1 D(y_m - y_m)$$
$$(d) = r\lambda_1 D(Ey_{m-1} - Ey_{m-2})(d) \leq \cdots \leq (r\lambda_1 D)^m (y_1 - y_0) \leq \beta r^m \lambda_1 \varphi^*(d) \quad (31)$$

Thus for $m, n \in M$,

$$| y_{m+n}(d) - y_m(d) |$$
$$= | y_{m+n}(d) - y_{m+n-1}(d) + \ldots. + y_{m+1}(d) - y_m(d) |$$
$$\leq | y_{m+n}(d) - y_{m+n-1}(d) | + \ldots. + | y_{m+1}(d) - y_m(d) |$$
$$\leq \beta \lambda_1 [r^{n+n-1} + \ldots + r^n] \varphi^*(d) = \beta \lambda_1 \frac{r^n(1-r^n)}{1-r} \varphi^*(d) \quad (32)$$

This shows that $\|y_{m+n} - y_m\| \leq \beta \lambda_1 \frac{r^n(1-r^n)}{1-r} \varphi^* \|\varphi^*\|$. So, $\{y_m\}$ Because A is a Banach space, there exists a Cauchy sequence in A. $y^* \in A$ such that $\lim_{m \to \infty} y_m = y^*$. Hence, y^* is a permanent point of E in Ω. We'll demonstrate that y^* is the only permanent point in this space. of E in Ω. Assume a

substance called $y \in T$ with $y = Ey$. By Remark 1, there exists $\beta_1 > 0$ such that

$$D(y - y_0) \leq \beta_1 \varphi^*, \quad (33)$$

and for any $m \in M$, we posses

$$y \geq y_m \geq y_o \quad (34)$$

Therefore,

$$y \geq y^* \geq y_m \geq y_0 \quad (35)$$

Then for all $m \in M$ and $d \in [0,1]$, and we possess

$$|y^* (d) - y(d)| \leq |y(d) - y_m (d)| + |y^* (d) - y_m (d)| \leq |(E^m y) (d) - (E^m y_0)(d)| + |(E^m y^*)(d) - (E^m y_0)(d)| \leq 2|(E^m y)(d) - (E^m y_0)(d)| \leq 2\beta_1 r^m \lambda_1 \varphi^* (d). \quad (36)$$

Thus, we get $= y^*$. This completes the proof. These findings are related to Theorem 2 and are similar.

Theorem 3: Let's pretend there is $y_0 \in A$, $r \in [0, 1)$ such that fulfils the following requirements:

$$T^b y_0(d) + o(d) + b(d)l(d, y_0(d))$$
$$\leq 0, d \in [0, 1), y_0'(0), y_0(0) = 0, y_0 (1) >= 0, \quad (37)$$

$$0 \leq l(d, w(d)) - l(d, c(d))$$
$$\leq r\lambda_1 (c(d) - w(d)), w(d) >= c(d), \forall d \in [0,1], w, c \in \Omega \quad (38)$$

In which $\Omega = \{y \in A \mid y \geq y_0\}$. Then (3) has a solution $**^*$in Ω.

5. Conclusion

This study examined the Analysis of solutions for non-linear boundary value class issues using equations of fractional differentials as part of this body of work. The fact that the Lipchitz constant is connected to the initial eigenvalues that correspond to the related function is the most significant contribution that this study makes to the body of knowledge. The existence-uniqueness theorem in this study enhances the findings of earlier research since it offers broader requirements for any continuous solution to the issue under investigation, not only for positive solutions.

References

1. Wang, J., Fečkan, M. and Zhou, Y., 2016. A survey on impulsive fractional differential equations. *Fractional Calculus and Applied Analysis*, 19(4), pp. 806–831.
2. Atangana, A. and Owolabi, K.M., 2018. New numerical approach for fractional differential equations. *Mathematical Modelling of Natural Phenomena*, 13(1), p.3.
3. Garrappa, R., 2018. Numerical solution of fractional differential equations: A survey and a software tutorial. *Mathematics*, 6(2), p.16.
4. Rani L., Srivastav A.L., Kaushal J. "Bioremediation: An effective approach of mercury removal from the aqueous solutions." Chemosphere. 2021; 280. Article No.: 130654.

5. Chhabra R., Sharma V. "Applications of blogging in problem-based learning."*Education and Information Technologies.* 2013; 18(1): 3–13.

6. Ahmad, B., Alsaedi, A., Ntouyas, S.K. and Tariboon, J., 2017. *Hadamard-type fractional differential equations, inclusions, and inequalities* (pp. 3–11). Cham, Switzerland: Springer International Publishing.

7. Kumar, D., Seadawy, A.R. and Joardar, A.K., 2018. Modified Kudryashov method via new exact solutions for some conformable fractional differential equations arising in mathematical biology. *Chinese Journal of Physics*, *56*(1), pp. 75–85.

8. Sun, Z.Z. and Gao, G.H., 2020. Fractional differential equations. In *fractional Differential equations*. De Gruyter.

9. Khan, A., Li, Y., Shah, K. and Khan, T.S., 2017. On coupled-laplacian fractional differential equations with nonlinear boundary conditions. *Complexity*, *2017*.

10. Qarout, D.A., Ahmad, B. and Alsaedi, A., 2016. Existence theorems for semi-linear Caputo fractional differential equations with nonlocal discrete and integral boundary conditions. *Fractional Calculus and Applied Analysis*, *19*(2), pp. 463–479.

Advancing Sustainable Science and Technology for a Resilient Future – Sai Kiran Oruganti et al. (eds)
© 2024 Taylor & Francis Group, London, ISBN 978-1-032-79020-6

Smart Cloud Data Management Based on Deep Reinforcement Learning with Spider Swarm Optimization Algorithm

3

Sandhya Rani Nallola*

Research Scholar, Department of Computer Science, GITAM School of Technology, GITAM Deemed to be University, Bengaluru, India,

Vadivel Ayyasamy[1]

Professor, Department of computer science, GITAM School of Technology, GITAM Deemed to be University, Bengaluru, 361203, India,

Omaia Mohammed Al-Omari[2]

Assistant Professor, Department of Information Systems, College of Computing and Information Technology, Shaqra University, Shaqra, Saudi Arabia,

A. Inbavalli[3]

Assistant Professor, Department of Computer Science and Engineering, IFET College of Engineering, Villupuram, India,

Abstract: More sophisticated techniques are now needed in order to manage and process the enormous volume of data coming from various sources. The disadvantages of traditional cloud data management techniques include poor performance, high costs, and a lack of adaptability to modify conditions. To manage cloud data, this research proposes revolutionary deep reinforcement learning with the Spider swarm optimization (DRL-SSO) approach. The SSO method is employed for optimizing the hyper-parameters that are part of the DRL approach, while a neural network is utilized for learning and making decisions depending on feedback from the environment. Using a real-world dataset, the efficiency of the suggested method is assessed and contrasted with that of different cloud data management strategies already used.

Keywords: Cloud computing, Data management, Deep reinforcement learning (DRL), Spider swarm optimization (SSO)

1. Introduction

The term smart cloud data management (SCDM) describes using cutting-edge tools and methods to manage data kept in the cloud. This strategy uses techniques and technologies that assist businesses in better managing, protecting, and using their cloud data assets (Zhou, et al., 2020). Compared to conventional power grids, smart cloud increases computer services' dependability, efficiency, and substantiality. SCDM comprises maximizing the available data space in the cloud, removing redundant or obsolete data, and employing compression and deduplication methods to reduce the amount of storage needed (Ramesh, et al., 2020). Due to the demand for massive data processing and administration in the cloud, reliable and fault-tolerant data management methods have garnered attention. Cloud data management handles data analysis. This involves using data analytics technologies to uncover patterns and insights and applying them to make business decisions. Cloud use has led to several approaches for handling duplicated, separated data sets (Maiyya, et al., 2019). SCDM works well for cloud data management. As companies use cloud-based technology to store, process, and analyse data, effective cloud data management is essential. Choose the correct cloud storage option, backup data, streamline processing times, and optimise processing times to keep data safe and compliant (Cavicchioli, et al., 2022). Cloud data management lets firms grow their data processing and storage capacities to meet changing business needs. Every firm that uses data to make decisions and understand its operations needs good cloud data management (Saqlain, et al., 2019).

2. Related works

Chen, et al., 2021 suggested Holistic Big Data Integrated Artificial Intelligent Modeling (HBDIAIM) to improve the

*Corresponding author: snallola@gitam.in

[1]vayyasam@gitam.edu, [2]omaiaomari@su.edu.sa, [3]inbavallikumaran.2010@gmail.com

DOI: 10.1201/9781003490210-3

privacy and security components of the data management interface in various smart city applications. Sinaeepourfard, et al., 2019 emphasized the benefits of this data management architecture, including data collecting and storage efficiency rates and data and network traffic reduction. The cloud data management framework may use both the cloud and fog technologies' potentials, including removing communication latency and creating different rules. Wang, et al., 2020 suggested Guard Health, a decentralized, effective Blockchain solution for sharing and protecting data privacy. With managing sensitive information, Guard Health maintains confidentiality, authentication, data preservation, and data exchange. Any cloud data management plan must include security and compliance since organizations must ensure their data is safeguarded from cyber threats and other security flaws. Kakkar, et al., 2019 offered a study of the IoT and cloud integration, highlighting the implementation difficulties and advantages of the connection. Due to the enormous amount of data being processed, cloud-based data processing might be complicated. Mittal, et al., 2022 offered a unique approach for an efficient and secure cloud-based E-health model that uses identity-based encryption, along with implementation details. Utilizing distributed processing methods to boost productivity and reduce processing times, smart cloud data management entails improving data processing in the cloud. This involves figuring out the ideal ratio of computing power, storage, and networking resources.

3. Methodology

Deep reinforcement learning with Spider swarm optimization algorithm (DRL-SSO): The reinforcement learning (RL) and DRL algorithms were utilized to enhance the associated variables for managed SLDM systems. The RL method is a methodology-free Q-learning method that uses a state action function named Q π (t_s, b_d) to provide values to select a decision in the setting of the present state st that conforms to a policy. The Q-table contains the compensation or storage for this function. The value that is the function of the condition and its action, that follows the Bellman condition, is shown in (1).

$$R^{\Pi}\left(t_s, b_d\right) = q_s + \gamma \max R^{\Pi}\left(t_{s+1}, b_{s+1}\right) \tag{1}$$

The learning rate β is added to the algorithm to improve performance, as shown in (2).

$$R^{\Pi}\left(t_s, b_d\right) = R^{\Pi}\left(t_s, b_d\right) + \beta((q_s + \gamma \max R^{\Pi}$$
$$\left(t_s, b_d, b_{s+1}\right) - R^{\Pi}\left(t_s, b_d\right) \tag{2}$$

The process will be repeated till the terminal condition is attained. These state action functions of value are consistently

stored in the Q-table as Q-values. That reveals a flaw in the Q-learning method. The Q-learning method in RL won't exhibit information efficiency, learning effectiveness, or stability as a consequence. Use Deep Q-network (DQN), a modified form of conventional Q-learning that uses experienced replay, targeted networks, investigation, and exploitation methods, to get around this problem. The parameters of Q-networks may be updated by reducing the loss function. The loss function may be created using equation (3), and the goal state-action value function can be established using equation (4).

$$K\left(\theta\right) = F[\left(target_s - R\left(t_s, b_s : \theta\right)\right)^2] \tag{3}$$

$$target_s = q_s + \gamma_{b_{s+1}}^{\max} R'\left(T_{S+1}, b_{s+1} : \theta\right) \tag{4}$$

where target s is a target state action-value function that is evaluated by the target Q-network using the parameters θ'. At each iteration, S, all the variables of the assessed Q-network are changed. Subsequently, the target Q-network's parameter θ' remains fixed and only changes throughout static periods. The desired Q-network current rate is thus lower than the assessed Q-network. Algorithm 1 describes the DRL algorithm's pseudo code. This technique is used by the data center controller to collect the crucial cloud data metrics that are updated as the environment changes. The loss function is developed using network parameters after the initial transition function (t_s, bs, q_s, t_{s+1}), and it is then ne-tuned.

Algorithm 1: DRL

---

1. *Restore memories D to its highest level N*
2. *Begin with assessed Q-network values $\varnothing$*
3. *Set up a goal Q-network with values θ*
   where $\theta' = \varnothing$
4. *Begin*
5. *Every episode f do*
6. *Restore the state with load*
7. *Execute every job in the assignment queues*
8. *if probability $\in$ then*
   Select a random action to
   Else
   Select at = s_{st}
   End if
9. *Apply action, calculate total reward*
10. *Translocation to the newly created state*
11. *A change in stores (t_s, $\square\square b_s$, t_{s+}) in memory C*
12. *Execute pseudo code*
13. *For Every T step, update target Q-network $\theta' = \theta$*
14. *End For*
15. *End For*
16. *Return the reward*
17. *End*

SSO are socially organized organisms that exhibit aggressive traits against other members of their species. Individuals work together to do everyday duties that contribute to the SSO. SSO is broken up into two distinct evolutionary operators. One of the two is responsible for web building. The other is in charge of predicting and improving the hyper-parameters. These evolutionary operators are based on female and male spiders. It is presumed that the search space is a spider and a community spider web that each prospective solution in the population has created. Solution fitness values define each spider's weight. The most hopeful particle's position throughout the swarm's migration and in its immediate environment affects each particle's position. If the particle swarm is around the particle, the surrounding particles' best position is the fundamental particle's best position. SSO refers to this method. The partial SSO approach considers nearby surroundings. SSO optimises the procedures above by using cloud-based resources and data.

4. Results and discussions

Deep reinforcement learning-based smart cloud data management can result in a number of advantages and advancements for cloud computing systems. It's vital to keep in mind that the precise outcomes and advantages could differ based on execution, the complexity of the cloud environment, the features of the dataset, and the effectiveness of the deep reinforcement learning algorithms used. To verify and quantify the precise results of a smart cloud data management system built on deep reinforcement learning, real-world deployments, and tests are required. To manage cloud data, this research proposes a revolutionary DRL-SSO approach. The efficiency and accuracy of a proposed method are compared to those of ways such as Logistic Regression (LR), support vector machines (SVM), and Linear Discriminant Analysis (LDA). These techniques are compared with previous methods using several parameters, including accuracy, precision, and recall. **Accuracy:** The accuracy that the data stored in the SCDM correctly represents the actual-world information it is supposed to represent might be referred to in this context as accuracy. It entails making sure the data is accurate, consistent, and fault-free.

$$(5)$$

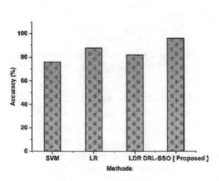

Fig. 3.1 Accuracy

Fig.1 shows the accuracy of the proposed and existing method. SVM has attained 76%, LR has attained 88%, LDA has attained 82%, whereas the proposed system reached 96% accuracy. It shows that the proposed approach has high accuracy than the existing one.

Precision: In the context of SCDM, precision refers to the accuracy and detail of the data saved in the cloud. It evaluates the degree that the recorded data corresponds to real-world values or measurements.

$$Precision = \frac{TP}{TP + FP} \qquad (6)$$

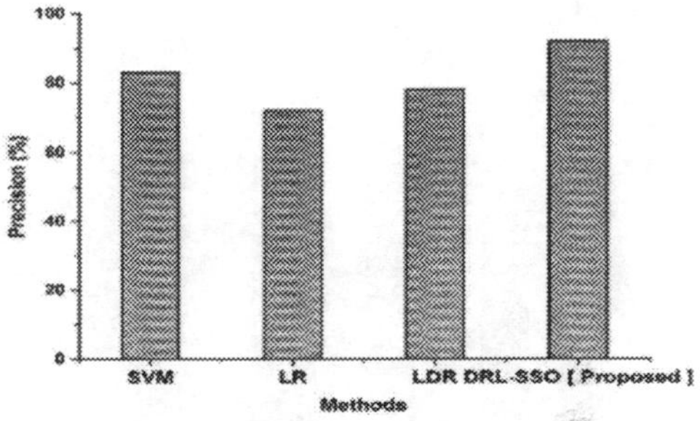

Fig. 3.2 Precision

Fig.2 shows the precision of the proposed and existing method. SVM has attained 83%, LR has attained 72%, and LDA has attained 78%, whereas the proposed system reached 92% of accuracy. It shows that the proposed approach has high precision than the existing one.

Recall: In the context of SCDM, recall refers to the capacity to access relevant data from the cloud as required. It evaluates whether the data can be called up or retrieved to respond to specific queries or demands.

$$Recall = \frac{TP}{TP + FP} \qquad (7)$$

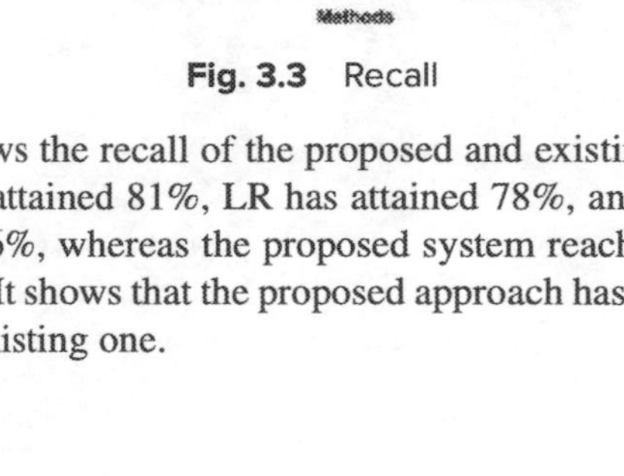

Fig. 3.3 Recall

Fig. 3 shows the recall of the proposed and existing method. SVM has attained 81%, LR has attained 78%, and LDA has attained 86%, whereas the proposed system reached 94% of accuracy. It shows that the proposed approach has high recall than the existing one.

Sensitivity: A statistic used to assess the efficiency of a binary classification model is sensitivity, sometimes referred to as recall or true positive rate. It counts the percentage of cases that the model actually identified as positive and did it accurately. The equation that follows is used to measure sensitivity:

$$Sensitivity = True\ Positives\ /\ (True\ Positives + False\ Negatives) \tag{8}$$

In this equation, "True Positives" denotes the percentage of positive cases that were accurately predicted, while "False Negatives" indicates the percentage of negative examples that were mistakenly anticipated.

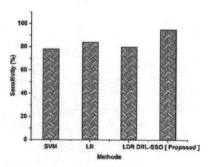

Fig. 3.4 Sensitivity

Figure 3.4 denotes the sensitivity of the proposed and existing method. SVM has attained 78%, LDA has attained 80%, and LR has attained 84%, whereas our proposed system reached 95%.

Specificity: Specificity is a statistic utilized to assess how well a binary classification model is doing. It gauges the percentage of actual unfavorable occurrences that the model properly predicted. The following formula is used to determine specificity:

$$Specificity = True\ Negatives\ /\ (True\ Negatives + False\ Positives) \tag{9}$$

In this equation, "True Negatives" refers to the percentage of cases that were accurately forecasted as negative, while "False Positives" refers to the percentage of instances that were mistakenly projected as positive.

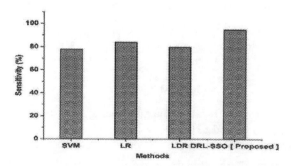

Fig. 3.5 Specificity

Figure 3.5 denotes the specificity of the proposed and existing method. SVM has attained 67%, LDA has attained 70%, and LR has attained 75%, whereas our proposed system reached 85% of specificity.

5. Conclusion

Advanced methods are needed to organise and analyse the expanding volume of data from varied sources. SCDM uses cutting-edge technologies and methodologies to manage, process, and store huge amounts of cloud data. DRL-SSO combines two cutting-edge methods to improve smart cloud data management. Our suggested method provides 96% of accuracy, 92% of precision, and 94% of recall, which is better than other methods. Neural networks learn to make decisions based on incentives and punishments in DRL. Spiders inspired evolutionary optimisation approach SSO. New encryption and authentication methods and blockchain data management may be explored in the future.

References

1. Zhou, T., Shen, J., Ji, S., Ren, Y., and Yan, L., (2020). Secure and intelligent energy data management scheme for smart IoT devices. Wireless Communications and Mobile Computing, 2020, pp.1-11.
2. Ramesh, D., Mishra, R., Edla, D.R. and Sake, M., (2020). Secure Identity-Based Proxy Signature With Computational Diffie-Hellman for Cloud Data Management. In Modern Principles, Practices, and Algorithms for Cloud Security (pp. 79-106). IGI Global.
3. Maiyya, S., Nawab, F., Agrawal, D. and Abbadi, A.E., (2019). Unifying consensus and atomic commitment for effective cloud data management. Proceedings of the VLDB Endowment, 12(5), pp.611-623.
4. Cavicchioli, R., Martoglia, R. and Verucchi, M., (2022). A novel real-time edge-cloud big data management and analytics framework for smart cities. Journal of Universal Computer Science, 28(1), pp.3-26.
5. Saqlain, M., Piao, M., Shim, Y., and Lee, J.Y., (2019). Framework of IoT-based industrial data management for smart manufacturing. Journal of Sensor and Actuator Networks, 8(2), p.25.
6. Chen, J., Ramanathan, L. and Alazab, M., (2021). Holistic big data integrated artificial intelligent modeling to improve privacy and security in data management of smart cities. Microprocessors and Microsystems, 81, p.103722.
7. Sinaeepourfard, A., Krogstie, J. and Petersen, S.A., (2020). D2C-DM: Distributed-to-Centralized Data Management for Smart Cities based on two ongoing case studies. In Intelligent Systems and Applications: Proceedings of the (2019) Intelligent Systems Conference (IntelliSys) Volume 2 (pp. 619-632). Springer International Publishing.

8. Wang, Z., Luo, N., and Zhou, P., (2020). GuardHealth: Blockchain-empowered secure data management and Graph Convolutional Network enabled anomaly detection in smart healthcare. Journal of Parallel and Distributed Computing, 142, pp.1-12.

9. Kakkar, L., Gupta, D., Saxena, S. and Tanwar, S., (2019). An analysis of the integration of Internet of things and cloud computing. Journal of Computational and Theoretical Nanoscience, 16(10), pp.4345-4349.

10. Mittal, S., Bansal, A., Gupta, D., Juneja, S., Turabieh, H., Elarabawy, M.M., Sharma, A. and Bitsue, Z.K., (2022). Using identity-based cryptography as a foundation for an effective and secure cloud model for e-health. Computational Intelligence and Neuroscience, (2022).

Note: All the figures in this chapter were made by the authors.

Advancing Sustainable Science and Technology for a Resilient Future – Sai Kiran Oruganti et al. (eds)
© 2024 Taylor & Francis Group, London, ISBN 978-1-032-79020-6

A Novel IoT Based Environmental Monitoring System Integrating Machine Learning Under the Background of Big Data

4

D. Saravanan*

Assistant Professor, Department of CSE, Sathyabama Institute of Science and Technology (Deemed to be University), Chennai, India,

Vijay Ramalingam[1]

Assistant Professor, Department of Computer Science and Engineering, Sathyabama Institute of Science and Technology (Deemed to be University), Chennai, India,

R. Parthiban[2]

Department of Computer Science and Engineering, IFET College of Engineering, Villupuram,

Manish Sharmad[3]

Associate Professor, Department of Computer Science & Engineering, Graphic Era Deemed to be University, Dehradun, Uttarakhand, India,

Abstract: In recent years, there has been a rise in environmental consciousness. This awareness motivates the effort to develop an efficient environmental monitoring system. Real-time water condition monitoring is essential for maintaining the water ecosystem in marine and archipelagic nations, which depend on the abundance of water resources. To enable real-time analysis, we use big data technology in conjunction with the water monitoring system. The purpose of this study is to develop an analytical system for the classification of water quality by using the Pollution Index approach. This system is an extension of the smart environment monitoring system (SEM). In addition, MQTT has replaced REST as the communication protocol. Real-time user interface implementation is included for visualization. The evaluations of performance metrics (Accuracy, Processing Time, and Mean Square Error) showed that the novel bird swarm optimization-based efficient SVM (BSO-ESVM) technique outperformed conventional approaches in analyzing the data.

Keywords: Smart environment monitoring (SEM), Internet of things (IoT), Water condition monitoring, Big data, Classification, Bird swarm optimization based efficient SVM (BSO-ESVM).

1. Introduction

The Internet of Things (IoT) is a new technology that makes it possible to connect intelligent gadgets to the Internet to form a network of interconnected items. This technology has a great deal of promise for ecological monitoring systems that can advance our knowledge of the natural world, offer early warning of environmental calamities, and allow us to take proactive steps to save our planet (Li, et al., 2021). A ground-breaking solution that utilises the power of both big data and machine learning to track environmental conditions in real time is an IoT-based ecological monitoring system that integrates machine learning while big data serves as the background. This system is made to gather and analyze information from various environmental sensors, such as those used to measure humidity, temperature, water quality, air quality, and the state of the soil (Hajjaji, et al., 2021). The system analyzes the data gathered from the sensors using a combination of machine learning techniques and data processing algorithms. This study aims to find trends and patterns in environmental information that may be utilized to forecast future circumstances (Climate Change, Biodiversity Loss, Pollution and Environmental Contamination). The system's scalability and flexibility enable simple adaptation to a variety of conditions in the environment and monitoring needs. Due to the system's scalability and use of the cloud for processing, it can analyze and store enormous amounts of data in real-time (Li, et al., 2022). One crucial component that makes it possible for the system to make exact and trustworthy forecasts about future environmental conditions

*Corresponding author: saravanan.d.cse@sathyabama.ac.in
[1]vijayrscs@gmail.com, [2]parthineyveli@gmail.com, [3]manishsharma.cse@geu.ac.in

DOI: 10.1201/9781003490210-4

is the incorporation of machine learning. The data gathered from the sensors is analyzed using machine learning algorithms to find trends and patterns that can be utilized to forecast future situations. The Decision tree method may be used, for instance, to forecast the probability of a forest fire based on information gathered from local temperature and humidity sensors. The system may utilize this information to build a predictive model that accurately predicts the chance of a forest fire developing by considering variables like the direction and speed of the wind, food volume, and other environmental parameters (Min, et al., 2019). The system is made to be highly accessible and simple to use. It can be accessed via a web-based interface from any gadget with a connection to the web. The interface enables users to view historical data, set up alerts for particular conditions, and monitor the surroundings in real time(Kukreja, et al., 2021).

2. Related works

Amanullah, et al., 2020explained the development of the Internet of Things (IoT), a platform for connectivity and interaction with many objects; technology has become ubiquitous in human existence. Atitallah, et al.,2020 described worldwide urban population growth is expanding quickly, posing new problems for inhabitants to deal with daily, such as pollution, public safety, traffic congestion, etc. Creating smarter cities, novel methods have been created to control this rapid growth. The Internet of Things (IoT) can be integrated into everyday life to develop new intelligence applications and services that benefit various city sectors, such as healthcare, security, agriculture, etc. IoT devices and sensors produce a significant amount of data that may be evaluated to learn essential facts and learn new things that improve the quality of life for citizens. IoT big data analytics may become more effective and efficient because of Deep Learning (DL), a recent development in artificial intelligence (AI). They review the literature on the application of the Internet of Things and DL to the development of smart cities in this study. Saheb, et al., 2019 Researchers telematics and health informatics have been significantly impacted by big datasets from IoT devices. They conducted a review of scientific papers and analyzed findings and trend on the Internet of Things Big Data Analytic paradigms (IoTBDA) in the field of healthcare. Identification of the effects of the IoT BDA paradigm on the conception, development, and implementation of IoT-based innovation in healthcare services is the aim of this work. In addition to reviewing 84 publications on cloud computing in the healthcare sector, they also did qualitative and quantitative reviews of 46 papers on IoTBDA. Jacob, et al., 2021 Explained the Internet of Things (IoT) is a network of interconnected devices and connections, A massive volume of data, and numerous consumers.

Machine learning is particularly suitable for these conditions because of its adaptability for "big data" difficulties and long-term concerns. For IoT management, however, maintaining security and privacy has become a severe issue. Recent studies show that deep learning algorithms are more effective at conducting security analyses for Internet of Things devices without the need for manually created rules. Sundas, et al., 2020 explored the cutting-edge technologies that have been used to handle trash in literary works. The authors also suggested an innovative waste disposal design that makes use of the Internet of Things and the processing of images.

3. Methodology

The Bird Swarm Algorithm

A brand-new global optimization algorithm called the Bird Swarm Algorithm (BSA) was inspired by the social iteration of bird behavior in the wild. The three primary activities of birds—are foraging, alertness. The following five steps serve as a summary of the algorithm's concept.

Step 1: Each bird may be in one of two states, either alertness or foraging.

Step 2: While the art of foraging, each bird keeps records of and remembers both its own and the swarm's finest experiences about food placements. This knowledge will influence how it moves and where it looks for food.

Step 3: Each bird in the alertness status makes a competitive effort to travel toward the central position of the flock, assuming that the birds with the highest reserves are located there.The likelihood of other predators attacking birds in the middle is lower.

Step 4: Birds continuously migrate from one location to another, alternately creating and scavenging food. The program makes the assumption that birds with higher reserves are producers and those with lower accounts are scroungers. Other birds, on opposite sides, are haphazardly considered to be providers or scroungers.

Step 5: Productive birds take the lead in looking for food, while scavengers sporadically lag.

4. SVM

The basic goal of the SVM method is to find a singular separating hyperplane (also known as the optimum margins hyperplane) that maximizes the difference between the two classes. Based on (1) training data points.

$$\left\{ (x_i, y_i) \right\}_{i=1,}^{l} x_i \epsilon R^N, y_i \epsilon \left\{ -1, 1 \right\}$$

The following optimization challenge needs to be resolved to use the support vector technique:

$$Minimize \, \Phi(\omega) = \frac{1}{2} w^T w + C \sum_{i=1}^{l} \xi i \quad (1)$$

$$yi \left(\omega, \phi(x_i) + b \right) \geq 1 - \xi i, i = 1, \ldots, l$$

$$\xi_i \geq 0, i = 1, \ldots, l \quad (2)$$

The function maps the training vectors xi onto a higher-dimensional space. The tradeoff between categorization violations and margin maximization is controlled by the positive user-specified parameter C. Its dual, the finite quadratic programming problem, is the most often used way to solve (1) today:

$$W(\alpha) = \sum_{i=1}^{N} \alpha_i - \frac{1}{2} \sum_{i,j=1}^{N} \alpha_i \alpha_j y_i y_j \left(\phi(x_i), \phi(x_i) \right)$$

$$= \sum_{i,j=1}^{N} \alpha_i - \sum_{i,j=1}^{N} \alpha_i \alpha_j y_i y_j K(x_i, x_j) \quad (3)$$

$$\sum_{i=1}^{N} y_i \alpha_i = 0, 0 \leq \alpha_i \leq C, i = 1, \ldots, l \quad (4)$$

where the Lagrangian argument is used. The absolute equality in (3) uses the technique known as the kernel as can be seen. The SVM's Kuhn-Tucker criteria are described by

$$\alpha_i \left[y_i \left(w, \phi(x_i) + b \right) - 1 + \xi_i \right] = 0, i1, \ldots, l$$

$$(C - \alpha_i) \xi_i = 0, i = 1, \ldots, l. \quad (5)$$

A support vector is a value xi with the matching αi> 0.

$$w_0 = \sum_{i=1}^{l} \alpha_i y_i \phi(x_i) = \sum_{i=1}^{l_s} \alpha_i y_i \phi(x_i),$$ The yields the superior value of the weight vector, where ls is the total amount of support vectors.

$$f(x) sign \left(w_{0,} \phi(x) b_0 \right) sign \left(\sum_{i=1}^{l} \alpha_i y_i \phi(x, x_i) + b_0 \right) \quad (6)$$

5. Result and discussion

The experiment is conducted using a MATLAB tool. In this section, the Accuracy, MSE, and Processing time of the suggested BSO-ESVM's performance are evaluated. The suggested BSO-ESVM is compared to existing RF, DT, and SVM techniques. The accuracy results for the guided and existing procedures are shown in Fig. 4.1. The percentage of the data for which the proposed BSO-ESVM correctly predicted the outcome is used to measure the model's effectiveness.

A measure of an outcome or results correctness or accuracy is called accuracy. It is frequently applied when assessing the

effectiveness of a model, system, or process by contrasting the projected or accomplished results with the actual or intended results. Figure 4.1 shows the accuracy of the suggested and accepted methods. While RF, DT and SVM only obtain 98.13%, 95.46% and 98.78% accuracy, respectively, the proposed approach BSO-ESVM receives 99.71% accuracy. BSO-ESVM proposed method is more accurate than traditional methods.

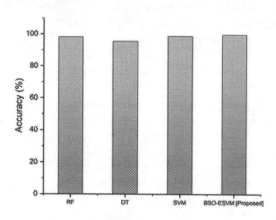

Fig. 4.1 Accuracy

Processing time is the length of time it takes for a system or process to finish a certain activity or function. It is the length of time taken from the start to the finish of a certain procedure or operation. Figure 4.2 illustrates the processing time of the suggested and accepted methods. The recommended method, BSO-ESVM, achieves a time of 0.248s, while RF, DT and SVM only achieve a processing time of 0.843s, 0.716s and 0.298s, respectively. BSO-ESVM procedures provide a lower time as compared to traditional methods.

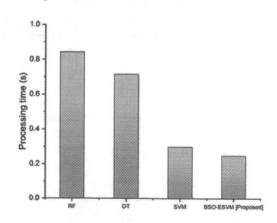

Fig. 4.2 Processing time

The average squared difference between a dataset's projected values and actual values is measured using the statistical metric known as mean squared error (MSE). Figure 4.3 shows the MSE for the suggested and common techniques. The methods mentioned While RF, DT and SVM only receive

0.076, 0.095 and 0.094 error, respectively; BSO-ESVM receives 0.071 error. The error of BSO-ESVM techniques is lower than that of conventional methods.

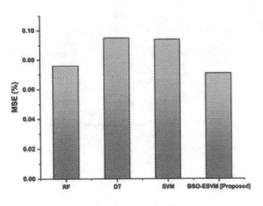

Fig. 4.3 MSE

5. Conclusion

The integration of IoT and big data has been used in this article to conduct a classification extension using IoT big data analysis for intelligent monitoring of the environment and real-time system. Evaluations proved that the information analytic function using the linear SVM was practical. An interesting field of study is improving the performance of support vector machines (SVM) for classification issues using the Internet of Things (IoT) and bird swarm optimization (BSO). BSO algorithms can enhance the characteristics of SVM to increase its precision as well as speed by leveraging IoT devices to gather data from numerous sources. Numerous industries, including healthcare, finance, and transportation, could benefit from the use of this strategy. However, additional study is required to comprehend the drawbacks and possibilities of this strategy as well as to improve the algorithms utilized for BSO and IoT data collecting. The values of performance metrics for our suggested method were obtained in terms of accuracy (99.71%), processing time (0.248s), and MSE (0.071). In summary, the application of BSO optimization for SVM using IoT has the potential to significantly increase accuracy and efficiency in classification tasks, leading to improved decision-making and results across various industries.

References

1. Li, W., Chai, Y., Khan, F., Jan, S.R.U., Verma, S., Menon, V.G. and Li, X., 2021. A comprehensive survey on machine learning-based big data analytics for IoT-enabled intelligent healthcare system. Mobile networks and applications, 26, pp.234-252.
2. Hajjaji, Y., Boulila, W., Farah, I.R., Romdhani, I. and Hussain, A., 2021. Big data and IoT-based applications in smart environments: A systematic review. Computer Science Review, 39, p.100318.
3. Li, X., Liu, H., Wang, W., Zheng, Y., Lv, H. and Lv, Z., 2022. Extensive data analysis of the internet of things in the digital twins of smart city based on deep learning. Future Generation Computer Systems, 128, pp.167-177.
4. Min, Q., Lu, Y., Liu, Z., Su, C. and Wang, B., 2019. Machine learning based digital twin framework for production optimization in petrochemical industry. International Journal of Information Management, 49, pp.502-519.
5. Kukreja V., Kumar D. "Automatic Classification of Wheat Rust Diseases Using Deep Convolutional Neural Networks."2021 9th International Conference on Reliability, Infocom Technologies and Optimization (Trends and Future Directions), ICRITO 2021. 2021
6. Amanullah, M.A., Habeeb, R.A.A., Nasaruddin, F.H., Gani, A., Ahmed, E., Nainar, A.S.M., Akim, N.M. and Imran, M., 2020. Deep learning and big data technologies for IoT security. Computer Communications, 151, pp.495-517.
7. Atitallah, S.B., Driss, M., Boulila, W. and Ghézala, H.B., 2020. Leveraging Deep Learning and IoT big data analytics to support the smart cities development: Review and future directions. Computer Science Review, 38, p.100303.
8. Saheb, T. and Izadi, L., 2019. Paradigm of IoT big data analytics in the healthcare industry: A review of scientific literature and mapping of research trends. Telematics and informatics, 41, pp.70-85.
9. Jacob, I.J. and Darney, P.E., 2021. Design of deep learning algorithm for IoT application by image based recognition. Journal of ISMAC, 3(03), pp.276-290.
10. Sundas, A. and Panda, S.N., 2020, June. IoT based integrated technologies for garbage monitoring system. In 2020 8th International Conference on Reliability, Infocom Technologies and Optimization (Trends and Future Directions)(ICRITO) (pp. 57-62). IEEE.

Note: All the figures in this chapter were made by the authors.

Advancing Sustainable Science and Technology for a Resilient Future – Sai Kiran Oruganti et al. (eds)
© 2024 Taylor & Francis Group, London, ISBN 978-1-032-79020-6

Facial Recognition Model using Hyper-Tuned ResNet50 in Localized Datasets

5

Shem L. Gonzales*

Surigao del Norte State University –
Del Carmen Campus, Philippines,

Jerry I. Teleron[1]

Surigao del Norte State University-
Main Campus, Philippines,

Rex Bomvet D. Saura[2]

Surigao del Norte State University –
Del Carmen Campus, Philippines

Abstract: This study developed a model that strengthens the framework for face recognition with the application of computer vision. ResNet50 from Deep Convolutional Neural Network Architecture was explored to create a model using transfer learning. Old photos were used as datasets in this study. Images were collected from seven (7) local personalities who served as classes. Each class contained thirty (30) images. The number of classes used to fit and train the model was used to fine-tune the output layer. The performances were evaluated using the different accuracy metrics generated from the learning curve. There was a 1% drop in training accuracy validation and testing accuracy with the hyper-parameter combinations and adjustments, yielding 73.68% and 68.42%, respectively.

Keywords: Computer vision, Deep learning, Face recognition, Hyper-parameter tuning, ResNet50

1. Introduction

Face Recognition is one of the highlights of biometric recognition (Wilmer, 2017). Its uniqueness provides a realization in many technology applications for a person's authentication (Boutet et al., 2015). Since the face is the most exposed physiological characteristic of humans, it provides an advantage and acceptance for providing relevant data that can be used in developing software applications for recognition and verification. Also, given its non-intrusive advantage, facial images can be quickly obtained without physical contact (Adiabi et al., 2020). The military is used for surveillance and security, schools and work areas incorporated for attendance monitoring, and business and online web applications for authentication (Bah et al.,2019).

I.   Several researchers proposed different methods of utilizing computer vision for face recognition. The study

of Dhawle et al (2020) used a cascade classifier from Open Computer Vision (OpenCV) to identify facial features in images and incorporated the Local Binary Pattern Histogram (LBPH) algorithm for recognizing both front and side faces in real-time (Deeba et al., 2019). Principal Component is another technique that was used for extracting facial features mathematically. This eigenfaces method is done by calculating the eigenvector to represent a large matrix for machine learning computation (Rowden & Jain, 2017). With the determined search for human-level accuracy, Deep Learning was established for enhancing the performance of classification models (Alzubaidi et al., 2021). The high computational cost of learning from scratch for training a model has been hurdled with the help of several pre-trained neural networks. The method integrates the concept of applying solutions learned from

---
*Corresponding author: j.doe@gmail.com.
[1]slgonzales@ssct.edu.ph

DOI: 10.1201/9781003490210-5

other similar problems. Due to the capability of feature learning, datasets that support the classes in classification problems are minimized (Setiowati et al., 2017). The consumed time and hardware resources are also lessened in processing with Deep Learning Convolutional Neural Network (CNN) Architectures.

Those mentioned methods computationally showed promising results. It was observed that in most study, the datasets that were used for training a model were from online image providers. These images provided by the web contain high resolution which gives less problem in terms of image quality. In this study, images from photo album collections were utilized as datasets. Seven (7) local personalities were selected as classes with 30 images per class. It contains a combination of low-quality images as to resolution, brightness, and contrast distribution, with a random age difference of 10 years from the class's present age. Further, (Alzubaidi et al., 2021) stated Deep Learning provides several CNN architectures that can be applied specifically in multiclass classification problems.

Thus, the outstanding performance of ResNet50 was employed in the study to attain the main goals of this paper as follows:

- To develop a face recognition model using ResNet50 using localized datasets and determined the training, validation, and testing accuracy.

- To describe optimal performance produced from harmonizing the hyper-parameter values of ResNet50

- To evaluate the training, validation, and testing accuracy of the hyper-tuned ResNet50 model.

2. Conceptual Framework

The paper adheres to the theory of human visual information processing which summarizes encoding visual images into neural patterns, detecting basic facial features, size standardization, dimensional reduction of the neural patterns, and correlation of the resulting pattern sequence with all visual patterns already stored in memory (Baron, 1981). Figure 1, showed the framework of the study to develop a face recognition model using ResNet50. An old photograph of a person in different age brackets (10, 20, 30 years) was experimented with to develop a face recognition model. Pre-training of the data set and creating a trained model hereafter undergone model evaluation through training accuracy, validation, and testing accuracy. Henceforward, hyperpara

meter tuning was done to produce optimal performance hereby rerun for model evaluation.

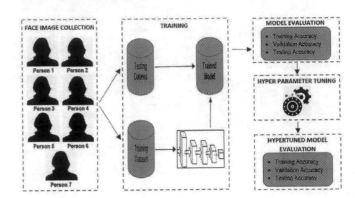

Fig. 5.1 Face recognition framework

3. Methodology

A. Materials

Computer hardware and software were used in this study. The development was based on software materials with a Windows 10-64 bit operating system, anaconda navigator, and installed Python, OpenCV, TensorFlow, and Keras. The Spyder was used for coding Python Script as Integrated Development Environment (IDE). Since the images were from photo album collections, Adobe Photoshop was utilized to convert the images into JPEG file format. For computer hardware specification, a desktop computer was used with a 2.80 GHz 8-core i7, 32-GB RAM, storage of 1 TB, and a graphical processing unit of 4 GB. To convert the images into digital format, an Epson 3110 was used.

B. Photo Image Collection, Preprocessing, and Training.

In this paper, the researchers identified seven (7) local personalities that serve as classes with a total of thirty images collected from photo albums to provide the datasets. The obtained photos were with the consent of the owner and the author's observed adherence to RA 10173, the Data Privacy Act. Each acquired face image was manually cropped and resized into 224x224 square pixels to satisfy the required input image of the neural network. Fig. 2 shows the sample of images that were used for training the model. Following the preparation of the datasets, they were fed into the neural network for training. The datasets were split into training and validation. 70% of the images were used as training datasets, with the remainder used for validation and testing. During the training process, images were augmented to keep the network from over fitting.

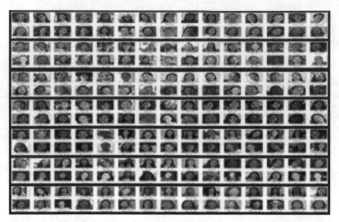

Fig. 5.2 Sample of collected face images

Note: if possible, please try to use composite figures instead of stand-alone figures.

C. Approach

ResNet50 was considered the convolutional neural network (CNN) architecture for this study. This CNN has the capability of dealing with training and overfitting problems in Deep Learning with the knowledge of residual learning. As the training developed, the deeper the network, the deeper the model will learn producing a degradation of image classification accuracy. This degradation is the increase of training error as more layers are added to the neural network (Baron, 1981). With the residual blocks design of the ResNet architecture, it can prevent degradation with the combination of multiple convolution filters resulting in the reduction of training time.

D. ResNet50 Hyperparameter Configuration

Table 1 displayed the values of the ResNet50 hyper-parameter by default and hyper-tuned. Adam optimization algorithm an optimizer used revealed the learning rate was 0.0001 considered optimized at the recommended range.

Table 5.1 Default and hyper-tuned values of ResNet50 hyper-parameter

Hyper-parameter	Default Values	Hyper-tuned Values
Batch size	5	32
Learning Rate	0.00001	0.0001
Optimizer	Adam	Adam
Dense	1024	512

3. Results and discussions

This section shows the learning curves and the generated training, validation, and training accuracy during the training process.

A. Learning Curves of ResNet50 in Default Values of Hyper-parameter

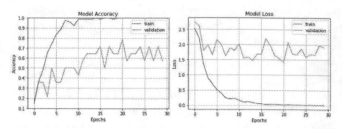

Fig. 5.3 Model and Loss Accuracy of ResNet50 using default given values.

B. Learning Curves of ResNet50 in Default Values of Hyper-parameter

The Fig. 5.4 provided an illustration of how the model learned and fitted during the training. As training accuracy increase, validation accuracy also increases along with the values of every epoch. The curve shown a best fit and overfitting was not observed.

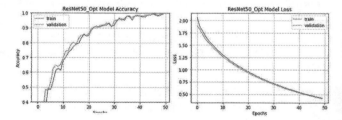

Fig. 5.4 Default and hyper-tuned values of ResNet50 hyper-parameter

C. Comparison of Accuracy

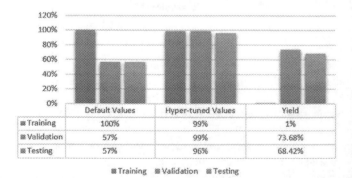

Fig. 5.5 Performance comparison of ResNet50

The different illustrations were presented in this section. The generation of the learning curve from Fig. 5.3 produced a 100% training accuracy, 57% validation accuracy, and 57% testing accuracy using the initial values of hyper-parameters in Table 5.1. To obtain optimal performance, the hyper-tuned values in Table 5.1 were applied, causing the training accuracy to be 99%, the validation accuracy to be 99%, and the testing accuracy to be 96%. Figure 5.5 provided a yield of 73.68% validation accuracy and 68.42% testing accuracy. The training accuracy was observed to have a drop of 1%.

4. Conclusion

The study's findings showed that ResNet50 obtained poor performance during the initial training. Learning rate was optimized to 0.0001 a significant improvement in accuracy metrics resulted during the hyper-parameter adjustment using Adam optimizer. Although there was a 1% drop in training accuracy, the hyper-tuned ResNet50 produced an outstanding model performance, yielding results of 73.68% for validation accuracy and 68.42% for testing. When the model's performance was assessed, the results were 100% Precision, 81.25% Recall, and 89.65% F1-score. Since overfitting did not occur in the performance, it is advised to use the model.

References

1. Wilmer, J. B. (2017). Individual Differences in Face Recognition: A Decade of Discovery. Current Directions in Psychological Science, 26(3),225–230. https://doi.org/10.1177/0963721417710693
2. Boutet, I., Taler, V., & Collin, C.A. (2015). On the Particular Vulnerability of Face Recognition to Aging. A Review of the Three Hypotheses. Frontiers in Psychology,6, Article 1139. https:// doi/10.3389/fpsyg.2015.01139
3. Adiabi, I., Ouahabi, A., Benzaoui, A. & Taleb-Ahmed, A (2020). Past, Present, and Future of Face Recognition: A Review. Electronics. https://doi.org/10.3390/electronics9081188
4. Bah, S.M., & Ming F. (2019). An Improve Face Recognition Algorithm and Its Application in Attendance Management System. Elsevier Inc. https://doi.org/10.1016/j.array.2019.100014
5. Dhawle T., Ukey, U., & Choudante, R. (2020). Face Detection and Recognition using OpenCV and Python. International. Research Journal of Engineering and Technology.
6. Deeba, F., Memon, H., Dharejo, F., Ahmed, A., & Ghaffar, A. (2019). LBPH-based Enhanced Real-Time Face Recognition. International Journal of Advanced Computer Science and Applications. http://dx.doi.org/10.14569/IJACSA.2019.0100535
7. Rowden, L. & Jain A.K. (2017) Longitudinal Study of Automatic Face Recognition. IEEE. https://doi.org/10.1109/tpami.2017.2652466
8. Alzubaidi, L. et al. (2021). Review of deep learning: concepts, CNN architectures, challenges, applications, future directions. J Big Data 8, 53. https://doi.org/10.1186/s40537-021-00444-8
9. Setiowati, S., Zulfanahri, Franita, E.L., & Ardiyanto, I. (2017). A Review of Optimization Method in Face Recognition: Comparison Deep Learning and Non-Deep Learning Methods. 9th International Conference on Information Technology and Electrical Engineering (ICITEE), 1-6.
10. https://doi.org/10.1109/ICITEED.2017.8250484

Note: All the figures and table in this chapter were made by the authors.

Advancing Sustainable Science and Technology for a Resilient Future – Sai Kiran Oruganti et al. (eds)
© 2024 Taylor & Francis Group, London, ISBN 978-1-032-79020-6

Mobile Application for Risk Assessment Using Naïve Bayesian Algorithm with Distance Notification against COVID-19

6

Aleta C. Fabregas*
Polytechnic University of the Philippines,
Philippines,

Carlo Inovero[1]
Polytechnic University of the Philippines,
Philippines.,

Ma. Leonila Amata[2]
Polytechnic University of the Philippines,
Philippines.

Armin S. Coronado
Polytechnic University of the Philippines,
Philippines.

Abstract: This research project developed a mobile application using Naïve Bayesian Algorithm that is able to perform self-assessment of any individual upon exposure to COVID-19 patients and recommended distance thereof as isolation intervention in combatting the spread of the virus using Received Signal Strength Indicator (RSSI). The target value for self-assessment is either Low or High risk with four selected features such as age, gender, pre-existing conditions and fever. The study evaluated the performance of the system by using the data from the university's medical clinic. There are 109 records used as training dataset to generate the model. The initial accuracy of the model is high which is 92.30. Distance Notification with RSSI, using BLE is functional but exact distance between contacts is still to improve due to effects of some environmental factors.

Keywords: Naïve Bayesian algorithm, Bluetooth, Bayesian model, RSSI

1. Introduction

Since January 2020, around 4 million cases of COVID-19 in the Philippines have been reported to the World Health Organization World Health Organization. (2022.). As of December 2021, the age group with the highest percentage of COVID-19 cases in the country were individuals aged between 20 and 39 years old, according to the Department of Health – Philippines. Department of Health (2021). Recently, the spread of COVID-19 infection in the University became very alarming. From 2020 until the first quarter of 2022, a total of 278 COVID-19 cases and 5 unfortunate deaths. According to the data from the University, 131 out of 278 confirmed cases are from the ages 20-39 years old, or the adults.

The Medical Services Department (MSD) lacks manpower in monitoring the increasing cases as wells as providing interventions to mitigate the escalation of the new sickness during isolation and/or quarantine of individuals contacted from the employees with confirmed infection of the virus. Therefore, a mobile application that can be used to assess the risk of vulnerable individuals would be of great help in mitigating the fast spread of the virus. The PUP Research Group in collaboration with the College of Computer and Information Sciences (CCIS) is proposing to develop a mobile application using Naïve Bayesian algorithm for self-assessment of contacted individuals with confirmed cases of COVID-19.

2. Related Literature

It was reported that Age, Gender, and Pre-Existing Medical Conditions are associated to the severity and mortality of confirmed cases. Jin, J.-M., Bai, P., He, W., Wu, F., Liu,

*Corresponding author:aletfabregas@gmail.com.
[1]cginovero@pup.edu.ph, [2]lanie.amata@gmail.com, [3]ascoronado@pup.edu.ph

DOI: 10.1201/9781003490210-6

X.-F., Han, D.-M., Liu, S., & Yang, J.-K. (2020). Moreover, the transmission of the virus is affected by the environmental conditions, which are favorable at temperatures between 21°c - 23°c with relative humidity of 40% over 7-day period and can survived up to 24 hours hanging on air and surfaces of various materials E. J. Anderson and Others, Gessain, A., & Others, M. B. and. (2020). Thus, these parameters will be explored to determine the risk level of an individual upon exposure to infected patients. Social distancing, approximately 2 meter to 3 meter apart, showed an effective way to stop the increase of COVID-19 Hans-Jürgen and Meckelburg (2020). It is then proposed that the mobile application can perform monitoring the distance of the contacted individuals within the workplace as well as during isolation in their respective homes to prevent its spread.

The contribution of this study is detecting risk level of Covid-19 with the distance notification using Bluetooth calculating Received Signal Strength Indicator (RSSI).

3. Methodology

The mobile application will evaluate the risk of the user based on the following parameters: (1) age; (2) gender; (3) pre-existing conditions; (4) body temperature; and (5) proximity of contact. Machine Learning Algorithm using Python will be used to generate the risk condition as to high and low risk. The researchers perform data cleaning by deleting records with blanks, inconsistent data type and labeling. Finally, the 278 was reduced to 109 records. The study used the Naïve Bayes model to determine the level of risk for COVID-19 infection.

Baye's Theorem Model

To calculate a posterior probability of A occurring given that B occurred:

$$P(A\#B) = P(B)P(B\#A) \cdot P(A)$$

where:

A,B – Events

P(B | A) – The probability of B occurring given that A is trueP(A) and P(B) – The probabilities of A occurring and B occurring independently of each other.P(A|B) - posterior probability is thus the resulting distribution, Naive Bayes was used because of its simplest probabilistic classifiers.

On the other hand, distance notification feature of the mobile application measures the distance between two individuals using the calculated Received Signal Strength Indicator (RSSI).

It measures how well a client device can hear (receive) a signal. Users are required to have a mobile device with android operating system and Bluetooth to determine the RRSI value. The system will notify the user if the required distance is within the required range of distance recommended during the isolation and/or quarantine. The RSSI value determines the distance between each user where: Distance = 10 ^ ((Measured Power — RSSI)/ (10 * N)). Measured power is set properly to the expected signal level at one meter which is -69 and it is often pre-configured into the device by the manufacturers but adjustable depending on the instances it would be used.

How the Bluetooth RSSI Works

For optimal performance, Bluetooth RSSI operates most effectively when the connection is robust. Familiarizing oneself with their device's Received Signal Strength Indicator (RSSI) is crucial due to variations in Bluetooth versions and connecting devices. The RSSI reading provides an indication of the Bluetooth signal strength, but it does not necessarily correlate with the speed of the connection. Its primary purpose is to help users assess the strength of the received signal.

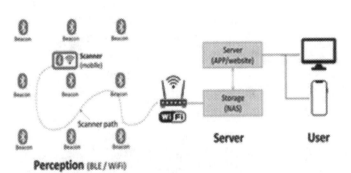

Fig. 6.1   RSSI workflow

The parameters of RSSI are the following: Distance, Measured Power,N (Constant depends on the Environmental factor. Range 2–4, low to-high strength.

A. System Architecture

The proposed system architecture of the research project shows the detailed process from the user wherein the input parameters will be evaluated by the Bayesian Naïve algorithm and the results will be tested to produce reliable model for predicting and assessing the probable risk. The application provides a distance alert feature that uses the Received Signal Strength Indicator (RSSI) value from the Bluetooth on the user's mobile phone to notify them if someone is not within the recommended government distance to stop the spread of the COVID-19 virus.

4. Results and discussions

A. Initial Findings of the Study

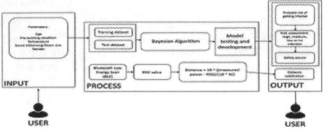

Fig. 6.2   System Architecture: PUP E-Guide: Mobile Application for Risk Assessment with Distance Safety Advice to Prevent the Spread of COVID-19

A. *Initial Findings of the Study*

The Initial Findings of the Study using Naïve Bayesian Algorithm in determining the Risk Level of 109 infected by COVID-19 employees/faculty in the state university are first determined by the frequency by age, gender, employee types, pre-existing condition, and body temperature.

The data were divided into age groups following the standard stages of life with the input from the INTEGRIS physicians Health, I. N. T. E. G. R. I. S. (n.d.)(2015).

The frequency and likelihood tables of all the patients with COVID-19 were presented below of how Naïve Bayesian algorithm works [8].

Table 6.I   Frequency and likelihood with the type of Risk for COVID-19 based on the Age Group

Age	Frequency	High Risk	Low Risk	Posterior Probability
Adult (20-39 years old)	61/109	33/57	28/52	.57
Middle Age Adult (40-59 years old)	42/109	23/57	19/52	.38
Senior (60 years old and above)	6/109	1/57	5/52	.05

Table 6.2   Frequency and likelihood table with the type of Risk by Gender for COVID-19 based on Gender

Gender	Frequency	High Risk	Low Risk	Posterior Probability
Female	66/109	37	29	.61
Male	43/109	20	23	.39

Table 6.3   Frequency likelihood table with the type of Risk by Gender for COVID-19 based on based on the Body Temperature

Body Temperature	Frequency	High Risk	Low Risk	Posterior Probability
Has Fever	55/109	54	1	.51
Normal	54/109	3	51	.49

Tables 6.1-6.3 shows that most of the employees who were tested positive of the virus are Adult, Female and experienced high fever. According to Mayo Foundation, the average body temperature is 98.6 F or 37.0 C but it can range between 97 F (36.1 C) to 99 F (37.2 C) [7]. The group with high fever is experiencing high risk of being infected by COVID-19.

Table 6.4   Likelihood table of the Employees with High and Low Risks

OVERALL FREQUENCY		
Risk	Frequency	Posterior Probability
High Risk	57/109	.52
Low Risk	52/109	.48

Based on the overall frequency of 109 covid-19 patients, the level of High risk infection is higher than low risk.

Python program is used to for learning and evaluation phases. There are 109 records trained using Naïve Bayesian algorithm model. Thirty percent (.30) of the trained records used for validation and testing. Model building and training is achieved using GaussianNB. Model Evaluation is achieved thru sample output simulation to test the model for prediction.

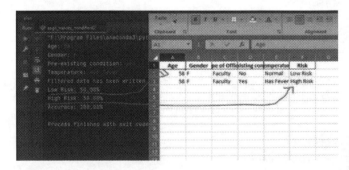

Fig. 6.3   Python simulation determining low risk and high risk using the four attributes (age, gender, pre-existing condition, Temperature

```
# Split the data into training and testing sets
    X_train, X_test, y_train, y_test = train_test_split(X_encoded,
y, test_size=0.3, random_state=70)
    # Create a Naive Bayes classifier (GaussianNB)
    clf = GaussianNB()
    # Train the classifier using the training data
    clf.fit(X_train, y_train)
    # Make predictions on the test data
    y_pred = clf.predict(X_test)
```

Fig. 6.4   Python simulation with the output of the dataset low and high risk

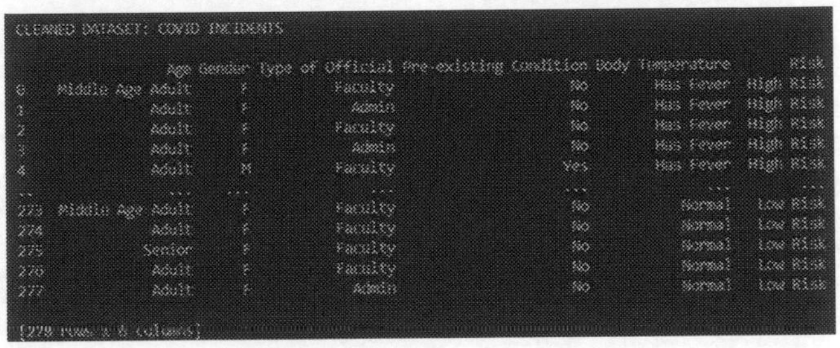

Fig. 6.5   Screenshot of the python program with learning and evaluation method

Measured Power	Trial 1	Trial 2	Trial 3	Trial 4	Trial 5	Trial 6	Trial 7	Trial 8	Trial 9	Trial 10
-60	3.1623 meters	7.9433 meters	3.1623 meters	6.3096 meters	4.4668 meters	10 meters	4.4668 meters	1.9953 meters	1.9953 meters	14.1254 meters
-65	3.9811 meters	0.7943 meters	1.122 meters	1.2589 meters	0.7079 meters	1.2589 meters	1.2589 meters	0.7943 meters	1.122 meters	1 meters
-70	0.3981 meters	0.2818 meters	0.3162 meters	0.4467 meters	0.4467 meters	0.4467 meters	0.3162 meters	0.4467 meters	0.3162 meters	0.4467 meters

Fig. 6.6   The data table below displays the corresponding distance for a certain measured power

The prototype using Naïve Bayesian algorithm is producing very high accuracy of 92.3 % in determining high and low level risk of COVID-19 infection. Still, an increase dataset from the university must be considered to make the model more consistent.

B. Bluetooth and RSSI for Distance Notification

Initial Findings using the Blue tooth and the RSSI. The distance between the smart fitness band and smartphone is 1 meter.

The measured power has a significant impact on the distance result. There are, however, gaps where the distance is too far from the desired distance. Humidity, air, and other interference that may alter the output are examples of environmental conditions that may be to blame for this issue with Bluetooth Low Energy (BLE)or the RSSI number. This means that, on this setting, a measured power of -65 is suitable. Additionally, the value of N is constant according to environmental factors. 2-4 range, and in this initial finding, N = 4 is the only value that was tested and has low-to-high strength. Notification for social distancing is improved by 50 % when the measured power is set to -65.

5. Conclusion

The prototype using Naïve Bayesian algorithm is producing very high accuracy of 92.3 % in determining high and low level risk of COVID-19 infection. Using Gaussian Naive Bayes (NB), the model predicts with high accuracy. Still,

there is the need to increase the dataset from the medical unit of the university to make very good model, since some predictions are not accurate. However, if the goal is to achieve a highly accurate distance, optimization is required.

References

1. World Health Organization. (2022.). [Philippines: Who coronavirus disease (covid-19) dashboard with vaccination data. World Health Organization.
2. Department of Health (2021). Philippines coronavirus disease (COVID-19) situation report #91, 6 December 2021 - Philippines. (2021, December 16).
3. Jin, J.-M., Bai, P., He, W., Wu, F., Liu, X.-F., Han, D.-M., Liu, S., & Yang, J.-K. (2020, April 29). Gender differences in patients with covid-19: Focus on severity and mortality. Frontiers in public health.
4. E. J. Anderson and Others, Gessain, A., & Others, M. B. and. (2020, May 14). Aerosol and surface stability of SARS-COV-2 as compared with SARS-COV-1: Nejm. New England Journal of Medicine.
5. Hans-Jürgen and Meckelburg,[April 2020] Contact Tracing Coronavirus COVID-19 -Calibration Method and Proximity Accuracy, DOI: 10.13140/RG.2.2.36337.22884
6. Health, I. N. T. E. G. R. I. S. (n.d.). Stages of life: Health for every age. Integris. Retrieved from https://integrisok.com/resources/on-your-health/2015/october/stages-of-life-health-for-every-age

Note: All the figures and tables in this chapter were made by the authors.

Advancing Sustainable Science and Technology for a Resilient Future – Sai Kiran Oruganti et al. (eds)
© 2024 Taylor & Francis Group, London, ISBN 978-1-032-79020-6

7

Physicochemical and Microbial Water Quality of the Coastal Waters of Sogod, Southern Leyte

Lelie Lou Bacalla*

Southern Leyte State University, Philippines

.**Abstract:** Marine water quality is important especially to communities that rely on marine waters for livelihood, economy and recreation. In the case of Sogod, Southern Leyte, its coastal waters are used for recreation, tourism and fishing. However, as it is the most commercially advanced community surrounding Southern Leyte's major water body, the Sogod Bay, its coastal waters are more susceptible to pollution. However, no past and present data on the physicochemical characteristics is currently available to determine the present health of the water body. In this study, a physicochemical and microbial water quality assessment of the coastal waters of Sogod Bay was carried out. Samples were taken within the coastal region of the most commercialized area of Sogod using direct sampling. The collected samples were analysed with methods mandated by the Water Quality Guidelines and General Effluent Standards (DAO 2016-08). Compared with the standards set in DAO 2016-08, the physicochemical characteristics of the coastal waters of Sogod, Southern Leyte are all acceptable and are waters suitable for the establishment of marine parks and sanctuaries. However, the microbial property showed elevated levels of fecal coliform. This makes the coastal waters of Sogod unsafe for the residents living around the coastal waters of the sampling area as they may carry water-borne diseases.

Index Terms: Coastal waters, Microbial analysis, Physicochemical analysis, Microbial analysis

1. Introduction

Water bodies are essential, especially in providing a living space for marine flora and fauna. In archipelagic countries such as the Philippines, besides being home to marine species, it also offers recreation (boating, swimming, water sports, etc.), livelihood, irrigation, transportation, livestock watering, tourism, and agricultural uses [1]. Coastal waters are used for small-scale and commercial fishing and as a food source through the fish and seafood that are consumed daily. In fact, the fisheries sector of the Philippines is valued at 273 billion pesos in 2020 [2]. Furthermore, it also plays a big part in tourism through diving and developing beaches all over the country. In fact, according to Zafra, in 2021 [3], coastal and ocean tourism is the most significant contributor to the Philippines' blue economy. PEMSEA has established this to have a value of around 884 billion pesos [4].

Different classifications of water bodies are for different uses. Marine waters, specifically, greatly influence the growth of fishes and marine organisms, which in turn are consumed by humans [5]. Eastern Visayas, otherwise known as Region VIII, is largely dependent on agriculture as its major source of economy. Although this is the case, aquaculture, commercial fishing, and municipal fishing are still valued at around 11.8 billion pesos in the year 2020, contributing to 4.32% of the country's total fisheries value [6]. Southern Leyte, particularly, contributed to around seven metric tons of fisheries production, equivalent to 5% of the fisheries sector of the region [6]

Once the water quality of a water body changes, its intended use cannot be served adequately. When water quality degrades, water that is previously classified to be healthy for livelihood fishing may not be used anymore for the purposes it should serve. Further, assessing the water quality of a

[1]lbacalla@southernleytestateu.edu.ph

DOI: 10.1201/9781003490210-7

water body is deemed essential to come up with necessary pieces of evidence to support policymaking that concerns the environment.

Sogod Bay, located on 9° 59.442'N, 125° 7.341'E, is one of the major water bodies found in Southern Leyte, Philippines. It is an unclassified water body and has not been classified by any government or private organization. Sogod is one of the coastal municipalities that is found within Sogod Bay, wherein according to the Department of Trade and Industry, it is a second-class municipality. In 2021, the municipality exhibited the biggest population among the municipalities around the bay, with 44,986 residents. It is located at the tip of the bay, as shown in Figure 1.

According to Tehreem et al., water pollution goes hand-in-hand with industrialization. Developed communities suffer water pollution due to industrialization [7]. Moreover, populated communities increase the risk of pollution due to increased waste and sewage. In the case of the municipalities around Sogod Bay, Sogod is the most populated (44,986 people) and is of the highest category (2nd class municipality) among all municipalities [8]; hence, the chosen for this study is Sogod, Southern Leyte.

Table 7.I   Coordinates of the sampling points

Points	Coordinates
Point 1	10.379757, 124.974664
Point 2	10.381097, 124.976971
Point 3	10.382453, 124.979960
Point 4	10.382353, 124.982655
Point 5	10.381037, 124.98581

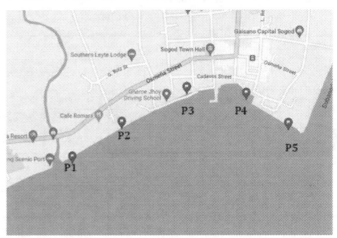

Fig. 7.2   Sampling points in the study

Table 7.2   Special Sampling and Handling Requirements

Analysis	Container	Sample Size (mL)	Preservation	Maximum Storage (hours)
BOD	Plastic	500	Cool; zero headspace	24
TSS	Plastic	500	Cool	168
Phosphate	Glass	100	Cool	48
Nitrate and Chloride	Plastic	100	Cool	48
Color	Plastic	100	Cool	24
Fecal Coliform	Sterilized Plastic or Glass	300	Cool	6

2. Methods

The study has made use of quantitative research methods. This research method is seen as the most fitting since the study involves the collection and analysis of water quality data. The area where the study is undertaken in the coastal waters of Sogod, Southern Leyte. Samples were taken from around the most commercialized area of the municipality of Sogod, Southern Leyte.

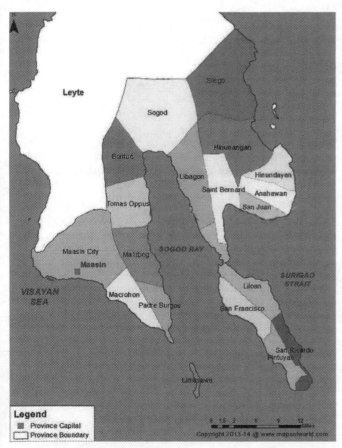

Fig. 7.1   Map showing Sogod Bay and the municipalities of Southern Letye

There were five points from where the samples were taken. The coordinates of the sampling points are listed in Table 7.1, all of which are within the municipality's center, as shown in Figure 7.2. These sample points are spaced evenly, at 300 meters apart within most commercial areas of Sogod, and represent the municipality's complete bathing area, as prescribed by the Water Quality Monitoring Manual of the DENR.

However, due to budgetary limitations, this study is limited to the five sampling points specified and is sampled at one time during the course of this study.

Sampling was done using direct sampling. The following steps were done in the following order: (1) the container and cap were rinsed with the water at least three times, and the rinse water was thrown in the direction of the waves; (2) The container was filled by first lowering it to the water face down to a depth of at least 4 inches below the surface; (3) the container was then slowly tilted to fill the container to the specified volume as prescribed by the laboratory, these volumes are laid out in Table 7.5 and; (4) The cap was then put back to avoid contamination and spillage.

Table 7.3 Analysis Methods as used in the EMB Regional

Parameter	Results					DENR Standards			
	P1	P2	P3	P4	P5	SA	SB	SC	SD
pH	7.57	7.54	7.56	7.54	7.71	7.0-8.5	7.0-8.5	6.5-8.5	6.0-9.0
Total Suspended Solids	7	7	8	4	6	25	50	80	110
True Color	5	5	5	5	5	5	50	75	150
Nitrate	8.25	8.28	8.29	8.28	8.29	10	10	10	15
Phosphate	0	0	0	0	0.002	0.1	0.5	0.5	5
Fecal Coliform	230	3300	1400	7900	2300	<1.1	100	200	400

After collecting the samples, they were stored and secured properly for transport, as prescribed by the laboratory. Table 7.2 shows the different preservation and storage recommendations made by the Regional Environmental Laboratory. The methods of analysis are all in accordance to the 23rd edition of the Standard Methods for the Examination of Water and Waste Water [9] as shown in Table 7.3.

Table 7.4 Analysis Methods as used in the EMB Regional

Parameter	Testing Method
Dissolved Oxygen	Azide Modification Method
pH	Electrometric Method
Total Suspended Solids	Gravimetric Method
True Color	Visual Comparison Method
Nitrate	Ion Chromatography with Chemical Suppression of Eluent Conductivity
Phosphate	Ascorbic Acid methods
Fecal Coliform	MTF Technique

3. Results and Discussion

According to the website of the EMB, Sogod Bay is an unclassified water body (Environmental Management Bureau, n.d.). This is an opportunity to compare results obtained during the analyses with the different standards that are set by the state according to water body classification. The results are summarized in Table 7.4.

As shown, all the samples sent for analysis are well within the set standards of DAO 2016-08, except for fecal coliform, where only the sample from P1 passed standard SD or the standard for navigable waters. These results may be a basis for the formulation of an Environmental Management Plan to reduce the elevated amount of fecal coliform. on the elevated amount of fecal coliform. An EMP is deemed beneficial to the Municipal Local Government Unit – the Municipal Environment and Natural Resources Office, in particular, to base activities on rehabilitation, monitoring and maintenance of the health of the environment, in this case, the coastal waters of Sogod, Southern Leyte, and further, the health and welfare of its citizens.

References

1. Water Quality Guidelines and General Effluent Standards of 2016. DAO 2016-08. § 6.1 (2016)
2. Philippine Statistic Authority, 2020, Fisheries Statistics Authority 2018-2020. ISSN 2012 – 0397 https://psa.gov.ph/

sites/default/files/Fisheries%20Statistics%20of%20the%20 Philippines%2C%202018-2020.pdf

3. Zafra, M.A. (2019). Developing the Philippine Blue Economy: Opportunities and Challenges in the Tourism Sector

4. Partnerships in Environmental Management for the Seas of East Asia. (2018). State of oceans and coast: Philippines. http://pemsea.org/sites/default/files/NSOC_Philippines_0.pdf

5. US EPA. (2021). Marine Water Quality. https://www.epa.gov/ salish-sea/marine-water-quality

6. Bureau of Fisheries and Aquatic Resources. (2018). Fisheries profile 2018. https://region8.bfar.da.gov.ph/ WebsiteNavigation?id=5b9b26fa-6288-42de-a6d5-915e0294 469a&pageAction=subContentFrame

7. Tehreem, H. S., Anser, M. K., Nassani, A. A., Abro, M. M., Zaman, K. (2020). "Impact of average temperature, energy demand, sectoral value added, and population growth on water resource quality and mortality rate: it is time to stop waiting around." Environmental Science and Population Research. https://doi.org/10.1007s11356-020-09822-w

8. Department of Trade and Industry. (2021). Province of Southern Leyte. https://cmci.dti.gov.ph/prov-profile. php?prov=Southern%20Leyte

9. Baird, R. B., Eaton, A.D., & Rice, E. W. (2017). Standard methods for the examination of water and wastewater (23rd ed.). American Public Health Association.

10. Environmental Management Bureau (n.d.). Region 8, List of water bodies. https://water.emb.gov.ph/?page_id=781

Note: All the figures and tables in this chapter were made by the authors.

Advancing Sustainable Science and Technology for a Resilient Future – Sai Kiran Oruganti et al. (eds)
© 2024 Taylor & Francis Group, London, ISBN 978-1-032-79020-6

K-Means Clustering on Spotify Music Data Using Supervised Feature Selection Techniques Enhancing Accuracy

8

Herbert Tagudin*

Polytechnic University of the Philippines, Philippines

Aleta Fabregas[1]

Polytechnic University of the Philippines, Philippines.

Abstract: Clustering, in machine learning, has been useful in identifying data with great similarities and combining them into related groups. Among several clustering algorithms, K-means clustering is widely used due to its simplicity and fast convergence. However, it shows certain limitations, as clusters produce multiple outliers if features are not selected properly. The objective of this paper is to explore how supervised algorithms can improve the accuracy of K-means clustering models. The study used the Philippines Top 200 songs dataset during five years, extracted from Spotify API. Three supervised algorithms, which are Decision Tree using Recursive Feature Elimination (RFE), Logistic Regression using RFE, and Random Forest, are used to perform feature selection with cross-validation to select the 6 most significant features. These features are then used to create the clusters as playlists containing songs of specific moods. The results showed that feature selection using Logistic Regression showed the highest improvement in clustering accuracy, with a silhouette score of 29.22 points, compared to the control model with 25.69 points.

Keywords: Clustering, Decision tree regression (DTR), Feature selection, K-means, Logistic regression (LR), Random forest (RF), Recursive feature elimination (RFE)

1. Introduction

In machine learning, clustering is an unsupervised technique in which data with similar attributes are partitioned into groups, referred to as clusters. (Madhulatha, 2012) Hence, data in these subsets exhibit some pattern of specific behavior that can be subjected to further analysis. Several clustering algorithms exist, such as Partition-based (e.g.: K-means, K-medoids), density-based (e.g.: DBSCAN, mean-shift), and hierarchy-based (e.g.: BIRCH, CURE), among others.

K-means is a popular choice for clustering algorithms due to its simplicity, speed, and efficiency. The algorithm works by grouping a data set into clusters of K. Each cluster must be non-empty and non-overlapping. Data shall belong to a cluster nearest to its corresponding centroid.

K-means is a common choice for implementing clustering machine models because of their wide usage. Several studies are available discussing this algorithm, and researchers continue to conduct improvements on how to maximize the power of k-means. Researchers and data science practitioners are inclined to use a more studied algorithm, even if it has its own set of limitations. Cluster elements created with the algorithm can produce outliers, particularly when multiple features are involved during the grouping process. Features in a data set do not always exhibit a direct relationship to which cluster it is expected to be grouped on.

---
*Corresponding author: hd.tagudin@iskolarngbayan.pup.edu.ph
[1]hd.tagudin@iskolarngbayan.pup.edu.ph

DOI: 10.1201/9781003490210-8

In the process of machine learning, data collected is preprocessed in preparation for feeding it to the model. Among several preprocessing techniques, feature selection can be performed to reduce the dimensionality of data. Nowadays, data sets can contain attributes with different types, and these can affect model training. Therefore, it is a good choice to ensure that features with high significance will be considered to get optimal results.

Different techniques can be done to implement feature selection. Commonly, machine learning models are run to perform a such selection. These models can be either supervised or unsupervised. This study combines the use of supervised and unsupervised algorithms, intending to achieve a more reliable result. A supervised machine learning algorithm is used to perform feature selection, while K-means, as an unsupervised algorithm, is used to perform clustering to playlist track data extracted from Spotify, a cloud-based music streaming service.

Spotify Audio Features and music data

In this day of the digital age, recreation with music consumption has evolved in the past years. As smart devices are becoming a more common staple, people are being reliant on these devices, including listening to songs. Music streaming apps allow users to easily browse and listen to their favorite tracks over the internet. One of the most popular music streaming platforms is Spotify, offering a community for both listeners and creators globally. As a community, it opens itself with a developer API. This allowed numerous developers to create apps integrated with Spotify. The API also made data science researchers interested to explore how the platform curates their playlists, chart rankings, and listener behaviors.

Spotify uses the concept of audio features, which are numeric values that represent a certain musical concept. The numerical interpretations of these attributes allow better measurements, especially when dealing with data science experiments. (Sciandra et al., 2020) emphasized the importance of popularity as an audio feature, as popularity defines the direction of whether a song is a top track on the platform.

Using feature selection techniques to enhance unsupervised prediction models

(Capo, et al., 2020) highlighted the limitations of K-means clustering, particularly when dealing with multi-attribute datasets. The study discussed using Lloyd's Algorithm to define the starting set of centroids and its dependence on the quality of initial data applied to K-means clustering. Data sets with high dimensionality tend to affect the performance of the algorithm, as well as the possibility of having irrelevant features that can negatively impact the clustering process.

Feature selection is one of the dimensionality reduction methods that can help mitigate these limitations of K-means. It is performed by selecting actual dimensions (i.e., features) among the data set of interest. (Alelyani, et al., 2018) discussed the technique and its application to clustering models, and they elaborated on how resulting clusters can be different based on the features included in the subset. Furthermore, the study enumerated approaches to feature selection. Generally, it is categorized into three (3) models: filter, wrapper, and hybrid. Filter-based models are more prominent in studies aiming to improve prediction (Rouhi, et al., 2018). Wrapper models tend to produce more precise results, although to be more expensive to use in terms of computational power. (Tadist, et al., 2019)

Among the wrapper methods, Recursive Feature Elimination (RFE) is the most common. (Bertoni, et al., 2021) formulated a feature bank from Brazillian songs on Spotify. The study considered RFE in reducing the dimensionality of the dataset for its efficiency. In other fields of study, (Mohamed, et al., 2020) also used RFE with Logistic Regression (LR) to reduce dimensionality in automated risk detection of Diabetes. This set of algorithms has been prominent in health-related diagnosis research, where data usually contains high dimensionalities, proving its great contribution to selecting attributes of high importance. (Reddy, et al., 2021)

Spotify audio data clustering approach

The experimental approach used in this study is divided into two stages. The first stage involves the data preprocessing execution. Data is subjected to techniques such as data cleaning, data reduction, and attribute selection. After applying the methods, features to be used for the clustering model are selected. The control setup does not have any feature selection techniques, thus applying all eleven (11) audio features *(i.e., acousticness, danceability, energy, instrumentalness, key, liveness, loudness, mode, speechiness, tempo, valence)*. On the other hand, experimental setups are formulated where supervised feature selection algorithms are applied. Attribute "popularity" from the dataset is used as the target to determine the six (6) most significant variables among the eleven Spotify audio features. The algorithms applied for the experiment are the following: (1) Recursive Feature Elimination (RFE) by Logistic Regression (LR), RFE by Decision Tree Regression (DTR), and Random Forest (RF).

The second stage involves the application of a clustering model to group the Spotify song tracks according to the values of their audio attributes. Based on the features selected by each algorithm, a sub-dataset was applied to the clustering machine learning model, using the K-means algorithm. The resulting clusters were observed and measured based on their clustering accuracy and performance. Results are compared to discuss any relevant conclusions.

Spotify Philippines Top 200 Tracks Dataset

The dataset used in this study is "Spotify Daily Top 200 Tracks in the Philippines", published by climate data scientist JC Albert Peralta. The dataset contains 4,467 records of playlist tracks extracted from the Top 200 songs in the Philippines from January 2017 to September 2020. Each track record contains basic information such as song titles, artist information, and album information. They also contain information about track audio features – these are numerical attributes that Spotify defined that provide in-depth audio analysis of tracks.

2. Data Preparation

The initial step towards data preparation is to download the Spotify track dataset from Kaggle. Rows with missing numerical values for audio features were removed, and non-numerical data such as artist names and track names were trimmed. Clean data extracted from the dataset were further analyzed before feeding to classification models for feature selection.

3. Tool

All experiments in this study are conducted using Python programming language. Since the dataset comes from Kaggle, the dataset can be easily linked to a Jupyter Notebook created to implement the models required in this study. Python libraries were utilized to seamlessly implement machine learning computational abilities, such as Numpy for data management, Scikit-learn for machine learning implementations, and PyPlot for data presentation.

Feature selection using supervised algorithms

Feature Selection is implemented by testing three (3) supervised classification algorithms: Recursive Feature Elimination (RFE) using Decision Tree Regression, RFE using Logistic Regression, and Random Forest algorithm. All audio attributes were used as features while using the popularity attribute as the target variable. A comparative experiment is performed on the feature selection models using cross-validation, setting up six (6) attributes as the target feature count. Results of the cross-validation are recorded, which include the Mean Absolute Error (MAE) and Standard Deviation. The feature selection algorithm with the most desirable result was considered. This will select the 6 features of the highest importance.

Table I presents a comparison matrix of supervised feature selection algorithms that are applied in this experiment using the Spotify Philippines Top 200 Tracks Dataset. Random

Forest resulted in the most desirable value, with a Mean Absolute Error of 25.99 and a standard deviation of 0.96.

Table 8.1 Comparison of mean absolute errors between feature selection supervised algorithms

Model	MAE	Standard Deviation
Decision Tree (RFE)	31.91	1.45
Logistic Regression (RFE)	46.38	1.55
Random Forest	25.87	0.87

Table II below shows the top six Spotify audio features selected by each algorithm using cross-validation. All three models exhibited a different set of selected features, except for danceability and valence being their common features.

Table 8.2 Comparison of selected clustering features

Model	Selected features
Decision Tree (RFE)	danceability, energy, key, loudness, valence, tempo
Logistic Regression (RFE)	danceability, energy, mode, acousticness, liveness, valence
Random Forest	danceability, loudness, speechiness, energy, valence, tempo

K-means clustering using selected Spotify features

The clustering model is constructed using the K-means algorithm. A Spotify data subset with the selected features is used to train the model. The elbow method is used to determine the ideal number of clusters based on this data set, using the Euclidean distance metric. It is determined that five (5) is the optimal number of clusters to use in this data set.

Table III shows the silhouette scores produced by creating the clusters based on the features selected by each supervised algorithm. Logistic Regression with Recursive Feature Elimination (RFE) resulted in 29.22, the highest among the experimental models measured. This is higher compared to the control variable, only having a score of 25.69, This is despite having the least desirable Mean Absolute Error (MAE) among the feature selection models.

Table 8.3 Comparison of silhouette scores with applied feature selection algorithms

Model	Silhouette Score (%)
Control	25.69
Decision Tree (RFE)	18.95
Logistic Regression (RFE)	29.22
Random Forest	22.52

4. Conclusion

The objective of this study is to enhance the accuracy of Spotify audio track clustering using supervised feature selection techniques. In this study, Spotify track data audio attributes are subjected to feature selection using different algorithms (Decision Tree - RFE, Logistic Regression - RFE, and Random Forest) with cross-validation to identify the most significant features before clustering. Selecting important features will reduce the complexity of the K-means model to be used for determining tracks to be grouped. The cross-validation results showed that the mean absolute error values of Random Forest had the highest desirable value, while the Logistic Regression had the least desirable value.

After including the most important features for each algorithm, the dataset is applied to the K-means clustering algorithm and compared to the control experiment without feature selection applied. Results showed that Logistic Regression got the highest silhouette score in creating five (5) clusters, despite having the least desirable MAE during cross-validation. This validates being useful for high-dimensionality datasets, such as audio features (Reddy, et al., 2021).

In this manner, this study suggests using feature selection models with the Logistic Regression algorithm in clustering Spotify tracks, as it surpassed the silhouette score of the control experiment and enhanced the clustering model accuracy.

For future studies, it is recommended to apply the same techniques with other music platforms, to open if such possibility of creating curated playlists and listening experiences for users.

References

1. Alelyani, S., Tang, J., & Liu, H. (2018). Feature selection for clustering: A review. Data Clustering, 29-60.
2. Bertoni, A. A., Lemos, R. P., Coelho, A. A., & e Silva, H. V. Three Feature Datasets extracted from Popular Brazilian Hit Songs and Non-Hit Songs from 2014 to 2019. International Journal of Computational Engineering Research (IJCER), vol. 11, no.1, 2021, pp 01-08.
3. Capo, M., Perez, A., & Lozano, J. A. (2020). A Cheap Feature Selection Approach for the K-Means Algorithm. IEEE Transactions on Neural Networks and Learning Systems, 1–14. doi:10.1109/tnnls.2020.3002576
4. Madhulatha, T. S. (2012). An overview on clustering methods. arXiv preprint arXiv:1205.1117.
5. Mohamed, A. (2020). Type 2 Diabetes Mellitus Automated Risk Detection Based on UAE National Health Survey Data: A Framework for the Construction and Optimization of Binary Classification Machine Learning Models Based on Dimensionality Reduction (Doctoral dissertation, The British University in Dubai (BUiD)).
6. Reddy, B. P. V., Alla, L. P., & Patil, H. Y. (2021, July). Parkinson's Disease Classification using Quantile Transformation and RFE. In 2021 12th International Conference on Computing Communication and Networking Technologies (ICCCNT) (pp. 01-05). IEEE.
7. Rouhi, A., & Nezamabadi-pour, H. (2018). Filter-based feature selection for microarray data using improved binary gravitational search algorithm. 2018 3rd Conference on Swarm Intelligence and Evolutionary Computation (CSIEC). doi:10.1109/csiec.2018.8405411
8. Sciandra, M., & Spera, I. C. (2020). A model-based approach to Spotify data analysis: a Beta GLMM. Journal of Applied Statistics, 1–16. doi:10.1080/02664763.2020.1803810
9. Tadist, K., Najah, S., Nikolov, N. S., Mrabti, F., & Zahi, A. (2019). Feature selection methods and genomic big data: a systematic review. Journal of Big Data, 6(1), 1-24.

Note: All the tables in this chapter were made by the authors.

Advancing Sustainable Science and Technology for a Resilient Future – Sai Kiran Oruganti et al. (eds)
© 2024 Taylor & Francis Group, London, ISBN 978-1-032-79020-6

9

Development of a Microcontroller-based Oil Spill Monitoring Buoy

John Francis R. Chan*
Department of Electrical Engineering,
University of the Philippines, Los Baños, Philippines

Melvin C. Ilang-Ilang[1]
Department of Electrical Engineering,
University of the Philippines, Los Baños, Philippines

Annie Liza C. Pintor
Department of Electrical Engineering,
University of the Philippines, Los Baños, Philippines

Abstract: Oil spill is a severe environmental problem that causes serious harm to marine environments and affects both locals and marine wildlife. This study aims to develop a microcontroller-based oil spill monitoring buoy capable of detecting the presence of crude oil. An oil sensor is developed based on the principle of capacitance sensing and frequency variation. A parallel plate capacitor acted as dielectric cell and was connected to a frequency generator circuit where its output frequency varies with the medium in the dielectric cell. Six cases of dielectric medium were prepared for the study namely crude oil, seawater, air, seawater-air, seawater-crude oil and air-crude oil. An Arduino UNO microcontroller was used to measure and to evaluate the output frequency while a GSM module was used for data transmission. Results showed that measurements from the oil sensor were more accurate with respect to the calibration dataset with a maximum percent error of 1.635% than the theoretical data with a maximum percent error of 2.293%. The response time of the monitoring buoy was also determined in this study. It takes 8.541 seconds for the system to measure the output frequency from the oil sensor and send an SMS message containing an oil spill confirmation if oil is detected.

Keywords: Oil spill, Oil spill detection, Environmental monitoring, Capacitance sensing, Microcontroller

1. Introduction

Oil spill is the release of hydrocarbon substances in both land and water environment that causes pollution. There are approximately 1.7 to 8.8 million tons of oil released in marine environments worldwide every year and 70% of these oil spill incidents are directly caused by human activities based on the study conducted by the National Academy of Science Fingas, M. (2011). Oil spill monitoring system plays a crucial role for contingency planning however, there is no sensor available that can give all the necessary information needed for oil spill surveillance as each method possess drawbacks at certain conditions Jha, M.N., Levy, J., & Gao, Y. (2008). For this reason, innovative techniques for environmental monitoring

and safeguarding in marine environments has shown a drastic increase for the past few years Moroni, D., Pieri, G., Salvetti, O., Tampucci, M., Domenici, C., & Tonacci, A. (2016). This study introduces a new method for oil spill surveillance through utilizing dielectric properties of crude oil and the change in frequency produced by a frequency generator circuit. The dielectric constant of water (80) is relatively high compared to that of oil (2) making dielectric constant a convenient way in identifying the two liquids apart Liptak, B. (2016). Through this study, oil spill response will be more proactive and effective making the people, who relied on maricultural industry for their livelihood, to have a more sustainable, wealthier, and easier life.

*Corresponding author: jrchan2@up.edu.ph
[1]mcilangilang3@up.edu.ph, [2]acpintor@up.edu.ph

DOI: 10.1201/9781003490210-9

2. Methods

A. System Design

The oil detection system is composed of power supply circuit, oil sensor, Arduino UNO, and a transmission module. The system was placed in a plastic buoy with a Styrofoam board was used as a floatation device. Fig. 9.1 shows the connection of the components of the system.

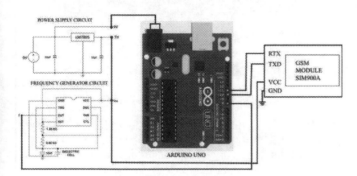

Fig. 9.1 Pin Diagram of the System

B. Oil Sensor

The developed oil sensor is made of dielectric cell and frequency generator. The cell is a parallel stainless-steel plate of dimension 100 mm × 100 mm × 0.5 mm with a 3 mm distance between the plates. This dielectric cell acted as capacitor while crude oil, air, and seawater served as dielectric medium.

C. Experimental Design

The dielectric cell served as electrodes while crude oil, seawater, and air acted as dielectric medium. The output frequency of NE555, which is used as frequency generator, was measured every second and the average of every five measurements was computed. The measured average output frequency from Arduino UNO was shown in the serial monitor. The obtained data were analyzed using percent error, Wilcoxon rank signed test, and Mann-Whitney U test.

D. Dielectric Medium Cases

Six cases of dielectric medium were prepared which were consisted of crude oil, seawater, air, seawater-air, seawater-crude oil, and air-crude oil. The seawater used is a replicate seawater which is based from Gillepsie, C. (2018). The visualization of the prepared dielectric cases is shown in Fig. 9.2.

E. Theoretical Capacitance

The theoretical value of capacitance for each case was determined in this study using (1).

$$c = \varepsilon_r \varepsilon_0 \frac{A}{d} \qquad (1)$$

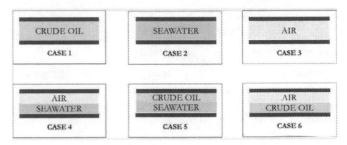

Fig. 9.2 Visualization of Cases 1 to 6

Where   C is the capacitance of the parallel plate (F)

ε_r is the dielectric constant of the material

ε_0 is the permittivity of free space (8.854×10^{-12} F)

A is the area of the parallel plate (m²)

d is the distance between the two plates (m)

The dielectric constants of crude oil, air, and seawater are 2.1, 1, and 80 respectively Musa, S.M. (2013) along the theoretical capacitance of each case are shown in Table I.

Table 9.1 Dielectric constants and theoretical capacitance involved per case

Case No.	1st Dielectric	2nd Dielectric Constant	Theoretical Capacitance (Pf) Constant
1	2.1 (Crude oil)	NA	61.979
2	80.0 (Seawater)	NA	2361.117
3	1.0 (Air)	NA	29.514
4	1.0 (Air)	80.0 (Seawater)	29.150
5	2.1 (Crude oil)	80.0 (Seawater)	60.394
6	1.0 (Air)	2.1 (Crude oil)	19.993

F. Oil Detection

The detection of oil spill was based on the evaluation of the measured frequency in the range of frequencies. Equation (2) was used to calculate the theoretical output frequency of the frequency generator.

$$f = \frac{1.44}{(1880 + 2(9900) \times C)} \qquad (2)$$

where f is the output frequency of NE555 (Hz)

C is the capacitance between TRG and GND (F)

Three frequency ranges were used for oil detection namely f_{Oil}, $f_{Oil\text{-}Seawater}$, and $f_{Air\text{-}Oil}$. The probable error, shown in (3), determined the allowable error that would set the upper and lower limit frequencies for Cases 1, 5, and 6.

$$PE = 0.6745\sigma \qquad (3)$$

where   PE   is the probable error of data
         σ is the standard deviation of data

G. Transmission System

The transmission system is composed of a GSM module and a phone, which is commonly used module for long range communications Mahmud, S., Alam, M., Abedin, J., & Roy, S. (2015). The algorithm for oil detection and data transmission is shown in Fig. 9.3.

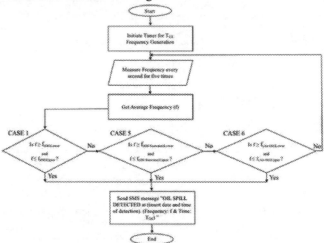

Fig. 9.3   Algorithm for the oil spill monitoring system

H. Data Collection

In this study, there were two sets of data obtained namely frequency measurement and response time. Fig. 9.4 shows the setup for the data of collection for frequency measurements using Arduino UNO.

Fig. 9.4   Frequency measurement set up

On the other hand, the response time of oil spill monitoring was only tested in cases with crude oil involved as dielectric medium. The total response time is equal to the sum of the time from frequency generation up to sending an SMS message (T_{GS}) and time after sending SMS message up to receiving the message by the phone (T_{SR}). Fig. 9.5 shows the setup for the data collection of response time.

Fig. 9.5   Set up of response time data collection

3. Results and Discussion

A. System Calibration

The output frequency in each case were gathered for comparison with the corresponding theoretical values. Table 9.2 shows the summary of the theoretical output frequency and the calibration dataset.

Table 9.2   Comparison of the theoretical data.

Case No.	Theoretical Output Frequency (Khz)	Calibration Dataset (Khz)	%Error
1	6.601	6.638	0.561
2	5.373	Inf	-
3	6.623	6.663	0.604
4	6.623	6.702	1.193
5	6.602	6.791	2.863
6	6.629	6.739	1.659

It can be observed that there were no frequencies measured in Case 2 since the dielectric was conductive, with 5.5 S/m conductivity based on Fondriest Environmental, as confirmed by continuity test.

Results from Table 9.2 showed that the initial measurement from the oil sensor closely matched the theoretical output frequency, with a maximum error of 2.863% in Case 5 and a minimum error of 0.561% in Case 1. It can observe that the oil sensor exhibits high precision and reliability. Small errors came from the estimations during implementation involving two substances in the dielectric cell and stray capacitance from the soldered board.

B. Accuracy of Measured Frequency

The accuracy of the oil sensor was verified using percent error and Wilcoxon signed rank test. Table 9.3 shows the comparison between the theoretical and average experimental output frequencies of each case.

Table 9.3 Theoretical and average experimental output frequency of each case

Case No.	Theoretical Output Freq. (Khz)	Experimental Output Freq. (Khz)	%Error
1	6.601	6.580	0.318
2	5.373	Inf	-
3	6.623	6.654	0.468
4	6.623	6.749	1.902
5	6.602	6.680	1.181
6	6.629	6.781	2.293

Table III showed that the experimental frequencies of the oil sensor were close to the theoretical frequencies in each case with maximum error of 2.293% in Case 6 and a minimum percent error of 0.318% in Case 1.

On the other hand, the result for the Wilcoxon signed rank test is shown in Table 9.4.

Table 9.4 Results for the Wilcoxon signed rank test in each case of the first test.

Case No.	Z-Value	P-Value	Decision
1	-0.976	0.329	Accept H_0
2	-	-	-
3	2.646	0.008	Reject H_0
4	2.850	0.004	Reject H_0
5	1.721	0.085	Accept H_0
6	2.871	0.004	Reject H_0

Results indicated that only the median of the experimental data from Cases 1 and 5 are equal to their corresponding theoretical value. While Cases 3, 4, and 6, did not match their corresponding theoretical value. Stray capacitance and estimations are some reasons for these results. The accuracy of the oil sensor is crucial in the oil monitoring process. Its reliability directly affects the effectiveness of oil spill surveillance. Both the percent error and Wilcoxon signed rank test showed that the measurements obtained from the oil spill sensor exhibit high level of accuracy.

C. Comparison Test

In this test, the calibration dataset was subjected to Mann-Whitney U test. The output frequencies of Cases 3, 4, 5, and 6 were compared to the output frequency of Case 1 to verify if there is a statistical difference between the data of Case 1 from the other cases. Table 9.5 shows the results for the Mann-Whitney U test of each comparison.

Table 9.5 Comparison of the theoretical data and calibration dataset of each case

Comparison No.	1st Mean (Khz)	2nd Mean (Khz)	U-Value	Decision
Comparison 1	6.601	6.580	0.318	
(Case 1 & 2)	6.638	inf	-	-
Comparison 2	6.623	6.654	0.468	
(Case 1 & 3)	6.638	6.63	30	Accept H_0
Comparison 3	6.602	6.680	1.181	
(Case 1 & 4)	6.638	6.702	30	Accept H_0
Comparison 4				
(Case 1 & 5)	6.638	6.791	24	Accept H_0
Comparison 5				
(Case 1 & 6)	6.638	6.739	30	Accept H_0

Results suggest that the medians for Cases 3, 4, 5, and 6 are all similar to that of Case 1. This likely because the theoretical capacitances of Cases 1, 3, 4, 5, and 6 are significantly small compared to the paralleled 10 nF capacitance. The effect of the external resistances to the frequency is less significant as to that of the external capacitor which was adjusted to limit the output frequency from the oil sensor.

D. Oil Detection of the System

The calculated range of frequencies, used for oil detection, are shown in Table 9.6.

Table 9.6 Range of frequencies for oil detection of Cases 1, 5, and 6

Case No.	Calibration Dataset (Khz)	Probable Error (Khz)	Upper Freq. Limit (Khz)	Lower Freq. Limit (Khz)
1	6.638	0.081	6.719	6.557
5	6.791	0.086	6.877	6.705
6	6.739	0.015	6.754	6.724

Table 9.6 shows that computed probable errors are very minimal. Also, there is an overlap in frequencieswhich means that cases with oil involved as dielectric medium have almost similar output frequencies.

Results showed that the setup was able to detect oil in Cases 1, 5, and 6. However, the system was still able to detect oil despite having no presence of oil in Cases 3 and 4. The main reason for this is that the output frequency from Cases 3 and 4, both involved air, coincide in the range of frequencies for oil detection. To solve this, the dielectric cell must be fully submerged in the seawater to avoid misdetection of oil.

E. Response Time of the System

Table 9.7 shows the obtained average T_{GS}, T_{SR}, and T_{TOTAL} of Case 1, 5, and 6.

Table 9.7 Average response time of the system for cases 1, 5, and 6

CASE NO.	T_{GS} (s)	T_{SR} (s)	T_{TOTAL} (s)
1	8.464	15.303	23.767
5	8.695	12.763	20.392
6	8.464	12.995	21.459

Results showed that T_{GS} of Cases 1, 5, and 6 are relatively identical with an average of time of 8.541 seconds. Meanwhile, Case 1 had a slightly higher value of T_{SR} that of Cases 5 and 6. One reason for this is due to unstable phone signal. The obtained minimal response time is favorable to monitoring systems.

4. Conclusion

Results showed that measurements from the oil sensor was more accurate with respect to the calibration dataset than the theoretical data. The main reason for this is that errors from stray capacitances and estimations were taken into consideration in the calibration dataset during data collection. For the response time of the system, it was found that the system measures the output frequency and sends SMS confirmation in about 8.542 seconds. While the receiving of SMS message from the monitoring system varies with the strength of mobile phone signal. This study recommends improving the quality of the buoy since the constructed buoy experienced an unsteady floatation during implementation.

The use capacitance sensing and frequency variation for oil spill surveillance presents a simple, fast, and cheap way of oil spill monitoring. The constructed oil spill monitoring buoy showed that it is capable of detecting the presence of crude oil and at the same time, it can also transmit necessary information needed for oil spill surveillance.

References

1. Fingas, M. (2011). Buoys and devices for oil spill tracking. International Oil Spill Conference Proceedings, 213-228. doi: 10.7901/2169-3358-2011-1-9.
2. Jha, M.N., Levy, J., & Gao, Y. (2008). Advances in remote sensing for oil spill disaster management: State-of-the-art sensors technology for oil spill surveillance. Sensors, 8(1), 236- 255. doi.org/10.3390/s8010236.
3. Moroni, D., Pieri, G., Salvetti, O., Tampucci, M., Domenici, C., & Tonacci, A. (2016). Sensorized buoy for oil spill early detection. Methods in Oceanography, 17(1), 221-231. doi: 10.1016/j.mio.2016.10.002.
4. Liptak, B. (2016). Detection of oil in or on water. Retrieved from https://www.controlglobal.com/articles/2016/detection-o f-oil-in-or-on-water.
5. Gillepsie, C. (2018). How to replicate seawater at home. Retrieved from https://sciencing.com/make-sea-water-home-6368912.html.
6. Musa, S.M. (2013). Computational nanotechnology using finite difference time domain (1st ed.) [Electronic version]. Boca Raton, Florida: CRC Press.
7. Mahmud, S., Alam, M., Abedin, J., & Roy, S. (2015). A GSM based intelligent wireless mobile patient monitoring system. International Journal of Research in Engineering and Technology, 4(4), 139-143. doi:10.15623/ijret.2015.0404024.
8. Fondriest Environmental. Conductivity, salinity& tital dissolved solids. Retrieved from fondriest.com/environmental-measurements/parameters/waterquality/conductivity-salinity-tds/

Note: All the figures and tables in this chapter were made by the authors.

Advancing Sustainable Science and Technology for a Resilient Future – Sai Kiran Oruganti et al. (eds)
© 2024 Taylor & Francis Group, London, ISBN 978-1-032-79020-6

10

Palayan: Forecasting Rice Crop Loss on Typhoons using PSO-SVR

Maria Sophia Balita*
Department of Computer Science, Polytechnic University of the Philippines, Philippines

Hannah Cailing - Carlos Joshua Elequin[1]
Department of Computer Science, Polytechnic University of the Philippines, Philippines

John Clarence Pagulayan - Jeremy Karl Santiagoa[2]
Department of Computer Science, Polytechnic University of the Philippines, Philippines

Ria Ambrocio Sagum[3]
Department of Computer Science, Polytechnic University of the Philippines, Philippines

Abstract: Typhoons are quite common in a tropical country like Philippines, and farmers and their crops are frequently affected by this natural disaster. The Support Vector Regression model is implemented with Particle Swarm Optimization for both the feature selection and parameter optimization which forecasts rice crop loss in Palayan City due to typhoons. This study was conducted to create a web-based decision support system for the local government units and provide guidance to insurance evaluators and humanitarian organizations in giving aid to the farmers pre- and post-disaster. Selected features were derived from municipal level, rather than provincial, for a more attuned forecasts and recommendations. The results indicate that PSO-SVR model yields an MSE forecast error of 87.64% and a Pearson's score of 1.0. This indicates poor forecasting accuracy of the model. This is attributed to the data quality despite implementing a data imputation method to fill in the incomplete records. With such results, it is suggested that future research gather quality data from more rice-producing municipalities with advanced record-keeping practices to train the model and yield better results.

Keywords: Crop yield, Crop loss, Support vector regression

1. Introduction

Emphasizing weather forecasts on its impact to the area suggests warnings to the people regarding the multiple hazards that a disaster may cause. (Global Facility for Disaster Reduction and Recovery, 2016) and Consequential events may lead to hazards to the public health, safety, and security. This would also help other sectors within the local community to act accordingly to mitigate the risks at an earlier time.

With the evolving technology, the application of Artificial Intelligence (AI) has been prevalent in the agriculture sector. According to Talaviya, Shah, D., Patel, Yagnik, and Shah, M.2, there will be 75 million connected devices which will be used by farmers at year 2020 and by 2050, an average farm can generate up to 4.1 million data points every day. (Talaviya et al., 2020) This has proven that AI in agriculture will continuously grow as time progresses. With this, weather prediction cannot suffice alone to ensure more production of goods. The implementation of AI in this sector has prevented a dramatic decline in yield production from factors such as climate changes and food security problems.

Existing software being utilized currently do not support the impact-based forecasting method. Harrowsmith, Nielsen, Jaime, de Perez, Uprety, Johnson, van den Homberg, Tijssen, Page, Lux, and Comment (Harrowsmith et al., 2020) have defined impact-based forecasting to be based on the projected impact severity caused by different weather phenomenon in contrast to the general weather forecasts where only the meteorological attributes are being forecasted.

To alleviate the problems concerning agricultural damages due to typhoons, the researchers will develop a tool that will forecast the rice crop loss for the impending typhoon

*Corresponding author: piangbalita@gmail.com
[1]hacailing@gmail.com, [2]carlos_elequin@yahoo.com, [3]rasagum@pup.edu.ph

DOI: 10.1201/9781003490210-10

based on past data of meteorological, geographical, and rice features of Palayan City, Nueva Ecija. To fulfil the goal of forecasting rice crop loss based on total area damaged due to typhoons, Support Vector Regression (SVR) will be utilized. For this, input variables are to undergo feature selection and parameter optimization using Particle Swarm Optimization (PSO) for reducing the noise in data to produce results of higher accuracy. PSO-SVR forecasting was implemented in the model to be developed since it is a powerful hybrid algorithm that leverages the global optimization capabilities of PSO and the non-linear modeling capabilities of SVR. It is particularly useful in solving regression problems with non-linear relationships between variables and can provide more robust and accurate results compared to traditional linear regression or standalone SVR approaches. (Alves et al., 2021; Adaryani et al., 2022; Huican et al., 2022; Xianting & Mao, 2022; Li et al., 2022) The goal of this study is to develop a model that will allow the local government to forecast in producing more efficient warning, focused on the farmers of rice crops to prevent significant losses in their yield production and avoid drastic consequences of rice loss.

2. Experimental Methods

A. Sources of Data

The data that will be used for meteorological features will be gathered from Climatology and Agrometeorology Division of Philippine Atmospheric, Geophysical and Astronomical Services Administration (PAGASA) and Electronic Freedom of Information (eFOI). The information will be utilized to determine which typhoons caused damage to rice production.

The geographical data will be from the City Government of Palayan, which is the setting of the study. The dataset will include information about the area's geographical characteristics, as typhoon damage to rice is correlated to its geospatial aspects. (Food and Agriculture Organization of the United Nations, 2015) The historical data for rice will be used as basis to predict its estimated damage due to typhoons. This includes statistical report of total rice farm area of damage. The data will come from the Philippine Rice Information System (PRiSM) of the Philippine Rice Research Institute (PhilRice).

The researchers have collected data from the website of the Philippine Statistics Authority (PSA) and some from the provincial office of Bataan. In total there were seven (7) years of data from January 2013 – December 2020 for meteorological, geographical, and rice dataset. These data will undergo feature selection to identify reliable variables before proceeding to parameter optimization of the model for forecasting.

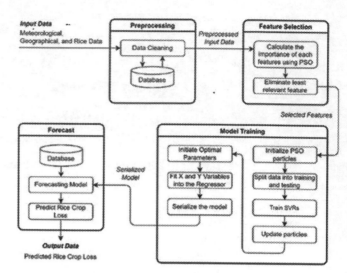

Fig. 10.1 System architecture

B. System Architecture

The study's system architecture encompasses the Input-Process-Output (IPO) model. The input data will include meteorological, geographical, and rice data, with the output data being forecasted rice crop loss. The tool developed has four (4) modules these preprocessing, feature selection, model training, and forecast.

The input data that are obtained through government agencies are cleaned and stored in the database during preprocessing stage. Once noise data are removed, most important features will be selected by computing the fitness value of a random velocity and position of a PSO particle. A best fitness value is returned once it meets termination requirements, and it will be handled as selected features. These selected attributes will be utilized to begin model training. Once again, PSO is employed for optimization of SVR parameters, and its particles are initiated. The data is divided into training and testing set. A different set of SVR parameters are trained and validation scores are obtained. These scores are compared to each particle, returning a set of optimal parameters by the end of the iteration. The parameters are then placed into the X and Y regressors and serialized to avoid training results from being overwritten, as the model will continue to train if the new input is introduced. A serialized model is used in the forecast module. A database, which contains the new input, is accessed by the forecasting model to predict the rice crop loss.

C. Statistical Data Analysis

This study aimed to obtain the accuracy and reliability of predicting rice damage due to typhoons. Forecast error measurement was used by researchers to increase overall forecasting accuracy and further refine forecasting procedures. (Chapman, 2021) Mean Squared Error (MSE) was employed to calculate the error of the study. The degree of inaccuracy in statistical models was gauged by the MSE. It evaluates the average squared difference between the observed and projected values.

$$MSE = \frac{1}{n}\sum_{i=1}^{n}\left(A_i - F_i\right)^2$$

Where:

n = number of observations

A_i = actual value

F = forecast value

Forecast accuracy KPI was used to obtain the forecast accuracy of the study. the equation is depicted below:

Forecast Accuracy KPI = 1 - % MSE Total Error

The stability or consistency of test scores is measured by reliability. The consistency of results when repeating the same test on the same sample at a different time is known as test-retest reliability. It is used when measuring anything that is expected to remain consistent in the sample. Hence, test-retest was utilized for determining the reliability of the system. The method consisted of administering the test to the same group of individuals on two different occasions. It was represented by Pearson's correlation coefficient equation:

$$r = \frac{n\left(\sum xy\right) - \left(\sum x\right)\left(\sum y\right)}{\sqrt{\left(n\left(\sum x^2\right) - \left(\sum x\right)^2\right)\left(n\left(\sum y^2\right) - \left(\sum y\right)^2\right)}}$$

Where:

n = total number of pairs of test and retest scores.

$\sum x$ = sum of x scores

$\sum y$ = sum of y scores

$\sum y^2$ = sum of the squared x scores

$\sum y^2$ = sum of the squared y scores

$\sum xy$ = sum of the products of the paired scores

Table 10.1 displays the interpretation of the reliability of test-retest measure or Pearson correlation coefficients.

Table 10.1 Interpretation of test-retest reliability coefficients

Pearson r	Interpretation
0.90 to 1.00	Excellent reliability
0.70 *to* 0.90	Good reliability
0.50 *to* 0.70	Acceptable reliability
0.30 *to* 0.50	Questionable reliability
0.00 *to* 0.30	Poor reliability

3. Results and Discussion

The accuracy of the developed tool was evaluated using the mean average error in percentage points. This metric was used to evaluate both the training and testing data and keep track of how well the model works at different stages of implementation.

Random data from the years 2013 to 2021 is used to train the impact-based forecasting model. In testing the performance of the accuracy of the model, forecast accuracy is obtained by subtracting the Mean Squared Error (MSE) from 1. Table 10.2 shows the summary of the mean average percentage error and accuracy of the forecast results in the testing phase.

Table 10.2 Summary of the accuracy of the system in forecasting rice crop loss in total rice area damage

Phase	MSE	Accuracy
Testing	0.8763	12.37%

In the training phase, fifty-eight (58) observations out of eighty-two (82) are employed. The remaining twenty-four (24) are fed during the testing phase, yielding a mean squared error of 0.8763 or 87.64%.

Figure 10.2 shows the graphical representation of the actual and predicted rice area damage of the system during testing phase. Twenty-four (24) observations ranged from 0.10 to 23.58 square kilometers are presented. It can be observed that lines that represent the actual and forecast fall along coordinates far from each other per instance. This indicates the poor accuracy of the model, as calculated through MSE.

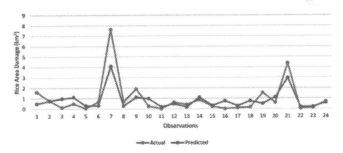

Fig. 10.2 Actual versus predicted rice area damage

Reliability measures the consistency of the system. To calculate the Pearson correlation coefficient, the researchers used a test-retest method. Table 10.5 displays the interpretation of the experiment administered to the same group of data on different occasions. The forecasted values of rice area damage produced a score of 1.0 indicating that the system has an excellent reliability in forecasting rice crop loss in terms of total rice area damage.

Table 10.3 Interpretation of test-retest reliability results

Attribute	Test 1 (Σx)	Test 2 (Σy)	Production of Tests 1 & 2 (xy)	Test-Retest Pearson (r)	Interpretation
Rice Area Damage	19.97	19.97	398.58	1	Excellent Reliability

4. Conclusion

The researchers have found that using particle swarm optimization in feature selection and parameter optimization for support vector regression to forecast rice crop loss was not very effective through analysis and experimentation, only yielding an accuracy of 12.37 percent. However, this finding can be attributed to several factors, such as: 1) high variance in data; 2) lack of datasets; and 3) quality of data.

Like with any data science problem, the computational costs can increase exponentially as more datasets are added to the system. The utilization of PSO for feature selection and parameter optimization can help lower this compared with other algorithms. Using this positively improved the accuracy rate of the SVR model, which is a renowned regression algorithm.

References

1. Alves, E., de Oliveira, J.F.L., de Mattos Neto, P.S.G., da Nobrega Marino, M.H., & Madeior, F. (2021). A Nonlinear Optimized PSO-SVR Hybrid System for Time Series Forecasting with ARIMA. *ChemBioChem*. https://doi.org/10.21528/cbic2021-54

2. Chapman, M. (2021). How to Calculate Forecast Accuracy and Forecast Error.

3. Huican Luo, Peijian Zhou, Lingfeng Shu, Jiegang Mou, Haisheng Zheng, Chenglong Jiang, & Yantian Wang. (2022). Energy Performance Curves Prediction of Centrifugal Pumps Based on Constrained PSO-SVR Model. *Energies*, 15(9), 3309–3309. https://doi.org/10.3390/en15093309

4. Li, Y., Lei He, He, L., Peng, B., Fan, K., & Tong, L. (2018). Remote Sensing Inversion of Water Quality Parameters in Longquan Lake Based on PSO-SVR Algorithm. *IEEE International Geoscience and Remote Sensing Symposium*.

5. Talaviya, T. et al. (2020). Implementation of Artificial Intelligence in Agriculture for Optimization of Irrigation and Application of Pesticides and Herbicides.

6. Xianting Yao, & Mao, S. (2022). Electric Supply and Demand Forecasting Using Seasonal Grey Model Based on PSO-SVR. *Grey Systems*. https://doi.org/10.1108/gs-10-2021-0159

Note: All the figures and tables in this chapter were made by the authors.

Advancing Sustainable Science and Technology for a Resilient Future – Sai Kiran Oruganti et al. (eds)
© 2024 Taylor & Francis Group, London, ISBN 978-1-032-79020-6

11

Artificial Intelligence as Support in Knowledge Management for Human Resource Management

Ria Ambrocio Sagum*

College of Computer and Information Sciences, Philippines,
Research Management Office, Polytechnic University of the Philippines

Abstract: This study examines the possibility of artificial intelligence as support for knowledge management. In order to identify the scope of AI support that may be used to knowledge management in human resource management, the application of knowledge management in human resource management was also explored. The benefits of combining AI and KM were discussed. A connection and incorporation of artificial intelligence in knowledge management and its different processes were highlighted and this literature review was able to probe the effects and issue of using knowledge management for human resource management, which is sometimes referred to as knowledge base human resource management. It also gathers the different achievements of AI when confronted with different km issues. The objective of the study was attained by reviewing different research papers in the field of artificial intelligence, knowledge management, and AI in HRM. It is concluded that AI systems can be built and deployed to aid KM processes, and it is suggested that a model for KM in HRM with AI help be constructed.

Keywords: AI support for KM, Knowledge management, Knowledge management for HRM, Knowledge

1. Introduction

Knowledge age is the term of how we use knowledge as the primary commodity. Human expertise is being used by people through their experiences and skills, and human interaction in its environment has become a source of knowledge. Knowledge is seen as the sole source of production of wealth in the contemporary era of globalization, and developing, utilizing, and creating knowledge is the primary source for a company in obtaining long-term benefits in the marketplace.

For enterprises, knowledge management has become a commercial need in terms of organizing and keeping their gained knowledge so that it may be conveniently accessed and employed later. In the study of Jia et al., (2018) their findings revealed that KM and HRM are inextricably linked. Humans are the major bearers of knowledge, according to their findings, and the knowledge management development concepts is based on a series of procedures and tasks, including knowledge acquirement or collection, knowledge preservation, knowledge sharing, and knowledge usage. Artificial Intelligence (AI) is now being used to support KM in a variety of disciplines, which is a big development in KM.

According to Rich (1985), the study of how to make computers execute things that people can currently do better is known as artificial intelligence (AI). Since AI is already employed everywhere, this definition has already been enlarged. It has been used in business for a variety of purposes, including enterprise management decision-making and assisting individuals not just in the operations side, but also in the middle hierarchy level, where managers and supervisors can complete their routine tasks quickly (Jia et al., 2018). Artificial intelligence can be used to improve efficiency and productivity in human resource management in a variety of ways. During the recruitment process, AI may be used to assist both the hiring company and the job applicants. It can also be used to improve employee retention and internal

*Corresponding author: rasagum@pup.edu.ph

DOI: 10.1201/9781003490210-11

mobility. Employee engagement and work satisfaction may now be assessed more effectively by human resources departments through customized feedback questionnaires and employee appreciation programs. Artificial intelligence (AI) software that automates administrative activities may be able to alleviate some of the stress (O' Connor, 2020).

This study will discuss the use of KM in HRM and the possibility of implementing AI in the knowledge management process.

2. Literature Review

A. Knowledge Management, needs, benefits, and its process

Knowledge Management is "the coordination and exploitation of organizational knowledge resources, in order to create benefit and competitive advantage" (Drucker, 1999). It plays a vital role in accelerating the wheels of information systems. Through knowledge management, companies have been provided with a strategic and sustainable competitive advantage. In other words, according to Davenport and Prusak (Davenport & Prusak, 1998), to enhance performance and create value, KM is a systematic and organizationally defined process for collecting, managing, maintaining, using, communicating, and renewing all personnel in a company's tacit and explicit knowledge.

It entails activities, procedures and systems that support and even improve knowledge storage, assessment, sharing, refining, and production in order to create value and meet tactical and strategic goals.

There are 4 main processes of the knowledge management for it to be successfully implemented: Knowledge Acquisition, Knowledge Storage, Knowledge Distribution, and Knowledge Use.

B. Human Resource Management and Knowledge Management

The term "human resource management" refers to a collection of policies and management tasks are relevant to a company's human resources. Jia et al. (2018), stated that there are six (6) dimensions of HRM, these are; Staffing and deployment, preparation and development, performance appraisal, pay management, and employee relationship management are all part of human resource management.

According to Noe et al. (2006), human resource planning assists a business in predicting future personnel demands as well as identifying the basic qualities of prospective hires. In human resource planning, it was mentioned that this solves the organization staffing and staff matching issues. Staff members are educated and updated through training and development. The most significant of the six

human resource management dimensions is performance management. It is the primary source of information for all other dimensions. Compensation management's goal is to incentivize employees to solve challenges at work. Management of employee relationship is the last of the six components, and its objective is to assist the organization in managing employees and creating a more efficient human resource allocation cycle. As a result, KM may be regarded a fundamental component of HRM, and to succeed in a highly competitive environment, any organization must ensure that its central skills, which for most companies is knowledge, is exploited in order to secure long-term viability. Zaim et al. (2018), present empirical evidence for KM's positive impact on HRM, demonstrating that knowledge usage performs an essential part in KM processes and mediates the relationship between KM and HRM.

C. Link between Artificial Intelligence and Knowledge Management

Knowledge management appears to have progressed to the next level in the twenty-first century. Artificial intelligence (AI) has started to make an impact on how knowledge is captured, developed, shared, and used (Murad and Kurdy, 2021) AI plays a critical role in the dissemination of knowledge. It enables machines to acquire, analyze, and apply knowledge in order to employ tasks, as well as to unlock knowledge that can be shared to humans so that making decision is improved. Artificial intelligence is poised to change knowledge management, particularly in terms of assisting enterprises in retaining knowledge, collaborating, and providing excellent customer service.

Knowledge management, allows for the understanding of knowledge, whereas artificial intelligence gives individuals the power to grow, utilize, and generate knowledge in ways they never dreamed. Knowledge management's interface with artificial intelligence is breaking new ground in cognitive computing, where computational models are utilized to replicate human brain processes. The approach employs deep learning artificial neural network software that mimics the human brain through data mining, pattern identification, and natural language processing. Big data can be analyzed to improve judgment in decision-making processes. Cognitive computing is paving the road for future artificial intelligence and knowledge management applications (Rhem, 2017).

For Murad and Kurdy (2020), knowledge management benefit artificial intelligence tools that are used to capture, filter, represent or apply knowledge. The existing knowledge management systems have a relationship with the key knowledge processes of acquiring, communicating, storing, codification, producing, applying and various sorts of innovation. To aid decision-making, Text selection, parsing, analysis, and categorization, as well as automated reasoning and visualizations, are all possible with AI technologies. The emergence of natural language processing and its vast and intelligent advances in the field of artificial intelligence provides ways to process human input such as handwriting and voice recognition. On other applications, AI systems can even manage large amounts of data while also providing a level of security through the use of new data storage methods.

D. Artificial Intelligence in Supporting Knowledge Management

AI techniques for Knowledge Elicitation as stated by Mohamed and Zaibon (2021), can be done through structured interviews, protocol analysis, concept sorting, and data mining. AI techniques for knowledge representation can be through production rules, frames, and semantic networks. They discovered that using these strategies, AI and knowledge-based systems knowledge management technologies are typically trustworthy in delivering the proper information at the right time.

Not only Mohamed and Zaibon established the use of AI as support in simple knowledge management, the use of AI in KM plays important roles in different areas like simplifying knowledge discovery, can aid employees in content creation, can strengthen the collaborations among employees, AI will amplify learning that may result in skills enhancement..

According to Jallow, Renukappa, and Suresh (2020), some organizations that implemented AI sort systems in a common data environment were able to assist employees in finding documents more easily, and AI systems can be built to help with knowledge management processes that businesses have already implemented. They suggested that a business model canvas for applying AI to benefit KM within enterprises be created in order to distinguish between organizations that are already using AI for KM and those who are not. According to their findings, other researchers should investigate the adoption of AI in at least one company as a case study to determine the results of hurdles and benefits on the use of AI to support KM within the business. As stated in the study conducted by Murad and Kurdy (2020), one possible approach to exchange information, queries, and requests with some other beneficiaries and agencies that share a common unified domain is through implementing an AI-based technique that uses automating knowledge. The impact of AI in KM can be seen in different business industries, it would be a long list if you are asked to enumerate all of them to summarize here are some of its known impact: AI makes huge data processing available and quicker in real time, which can substantially improve customer relationships, AI aids in the upkeep of knowledge base content, it can help a company's

management and organizational capability by going beyond human capabilities, by identifying latest patterns in data and maximizing the use of existing knowledge assets, and enhances knowledge management by giving real-time measurements and data generation tracking tools. These proofs of AI's impact on knowledge management suggest that AI has already created a key contribution to the quality and utility of knowledge management, not only in terms of thinking and methods in solving problems, but also in terms of knowledge acquisition, management and exhibition, decision support systems, smart tutors, planning, making schedules, and optimization structures. (Net al, 2020) (Mercier-Laurent, 2015). As Akerkar (2019) said, because of the value provided in assisting humans to learn and build innovation quicker, Artificial intelligence (AI) is expected to usher in a new era of business efficiency, competitive advantage, and even economic development.

E. AI Application Framework in Human Resource Management

The expansion of a Human Resource Information System (HRIS) to help in HRM marks the support given by ICT in the field of HRM. HRIS assists in strategic planning, development of necessary training for the company, and helps in the evaluation of employees' performance. At this time the implementation of AI in some human-computer interaction functions was developed.

The "AI+HRM" model framework is designed to assist human resource managers in making better, more efficient decisions in the face of large amounts of data. Jia et al (2018) developed a paradigm that explains how AI can be combined with HRM. It explains the connections between human resource management, AI technology, and intelligent systems that have been established.

3. Methodology

This study made use of related reviews to identify the possibility of artificial intelligence support in different phases of knowledge management for human resource management. The researcher made use of different literature reviewed and able to discuss different research made to define Knowledge Management, its functions and it benefits, as well as Human Resource Management, and the link of KM with HRM. Implications of using artificial intelligence, AI, in knowledge management and AI in different HRM systems. Researches that were used in this review article came from international refereed journal and some in the SCOPUS indexed journal. Some models for Knowledge Management were reviewed together with conceptual framework of HR activities. The models were analyzed and observed how one model may contribute to the other.

4. Results and Discussion

The use of KM in HRM was clearly seen in different reviews, focusing on the connection between KM and HRM, the reason why KM can be used in HRM, and its benefits. A study gave HRM processes were also discussed, and the use of Artificial intelligence can be seen as a support in the development or designing a knowledge management, as well as the use of AI in some of HRs functions. The researcher was able to see a model (structural model), see Fig. 3, that summarizes the positive effect of KM and HRM. The first outcome of their study reveals that KM has a favorable impact on HRM practices, which is in line with previous. According to Zaim et al. (2018), future researchers should look at which components of HRM are more closely tied to KM and how implied knowledge will become more prevalent in HRM practices. As a result, the researchers are considering introducing artificial intelligence into the knowledge management process. According to the research, the precise contribution of IT systems to the transmission and development of tacit knowledge is a tough matter. Some academics have proposed a new trend in AI termed cognitive computing, which let this to take place (Rhem, 2017).

5. Conclusion

The goal of this study was to look over the existing literature on the use of artificial intelligence as a knowledge management aid. In order to identify the scope of AI support that may be used to knowledge management in human resource management, the application of knowledge management in human resource management was also explored. The advantages of merging AI with knowledge management were highlighted. A connection and incorporation of artificial intelligence in knowledge management and its different processes were highlighted and this literature review was able to probe the effects and issue of using knowledge management for human resource management, which is sometimes referred to as knowledge base human resource management. It also gathers the different achievements of AI when confronted with different km issues. The objective of the study was attained by reviewing different research papers in the field of artificial intelligence, knowledge management, and AI in HRM. There exist a model stating the positive effects of KM in HRM, positive effects of using AI in HRM, studies show the possible effect of AI support in KM. Through some readings there are still recommendations to look on what dimensions of HRM KM is more aligned and the conversion/interpretation of tacit knowledge is unclear. The emerging trend cognitive computing can be seen as a potential solution, and further studies still needs to be done.

References

1. Davenport, T., & Prusak, L. (1998). *Working Knowledge: How Organizations Manage What They Know*. Retrieved from https://www.researchgate.net/
2. Guo, Y., Jia, Q., Li, R., Li, Y. R., & Chen, Y. W. (2018). A Conceptual Artificial Intelligence Application Framework in Human Resource Management. In *Proceedings of the 18th International Conference on Electronic Business* (pp. 106-114). ICEB.
3. Jallow, H., Renukappa, S., & Suresh, S. (2020). Knowledge Management and Artificial Intelligence (AI). *21st European Conference on Knowledge Management (ECKM 2020)*.
4. Jaradat, A., Kastrati, S., Keceli, Y., & Zaim, H. (2018). The Effects of Knowledge Management Processes on Human Resource Management: Mediating Role of Knowledge Utilization. Retrieved from https://www.emerald.com/.
5. Mohameda, S., & Zaibon, S. (2021). Artificial Intelligence Support for Knowledge Management in Construction. Retrieved from http://repo.uum.edu.my/
6. Murad, B., & Kurdy, M. (2020). Knowledge Management Referral System Using Artificial Intelligent Techniques. *Journal of Engineering Sciences and Information Technology*, 4(3), 117-144.
7. Noe, R. A., Hollenbeck, J. R., Gerhart, B., & Wright, P. M. (2006). *Human Resource Management*.
8. O'Connor, S. (2020). Artificial Intelligence in Human Resource Management. Retrieved from https://www.northeastern.edu/
9. Rich, E. (1985). Artificial Intelligence and the Humanities. Retrieved from https://www.jstor.org/.
10. Rhem, A. (2017, July 18). The Connection between Artificial Intelligence and Knowledge Management. Retrieved from https://www.kminstitute.org/.

Advancing Sustainable Science and Technology for a Resilient Future – Sai Kiran Oruganti et al. (eds)
© 2024 Taylor & Francis Group, London, ISBN 978-1-032-79020-6

12

Sentiment Analysis on Product Review Using Sarcasm, Emoticon, and Internet Slang Detection

Ria Ambrocio Sagum*

College of Computer and Information Sciences, Philippines.
Research Management Office, Polytechnic
University of the Philippines,

Abstract: This research aims to improve the accuracy sentiment analysis by adding one variable, internet slang detection. The system made use of Backward Chaining algorithm to identify each internet slang given a sentence, give its valuable meaning and infer the internet slang for later usage in the final polarity. The system determines if the review is sarcastic, then integrate the internet slang as a new variable. The study showed that adding an internet slang detection intensified the input to its polarity as compared to a positive or negative message with emoticon. The system gained 92% of accuracy, 80% for precision, recall of 100% and 88.88% for F1-Score. The result shows that there is an increase in accuracy of the Sarcasm Detection gained with the integration of Emoticon and Internet Slang Detection. It means that product review with Sarcasm, Emoticon and Internet Slang Detection Module successfully implemented. This research, sentiment analysis with internet slang detection, that has a 90% accuracy rate is doing well compared to sentiment analyzing with emoticon detection alone with the accuracy of 82% but this research has a lot of rooms for improvement.

Keywords: Sentiment analysis, Internet slang detection, Emoticon detection, Sarcasm detection, Backward chaining, Product review

1. Introduction

Sentiment analysis is one of the emerging research areas in computing science. The studies of public opinion at the beginning of 20th century is the root of sentiment analysis began in the. However, it was not until the availability of the subjective texts on the Web when computer-based sentiment analysis began (Mäntylä et al.). As observed by the study of Hussein et.al, in the past few years, sentiment analysis became very beneficial for different areas of society such as businesses, governments, and individuals. Some sentiment analyzer are being used to understand the needs of the customer (Nagaraj et al., 2020; Shashkova et al., 2020; Srinivas & Ramachandiran, 2020), their behaviors (Nagarit et al., 2018), and also the opinion of people (Kochuieva, 2021). Though, there are numerous difficulties being faced in the evaluation process in sentiment analysis.

One of the known challenges in sentiment analysis researchers faces, is the problem with sarcasm detection.

Sarcasm happens when a person used a word that conveys implicit information. These words being used is usually the opposite of what is said. Recognizing sarcastic statements can be useful in improving automatic sentiment analysis most especially the data to be collected is from social networks.

Studies states that emoticon detection is one good approach for detecting a sarcastic statement. Emoticon, as defined by DataGenetics in 2013, can be seen representation of a facial expression using characters. (Gael et al., 2016) In a recent report the Emoji Research Team of Emoji has shown that 92% of the online population are using emojis.

Moreover, another variable that could be used in sentiment analysis is the widely used internet slang. It was used by some people during Internet's early (Dear, 2017). Internet slang are sometime being used in chat rooms, social networking services, online games, video games and in the online community. (Manning, 2018) The creation of this shorthand (Internet slang) can be rooted in 1979 when users of communication networks made shortcuts for their easy

*Corresponding author: rasagum@pup.edu.ph

DOI: 10.1201/9781003490210-12

communication. From the use of these shorthand emerge and later on referred to as Internet slang. (Data Genetic, 2013).

This research aims to solve the problem with the lack of sufficient variables that could help in enhancing the accuracy of detecting the sentiment analysis. To tackle this problem, the researcher added another variable, internet slang, with the existing variables (sarcasm and emoticon analysis). Experiments were made to inspect if there will be improvement in accuracy with the new system.

2. Related Works

One of the definitions of Internet Slang (IS) was provided by Urban Dictionary where it was stated that, Internet Slang is a kind of language that is commonly use on internet, it includes letter homophones (b r b), punctuation (!!!), capitalization (STOP!), onomatopoeic (hehehe) and emoticons (:3). In line with this, it was also stated that, Internet slang is commonly used in websites that have online communities such as forum boards or community irc lines. Such words often contain 3 letter, (a common pattern noticed).

"LOL" meaning "laugh out loud" is an example of IS. IS cannot be standardized due to the constant changes made. Sometimes it is how it became popularized in a locality or country. IS was popular since it made easier for the people to keyin in a keyboard or in cellular phones. It became mostly used for text messaging, or chats in social networks. Some uses it just for uniqueness and fun on using it. Others find it their way not to be understood intentionally some sort of coding of a new generation that cannot be understood by their parents (older generations).

Methods for Internet Slang Extraction

Detection of Internet Slang and scoring shows a promising result in the detection of sentiment. (Manning, 2018). Backward chaining with its counterpart forward chaining shows promising result when dealing with Artificial Intelligence (AI) detection and extraction.

Sarcasm

These words are being used to criticize someone or something in a way that it sounds positive to others yet in actuality it is the exact opposite to the person criticized. This definition helped to connect internet slang in sarcasm detection where in through the help of internet slang, the accuracy of sarcasm detection will be enhanced.

Sentiment Analysis

Sentiment can easily be recognized in a face-to-face communication by merely looking at the visual feature like frowning or smiling. This will be hard to detect in a plain text since no visual emotion can be seen. To date people have embraced the use of emoticons as substitute to illustrate visual emotion in communication via computer it is very popular as a substitute emphasizing opinions in the social web. The use of emoticons allows the writer of opinion and state of emotion in messages, chatting in the social media, even in giving feedbacks and comments. It is being written typically as sequences of typographical symbols such as ":3", ";(", ":(", or "(". It is being understood as facial expressions and being read sideways. (Kolowich, 2018)

3. Discussion of the Research

Figure 12.1 shows the system architecture of the study. the research composed of three important parts which are the implementation of backward chaining, the computation, and the input of language model.

The Language Model will be constructed using set of training data and scoring of the N-Gram that assigns a probability to that set of string base on its incident in text before processing.

Backward Chaining is the algorithm is being used to verify the internet slang word, evaluate the slang word using the rule-based conditions, and inferring the internet slang words.

Upon verifying, the system will check if an internet slang word exists in the given sentence. The evaluation of the internet slang words follows, where the predicted values of the verified words will be assessed based on the rule base condition of the backward chaining algorithm. Lastly, the system will infer the initial polarity of the evaluated internet slang that will become the input for the computation of the final polarity of every variable which are internet slang, emoticon, and sarcasm.

In the computation part, systems will conjunct all the variable polarity at one. This is the part of the program where the sentiment value of the system will come up.

Computational part plays the role of the output details. It handles the combination logic rule of the system detecting the true meaning of the given product review sentence. Inferred inter slang will associate in the variable of emoticon and sarcasm with its perspective polarities. Upon associating, having the final polarity of the sarcasm and emoticon detection, the infer value of internet slang will now contradict or give more weight in the sentiment of the given product review sentence through the rules that is implemented in the system.

Data Generation

The reviews that will serve as a training data for sarcasm, emoticon and internet slang detection will be gathered from an online shopping website named Lazada.ph The samples

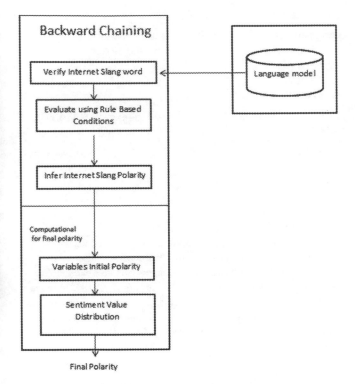

Fig. 12.1 System architecture

Source: Author

collected from this kind of platforms is known for having reviews regarding a specific product. It is also one of the useful sites for conducting sentiment analysis because reactions and opinions of people are often expressed in certain discussions like the product reviews.

Reviews that were gathered contain variables like emoticon and internet slang for better verification and testing of the functionality of the system. Sixty percent (60%) of the gathered data served as training data for the sarcasm, emoticon, and internet slang detection, twenty percent (20%) was used for will be collected, while the 20% of data will be tested by a natural language processing expert and for the evaluation, 20% of the data gathered .

Data Analysis

The overall performance of the system was measured using accuracy. Percentage of correctly identified results consists of true positive and true negative, amongst the entire set of reviews fed to classified. It is computed as the total of TP and TN divided by the total of TP, TN, FP, and FN.

F1 score was also computed to further see how the systems performs. This is said to be the harmonic mean of the computed precision and recall.

True Positive (TP) is the total sentiments classified as positive both system and human while True Negative (TN) is the number of sentiments that were classified as negative by both system and human. Furthermore, False Positive (FP) is calculated as the number of sentiments classified as positive by system but in reality, it is a negative sentiment as its true value, while False Negative (FN) is the number of sentiments classified as negative by the system but assessed by human as positive sentiments.

For the evaluation of the two systems, sentiment analyzer and sarcasm detection module of the system, the researcher experiment paper to be checked by the administrators that also tag the sentiment seed. As for the accuracy of the system, the researcher would be using an experiment paper to be answered by different customers.

4. Result

The evaluation of the system's accuracy was compared to other three criteria, where the first one is an ordinary sentiment analysis, the next is a sentiment analysis adding variables with sarcasm and emoticon, the third one is sentiment analysis with additional variable sarcasm and internet slang and of course the developed system, a sentiment analyzer with sarcasm, emoticon, and internet slang were identified and used for sentiment analysis. The computed accuracy were 24%, 56%, 60%, and 92% respectively. By this result it was seen that as another variable were added to the sentiment analyzer the more accurate the analyzer is.

The system's performance was also evaluated by computing its F-Score. It can also be noted that as more variable were used to classify the sentiment given a product review the more accurate the result is. Take note that three criteria were also used and the system's developed. The four systems precision were 20%, 35.71%, 34.86%, and 80%. Likewise the recall for the systems were 30%, 71.42%, 71.42%, and 100%, respectively.

The criteria for Product Review with Sarcasm, Emoticon and Internet Slang garnered 80.0% in terms of precision, a perfect 100% performance in terms of recall and a promising performance of 88.88% in terms of F1-Score that shows that the utilization of variables which are sarcasm, emoticon and internet slang garnered the highest system performance based precision, recall and F1-score. For the comparison of accuracy of in previous research's system, Sentiment Analysis Using Sarcasm and Emoticon, has an accuracy of 52% while this study Sentiment Analysis Using Sarcasm, Emoticon and Internet Slang has 92%. From the formula of Percentage, it increased by 64.2857%, this means that this research has more potential in detecting the correct sentiment of the product review than the previous research.

5. Conclusion

It can be stated that the developed system lives to the expectancy on detecting sarcasm, emoticons, and internet slang. This study used backward chaining algorithm for the language model, the system's precision and recall were quite higher than existing papers. However, there are some words that the program was not able to detect that may be used in the identification of the sentence's sentiments. The result of the study showed that the system was able to correctly determine the detection in review on the product production. Product Review with Sarcasm, Emoticon and product review with Sarcasm, Emoticon and Internet Slang gaining 56% and 92% consecutively were tagged with the high accuracy and gaining excellent rate in the interpretation data. The quantity of the testing set for both Product Review with sarcasm, emoticon and Internet slang possibly affected the respective accuracies.

The F1-score showed a dense result. F1-Score from ranged 80-84%, is considerably satisfactory to excellent rate, but can be improved through a larger set of implementation data. It was observed that the F1-Score and Accuracy of the criteria product review with sarcasm, emoticon and internet slang is higher than the criteria of product review with sarcasm and emoticon. The goal to integrate the internet slang detection to sarcasm and emoticon for better recognition of sarcasm and improved the previous study yield to a better accuracy of sentiment analyzer.

6. Recommendations

The system will be more accurate based on the regular expression that is in the developed system. This system will be accurate if the evaluation data will follow the construction of a sentence which is the sentence/phrase that will be the first one to compute the polarity by the system, next is the emoticon and internet slang word for the last component of the sentence.

Despite the promising results that the system with internet slang detection provided, the system still lacks sufficient polarity detection especially in terms of detecting the neutral polarity of a given statement therefore, affecting the final polarity and accuracy of the system. That is why, it is recommended that the future researchers should focus on adding the detection of neutral polarity. Future researchers can also use the approach of Fuzzy Logic Algorithm in predicting the correct detection of the polarity of a given statement.

Acknowledgment

The author would like to acknowledge group 4 of BSCS 3-FS2 batch 2019, for the help in programming and in the experimental stage of the research.

References

1. Bautista, S. C., Bautista, S. C., Bermudo, P. J. V., Yango, A. D., Yango, A. R., Galicia, L. S., & Nagarit, N. B., EdD Susana C. (2018). Transforming Online Negative Blogs in the Use of Credit Cards in Electronics Transactions into Constructive Action: Basis of Creating Business Spend Analyzer Model. *Journal of Business Management*, 6(1), 66–83. https://doi.org/10.25255/jbm.2018.6.1.66.83
2. Data Genetics. (2013). "Emoticon Analysis in Twitter."
3. Dear, Brian (September 19, 2012). "PLATO Emoticons, revisited." *PLATOHistory*.
4. Gaël Guibon, Magalie Ochs, Patrice Bellot. (2016) "From Emojis to Sentiment Analysis."
5. Kochuieva, Z., Borysova, N., Melnyk, K., & Huliieva, D. (2021). Usage of Sentiment Analysis to Tracking Public Opinion. *International Conference on Computational Linguistics and Intelligent Systems*.
6. Kolowich, Lindsay (March 2018). "The Evolution of Language: How Internet Slang Changes the way we Speak." Retrieved from https://blog.hubspot.com/marketing/how-internet-changes-language
7. Manning, Christopher. (2012). Evaluation of Text Classification. Stanford University Coursera.
8. Nagaraj, M., & Ruba, G. (2020). Big Data – Concepts, Applications, Challenges and Future Scope. *Software Engineering and Technology*, 12(1), 10–18.
9. Shakhovska, K., Shakhovska, N., Veselý, P., & Veselý, P. (2020). The Sentiment Analysis Model of Services Providers' Feedback. *Electronics*, 9(11), 1922. https://doi.org/10.3390/electronics911192
10. Srinivas, S., & Ramachandiran, S. (2020). Discovering Airline-Specific Business Intelligence from Online Passenger Reviews: An Unsupervised Text Analytics Approach. *ArXiv: Information Retrieval*.

Advancing Sustainable Science and Technology for a Resilient Future – Sai Kiran Oruganti et al. (eds)
© 2024 Taylor & Francis Group, London, ISBN 978-1-032-79020-6

13

Antonym and Meronym Identification as an Aid for Filipino Corpus Builder

Ria Ambrocio Sagum*

College of Computer and Information Sciences, Philippines.
Research Management Office, Polytechnic University of the Philippines,

Abstract: This project aims to add features to the existing FiCoBu System by adding antonym and meronym. These added to a Filipino Wordnet system can enhance its usefulness, enabling a more comprehensive understanding of the language and facilitating various natural language processing and knowledge representation tasks. The accuracy of the system in terms of identifying antonyms given Filipino word as an input is 82.24%. For the meronyms accuracy the system was evaluated to have 85.17% accuracy given a Filipino word. The project can be extended by adding of different synset relation like holonyms, hypernyms and hyponyms for noun words and troponyms and entailment for verb.

Keywords: Natural language processing, Corpus building, WordNet, Filipino language, Language modeling

1. Introduction

A collection of documents of a language or languages that are to be kept, managed, and evaluated in digital form is said to be a corpus (Dita et al., 2009). McEnery and Wilson (2001), indicated that "any collection of more than one text can be called a corpus". A corpus builder is an automatic system that constructs a corpora for a certain language. WordNet is an example of a result of a corpus builder.

Different methods and tools were being used for a machine to recognize, understand, and generate human language. One of these is Natural Language Toolkit (NLTK). NLTK is a toolkit that can be used with natural language, this toolkit makes use of a corpora as its component to deal with processing of human language (Bird et al., 2009), WordNet is one of the corpora it uses.

There are seventy-two existing Wordnets available today (Weischer, 2016). To name a few we have AlbaNet which is intended for the Albanian language, BulNet for Bulgarian Language, DanNet which is developed for Danish Language and EuroWordNet made for English Language are just some of the existing WordNet that qualified the WordNet Organization.

Building language corpora set to serve as Wordnet is a challenge for researchers most especially for small resource language. A language corpus like Filipino corpus, can help preserve the existence of the language, language reservation, and language research. Challenges and limitations can be encountered in building one, sample size, bias in data selection such as over representing of some genre which may cause invalidity for the language data set. An available tool like this is useful in the field of Natural Language Processing (NLP). The researchers aim is to improve and add features of the existing FiCoBu (Sagum et al., 2015), system by including antonymy and meronym words for Filipino language.

2. Related Works

Constructing a WordNet for Filipino language will not only provide an important lexical-semantic resource for this language but may be used for that language in several applications that employ WordNets.

FilWordNet, discussed that morphology is needed to ascertain analyzers and generators to support root word entries in the WordNet as well as synset entries in root word form (Borra et

*Corresponding author: rasagum@pup.edu.ph

DOI: 10.1201/9781003490210-13

al., 2010). A study called Palito, a Tagalog term which literally means 'stick', uses a modified Morphological Analyzer and Generator for Tagalog or MAGTAG to suit the synset creation form (Bondoc et al., 2009). The process becomes complicated when the original word and the translated word are both nouns. The definition of the Filipino WordNet synset become more complex once translated word is inflected and becomes and adjective or a verb. Morphological information is important to know the word's meaning and sense. It is suggested to establish a morphological analyzer and put more synsets in the existing research.

FiCoBu research states that the system (which is made at the time when the paper was published) had issues in detecting uncommon and deep words, as well as words with several affixes as a Filipino word. Improving the system to be able to detect those kinds of words and crawling more Filipino websites and dictionaries to further build the FiCoBu Word Archive.

We recommend improving the system on how to detect deep Filipino words and if possible, to improve the FiCoBu training data to be able to improve existing capability of detecting Filipino words and put a capability to enter texts ourselves on the program. An option should be present as well to let users enter Filipino words, through it should be still supervised in a manner that will be controlled by an arbitrary or a ruling party based on the recommendation of the research.

Of the many years that have passed, Wordnets for different languages have been developed. Of all the attempts in developing the said system, only 72 met the standards of the Global Wordnet Association in qualifying as an official Wordnet for a particular language. The AlbaNet which is intended for the Albanian language, BulNet for Bulgarian Language, DanNet which is developed for Danish Language and EuroWordNet made for English Language are just some of the existing WordNet that qualified the WordNet Organization. Some added different relation to its word like meronyms and antonyms (Sinaga et al., 2022, Rahman, 2021).

In this project, the researcher focused on the development on the identification of antonyms and meronyms of words as an additional feature for the existing FiCoBu system. Antonyms are words opposite in meaning to another (e.g., Mabait and Masama). Meronyms a term which denotes part of something, but which is used to refer to the whole of it (e.g., "Daliri" is meronym of "Kamay").

3. Methodology

Experimental method was used to assess the performance of the developed system. The system architecture depicts the design of the software.

Figure 13.1 illustrates the flow of the proposed system. There are two major components in the system, Corpus Builder that deals with collecting articles, and building a corpus, and the API Block that deals in using the created corpus to recognize Filipino words in an article.

The system made is based on FiCoBu System that made use of online dictionary (KATIG.COM and tagalog.pinoydictionary.com) the words on this dictionary will be the initial words (keywords) of the system. Keywords were set to be the first piece to make the crawler does its work.

In building the systems's corpus builder, a Crawler looks for Filipino articles in the world wide web and will gather these articles. The link crawler will automatically find links on its own. Then, the links gathered will be added to the database. The user can also enter specific link to be stored in the database, assuming that the link is not yet stored in the database. Content crawler, will automatically fetch the contents of every link that is in the db and store the content in a textfile. The user can enter a specific link to get its content. In tokenizer module, inputs are the textfiles from the article storage and will simply recognize the words in the input textfile and its output will become the input in the language model.

In language modelling part, the inputs are from the tokenizer. The language model classifies the Filipino and non-filipino words from the input textfile.

The stemmer part will reduce derived words to their stem words, base, or root form (Sagum et al., 2014). In this part,

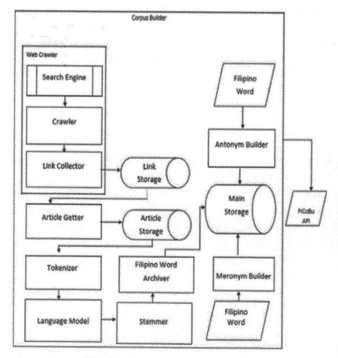

Fig. 13.1 The proposed system

the inputs are the classified Filipino word from the language model. Filipino language has several ways of placing an affix to a word such as prefix (unlapi), suffix (hulapi), infix (gitlapi), circumfix (kabilaan), and partial or full reduplication. (Noche, 2012).

a. Prefix, an affix that is placed before a rootword to form new words. Valid affixes includes pinag-, pina-, nag-, pa-, um-, mag-, ma- (Ebona et al., 2013).

b. Suffix, a type of affix where a new word is being formed once a root word was added in the word. Root words like an, -in, -hin are examples of this.

c. Infix, when an affix is inserted within the root word and form a new word, it is called as infix. Some valid infixes are –um-, -in-.

d. Circumfix, if a word has two or more type of affixes, it is called a circumfix.

e. Reduplication, it can be categorized into two partial or full. Full reduplication exists when the root word was entirely duplicated to form a new word. A partial reduplication however is when a particular syllable in a word is being duplicated.

In this part, the system will collect the classified Filipino words, that is reduced in its root or base form, from the input textfile and store it in the main storage. An online dictionary was utilized to get the antonym of a word.

The meronym of the word is manually typed by the user. Once word were typed APIs were used to help in corpus building. These are: isFilipino- identify if the word is a Filipino or not, antonym- returns the antonym of the word, and .meronym- returns the meronym of the word

4. Data Gathering

Filipino articles from online sites with .txt format will be used as the objects for this project. These files will be used for the training of the system.

One thousand words were used as test data for the system. 500 words for testing the retrieving of antonyms and the other 500 words is used for testing the retrieving the meronyms.

This section discusses the step-by-step procedure in gathering data.

1. The system will be trained on the domain chosen by the researcher. In this case, Filipino text.

2. After training, actual testing will proceed. The API created after and/or during the building of system will be used in experimentation. The system that uses the API will accept user input by opening a file. The input will be processed by the API and output will be displayed by the system that used the API.

The researcher of the study will record the following in the test plan:Number of words with correctly tagged antonym/meronym (TP), number of words that is not tagged because there is no antonym/meronym (TN), number of words that is wrongly tagged that it have a antonym/meronym(FP), and number of words that is not tagged but the word has a antonym/meronym(FN)

3. The Precision, Recall, and Accuracy will be computed using the formula provided, and will also be recorded in the test plan.

5. Evaluation Methodology

Quota sampling is used in this project. It is a non-probability sampling for there is no knowledge of the total count of Filipino words found on the World Wide Web. This sampling technique does not require the population lists, though as limitation for this technique it may not be good to represent the entire population list and might result to bias.

A tool in Java programming Language was provided that will test and train all the data and sample Filipino article in the text file to check the accuracy of the system.

The accuracy of the system used in this research was from the computed harmonic mean of the system's precision and recall.

Recall will be calculated by number of words with correctly tagged antonym/meronyn by the system(TP), and dividing it by the sum of total words with correctly tagged antonym/meronym(TP) and total words not tagged but the word has a antonym/meronym(FN).

Precision was calculated by the number of words and divide it with the correctly tagged antonym/meronyn by the system(TP), by the sum of total words with correctly tagged antonym/meronym(TP) and total words that is wrongly tagged that it have a antonym/meronym(FP)

6. Results and Discussion

The system got a total recall of 69.83% and precision of 100% and an accuracy of 82.24% in identifying the antonym of the words in the input files. The system got a low score in terms of recall or retrieving words with antonyms because there are still Filipino words that has no antonym, and the system got a good score in terms of precision since all the words with retrieved/tagged antonym are all correctly identified and total number of words that were not tagged because there is no antonym (**TN**) is 195 among 500 words because not all Filipino words has an antonym.

Table 13.1 Summary of 500 words for antonym identification

RECALL	PRECISION	ACCURACY
69.83%	100%	82.24%

The system got a total recall of 69.83% and precision of 100% and an accuracy of 82.24% in identifying the antonym of the words in the input files. The system got a low score in terms of recall or retrieving words with antonyms because there are still Filipino words that has no antonym, and the system got a good score in terms of precision since all the words with retrieved/tagged antonym are all correctly identified and total number of words that were not tagged because there is no antonym (**TN**) is 195 among 500 words because not all Filipino words has an antonym.

Table 13.2 Summary of 500 words for meronym identification

RECALL	PRECISION	ACCURACY
74.18%	100%	85.17%

The system got a total recall of 74.18% and precision of 100% and an accuracy of 85.17% in identifying the antonym of the words in the input files. The system got a low score in terms of recall or retrieving words with meronyms because there are still Filipino words that has no meronym, and the system got a good score in terms of precision because all the words with retrieved/tagged meronym are all correctly identified. The total count of words that is not tagged because there is no meronym (**TN**) is 317 among 500 words because not all Filipino words has an antonym.

Table 13.3 Summary of the system's performance

	Recall	Precision	Accuracy
Antonym	69.84%	100%	82.24%
Meronym	74.18%	100%	85.17%
TOTAL	**71.46%**	**100%**	**83.35%**

In Table 13.3, it shows that the system got a low score of 71.46% in recall for both antonym and meronym identification for the reason that not all Filipino words has an antonym or a meronym. The system got a good score, 100%, in terms of precision because words with retrieved antonym and meronyms were correctly identified. The harmonic mean is 83.35%.

7. Conclusion and Recommendations

This paper aims to improve the existing features of FiCoBu, by adding different synset relation which are antonym and meronym. New entities for the antonyms and meronyms of Filipino words were added to the database of the existing FiCoBu. The accuracy of the developed system in terms of identifying the antonym and meronym of a word was evaluated after experimentation was done.

For the identification of antonyms, the system got a precision rate of 100%, its recall is 69.84% and an accuracy of 82.24%. With this evaluation the systems performance is satisfactory for accuracy, a good rate in precision, and a bad rate for recall. (Sagum et al., 2014).

The degree of accuracy of the systems in identifying the meronyms are, a precision of 100%, recall of 74.18%, and its harmonic mean is 85.17%.

The mean of the scores for both antonyms and meronyms were 71.46%, 100%, and 83.35% for precision, recall, and harmonic mean respectively.

Based on the obtained findings, it is recommended that in order to improve the language model by adding some characteristics that will help identify both meronyms and antonyms. The content of the corpora can be extended to include different synset relations including holonyms, hypernyms, and hyponyms for noun word and troponyms and entailment for verbs. The methodology of the system may be extended to use special algorithm to easily classify antonyms and meronyms given Filipino words.

References

1. Bird, S., Loper, E., & Klein, E. (2009). *Natural Language Processing with Python*. O'Reilly Media Inc.
2. Borra, A., Pease, A., Roxas, R. E. O., & Dita, S. (2010). Introducing Filipino WordNet. (Manila).
3. Bondoc, R. J., Garcia, A., Lacaden, J. B., Ping, Y. H., & Borra, A. (2013). The Filipino Wordnet Construction. (DSLU).
4. Dita, S. N., Roxas, R. E. O., & Inventado, P. (2009). *Building Online Corpora of Philippine Languages*. DSLU.
5. McEnery, T., & Wilson, A. (2001). *Corpus Linguistics: An Introduction* (2nd ed.). Edinburgh University Press.
6. Rahman, A. A. (2021). Multi-Relational Latent Morphology-Semantic "MOR-PHOSEM" Analysis Model For Extracting Qura'nic Concept: A New Innovative For Sustainable Society, 12(8), 1967–1977. https://doi.org/10.17762/turcomat.v12i8.3402
7. Sagum, R. A., Barcelona, A. C., Briore, J., et al. (2014). Stemmer. Unpublished Paper, Manila. 4000

Note: All the figure and tables in this chapter were made by the authors.

Advancing Sustainable Science and Technology for a Resilient Future – Sai Kiran Oruganti et al. (eds)
© 2024 Taylor & Francis Group, London, ISBN 978-1-032-79020-6

Multi-staged Bird Species Classification Through CNN using Bird Vocalizations

14

Albert John De Vera

Department of Electrical Engineering, University of the Philippines, Los Baños, Philippines

Rock Christian Tomas

Department of Electrical Engineering, University of the Philippines, Los Baños, Philippines

Jabez Joshua Flores

Faculty of Management and Development Studies, University of the Philippines Open University, Philippines

Anton Domini Sta. Cruz

Department of Electrical Engineering, University of the Philippines, Los Baños, Philippines

Abstract: The biodiversity of birds may serve as a proxy for the Gini index of a permaculture landscape. Manual biodiversity monitoring can be expensive, time-consuming, and requires highly trained ornithologists. To mitigate these problems, an automated solution is proposed using digital signal processing (DSP) and image classification using deep learning. In this study, a two-stage approach was applied using bird vocalization recordings from public platforms to classify 20 native bird species. The audio data was pre-processed by a bandpass filter to remove noise, then the respective mel spectrograms extracted from the audio recordings were fed to a convolutional neural network (CNN) for classification. A custom CNN model with 4 convolutional layers and 1 fully connected layer was used. A pre-trained model of EfficientNetV2 was used to benchmark the custom CNN model. The custom CNN model was able to achieve top 1 training and validation accuracies of 98.90% and 86.45%, respectively. On the other hand, the EfficientNetV2 model was able to achieve top 1 training and validation accuracies of 79.56% and 75.00%, respectively. While the custom CNN model was able to achieve better validation performance, the large discrepancy between its training and validation metrics implied overfitting. Further investigation revealed that the similarity of mel spectrogram patterns and frequencies of the vocalization among bird species are possible reasons for misclassifications. The proposed method may serve as a benchmark in the deployment of onsite classification systems in the future.

Keywords: Automation, Biodiversity monitoring, Bird vocalization convolutional neural network, Digital signal processing

1. Introduction

Around half of the bird species around the world are at risk where intensification of agriculture is one of the primary factors (Conde and Choi, 2022). Although it maximizes the yield of crops for food security, it also destroys the habitats of birds and other animals that depend on the displaced landmass. One such solution that can balance the trade-off between these two aspects is through permaculture. Permaculture, or permanent agriculture, is a regenerative, closer-to-nature type of agricultural practice aiming for food security and preservation of nature and the biodiversity of animals by replicating the system of natural landscapes (Flores and Buot,

2021). In agricultural landscapes, birds play a vital role in the ecosystem of the area. Birds perform various activities such as seed dispersal, pest control, and provide additional fertilizer components that are beneficial to the permaculture site (Flores, 2018). Different species impart different benefits; thus, the biodiversity of bird species in the landscape may serve as a proxy for the Gini index of the location providing an insight into the overall health of the site.

Monitoring birds manually, however, is an expensive, laborious, and time-consuming task that may be erroneous. Inconsistencies such as double counting and biases from the observer are present when conducting field surveys (Piczak, 2016). It also requires intervention and validation from a bird

[1]acdevera1@up.edu.ph, [2]rvtomas1@up.edu.ph, [3]jabezjoshua.flores@upou.edu.ph, [4]acstacruz2@up.edu.ph

DOI: 10.1201/9781003490210-14

expert which requires additional costs. To alleviate some of these difficulties, automation of bird classification can be implemented to monitor the diversity of bird species in the area. Since it is easier to track birds through the sounds they make than through visual confirmation, monitoring them through audio recordings and the use of deep learning is the preferable method for the classification task (Puget, 2021).

2. Literature Review

Early solutions use mel-frequency cepstral coefficients as the input for various machine learning algorithms for bird classification (Sprengel et al., 2016). In 2016, the winning solution in the BirdCLEF challenge, a yearly challenge of identifying bird species in audio recordings using machine learning, was using convolutional neural networks (CNN) with spectrograms as its input data (Stowell and Plumbley, 2014). Since then, it was the state-of-the-art solution for bird classification problems infused with deep learning algorithms. Converting the audio recordings to mel-scaled spectrograms, or simply mel spectrograms, was the common method in audio classification (Tivirkin et al., 2022). Piczak (2016) concluded that among the four different types of spectrograms, the mel spectrogram achieved the best results in all the performance metrics used. Weston (2022) presented a solution to address the challenges of long-tailed distribution and few-shot training for the imbalance of data in classes. Although various studies have been conducted regarding bird classification using deep learning, none used species endemic to the Philippines or in a permaculture setting.

With these findings, a bird classification system can be created in the context of permaculture in the Philippines. Birds are essential in permaculture and monitoring its diversity is an important factor in monitoring the health of the landscape, providing the practitioner an insight into it (Flores and Buot, 2021). Automating this task eliminates the involvement of an expert while removing the time-consuming and laborious aspect of it.

3. Methodology

A. Data Collection

The audio recordings were obtained from Xeno-Canto and Avian Vocalization Center (AVoCet), websites dedicated as a repository for bird audio around the world. Audio recordings of twenty (20) bird species that are present in permaculture sites in the Philippines were downloaded from these public platforms to build the dataset for training the inference model. The Xeno-Canto API wrapper (Xie et al., 2019) was used to extract relevant audio from the recordings. Each extracted audio was labeled according to species, regardless of the type of call or vocalization, i.e., whether the call was that of

a mating call, a contact call, a fight call, etc. The extracted dataset, X_{raw}, comprised a total of 829 audio recordings as shown in Table 14.1.

Table 14.1 List of bird species and the number of audio recordings obtained

Bird Species	No. of Audio Recordings
Asian Glossy Starling	42
Black-crowned Night Heron	34
Black-naped Oriole	46
Blue-headed Fantail	30
Blue-tailed Bee-eater	43
Brown Shrike	50
Chestnut Munia	40
Collared Kingfisher	49
Eurasian Tree Sparrow	51
Grey-backed Tailorbird	36
Grey Wagtail	41
Mangrove Blue Flycatcher	51
Olive-backed Sunbird	79
Philippine Magpie-Robin	34
Philippine Pied Fantail	32
Pied Bush Chat	38
Red-keeled Flowerpecker	45
Rufous-crowned Bee-eater	12
White-breasted Waterhen	43
Yellow-vented Bulbul	33

B. Preprocessing

The collected audio recordings were first split into 5-second clips. Waveforms from frequencies less than 500 Hz were filtered out from each element in X_{raw}. Furthermore, waveforms with frequencies greater than 11 kHz were also filtered out. A bandpass filter with cutoff frequencies of 500 Hz and 11 kHz was used to process the signals. After denoising X_{raw}, the mel spectrogram of each element in X_{raw} was then obtained ($X_{raw} \rightarrow X$) with a frame size of 2048, a hop size of 512, a sampling rate of 22050 Hz, and mel bands of 128 units. Additionally, the mel spectrograms generated were in the RGB color space. All preprocessing was done in Python with SciPy for the creation of the bandpass filter and the Librosa package for the mel spectrogram generation.

C. Data Augmentation

Data augmentation was used to artificially populate the dataset for training the inference model. Implementing this method helps the model prevent overfitting and increase its robustness (Weston, 2022). Three data augmentation techniques were used – Noise injection, pitch shifting, and time stretching.

Noise Injection. To virtually create a noisier audio, noise injection was used which can be considered a unique entry to the dataset. Gaussian white noise was inserted into each audio with a mean of zero and noise factor of 0.1.

Pitch Shifting. Pitch Shifting altered the pitch of an audio recording without changing the time domain. The Librosa package was used to implement this technique with a mean value of 1.5 and a standard deviation of 0.25.

Time Stretching. In contrast with pitch shifting, time stretching alters the time domain without changing the pitch of the audio. This causes the data points in the mel spectrogram to stretch to the horizontal axis. It was also implemented using Librosa with a mean of 1.5 and a standard deviation of 0.05.

D. Custom CNN Model Architecture

A custom CNN model was trained to classify the 20 bird species according to their vocalizations using X. The architecture of the custom CNN model consists of three convolution layers each followed by a max pooling layer as shown in Fig. 14.1. The last convolution layer is then followed by a flattening layer to convert the output matrix into a one-dimensional vector for the feed forward layer of 512 nodes, then the output node of 20 nodes corresponding to each focal species of the dataset. Additionally, a dropout of 50% was utilized at the feed forward layer to prevent overfitting. Rectified Linear Unit (ReLU) activation functions were used for all layers except for the output layer where a softmax function was used.

E. Training and Validation

To prevent an imbalance of each class in the dataset, stratified random sampling was used to remove sampling bias and create a balanced dataset for model training. Each species was considered as a stratum wherein an equal number of samples were randomly selected with 80% and 20% of the total dataset allocated to the training and validation sets, respectively. The performance of the custom model was benchmarked using a trained EfficientNetV2 model utilizing transfer learning for the specific dataset. All models were designed, trained, and evaluated in Google Colaboratory using Python and TensorFlow with training parameters of 25 epochs, categorical cross-entropy loss function, and using Adam as the optimization algorithm.

F. Analysis

To present further insight on the performance of each model, a confusion matrix was generated which gives information on the performance for each evaluated species. Misclassified samples were also analyzed and possible reasons were provided as to why the model incorrectly predicted some samples.

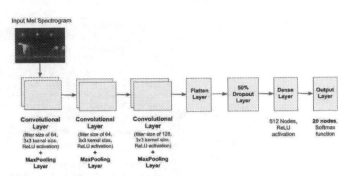

Fig. 14.1 Architecture of the custom CNN model

4. Results and Discussion

The total mel spectrograms generated vary from 300 to 1000 images depending on the number of recordings from each species and its length. After splitting the entire generated dataset, the training set contained 6980 mel spectrogram images (349 for each species) while the validation set contained 1740 mel spectrogram images (87 for each species). These two sets were used for training the custom CNN model and the EfficientNetV2 model.

A. Custom CNN Model

Training the custom CNN model for 25 epochs, it achieved 98.90% training accuracy and 86.45% validation accuracy. Although it resulted in reasonably high training and validation accuracies, the large discrepancy of 12.45% between the two implied overfitting. This means that the model was trained too much using the training dataset such that it predicted almost all its samples correctly; however, when inferring new samples presented by the validation set, its performance was relatively worse. The confusion matrix of the validation dataset is shown in Fig. 14.2 wherein the custom CNN model performed well on nine of the twenty species with accuracies equal to 90% or higher.

The model predicted Eurasian Tree Sparrow, Philippine Magpie-Robin, and Rufous-crowned Bee-eater exceptionally well, achieving significantly high accuracy. On the other hand, it predicted Olive-backed Sunbird relatively poorly with an accuracy of 69% on its validation samples while the model correctly predicted less than 80% of the samples from three other species.

B. EfficientNetV2

In contrast, the EfficientNetV2 achieved relatively lower accuracies, having 79.56% training accuracy and 75.00% validation accuracy, but the deviation between the two parameters was much lower compared to the custom model. Hence, the model did not underfit nor overfit. The resulting confusion matrix of the validation dataset using this model is shown in Fig. 14.3. The EfficientNetV2 model predicted

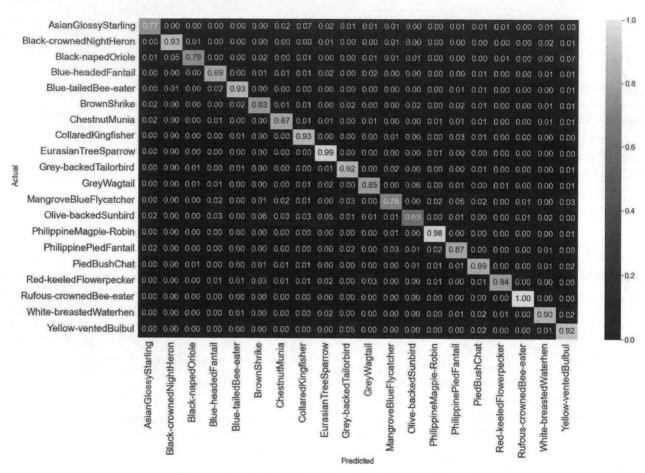

Fig. 14.2 Confusion matrix of the custom CNN model

Eurasian Tree Sparrow, Grey-backed Tailorbird, and Rufous-crowned Bee-eater well with accuracies over 90%. In contrast, the Philippine Pied Fantail was the least accurately predicted species with only 46% of the validation samples being correct. Twelve of the twenty species obtained an accuracy lower than 80%.

Based on these results, the custom CNN model was more suitable for bird classification with the given dataset despite the overfitting of the model

C. Misclassification

One of the reasons why the CNN models misclassified some samples was due to the similarity in structure and frequency of the bird vocalization to other species present in the dataset. For example, the bird vocalization in Fig. 14.4 belonged to an Olive-backed Sunbird but it was misclassified as vocalization of a Philippine Pied Fantail. The custom CNN model predicted as such with a confidence level of 51.17% while the correct species, Olive-backed Sunbird, was at 47.52%. Fig. 14.5 shows the difference between the mel spectrograms of the two species. The spectrogram of the Olive-backed

Sunbird was similar to some portions of the Philippine Pied Fantail specifically along the y-axis, which corresponds to the frequency. Hence, some frequencies of Olive-backed Sunbird and Philippine Pied Fantail vocalizations were similar. Additionally, the upward curve structure of some vocalizations of the Philippine Pied Fantail resembled that of the test sample in Fig. 14.4, causing misclassification of the model.

Another instance of misclassification is shown in Fig. 14.6 wherein the vocalization from a Grey Wagtail was predicted as a Blue-headed Fantail due to the similar range of frequency from both species. The validation sample was predicted as such with a confidence level of 75.73% while its prediction as the correct species was only 0.35%.

5. CONCLUSION

Fusing deep learning with bird classification tasks through its vocalizations has been studied recently to predict the species present in an area. Collecting audio recordings of bird species found in permaculture settings in the Philippines and training an inference model proves to be a viable automated solution

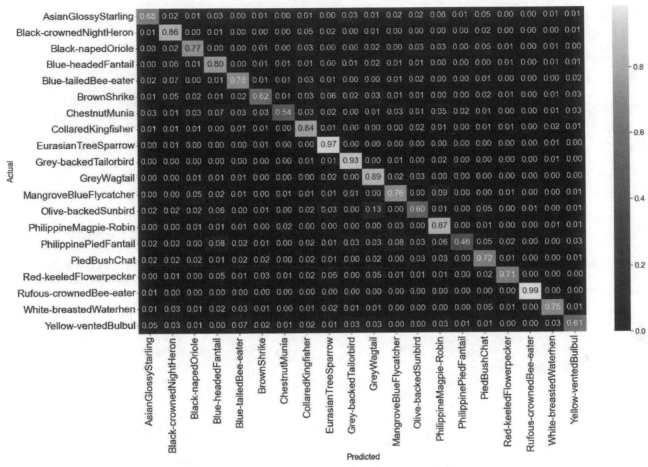

Fig. 14.3   Confusion matrix of the EfficientNetV2 model

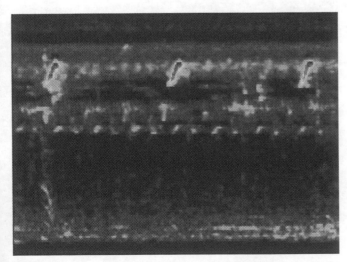

Fig. 14.4   Vocalization of an Olive-backed Sunbird predicted incorrectly

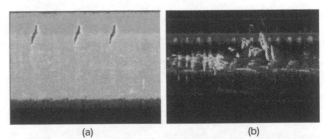

(a)   (b)

Fig. 14.5   Vocalization of (a) Olive-backed Sunbird and (b) Philippine Pied Fantail.

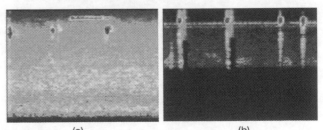

(a)   (b)

Fig. 14.6   (a) Sample from Grey Wagtail test set and (b) vocalization of a Blue-headed Fantail

for monitoring the diversity of species in a permaculture site. With the obtained dataset, the custom CNN model performed better compared to the EfficientNetV2 model with transfer learning in this bird classification task. This study may serve as a benchmark for a deployable bird classification solution that may provide benefits to permaculture practitioners in the Philippines aiding them in monitoring the diversity of bird species in their landscape giving insights into the health of the landscape.

REFERENCES

1. Conde, M. V. & Choi, U. J. (2022). Few-shot long-tailed bird audio recognition, (arXiv:2206.11260). arXiv. http://arxiv.org/abs/2206.11260.
2. Flores, J. J. & Buot, I. (2021). The structure of permaculture landscapes in the Philippines. Biodiversitas Journal of Biological Diversity, 22(4). https://doi.org/10.13057/biodiv/d220452.
3. Flores, J. J. (2018). Designing food security: The applications of permaculture in sustainable agriculture – Case studies in the Philippines. University of the Philippines.
4. Piczak, K. J. (2016). Recognizing bird species in audio recordings using deep convolutional neural networks. In: Working Notes of CLEF 2016 - Conference and Labs of the Evaluation forum, Evora, Portugal, 5-8 September, 2016. CEUR-WS Proceedings Notes, vol. 1609, pp. 534–543.
5. Puget, J. F. (2021). STFT transformers for bird song recognition. In CLEF 2021 – Conference and Labs of the Evaluation Forum in Bucharest, Romania.
6. Sprengel, E., Jaggi, M., Kilcher, Y., & Hofmann, T. (2016). Audio based bird species identification using deep learning techniques. In: Working notes of CLEF 2016.
7. Stowell D. & Plumbley, M.D. (2014). Audio-only bird classification using unsupervised feature learning. In: CLEF (Working Notes), 1180.
8. Tivirkin, N., Davison, D., & Bilzard. (2022, Aug. 10). Xeno-canto API wrapper. GitHub Repository. https://github.com/ntivirikin/xeno-canto-py.
9. Weston, P. (2022, September 28). Half of world's bird species in decline as destruction of avian life intensifies. The Guardian. https://www.theguardian.com/environment/2022/sep/28/nearly-half-worlds-birdspecies-in-decline-as-destruction-of-avian-life-intensifies-aoe.
10. Xie, J., Hu, K., Zhu, M., Yu, J., & Zhu, Q. (2019). Investigation of different CNN-Based models for improved bird sound classification. IEEE Access, 7, 175353–175361. https://doi.org/10.1109/ACCESS.2019.2957572.

Note: All the figures and table in this chapter were made by the authors.

Advancing Sustainable Science and Technology for a Resilient Future – Sai Kiran Oruganti et al. (eds)
© 2024 Taylor & Francis Group, London, ISBN 978-1-032-79020-6

Performance of ArUco Detector in ArUco Marker Detection Using Non-GPS Drone

15

Anton Domini Sta. Cruz[1]

Department of Electrical Engineering, University of the Philippines, Los Baños, Philippines

Jayson Osayan[2]

Department of Electrical Engineering, University of the Philippines, Los Baños, Philippines

Luis Rafael Moreno[3]

Department of Electrical Engineering, University of the Philippines, Los Baños, Philippines

Rob Christian Caduyac[4]

Department of Electrical Engineering, University of the Philippines, Los Baños, Philippines

Abstract: Autonomous drones use Global Positioning System (GPS) for flight control, however; GPS-equipped drones are costly, hence the use of non-GPS drones. Autonomous flight control for non-GPS drones is achieved using computational-heavy algorithms or expensive components. This study examines the performance of detecting ground control points (GCP) at different altitudes. Understanding the detection performance can be leveraged for developing non-GPS autonomous flight control. ArUco Detector and six 10.5x10.5 inch ArUco Markers were used as the GCP detector and GCPs, respectively. The experiment was carried out at different times of the day in a grass-covered area. Detection rates from 2-8m altitude levels with 0.1m increments were examined. It was observed that the ArUco Markers should be placed in a white material for easier ArUco Marker boundary detection. At least a 70% detection rate was obtained at 2-3.6m altitude levels with a 92.49% maximum detection rate at 2.1m. The average detection rate follows a linear trend and decreases by 2.26% per 0.1m with a 94.5% coefficient of determination at 3.7-6m altitude levels. For beyond 6m altitude level, the detection rate is less than 10%. ArUco marker pixel area relative to altitude levels was also measured using a 300x300 resized image from the drone and it was observed that the pixel area follows an exponential decay trend with 2388.58 and 0.9954 as its first term and common ratio respectively.

Keywords: ArUco, Automation, Computer vision, Non-GPS drone

1. Introduction

Drones are being utilized in a vast array of applications, one of which is monitoring ecological landscapes. Some monitoring approaches, such as generating orthomosaics and 3D point clouds, may require multiple overlapping and consistent images. Experts are capable of precisely operating drones using Global Position System (GPS), however; there are still certain regions wherein GPS data cannot be relied upon for mission control. In addition, ecological analysis, such as done by Flores et al. (2020), manually pilot non-GPS drones to monitor developments in farms. Aside from being tedious and laborious, manually operating non-GPS drones might significantly affect the precision of the images taken.

Autonomous flight control is an efficient drone navigation method for automated monitoring. However, precise navigation comes with the price of using expensive GPS-enabled drones. Autonomous flight control may utilize Ground Control Points (GCPs) – a pattern marked on the ground which can be used as a visual marker. In recent years, there has been a growing interest in integrating autonomous flight control for non-GPS drones brought about by the development of computer vision and data fusion. Recent methods for non-GPS autonomous flight control include optical flow navigations (Aasish et al., 2015), visual odometry (Romero et al., 2013), multi-sensors (Ashraf et al., 2019, and Zahran et al., 2019), and ultrasonic signals (Famili et al., 2022). These techniques demonstrated successful non-GPS automation through the use of additional components,

---

[1]acstacruz2@up.edu.ph, [2]jmosayan@up.edu.ph, [3]lpmoreno@up.edu.ph, [4]rmcaduyac1@up.edu.ph

DOI: 10.1201/9781003490210-15

data computations, in exchange for additional costs. On the contrary, by utilizing GCPs to control non-GPS drones, a potential outcome could be the simplification of detection and data feedback through computer vision.

Planar markers such as ArUco markers are being utilized in robot navigation and augmented reality (Garrido-Jurado et al., 2014). These are binary square fiducial markers with an external black border and an inner matrix of black and white squares. The inner matrix contains different patterns which are associated with marker ids. Existing literature further demonstrates the leveraging of ArUco markers in drone flights: Miranda et al. (2021) and Lebedev et al. (2020) on improving the GPS-enabled multi-sensor drone landing using markers, and a comprehensive study by Zoltan Siki and Bence Takács (2021) of drones detecting ArUco markers showed the performance of OpenCV-based detectors at different marker sizes and flight altitudes. However, it was observed that there was a limited number of images that were considered. In addition, performance at different times of the day should be explored as it may affect the detection due to lighting conditions. Understanding the performance of ArUco marker detection at different light conditions may be used for designing vision-based autonomous flight control of non-GPS drones.

This study aimed to benchmark the performance of an OpenCV-based ArUco marker detection from image streams that were captured by a non-GPS drone. Different altitudes and times of day were considered. The results of the study may aid in designing autonomous flight control for non-GPS drones with the use of ArUco markers, especially in validating if the ArUco detection would be relevant to the flight altitude of the use case.

2. Methodology

The performance analysis involved: A) the development of ArUco markers which served as visual targets for detection; B) the test system development which covers the ArUco

Fig. 15.1 Actual drone-captured markers in a grass field

inference and drone control; and C) the design of experiments for analysis.

A. ArUco Markers

Six ArUco markers were placed on a grass field following the placement captured by the non-GPS drone illustrated in Fig. 15.1. With this setup, it is expected to detect 6 markers per image, given a 100% detection rate. For the construction of the markers, the 10.5 in x 10.5 in ArUco patterns were printed on bond papers which were then affixed onto heavier illustration boards to prevent them from being blown by the wind. The ArUco marker ids used are ids 1 to 6. The approximate horizontal and vertical distance of adjacent ArUco markers were 50 cm and 60 cm, respectively.

B. Test System Development

The test system included a non-GPS drone and computer which were configured to detect ArUco markers at different altitudes (Fig. 15.2). The non-GPS drone was used to capture real-time image streams. On the other hand, the computer was provisioned for the detection of ArUco markers and flight control.

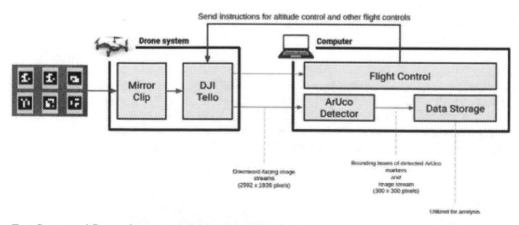

Fig. 15.2 Test System: a) Drone for capturing images and b) Computer for flight control and ArUco marker detection

Fig. 15.3 Actual drone-captured markers in a grass field

A DJI Tello drone was used as the non-GPS drone. As shown in Fig. 15.3, a mirror clip was installed on the drone since the built-in camera is forward-facing. Together with the drone camera, 5 MP (2592x1936) images with a downward field of view can be recorded. These image streams were sent to the computer through WiFi in real time.

Image streams were passed through the detector pipeline and flight control pipeline (Fig. 15.2b). The flight control pipeline simply sends instructions to the drone to change altitude. On the other hand, the detector pipeline consists of preprocessing and inference.

Images were initially resized to 300x300 pixels and then used for inference. An OpenCV-based ArUco detector proposed by Zoltan Siki and Bence Takács (2021) was utilized for marker recognition.

C. Design of Experiment and Performance Analysis

Shown in Fig. 15.4 is the overall workflow of the design experiment and the performance analysis. The non-GPS drone was programmed to capture images for 5 minutes from altitudes of 2 m to 8 m with 0.1 m intervals. Measurements

of percent detection rate and pixel area were observed for all 6 ArUco markers in the setup. The drone collects data every 0.05 seconds, therefore, around 10 observations on average were observed for every altitude measurement. For analysis of the effects of different illumination, three trials were performed during the morning (~8 AM), noon (~12 NN), and afternoon (~5 PM). These trials were performed on 2 consecutive days. This resulted in ~60 images per altitude.

The average detection rate was calculated by taking the averages of the detection rates of all images captured at the same altitude level. The trend of the average detection rate is expected to follow a linear trend with altitude (centimeters) as the independent variable and detection rate as the dependent variable since the number of ArUco markers are fixed and only the height (one-dimensional) is varied.

Unlike the average detection rate, the average pixel area was also calculated by taking the averages of the bounding box areas of all images captured at the same altitude level. The trend of the average pixel area is expected to follow an exponential decay trend with the general equation:

$$y = AB^x \tag{1}$$

wherein x is the altitude and y is the average ArUco pixel area. This is expected since the variation of the height versus the area (two-dimensional) was considered. While quadratic trend is also a reasonable trend, it is not considered since height and pixel area are impossible to be negative. Applying logarithm on both sides of equation 1:

$$lm\ y = ln\ a + x\ ln\ b \tag{2}$$

The best-fit equation of the exponential decay can be obtained by using least squares regression line with altitude (centimeters) as the independent variable and logarithm of the pixel area as the dependent variable.

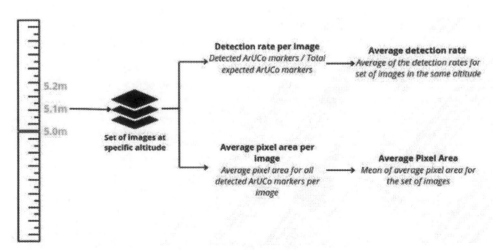

Fig. 15.4 Workflow of the design experiment and performance analysis

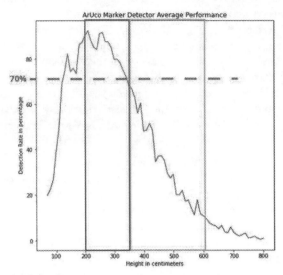

Fig. 15.5 Average detection rate performance

3. Results and Discussion

A. Overall detection rate

The average detection rate relative to the altitude of the drone shows a quadratic trend as expected (Fig. 15.5). At least a 70% detection rate was observed from 2 m to 3.6 m. The highest detection rate is 92.49% which was observed at 2.1 m altitude. Therefore, it is advisable to operate the non-GPS drone at a 3.6 m maximum altitude for a better detection rate.

At altitudes from 3.7 m to 6 m, the detection rate follows a linear trend and it decreases by 2.26% per 0.1 m altitude increments with a 94.06% coefficient of determination. Beyond the 6 m altitude level, the detection rate is less than 10%.

B. Detection rate at different times of the day

Table 15.1 summarizes the percent detection rate at different times of the day. There is an observed similarity in terms of detection rate in the morning and noon. The maximum altitude of operation for the drone to have at least a 70% detection rate is around 3.65 m. Meanwhile, the drone is advised to operate at a maximum altitude of 3.1 m only in the afternoon to have at least a 70% detection rate. All trials are observed to have a negative linear trend in detection rate as altitude increases.

C. Observed Pixel Area

Shown in Fig. 15.6 is the trend of the average pixel area relative to the altitude of the drone. On average, at least 100 square pixels were covered by the ArUco marker at 2 m to 6.6 m using a 300x300 square pixel drone camera.

After least squares fitting, the coefficient of determination is 97.86% and the best fit exponential decay equation is given by:

$$y = 2388.58(0.9954)^x \qquad (3)$$

D. Observations on ArUco Marker Material

During the first attempts of the experiment, the 18 in x 18 in markers were printed on PVC-Flex (as shown in Fig. 15.7).

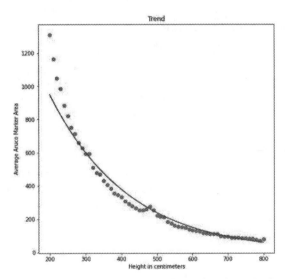

Fig. 15.6 Average ArUco marker bounding box pixel area

Table 15.1 Summary of detection rate per time of the day

Observation Parameter		Morning	Noon	Afternoon
At least 70% average detection rate		2m to 3.7m	2m to 3.6m	2 to 3.1m
Highest detection rate		96.9% at 2.1m	91.06% at 2.6m	99.97% at 2m
Decrease in detection rate as altitude increases	Linear trend (per 0.1m increase in altitude)	-2.4%	-2.56%	-2.4%
	R²	0.79	0.94	0.94
	Altitude range	3.8m to 6m	3.7m to 6m	3.2m to 6m

Fig. 15.7 ArUco markers printed on different materials: PVC-Flex (left) and bond paper (right)

The setup was initially tested on a concrete parking space. The markers were detected up to 5 m. However, when placed on the grass field, it was observed that the markers were not detected even at a 2 m altitude. The type of material was suspected to be the main cause of nondetection – possibly due to the reflectance of the material.

4. Conclusion

The performance of the ArUco marker detection using a non-GPS DJI Tello drone was performed. On average, it was found that non-GPS drones operate best at a maximum altitude of 3.6 m to have at least a 70% detection rate. Marginal differences in performance were observed when comparing the detection at different times of the day. Moreover, at least 10x10 pixel is expected to be observed from a resized 300x300 pixel image from the drone when operated on or below 6.6 m. To satisfy at least a 70% detection rate and obtain at least a 10x10 ArUco pixel area, non-GPS drones must be operated on or below an altitude of 3.6 m. Lastly, bond paper and illustration board were used for printing and affixing ArUco markers to minimize reflectance and to attain better detection.

References

1. Aasish, C., Ranjitha, E., Razeen, R.U., Bharath, R.S., & Angelin Jemi L. (2015). Navigation of UAV without GPS. 2015 International Conference on Robotics, Automation, Control and Embedded Systems (RACE). https://doi.org/10.1109/race.2015.7097260

2. Ashraf, S., Aggarwal, P., Damacharla, P., Wang, H., Javaid, A. Y., & Devabhaktuni, V. (2018). A low-cost solution for unmanned aerial vehicle navigation in a Global Positioning System–denied environment. International Journal of Distributed Sensor Networks, 14(6), 155014771878175. https://doi.org/10.1177/1550147718781750

3. Famili, A., Stavrou, A., Wang, H., & Jung-Min, P. (2022, January 25). Pilot: High-precision indoor localization for autonomous drones. arXiv.org. Retrieved April 11, 2022, from https://arxiv.org/abs/2201.10488

4. Flores, J. M., Bagunu, A. K. , & Buot, I. E. (2020). Documenting permaculture farm landscapes in the Philippines using a drone with a smartphone. In Methodologies Supportive of Sustainable Development in Agriculture and Natural Resources Management (pp. 71–86). SEARCA.

5. Lebedev, I., Erashov, A., & Shabanova, A. (2020). Accurate Autonomous UAV Landing Using Vision-Based Detection of ArUco-Marker. International Conference on Interactive Collaborative Robotics.

6. Miranda, V. R. F., Rezende, A. M. C., Rocha, T. K., Azpurua, H., Pimenta, L. C. A., & Freitas, G. M. (2021). Autonomous Navigation System for a Delivery Drone

7. Romero, H., Salazar, S., Santos, O., & Lozano, R. (2013). Visual Odometry for Autonomous Outdoor Flight of a quadrotor UAV. 2013 International Conference on Unmanned Aircraft Systems (ICUAS). https://doi.org/10.1109/icuas.2013.6564748

8. S. Garrido-Jurado, R. Muñoz-Salinas, F.J. Madrid-Cuevas, & M.J. Marín-Jiménez. (2014). Automatic generation and detection of highly reliable fiducial markers under occlusion. Pattern Recognition, 47(6), 2280–2292. https://doi.org/10.1016/j.patcog.2014.01.005

9. Zahran, S., Moussa, A., & Naser El-Sheimy. (2019). Enhanced drone navigation in GNSS denied environment using VDM and hall effect sensor. ISPRS International Journal of Geo-Information, 8(4), 169. https://doi.org/10.3390/ijgi8040169

10. Zoltan Siki, & Bence Takács. (2021). Automatic recognition of ArUco codes in land surveying tasks. Baltic Journal of Modern Computing, 9(1). https://doi.org/10.22364/bjmc.2021.9.1.06

Note: All the figures and table in this chapter were made by the authors.

Advancing Sustainable Science and Technology for a Resilient Future – Sai Kiran Oruganti et al. (eds)
© 2024 Taylor & Francis Group, London, ISBN 978-1-032-79020-6

A comparative analysis of three classification algorithms to predict students board performance

16

Marienel N. Velasco*
Polytechnic University of the Philippines, Philippines.

Jayson M. Victorianob[1]
Bulacan State University, Philippines.

Abstract: Educational Data Mining refers to the process of gathering and analyzing various raw data taken from an academic setting that employs different Data Mining techniques. EDM aids any academic institution in making better decisions for its success and its students. This study aims to show a comparative analysis among the three data mining algorithms, namely; Naïve Bayes, Random Forest, and Decision Tree using WEKA software to predict students' board examination performance on Licensure Examination for Teachers. The data mining algorithms were validated using different measurements such as recall, precision, kappa statistics, and f- measure. Data were taken from the Bachelor in Business Teacher Education (BBTE) program from 2017-2019. The attributes used for the prediction model are the results of the LET passers posted on the PRC websites and student records taken from the registrar's office and the BBTE program head, consisting of the student's General Weighted Average (GWA), scholarship, review center records, and honor received. The study's findings revealed that Decision Tree had an accuracy percentage of 96.2264%, whereas both Naïve Bayes and Random Forest had a percentage of 94.3396%. Moreover, students' GWA, review center, and scholarship records influenced their board examination performance. This study will help the university in making a strategic plan to increase the percentage of their board passers and make an early intervention for students who will likely fail the board exam.

Keywords: Board examination performance, Educational data mining, Decision tree

1. Introduction

Data mining of educational data is a method of gathering and analyzing data taken from an academic setting (Al Breiki et al., 2019) that employs various techniques of Data Mining, Machine Learning Algorithms, as well as statistical approaches (Jalota & Agrawal, 2019) to predict students' academic success (Ajibade et al., 2019). Today, EDM has gained popularity among many researchers because of its importance in higher education institutions in making better decisions and bringing innovation to educational systems in general. However, because educational data sets contain so much information, predicting a student's progress at any higher education institution is challenging and time-consuming (Arcinas et al., 2021).

Many researchers conducted various studies in the field of EDM. For instance, one study used regression analysis and the WEKA platform to predict the number of possible board passers (Abaya et al., 2019). Another paper focuses on creating a system for analyzing and forecasting the results of the teacher's board examination using only a limited number of data mining classifiers (Tarun, 2017).

In addition, one study used a decision tree and WEKA but failed to build a system model capable of predicting student licensure examination performance (Rustia et al., 2018).

While another study used Naïve Bayes and Decision tree algorithms to analyze the student's board examination performance (Polinar et al., 2020).

Furthermore, several studies suggested the following; 1) to use additional data mining techniques, 2) to include more datasets and attributes, 3) to improve the machine learning algorithm, and 4) to create a system that properly predicts a student's outcome on a board examination.

*Corresponding author: mnvelasco@pup.edu.ph

DOI: 10.1201/9781003490210-16

This study aims to; 1) perform a comparative analysis on the accuracy percentage of different data mining or classification algorithms such as Naïve Bayes, Decision Tree, and Random Forest using WEKA to forecast students' performance on board examinations for teachers, and 2) validate their accuracy using precision, recall, kappa statistics, and f-measure.

A total of 106 datasets gathered from the Bachelor in Business Teacher Education (BBTE) program from 2017-2019. The attributes used for the prediction model are the results of the LET passers posted on the PRC website and student records gathered from the registrar's office and the BBTE program head, consisting of the student's General Weighted Average (GWA), scholarship, review center records, and honor received.

This research is valuable because Educational Data Mining (EDM) is becoming increasingly popular in predicting students' board examination performance, which will aid many higher education institutions in decision-making.

Moreover, this study will help the university in making a strategic plan to increase the percentage of their board passers and conduct an early intervention for students who are most likely to fail the board exam. The study will also guide other researchers interested in conducting a similar study in Educational Data Mining (EDM), and Machine Learning Algorithms (MLA).

2. Literature Review

In order to benefit from technological advancements, higher education institutions must learn to be adaptable. Data mining of educational data and machine learning algorithms aid in accurately forecasting a student's board test performance

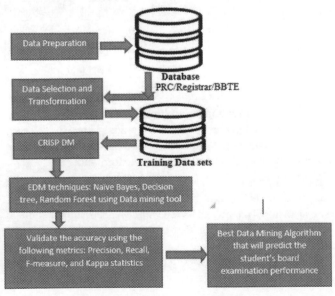

Fig. 16.1 Framework of the Study

which will enable educational institutions in formulating ways to improve the quality of education provided to all students.

CRISP-DM, or Cross Industry Standard Procedure for Data Mining, is the most commonly used approach for data mining and science projects (Villarica, 2019).

CRISP-DM framework was utilized in the paper, and it consists of various data mining stages: the first part is the data preparation or the process of collecting data from the PRC website, registrar's office, and BBTE program. Next is the data selection and transformation.

The Data mining stages were based on CRISP-DM, a data science process model consisting of several stages such as data analysis and preparation, model construction, assessment, and distribution.

Finally, all datasets and attributes were tested using various data mining approaches: Random Forest, Naïve Bayes, and Decision Tree. The three data mining techniques were analyzed using WEKA, a data mining tool, and validated using various measurements such as f-measure, kappa statistics, recall, and precision. Furthermore, the most accurate model was chosen based on how well it predicted the study's outcomes.

3. Methodology

CRISP-DM approach was employed in the study, which consists of the following phases: 1) Collecting data from the PRC website, registrar's office, and BBTE program student's records from the years 2017 to 2019, 2) the data selection and transformation using the student's board performance, GWA, honor received, review class, and scholarship as the attributes needed on the study, 3) testing of 106 datasets and attributes using WEKA to analyze various classification algorithms such as Random Forest, Naïve Bayes, and Decision Tree, and 4) comparative analysis of the accuracy percentage of the three algorithms validated using several validation criteria such as kappa statistics, precision, recall, and f-measure.

WEKA

The study used WEKA, which refers to Waikato Environment for Knowledge Analysis, an open-source or free software platform that consists of multiple machine learning techniques for data mining applications. Its name was inspired by a flightless bird that can only be found on the islands of New Zealand and was developed by a team of researchers at the University of Waikato in that country. Moreover, WEKA allows users the option of interacting with its platform via a command-line interface or a Graphical User Interface (GUI). It processes and stores the data in Attribute-Relation File Format (ARFF), consisting of a list of instances. It is compatible with different operating systems.

Naïve Bayes

It is a straightforward classification technique that treats all provided attributes in a dataset separately. This classification algorithm is also effective and accurate based on Bayes' theorem. Naïve Bayes has the advantage of dealing with large sets of data and its effortless implementation. This algorithm uses a method to predict the probability of unrelated actions on several attributes.

Decision Tree

A decision tree-based algorithm requires developing a tree model classification process that involves two significant steps; build a decision tree and apply that decision tree to a database (Jalota & Agrawal, 2019). It is a flowchart-like structure or graph composed of leaf nodes representing a decision and a branch representing a value that leads to those decisions. Decision tree such as J48 is widespread and commonly used by various researchers because of their simplicity in which a set of conditions are hierarchically arranged.

Moreover, the decision tree algorithm offers several advantages, including the ability to handle a wide range of input data, platform compatibility, and excellent handling of missing values in the data set.

Random Forest

It was given the name "forest" because of its capacity to efficiently assemble and analyze several decision trees. It was included in numerous data mining software packages, including WEKA.

Table 16.1 The study's dataset

Attributes	Values
GWA	Numeric
Review Center	Yes or No
Scholar	Yes or No
Honor Received.	Yes or No
Performance	Passed or Failed

Table 16.1 shows the different attributes and its corresponding values used in the study which includes; student's board performance (passed or failed), GWA (numeric), honor received (yes or no), review class (yes or no), and scholarship (yes or no). Moreover, the study includes the entire population of 106 students of the Bachelor in Business Teacher Education (BBTE) program for the years 2017 to 2019 to avoid biases or wrong interpretations of data.

4. Results and Discussions

The collected student's GWA records from the years 2017 to 2019 from the registrar's office include the following values: a minimum of 1.28, a maximum of 2.29, a mean of 1.708, and a standard deviation of 0.812. (Table 16.2).

Table 16.2 General weighted average (GWA)

Statistics	Value
Minimum	1.28
Maximum	2.29
Mean	1.708
StdDev	0.812

Table 16.3 shows the study's attributes; 86 students passed the board test, 67 were scholars, 45 were honor students, and 77 registered in a review center based on their academic records from the years 2017 to 2019.

Table 16.3 Attributes of the study

Review Center	Scholar Honor		Passed
Yes 77	67	45	86
No 29	39	61	20

Table 16.4 illustrates the detailed accuracy of each data mining technique based on the various metrics; precision, recall, and f-measure were based on the weighted average. In addition, the kappa value displays the level of agreement based on the data reliability percentage. Comparative analysis reveals that the Decision Tree achieved the highest accuracy.

Table 16.4 Metrics

Techniques	Precision	Recall	F-Measure	Kappa
Naïve Bayes	0.951	0.943	0.945	(Almost Perfect)
Decision Tree	0.969	0.962	0.964	(Almost Perfect)
Random Forest	0.946	0.943	0.945	(Almost Perfect)

In addition, a number of studies perform a comparative analysis of various classification algorithms utilizing WEKA software and validated by different metrics, including precision, recall, f-measures, and kappa statistics, to find the model that is most accurate at predicting student board performance.

Table 16.5 summarizes the comparison of various classification algorithms based on their accuracy percentages. The Decision Tree model has the highest accuracy percentage of 96.2264%, indicating that it is the most accurate model used in the study. Due to its high accuracy rate, a similar study (Velasco et al., 2023) also employed the Decision Tree model to predict a student's success on the licensing exam.

Table 16.5 Accuracy percentage

Techniques	Accuracy Percentage (%)
Decision Tree	96.2264%
Naïve Bayes	94.3396%
Random Forest	94.3396%

5. Conclusions

According to the study's findings, the Decision Tree has the best accuracy percentage of any prediction model tested. In addition, precision, recall, f-measure, and kappa metrics were used to properly validate the accuracy of the three classification algorithms. Using this information, the university can make a strategic plan to increase the percentage of their board passers and make an early intervention for students who will likely fail the board exam.

The study is very beneficial to the university. However, the following are highly recommended: 1) the use of additional or bigger datasets, 2) the use of different attributes, and 3) the implementation of different data mining algorithms and comparison of these techniques based on their accuracy percentage when it comes to predicting the student's board performance.

References:

1. Abaya, S. A., Orig, D. A. D., & Montalbo, R. S. (2019). Using Regression Analysis in Identifying The Performance of Students in The Board Examination. Online Journal of New Horizons in Education, 6(4), 290–296.
2. Ajibade, S. S. M., Bahiah Binti Ahmad, N., & Mariyam Shamsuddin, S. (2019). Educational Data Mining: Enhancement of Student Performance model using Ensemble Methods. IOP Conference Series: Materials Science and Engineering, 551(1). https://doi.org/10.1088/1757-899X/551/1/012061
3. Al Breiki, B., Zaki, N., & Mohamed, E. A. (2019). Using Educational Data Mining Techniques to Predict Student Performance. 2019 International Conference on Electrical and Computing Technologies and Applications, ICECTA 2019. https://doi.org/10.1109/ICECTA48151.2019.8959676
4. Arcinas, M. M., Sajja, G. S., Asif, S., Gour, S., Okoronkwo, E., & Naved, M. (2021). Role of Data Mining in Education for Improving Students Performance for Social Change. Turkish Journal of Physiotherapy and Rehabilitation, 32(3), 6519–6526.
5. Jalota, C., & Agrawal, R. (2019). Analysis of Educational Data Mining using Classification. Proceedings of the International Conference on Machine Learning, Big Data, Cloud and Parallel Computing: Trends, Prespectives and Prospects, COMITCon 2019, 243–247. https://doi.org/10.1109/COMITCon.2019.8862214
6. Polinar et al. (2020). International Journal of Advanced Trends in Computer Science and Engineering Available Online at http://www.warse.org/IJATCSE/static/pdf/file/ijatcse102912020.pdf Big Data Management Challenges. International Journal of Advanced Trends in Computer Science and Engineering, 8(June), 1965–1968.
7. Rustia, R. A., Cruz, M. M. A., Burac, M. A. P., & Palaoag, T. D. (2018). Predicting student's board examination performance using classification algorithms. ACM International Conference Proceeding Series, 233–237. https://doi.org/10.1145/3185089.3185101
8. Tarun, I. M. (2017). Prediction Models for Licensure Examination Performance using Data Mining Classifiers for Online Test and Decision Support System. Asia Pacific Journal of Multidisciplinary Research, 5(3), 20. http://www.apjmr.com/wp-content/uploads/2017/06/APJMR-2017.5.3.02.pdf
9. Velasco, M. N., Malabuyoc, A. A., dela Cueva, G. V., & Enriquez, K. L. (2023). Predicting Licensure Examination Performance Using Data Mining Techniques. 2023 8th International Conference on Business and Industrial Research (ICBIR), May, 01–04. https://doi.org/10.1109/icbir57571.2023.10147475
10. Villarica, M. (2019). the Use of Data Mining To Model Personalized Learning Management System. International Journal of Advanced Research, 7(3), 1191–1200. https://doi.org/10.21474/ijar01/8753

Note: All the figure and tables in this chapter were made by the authors.

Advancing Sustainable Science and Technology for a Resilient Future – Sai Kiran Oruganti et al. (eds)
© *2024 Taylor & Francis Group, London, ISBN 978-1-032-79020-6*

Assessment of Neighborhood Watch System for Crime Prevention

17

Marienel N. Velasco*

Polytechnic University of the Philippines

Abstract: the aim of the study is to develop an electronic neighborhood watch system in the community of Le Moubreza, San Antonio, Sto. Tomas, Batangas. The system will act as a tool for "warners" or concerned citizens, and "responders," or the respective authorities, in their joint effort to lower the crime rate in the community. Waterfall Model was used for the system's development using several technologies like JOOMLA 3.2, and WAMP. The respondents assessed the system based on ISO/EIC 25010 standard; functionality, usability, dependability, and maintainability. Results showed that the system was broadly accepted by the respondents, who gave it a weighted mean rating of 4.74 on a five-point, which is interpreted as "strongly agree". In addition, the system had an appropriate framework for handling the distribution method of reports and evidence such as pictures and videos. Furthermore, there are different user's levels in accessing the system.

Keywords: Crime prevention, Neighborhood watch system, Peace, Security

I. Introduction

The neighborhood reporting the incident or suspicious activity directly to the authority has been the conventional approach to crime prevention (Bennett et al., 2008). But today, the emergence and use of ICT have proven to be one of the most effective and fastest schemes in reducing crime rates than the traditional method. For instance, several technical advances for crime prevention have been developed and are classified as either hard or soft technology (Byrne & Marx, 2011).

Finding a workable way to reduce the crime rate in a neighborhood or community has been a crucial societal issue because crime has a detrimental impact on people's lives (Moon et al., 2014). One study in the Philippines examined the effectiveness of certain barangays' crime reduction initiatives (Ayeo-eo, 2020). Another study examined the various measures taken by the Philippine National Police and the barangay officials to prevent crime in their community. It was found that these measures were difficult to implement because of a lack of staff, insufficient equipment, and instances of unreported crimes (Ammiyao et al., 2020).

The neighborhood of Le Moubreza in Barangay San Antonio, Sto. Tomas, Batangas had several numbers of crimes based on barangay blotters including theft, robbery, and physical assault. In particular, "Akyat Bahay" crimes—where people trespass onto neighbors' premises and steal expensive items from them—were the most often reported cases in the neighborhood. At present, the neighborhood's only means of crime prevention is by reporting it directly to the authority. They had little joint effort in maintaining peace and security within their community, which makes it difficult to report and respond to issues and concerns.

For this reason, the researcher attempts to develop a system that will help the neighborhood in reducing crime.

The community can use this system to report crimes, unlawful activity, and other issues. They can also request assistance and support from the appropriate authorities. e.g., reporting a missing person or asking for financial assistance for sick family members, etc. Each problem is logged on the site, making it easier for the authorities to respond to issues of importance to the community. This report will help to warn the public and the citizens to become aware of the issues. In addition, they can also upload pictures or videos as supporting evidence of the reported incident. The message posted will be sent to the respective local authority in Le Moubreza. Furthermore, citizens can share their experiences and learn

*Corresponding author: mnvelasco@pup.edu.ph

DOI: 10.1201/9781003490210-17

from one another about conflict prevention and peace building. Local news and current events are also allowed to be shared on the site. The system can be accessed 24/7 so that citizens can report and post their concerns anytime.

Specifically, the system lets a person keeps an eye on his neighbors (using a mobile device or other gadgets to record an image or video as evidence), logs in to the system right away, reports the issue and uploads evidence, then sends the report to the appropriate authorities, and finally, lets them fix the reported concerns. The system will act as a tool for "warner" (the concerned citizen) and "responders" (respective authorities), in their joint effort to reduce the crime rate in the community.

Generally, the aim of the study is to develop a neighborhood watch system to lessen the crime rate in the neighborhood of Le Moubreza, a residential community in Barangay San Antonio, Sto. Tomas, Batangas.

Specifically, it aims to answer the following problems:

1. What is the system acceptance based on ISO/IEC 25010 standard; functionality, usability, dependability, and maintainability?

2. What are the system's frameworks in handling the dissemination mechanism of reports and evidence?

3. What are the different user's levels in accessing the system?

2. METHODOLOGY

The study used a quantitative research method since it involves the process of gathering and interpreting data. A descriptive-developmental research approach was used to meet the needs of the study's investigation; with this methodology, the researcher preeminently benefited from system development.

In this study, the population and the respondents of the study are the neighborhood of Le Moubreza located in San Antonio, Sto. Tomas, Batangas. Using purposive sampling, there were a total of fifty respondents as the study does not require the participation of the entire population.

The main research tool for this study is a survey questionnaire with a five-point Likert scale. A survey research approach is appropriate because the researcher aims to collect information that will aid in the evaluation of the system using the ISO/IEC 25010 standards; system functionality, usability, dependability, and maintainability.

Some of the study's ethical considerations are as follows: the information obtained from this study was kept confidential, the research did not do any harm to any person or organization, the study was free from plagiarism or research misconduct, and the results were presented honestly.

A Likert scale was used to measure approach, preference, and subjective solutions. The scale is designed to evaluate the system in terms of the characteristics of quality software. For interpretation of the computed weighted mean, the researcher set different scales as shown in Table 17.1.

Table 17.1 The five-point likert scale

Scale	Range	Interpretation
5	4.21–5.0	Strongly Agree
4	3.41–4.20	Agree
3	2.61–3.40	Normal
2	1.81–2.60	Disagree
1	1.0–1.80	Strongly Disagree

In addition, the researcher used informal interviews using a social site like Facebook, and references like books, journals, magazines, and internet articles were also utilized for other information needed in the study.

For the statistical treatment of data, the researcher used the Weighted Mean (WM), with the following formula:

$$\text{Weighted Mean (WM)} = \frac{\sum f(x_1 + x_2 + x_3 + \ldots + x_n)}{N}$$

Where:

$$\sum f(x_1 + x_2 + x_3 + \ldots + x_n) = \text{sum of all Means (M) of each criterion}$$

The Waterfall model is one of the most common methods in system development, which presents the different phases of software development life cycle in a sequential order starting with analysis, design, code implementation, testing, and maintenance (Herawati et al., 2021).

The Waterfall Model employed in the system's development is shown in Figure 17.1. The researcher developed the system in accordance with the model's six-steps processes, which are as follows: 1) requirement gathering, 2) analysis of the provided data, 3) system designing, 4) coding and system's

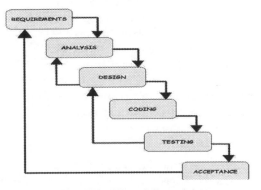

Fig. 17.1 Waterfall model

development using a variety of technologies, including Joomla 3.2, WAMP, and Adobe Photoshop, 5) testing the system for potential errors, and 6) system deployment for user acceptance. Generally speaking, the Waterfall Model offered a helpful framework for the system's development.

3. RESULTS AND DISCUSSIONS

Table 17.2 System Acceptance based on ISO/IEC 25010 standard

ISO/IEC 25010 standards:	WM	Interpretation
Functionality The system can perform the required task and has a complete function	4.82	Strongly Agree
Usability The system is user-friendly.	4.81	Strongly Agree
Dependability The system does not often fail or produce errors and is capable of handling errors.	4.49	Strongly Agree
Maintainability The system is easy to modify or extend.	4.84	Strongly Agree
Average	**4.74**	**Strongly Agree**

Legend: WM- Weighted Mean

The respondents evaluated the system based on ISO/IEC 25010 standard such as functionality, usability, dependability, and maintainability. The weighted mean for functionality, which was 4.82 and interpreted as "strongly agree," indicates that the system can carry out the necessary task and has a comprehensive purpose. While a weighted mean of 4.81 interpreted as "strongly agree" for the usability of the system because the user thought it was user-friendly. The respondents provided a weighted mean score of 4.49, or "strongly agree," for the system's dependability, which is the ability of the system to handle faults and not frequently fail or cause problems. In addition, the user rated the system's maintainability with a weighted mean of 4.84 and "strongly agree" because they thought it was easy to modify. In general, the respondents gave the system an average score of 4.74 on a scale of 1 to 5 for all factors, indicating that they generally found it to be extremely acceptable.

Figure 17.2 shows how the system framework handles the dissemination mechanism of reports and evidence such as photos and videos. Only logged-in users are permitted to

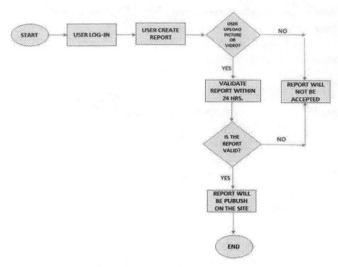

Fig. 17.2 System's framework in handling the dissemination mechanism of reports and evidence

access the system and submit their reports. The user will click "Create post option" to begin entering the report. The user is recommended to submit a photo or video as verification of the report's veracity in the event that it involves a criminal offense. However, a notification message will be given to the user if they neglected to submit the required proof.

The administrator validated the report and evidence within 24 hours after the user uploaded the necessary proof. If the information is confirmed to be true, the administrator will post it online and notify the respective authorities. However, if it is false, the report will be removed due to lack of evidence or credibility.

Figure 17.3 shows the system's different level of access; administrators, and registered users. The privilege of validating and approving a user's report is reserved for the

Fig. 17.3 System's level of access

administrator, whereas a registered user is only permitted to post reports. In addition, the system's interface, consists of Home, About, Related link, Rules, Contact Us, Create a Post and Change Password menus.

4. CONCLUSION

In light of the results of the study, the researcher concluded that the system was broadly accepted by the respondents, who gave it a weighted mean rating of 4.74 on a five-point, which is interpreted as "strongly agree". In addition, the system had an appropriate framework for handling the distribution method of reports and evidence such as pictures and videos. Furthermore, there are different user's levels in accessing the system.

The study is very beneficial to the community however, it is highly recommended that the implementation of the system via Internet technology should be supported and maintained by the community and that there should be trusted individuals who are willing to administer and maintain the system. Moreover, a follow-up study should be conducted to strengthen and enhance the specific features of the system.

References

1. Ammiyao, G. D., Arud, P. M. T., Asiaben, C. P., Balunos, N. R., Bangsara, U. D., Cacap, A. B., Dogui-is, C. M., Fianitog, J. J., Jagmis, M. D., Lagan, J. C., Leocadio, M. V, Liwayan, F. K., Mallari, V. Y., Nayosan, V. P., Pagulayan, M. L. M., Refuerzo, R. G., & Rufino, C. D. (2020). Community Crime Prevention: the Case of a Barangay in the Northern Philippines. International Journal of Advanced Research, 9(6), 98–126. https://garph.co.uk/IJARMSS/June2020/G-2815.pdf
2. Ayeo-eo, S. P. (2020). Assessment of Crime Prevention Programs of Selected Barangays in Cabanatuan City. 5(12).
3. Bennett, T., Holloway, K., & Farrington, D. (2008). The Effectiveness of Neighborhood Watch. Campbell Systematic Reviews, 4(1), 1–46. https://doi.org/10.4073/csr.2008.18
4. Byrne, J., & Marx, G. T. (2011). Technological innovations in crime prevention and policing. A review of the research on implementation and impact. Journal of Police Studies, 3(20), 17–40. https://www.ncjrs.gov/pdffiles1/nij/238011.pdf
5. Herawati, S., Negara, Y. D. P., Febriansyah, H. F., & Fatah, D. A. (2021). Application of the Waterfall Method on a Web-Based Job Training Management Information System at Trunojoyo University Madura. E3S Web of Conferences, 328. https://doi.org/10.1051/e3sconf/202132804026
6. Moon, T.-H., Heo, S.-Y., & Lee, S.-H. (2014). Ubiquitous Crime Prevention System (UCPS) for a Safer City. Procedia Environmental Sciences, 22, 288–301. https://doi.org/10.1016/j.proenv.2014.11.028

Note: All the figures and tables in this chapter were made by the authors.

Advancing Sustainable Science and Technology for a Resilient Future – Sai Kiran Oruganti et al. (eds)
© 2024 Taylor & Francis Group, London, ISBN 978-1-032-79020-6

The Effects of E-Machine Shorthand Online Teaching Practices to Students' Performance: Evidence from Rizal Technological University

18

Jayvie O. Guballo[1]
Rizal Technological University

Ely B. Marbella[2]
Rizal Technological University

Marvin A. Fuentes[3]
Rizal Technological University

Abstract: The study's goal is to determine the relationship between online e-machine shorthand teaching practices and student performance. Non-parametric statistics was employed to test the relationship of the independent and dependent variables. There were 328 student participants who took up e-machine stenography course. The respondents assessed e-machine stenography online teaching practices in terms of resources, student engagement, and faculty presence as extremely effective while orderliness as effective. In terms of students' performance, the respondents assessed examination and recitation as "very often." The study's results also revealed that there is a significant but weak association between e-machine stenography online teaching practices and students' performance. This study validates the importance of upholding and maintaining excellent online teaching practices in order to continuously improve students' academic performance. It is suggested that to engage students in a more enjoyable and conducive e-learning environment, the use of gamification, social media integration, reflective thinking, and high quality educational resources should be explored. Future research in other contexts may be conducted to back up or bolster the study's findings. Other variables evaluating online teaching and learning practices could also be used to test other factors influencing students; performance in e-machine stenography teaching courses.

Keywords: e-machine stenography, Online teaching, Students' performance

1. INTRODUCTION

In the Philippines, the first quarter of 2020 saw an unexpected challenge in the field and system of education, as the SARS COV-19 global pandemic became an issue, bringing many challenges, particularly to educational institutions, faculty, parents, and students. The pandemic imposed many constraints, particularly on the operations, systems, and even strategies of the educational system; as a result, the majority of educational institutions, both private and public, shifted to a new paradigm shift from traditional to online platform style of teaching. The term "online class" or "classroom" refers to the use of a specific e-learning management system that allows students and teachers to connect, participate, and interact with one another both synchronously and asynchronously. Because

there are some limitations in preparation and the landscape of education has been altered due to the SARS COV19 global pandemic, numerous educators are working hard to find ways to maximize education training, particularly in the online space. E. A. Sharp, M. K. Norman, C. L. Spagnoletti, and B. G. Miller. (2021).

As the Global Pandemic arose, faculty members teaching e-machine shorthand faced a difficult situation in terms of how to teach the subject online. Because the majority of faculty members lack sufficient training in conducting online classes. As a result, faculty members devise various teaching strategies such as providing open access resources, utilizing social media outlets as a medium for engaging students, and maintaining a constant presence of the lecturer to continuously guide and monitor the students' performance.

[1]joguballo@rtu.edu.ph, [2]ebmarbella@rtu.edu.ph, [3]mfuentes@rtu.edu.ph

DOI: 10.1201/9781003490210-18

Before the pandemic, the practice of teaching e-machine shorthand is simply by presenting the theory of its usage, followed by hands-on calibration and performance assessments. However, when the pandemic struck, major shifts and adjustments in teaching practices were made by using teleconference discussions, free online applications, and the creation of a crude prototype of e-machine shorthand in order to become familiar with correct finger position and navigation during hands-on practice.

The CHED Memorandum Order No. 4 Series of 2020 was issued by the Commission of Higher Education in the Philippines, whose main role is to make and create appropriate steps or actions to ensure that quality education is accessible to all. The circular focused on that during the time when face-to-face classes were disrupted due to the community quarantine or lockdown classification was imposed, Higher Educational Institutions (HEIs) were directed to use a flexible style platform to deliver and facilitate a continuous learning scheme in order to continue providing services in difficult times. According to Gerio, Eloisa Sandra B., et al., (2021) As a result of the abrupt paradigm shift, students are enforced to cope with the new method of instruction and learning landscapes. During the time of contagion, the underlying circumstance leads to determining the Academic Learning Process between face-to-face and online Medical Technology students.

E-machine shorthand, also known as steno machine, is a specialized typewriter or chorded keyboard used by stenographers in court proceedings. Under the CHED Memorandum No. 19 Series of 2017, e-machine shorthand is offered as part of the Bachelor of Science in Office Administration. E-machine shorthand teaching practices, particularly on online learning, were integrated as a response to emerging trends and the CHED memorandum circular. Using traditional teaching styles was integrated into various modalities, and even online lectures were introduced. One of the most important applications of Learning Analytics is providing an Institutions now have the ability to track students' academic activities and provide them with real-time adaptive consultations about their academic performance. Khan, I., Ahmad, A. R., Jabeur, N., & Mahdi, M. N. (2021)

In the Philippine context, there were no existing studies that specifically focused on e-machine shorthand online teaching practices, and the majority of existing researches focused only on the traditional style of teaching practices rather than the embraced new better normal strategy. Researchers were prompted to investigate and explore the relationship between e-machine shorthand online teaching practices and student performance in the context of the new normal in education system and setup.

2. Literature Review

E-Machine and Steno Machine

E-Machine Shorthand, also known as Steno Machine, was used extensively in industries such as corporate and court proceedings. Computer stenography is also considered an input method; however, in order to master the vital role of the input method of stenography, the application of broad tasks is required. It was also determined that there is a need to improve machine shorthand training techniques, as PTL can be considered a useful technique for building fundamental blocks of stenography T. T. Aveni, C. Seim, and J. Aveni a celebrity (2019). In addition, according to T. J. Aveni (2019) stenography teaching foundation must be established, because the role of teaching fundamental of the stenotype keyboard primer is part of the primary introduction, it validates that teaching writing full words in stenography but also identifying its parts and function is very important.

Mu, X. (2016). Their findings concluded that universities and academic institutions must begin identifying social needs for establishing training objectives, curriculum systems, adequate curriculum design, and establishing professional knowledge through enhanced teaching methods of styles in order to groom or develop potential high caliber talents in the technical setting. In the study conducted of D. Han, J. Yang, and W. Summers (March 2017) noted in their study that steganography is becoming more popular as the techniques pertaining to steganography in cybersecurity education are becoming more popular, as many institutions offer it as part of the program. It was also suggested that since then, steganography has played an important role in incorporating the aforementioned subjects into the curriculum.

Online Teaching Practices

In the study conducted of S. Rakic, N. Tasic, U. Marjanovic, S. Softic, E. Lüftenegger, and I. Turcin. (2020). Their study concluded that there is a significant correlation between the use of e-learning digital resources and student academic performance. On the other hand, self-efficacy and feedback from the course or subject lecturer were found to be positive predictors in all factors and aspects related to student engagement. While the shared control predicted the entire set of factors related to student engagement, it did not include emotional engagement. R. Truzoli, V. Pirola, and S. Conte (2021).

C. Tan (2021). According to their findings, there is a significant difference in the learning process in the implementation of recovery movement control order and movement control order. The results revealed that while the implementation of movement control order is taking place, the most valuable factor identified in the study that establishes influence on the learning performance of the student is the subject lecturer's social presence. Furthermore, it was discovered that there is a significant relationship between the academic standing of the student and the involvement of parent guidance. J. Tus (2021).

In the study completed by S. Chaturvedi, S. Purohit, and M. Verma (2021). It was noted and highlighted the primary teaching strategies or practices in order to create an efficient blended style of learning environment for learners and lecturers, which include: redesigning or revitalizing virtual spaces by providing online learning kits and virtual incubation rooms, utilizing reflective manner of thinking by applying pedagogic and andragogic approaches, and empowering lecturers to offer a variety of timely and relevant courses.

Based on Budhai's research, S. S. (2021) that online teaching practices played an important role in students' learning experiences, as well as establishing student engagement and making learning environments more enjoyable and inclusive. The study also noted that the integration of gamifications and social media has benefits on learners' critical thinking, communication, discovering new skills, and enabling students to handle problem-solving situations.

Students' Academic Performance

Academic performance of students is important to academicians and students because it is regarded as a pre-determinant or measurement of students' achievement across academic subjects. In the study conducted by E. Galy, C. Downey, and J. Johnson (2011). It was discovered that there were significant differences in learners' perceptions of e-learning tools between those who chose to take the online course and those who chose to take the on-campus learning courses. Furthermore, the learners' ability to work independently, ease of use, and usage perception are found to be statistically significant as student final grade predictors. In the context of academic measurement particularly the course outline assessment designs such as rubrics in the study of S. S. Jaggars and D. Xu (2016) It was discovered that the importance and excellence of interpersonal interaction among learners and course delivery have a significant and positive relationship on student ratings. Furthermore, it was suggested in the analyzed course observation and data interviews that excellent and continuous student and lecturer interaction can establish an online environment that can push learners to perform and commit to the subject course at a high level of academic performance.

Furthermore, there were no significant differences in assessment or measurement techniques for the end of course, as well as the multiple-component learning assessment and the one-moment assessment. The discovery confirms that blended learning is significantly related to the learning performance of STEM learners rather than the traditional

style of on-site classroom practice H. M. Vo, C. Zhu, and N. A. Diep (2017). In connection to student's assessment performance, according to C. Korkofingas and J. Macri (2013) it was highlighted that there is a significant relationship between a student's time spent on the course website and the student's assessment performance for university business forecasting course.

According to the study of S. S. Jaggars and D. Xu (2016). Students access on the online course materials is greatly associated in improving the learner's performance and can create added value on the learning experience. The study conducted by. Joksimovi, D. Gaevi, T. M. Loughin, V. Kovanovi, and M. Hatala. (2015) it was discovered that the number of time utilized on the system student interactions and connection has a continuous positive impact on the learner's learning outcome, however, the quantity of student related content interactions was not directly associated on the final course rating. In the issue of internet connectivity, it was revealed by L. D. Lapitan Jr., C. E. Tiangco, D. A. G. Sumalinog, N. S. Sabarillo, and J. M. Diaz (2021) Subject lecturers should identify some ways to improve their learners' engagement and commitment in online classes by investing in internet-based teaching kits and training on e-learning familiarity. Lastly, A. Crampton, A. T. Ragusa, and H. Cavanagh (2012) It was discovered that learners who continuously accessed the readily available online learning resources could achieve a higher score.

There are no studies that deal with managing both online and in-person classes in the context of teaching e-machine shorthand. The researcher carried out this study in order to fill a gap in the literature by presenting best practices and sufficient support for instructing e-machine shorthand in an online setting.

3. Research Method

After gathering and cleaning the data, the researchers made used of correlational research design to determine the significant and positive relationship of the independent variables which pertains to e-steno online teaching practices and the dependent variables under students' performance. The researcher employs WarpPLS to assess the statistical power of the sample size using the Gamma-Exponential Method and the Inverse Square Root Method.

Before distributing the questionnaire to the identified respondents, the researcher created a self-made survey in which factors analysis and identifying existing practices were identified in order to devise a questionnaire. The questionnaire also underwent content validation and a reliability test. The researcher used Google Sheets to distribute an online survey to the respondents as part of the data collection process.

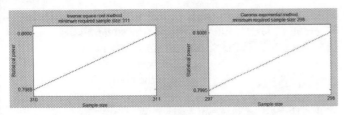

Fig. 18.1 Sample Size Computation

Apart from being one of the best tools in the emerging research trends, WarpPLS was chosen as the analytical tool in this study because it is ideal for conducting an accurate analysis of the data without bias and the best software for calculating p values.

According to the calculated sample size, the minimum requirements are 298 to 311, with a statistical power of 80% interpreted as acceptable. The study's total sample size is 328, which is greater than the required minimum number of survey takers.

Table 18.1 Profile of respondents

Age	Frequency	Percent
21 to 26 years	311	94.8
27 to 32 years	17	5.2
Total	328	100
Gender		
Male	45	13.7
Female	252	76.8
Lesbian	3	0.9
Bisexual	20	6.1
Gay	8	2.4
Total	328	100
Status		
Single	326	99.4
Married	2	0.6
Total	328	100
Learning Modality		
Google Classroom	35	10.7
Moodle	293	89.3
Total	328	100

Table 18.1 shows the profile of the survey takers of the study. It depicts that most of the respondents are aged 21 to 26 years of age with a frequency of 311 and percent of 94.8, female (F=252; Percent=76.8), single (F=326, Percent=99.4), and using Moodle (F=293; Percent=89.3) as their learning modality platform.

Table 18.2 shows the validity and reliability tests of the constructs used in the study. Average Variance Extracted is

Table 18.2 Validity and Reliability of the Constructed Survey Questionnaire

Construct	Indicator Loading
Resources (AVE = 0.637, CR = 0.898, CA = 0.857)	
2.1.1 A video presentation on a certain topic gives a clear and straightforward explanation.	0.808
2.1.2 Lessons are being updated and are being linked with learning objectives and goals.	0.81
2.1.3 The course facilitator provides references to meet learning objectives.	0.807
2.1.4 The course facilitator delivers easy-to-understand and navigate instructional tools.	0.835
2.1.5 The online course materials are simple to use and constantly available.	0.728
Student Engagement (AVE = 0.678, CR = 0.913, CA = 0.881)	
2.2.1 The course facilitator provides opportunities for students to interact with their peers.	0.806
2.2.2 The course facilitator citing relevant practical examples that are linked to the lesson.	0.806
2.2.3 The course facilitator provides students with educational experiences that are challenging, enriching, and enhance their academic ability.	0.85
2.2.4 The course facilitator encourages students to routinely present and discuss their work.	0.817
2.2.5 The course facilitator fosters a welcoming environment.	0.837
Orderliness (AVE = 0.642, CR = 0.899, CA = 0.859)	
2.3.1 Expectations and objectives are communicated by the course facilitator.	0.828
2.3.2 The course facilitator provides updates in connection to school calendar.	0.843
2.3.3 The course facilitator ensures course information and resources are easy to find.	0.846
2.3.4 The course facilitator provides explicit directions and reliable clear due dates.	0.786
2.3.5 The course facilitator set up the Moodle learning modality for easy access and navigation.	0.695
Faculty Presence (AVE = 0.661, CR = 0.907, 0.871)	
2.4.1 The course facilitator post announcements regarding important things about the course subject.	0.795
2.4.2 The course facilitator appears through video conferencing and takes part in the conversation.	0.801
2.4.3 The course facilitators demonstrate enthusiasm and skill in teaching the course.	0.436
2.4.4 The weekly synchronous and asynchronous session is led by the course facilitator.	0.773
2.4.5 The course facilitator offers timely feedback on student work on a regular basis.	0.803

Construct	Indicator Loading
Examination (AVE = 0.541, CR = 0.850, CA = 0.776)	
3.1.1 I pass the quiz	0.818
3.1.2 I pass the major examination	0.852
3.1.3. I complete and summit the assignment on time	0.813
3.1.4 I pass the typing drilling examination	0.845
3.1.5 I pass the timed writing examination	0.733
Recitation (AVE = 0.705, CR = 0.923, CA = 0.895)	
3.1.6 I pass the transcribing examination	0.842
3.3.1. I pass the manipulation of machine shorthand keyboard recitation	0.831
3.3.2 I pass the reading recitation	0.807
3.3.3 I pass the transcribing recitation	0.877
3.3.4 I pass the typing drilling recitation	0.839

Note: AVE = Average Variance Extracted, CR = Composite Reliability, CA = Cronbach's Alpha

Table 18.3 Normality Test of the Distribution of Questionnaire

Learning Modality		Kolmogorov-Smirnov[a]			Shapiro-Wilk		
		Statistic	df	Sig.	Statistic	df	Sig.
ORES	Google Classroom	0.179	35	0.006	0.861	35	0.000
	Moodle	0.115	293	0.000	0.941	293	0.000

being used to tests the convergent validity of the constructs. In the study the AVE of the latent variables are ranging from 0.541 to 0.705. According to Fornell and Larcker (1981) criterion, the average variance extracted should be greater than 0.50. Also, the indicator loading of each construct should be greater than or equal to 0.50. Since all the indicator loading and AVE is higher than 0.50 it means that the constructs of the survey questionnaire passed the validity test. Further, The internal consistency of all constructed statements in the survey was determined using composite and Cronbach's Alpha reliability tests. According to Fornell and Larcker (1981), composite and Cronbach's alpha values should be equal to or greater than 0.70. The composite and Cronbach's alpha values in the study were greater than 0.70, indicating that all of the constructs used in the survey had internal consistency.

The researchers test the normality of the distribution of the questionnaire to determine whether there are outliers in the study and to identify the proper statistical tools to be used in the study. According to Lacap (2021), to use parametric statistics, the significant value should be greater than 0.05, otherwise used non-parametric if the significant value is 0.05 and below. Results shows that the significant value is less than 0.05, therefore the researcher used non-parametric statistics to test the relationship of both independent and dependent variables which is the Spearman Rho Correlation.

4. Results

Table 18.4 Respondents' Assessment on the e-machine shorthand online teaching practices

Table 18.4 shows the respondents' assessment on e-machine stenography online learning practices. The table reveals that in terms of resources the respondents assess "Lessons are being updated and are being linked with learning objectives and goals" and "The course facilitator delivers easy-to-understand and navigate instructional tools" as extremely effective with mean both mean of 4.2744 and SD 0.74962 and 0.71955. In terms of student engagement, the respondents assess "The course facilitator citing relevant practical examples that are linked to the lesson" as extremely effective with mean of 4.3079 and SD of 0.74583. While in terms of orderliness, the respondents assess "The course facilitator ensures course information and resources are easy to find" and "The course facilitator provides explicit directions and reliable clear due dates" both got a mean of 4.2317 and with SD of 0.75092 and 0.71760 respectively which is verbally interpreted as extremely effective. Finally, in terms of faculty presence, the respondents assess "The weekly synchronous and asynchronous session is led by the course facilitator" with a mean of 4.3963 and SD of 0.73139 interpreted as extremely effective. Overall, the four variables were assessed effective to extremely effective by the respondents.

Table 18.5 shows the respondents' assessment on their performance. The results revealed that "I complete and summit the assignment on time" has the highest mean of 4.3476 and SD 0.77099 interpreted as always. While "I pass the manipulation of machine shorthand keyboard recitation" has the highest mean of 4.0732 with SD 0.77073 interpreted as very often.

It was concluded based on the research of Liang, G., Jiang, C., Ping, Q., and Jiang, X. (2023). It was stated that completed assignments submitted by students are one of the most important factors influencing final exam scores as an output as part of academic performance prediction associated with online learning set up.

According to the research of Patel, H. R., and Singh, S. D. (2023). The availability of self-study learning materials is critical, as students rely on various modes of communication and learning. Instruction materials are created in a variety of ways to provide information to students and help them fully understand it while learning and teaching.

Table 18.6 depicts the relationship of e-machine stenography online teaching practices and students' performance. It

Table 18.4 Respondents' Assessment on the e-machine shorthand online teaching practices

Resources	N	Mean	SD	Interpretation
RESOURCES 1	328	4.2043	0.82268	Effective
RESOURCES 2	328	4.2774	0.74962	Extremely Effective
RESOURCES 3	328	4.3598	0.70287	Extremely Effective
RESOURCES 4	328	4.2744	0.71955	Extremely Effective
RESOURCES 5	328	3.9939	0.91536	Effective
Over all Mean	**328**	**4.222**	**0.62232**	**Extremely Effective**
Student Engagement	N	Mean	SD	Interpretation
SE1	328	4.1463	0.79558	Effective
SE2	328	4.3079	0.74583	Extremely Effective
SE3	328	4.2683	0.7386	Extremely Effective
SE4	328	4.2226	0.77174	Extremely Effective
SE5	328	4.1951	0.71573	Effective
Over all Mean	**328**	**4.228**	**0.62011**	**Extremely Effective**
Orderliness	N	Mean	SD	Interpretation
ORDER1	328	4.1921	0.77202	Effective
ORDER2	328	4.2073	0.754	Effective
ORDER3	328	4.2317	0.75092	Extremely Effective
ORDER4	328	4.2317	0.7176	Extremely Effective
ORDER5	328	4.0427	0.94724	Effective
Over all Mean	**328**	**4.1811**	**0.628**	**Effective**
Faculty Presence	N	Mean	SD	Interpretation
FACULTY1	328	4.3384	0.67147	Extremely Effective
FACULTY2	328	4.3445	0.68194	Extremely Effective
FACULTY3	328	4.314	0.72241	Extremely Effective
FACULTY4	328	4.3963	0.68732	Extremely Effective
FACULTY5	328	4.2348	0.73139	Extremely Effective
Over all Mean	**328**	**4.3256**	**0.56705**	**Extremely Effective**

Table 18.5 Respondents' Assessment on students' performance

Quiz, Major Exam, Assignment	N	Mean	SD	Interpretation
EXAM1	328	3.9055	0.77038	Very Often
EXAM2	328	4.0457	0.72592	Very Often
EXAM3	328	4.3476	0.77099	Always
EXAM4	328	3.8171	0.76431	Very Often
EXAM5	328	3.7927	0.74993	Very Often
EXAM6	328	3.8841	0.74931	Very Often
Overall Mean	**328**	**3.9817**	**0.54848**	**Very Often**
Recitation	N	Mean	SD	Interpretation
RECIT1	328	3.7866	0.76037	Very Often
RECIT2	328	4.0732	0.77073	Very Often
RECIT3	328	3.9787	0.76796	Very Often
RECIT4	328	3.8598	0.76135	Very Often
Overall Mean	**328**	**3.9165**	**0.63936**	**Very Often**

Table 18.6 Significant Relationship of the Respondents' Assessment on e-machine shorthand online teaching practices and students' performance

Aspects of Teaching Practices		Student Performance	Spearman's Rho Value	p-value	Decision	Remarks
Resources	vs	Examination	0.371	<0.001	Reject H$_0$	Significant
		Recitation	0.337	<0.001	Reject H$_0$	Significant
Student Engagement		Examination	0.349	<0.001	Reject H$_0$	Significant
		Recitation	0.344	<0.001	Reject H$_0$	Significant
Orderliness		Examination	0.377	<0.001	Reject H$_0$	Significant
		Recitation	0.36	<0.001	Reject H$_0$	Significant
Faculty Presence		Examination	0.366	<0.001	Reject H$_0$	Significant
		Recitation	0.345	<0.001	Reject H$_0$	Significant

shows that all four aspects of online teaching practices have a significant and positive but weak relationship with students' performance. This means that teaching practices has a direct effect with students' performance. It also shows that if the value of teaching practices increases (decrease), the value of students' performance also increases (decreases). It implies that teachers should always uphold good practices to boost students' performance.

According to the study of Ghosh, R., Khatun, A., and Khanam, Z. (2023), it was stated that using online applications such as social media apps has a significant relationship with the academic standing of currently enrolled students during the pandemic.

5. Discussion

In this study, in the aspects and factors identified to be measured in this study such as resources, student engagement, orderliness and faculty presence has significant relationships with the academic performance of the students.

In earlier research, like the study of S. Rakic, E. Lüftenegger, S. Softic, U. Marjanovic, N. Tasic, and I. Turcin. (2020). which highlighted the important connection between using digital resources for e-learning and students' academic performance while studying S. Jaggars, S., and D. According to Xu (2016), making online learning materials more accessible has significantly enhanced students' academic performance. In which the findings of this study support the association between students' academic performance and online teaching strategies. This study contributes to the realization that an academician must constantly improve and nurture the academic performance of the student. Online resources should be available to everyone, wherever they are. Student involvement is also essential for connecting with and assisting the students, and it is important to give clear instructions to prevent potential obstacles from impeding the learner's ability to learn. Last but not least, it's crucial that the subject facilitator is present online so that students can feel as though their queries and worries are being addressed appropriately and that someone is still there to help them even after online setup is complete.

The study filled a knowledge gap by addressing the factors that should be taken into consideration in order to give students an excellent and remarkable learning experience in the field of technical and skill reference subjects. The factors identified in this study that are related to the academic performance of the students were also added value to the body of knowledge.

According to the findings of this study, all aspects of online teaching practices, including online resources, student engagement, orderliness, and faculty presence, have a significant and positive but weak relationship with student performance. In the study, online teaching practices had a direct effect on the performance of the student learners, and it was also demonstrated that as the value of teaching practices increased (decreased), so did the value of students' performance (decreases). It also implies that academicians should always uphold good practices in order to improve student performance.

6. Recommendations and Future Research Directions

Since the SARS COVID-19 global pandemic imposed many constraints and dilemmas on the operations and strategies of higher education institutions, the Philippine government particularly IATF (Inter-Agency Task Force for COVID-19) has ordered and directed that no 100% on site classes be held for the 2020 – 2021 School / Academic Year until the vaccine drive has achieved 100% herd immunity Ancheta, R., & Ancheta, H. (2020). As the current study suggests that e-machine stenography online learning practices have a significant effect on students' performance, teachers must always look for additional ways to improve course materials or resources, as well as student interaction, to improve students' performance. To engage students in a more enjoyable and conducive e-learning environment, the use of gamification, social media integration, reflective thinking, and high-quality educational resources should be explored.

It is also recommended for the provision of online applications that can be used for gamifying online lectures, such as video editors, design tools, grammar checkers, and plagiarism checkers. Consider institutionalizing public-private partnerships to fund and improve existing online learning materials.

Further research in other contexts may be conducted to support or repute the study's findings. Other variables assessing online teaching and learning practices may also be used to test other factors influencing students' performance in e-machine stenography teaching practices.

REFERENCES

1. A. Crampton, A. T. Ragusa, and H. Cavanagh, "Cross-discipline investigation of the relationship between academic performance and online resource access by distance education students," Research in Learning Technology, 20(1), n1, 2012.

2. C. Fornell, and D. F. Larcker, Structural equation models with unobservable variables and measurement error: Algebra and statistics," Journal of Marketing Research, 18(3), 382-388, https://doi.org/10.2307/3150980, 1981.

3. E. Galy, C. Downey, and J. Johnson, "The effect of using e-learning tools in online and campus-based classrooms on student performance," Journal of Information Technology Education: Research, 10(1), pp. 209-230, 2011.

4. Ghosh, R., Khatun, A., & Khanam, Z. (2023). The relationship between social media based teaching and academic performance during COVID-19. Quality Assurance in Education, 31(1), 181-196.

5. I. Galikyan, and W. Admiraal, "Students' engagement in asynchronous online discussion: The relationship between cognitive presence, learner prominence, and academic performance," The Internet and Higher Education, 43, 100692, 2019.

6. Liang, G., Jiang, C., Ping, Q., & Jiang, X. (2023). Academic performance prediction associated with synchronous online interactive learning behaviors based on the machine learning approach. Interactive Learning Environments, 1-16.

7. Patel, H. R., & Singh, S. D. (2023). The Significance of Instructional Materials in Enhancing Students' Academic Performance. Journal of English Language Teaching, 65(2), 28-36.

8. R. Ancheta, and H. Ancheta, "The new normal in education: A challenge to the private basic education institutions in the Philippines," International Journal of Educational Management and Development Studies, 1(1), 2020

9. R. Y. Chan, K. Bista, and R. M. Allen (Eds.), "Online teaching and learning in higher education during COVID-19: International perspectives and experiences," Routledge, 2021.

10. S. Chaturvedi, S. Purohit, and M. Verma, Effective Teaching Practices for Success During COVID 19 Pandemic: Towards Phygital Learning," Front. Educ. 6: 646557. doi: 10.3389/feduc., 2021.

11. S. S. Budhai, "Best practices in engaging online learners through active and experiential learning strategies," Routledge, 2021.

12. T. J. Aveni, C. Seim, and T. Starner, "A preliminary apparatus and teaching structure for passive tactile training of stenography," In 2019 IEEE World Haptics Conference (WHC), pp. 383-388, IEEE., July 2019.

Note: All the figure and tables in this chapter were made by the authors.

Advancing Sustainable Science and Technology for a Resilient Future – Sai Kiran Oruganti et al. (eds)
© 2024 Taylor & Francis Group, London, ISBN 978-1-032-79020-6

19

Digital Wallet: Usage and its Security for Cashless Society

Melani L. Castillo*

Polytechnic University of the Philippines -
Sto. Tomas Branch, City of Sto. Tomas,
Batangas, Philippines.

Abstract: Digital transactions became enormous when the COVID-19 pandemic hit the country. Many transactions are also paid for using digital wallets primarily for safety from being exposed to viruses when using cash. Increasing use of these applications also raises issues on cybersecurity and their impact on overall end-user confidence. Thus, this study discussed the features and uses of digital wallet applications and explores the awareness of end-users on cybersecurity issues. Quantitative research methods were utilized in this research work. The study found out that the respondents are aware of the various cybersecurity measures that can be used. The strong positive feedback of the users regarding the use of digital wallet for online transactions would conclude their overwhelming trust and support on these applications. It only proves that more people are convinced with the ease of using cashless payments, paving the way for technology to advance in society.

Keywords: Digital wallet, e-wallet, Online transactions, Cybersecurity, Cashless society

1. Introduction

The increase of digital users in the Philippines is undeniable massive. With the outbreak of the COVID-19 pandemic, most individuals, businesses, and large institutions, including the government, have gradually shifted their transactions to digital to keep up with the changing times and the challenges of current crisis. Consumer transactions such as merchant payments, utility payments, and e-commerce transactions have contributed to this increasing use and adoption of digital payments.

While moving into digital ecosystem, the cashless society faced the one of the most critical challenges in their transaction which is security. There is a high chance of getting subjected to cybersecurity risks such as online fraud, information theft, and malware or virus attacks. It is indeed that technology is making merely every aspect of life easier but there are factors that might be a disadvantage for its users. Thus, the main objective of the study is to determine the usage of electronic wallet among participants and the cybersecurity measures they use on their day-to-day transactions. Specifically, the study aims to answer the following questions:

1. What are the digital wallet applications do the participants used?

2. What transactions do participants used their digital wallet?

3. What are the common features of their digital wallet accounts?

4. What is the level of awareness of the participants on the cybersecurity threats on their digital wallet accounts?

5. What security measures do they observe to protect their digital wallet?

[1]mlcastillo@pup.edu.ph. (Paper ID :1208)

DOI: 10.1201/9781003490210-19

2. Research Methods

A quantitative research method was used in the study as it was considered the most appropriate method for this study. A total of eighty-three (85) respondents participated in this study. Most of the respondents who participated in the study are female while most of them are between 18 and 24 years old. The researcher used an online questionnaire constructed via Microsoft Forms. It was validated by ten (10) initial respondents and a statistician and two (2) IT experts. After, it was distributed via the researcher's social media accounts and posted online to ensure the randomness of the respondents who will voluntarily fill in the questionnaire. The questionnaire used a Likert scale with different response anchors depending on the type of question asked to the respondents. The results are automatically generated from the Microsoft Form. The consolidated data was downloaded and then computed using SPSS and was analyzed using statistical software.

3. Results and Discussion

After tabulating and analyzing the data gathered, the following are the general assessment of the respondents regarding the objective questions.

The Digital Wallet Applications used by the Respondents.

The first objective of the study is to determine the most preferred digital wallet application of the respondents. Table 19.1 shows the results of this inquiry.

Table 19.1 Most preferred digital wallet applications

Digital Wallet Applications	Frequency	%	Rank
Lazada Wallet	7	8	4
GrabPay	3	3.5	6
ShopeePay	12	14	3
GCash	29	34	1
PayMaya	22	26	2
Dragonpay	1	1.17	10
Coins.ph	5	5	5
CLiQQ PAY	2	2.35	7.5
ML Wallet	1	1.17	10
PayPal	2	2.35	7.5
BPI Direct BanKo	1	1.17	10
Total	85	100%	

Table 1 shows that the GCash is the most preferred digital wallet which comprise 34% of the total respondents. It is commonly called as GCash which is managed by Globe Telecom, the leader in the mobile phone service in the country. It allows their users to send money, remittances and do real-time transactions to more than 40 banks and billers nationwide. The results indicate the preference of the respondents of their digital wallet applications and implies their awareness and popularity. This leads to familiarity of the users to the benefit of the applications and gain popularity among others.

Digital Wallet Usage

Table 19.2 shows the usage of digital wallets. The results show that the participants always use their digital wallet for digital purchases and for their bills and loans payment. While other participants use their digital wallets for physical store purchases, money transfer, cash withdrawal, collecting coupons & offers and for virtual cash savings.

Table 19.2 Digital wallet usage

Indicators	Mean	V.I.	Rank
Bills and loans payment	4.45	A	2
Digital purchases	4.57	A	1
Physical Store purchases	4.22	O	4
Cash Withdrawal	4.12	O	5
Money Transfer	4.38	O	3
Virtual cash savings	3.87	O	7
Collect coupons or promotional offers	3.93	O	6
Overall Assessment	4.21	O	

VI- Verbal Interpretation, Scale: 4.25 - 5.00, Always (A), 3.45 - 4.24, Often (O), 2.55 - 3.44, Sometimes (S), 1.75 - 2.54, Rarely (R), 1.00 - 1.74, Never (N)

In the most recent study of Bangko Sentral ng Pilipinas State, it has revealed that "for 2022, the share in terms of volume of digital payments over total retail payments considerably grew to 42.1%, supported by an increase of 611.7 million retail payments transactions from the previous year". (BSP Status of Digital Payment Report, 2022). Since digital payment methods allow consumers to purchase goods and services conveniently, it leads to more consumers relying on this online method. In addition, users gain equability knowing that their funds are secured in their digital wallets.

Digital Wallet Features

Table 19.3 show that most of the respondents strongly agree that their digital wallet is user friendly and convenient to use, offers a wide range of features and provides fast and reliable processing of transactions. Based on the results of study of Ching (2017), doing transactions online is much convenience for Filipinos since mostly they have limited time to do their transactions physically such as paying utility bills and transfer money to their relatives.

Table 19.3 Assessment of digital wallet features

Indicators	Mean	V.I.	Rank
My digital wallet offers wide range of features	4.32	SA	3
My digital wallet is user-friendly and convenient to use.	4.68	SA	1
My digital wallet provides fast and reliable processing of transactions.	4.36	SA	2
My digital wallet is trusted by most users.	3.88	A	6
My digital wallet stores the history of my transactions.	3.93	A	5
My digital wallet provides privacy features.	3.87	A	7
My digital wallet provides security functions.	4.09	A	4
Overall Assessment	4.16	A	

Scale: 4.25 - 5.00, Strongly Agree (SA), 3.45 - 4.24, Agree (A), 2.55 - 3.44, Neutral (N), 1.75 - 2.54, Disagree (D), 1.00 - 1.74, Strongly Disagree (SD)

The listed digital wallet features in the table indicate that the companies must identify and understand the features valued by users through understanding their standpoints, so that they will be able to help them and increase their loyalty resulting to more sales and profits in the future. According to Ching (2017), having diverse payment method encourage more Filipinos to do their transactions online. They also considered their chosen method in their trades based on their "general, privacy, security, and trust perceptions". Thus, the given digital wallet applications provided some similar functions and features while other have different features that can suit the needs of their users.

Awareness on Cybersecurity Threats on Digital wallet

Table 19.4 below shows the overall weighted mean of 4.20 which indicate that majority of the respondents are Aware of the cybersecurity threats.

Table 19.4 Awareness on cybersecurity threats on digital wallet

Indicators	Mean	V.I.	Rank
Online Fraud	4.84	HA	1
Malware attacks	4.22	A	3
Identity theft	4.73	HA	2
Hacking	4.07	A	5
Illegitimate/unauthorized payments	4.11	A	4
Exploit the vulnerabilities of the other application	3.67	A	7
Attempt to tamper with other application	3.80	A	6
Overall Assessment	4.20	A	

Scale: 4.25 - 5.00, Highly Aware (HA), 3.45 - 4.24, Aware(A), 2.55 - 3.44, Neutral (N), 1.75 - 2.54, Somewhat Aware (SA), 1.00 - 1.74, Not Aware (NA)

The respondents are highly aware (HA) of the higher possibility for online fraud and identity and information theft. Lack of security in acquiring personal sensitive information is one of the main issues and challenges in electronic payment system (Rachna and Singh, 2013). While in the study of TransUnion in 2021, they found out that 48% of consumers in the Philippines is being targeted by digital fraud in the last three months while, various business groups in the Philippines have projected that the cost of cybercrime will be around $6 trillion in 2021 increasing to $10.5 trillion annually by 2025 (Balinbin, 2021). The results imply that despite the security features imposed by digital wallet providers, threats, and risks in the use of the digital wallet is still more likely to occur. Thus, users must be cautious with these threats to avoid bigger problems in the future.

Security Measures to Protect their Digital Wallet

The Table 19.5 below shows that the overall assessment of 3.71 mean "observe" which implies that the participants apply these measures to protect their digital wallet.

To adequately prevent and mitigate these evolving threats in the use of digital wallet, established digital wallet and e-money providers also take the security of their customers very seriously. For instance, GCash provide authentication features for their digital transactions using the 'GCash Security Code (GCSC).' This security element uses code/PIN that will be sent to the registered mobile number of the customer via their Short Messaging System (SMS) once they make digital transaction using their GCash account (Gcash Website). On the other hand, other e-wallet companies such as Paymaya and Coins.ph also have different support channels to cater their respective customers about their security concerns.

Table 19.5 Security measures to protect their digital wallet

Indicators	Mean	V.I.	Rank
Use of strong PIN and Password	4.45	HO	2
Keep login credential secure	4.67	HO	1
Securing OTP codes	4.38	HO	3
Checking URL links before log-in	3.80	O	5
Register for alerts through SMS and emails	4.22	O	4
Use secure network connections	3.16	O	6
Finger-print login	2.52	N	7
Dual verification	2.45	N	8
Overall Assessment	3.71	O	

Scale: 4.25 - 5.00, Highly Observed (HO), 3.45 - 4.24, Observed(O), 2.55 - 3.44, Neutral (N), 1.75 - 2.54, Not Observed (NO), 1.00 - 1.74, Highly Not Observed (HNO)

4. Conclusions

Based on the findings of the study, among the top three digital wallets preferred by the respondents include GCash, Paymaya and ShopeePay. The respondents found digital wallets convenient and easy to use in terms of digital payment, money/bank transfer, and paying loans and bills/utility. While digital wallet users convincingly agreed that their applications provide convenience, ease of use and fast transactions. The study also revealed that most of the respondents are aware of these cybersecurity threats they may encounter while using their digital wallets. The results also indicated that respondents generally observed cybersecurity measures when using their digital wallet applications.

5. Recommendations

Considering the results of this study, the following recommendations were made. First, the users must be aware of the features and advantages of their digital wallet to utilize its full potential. Investigating the user's feedback towards digital payments and transactions generally would give ideas to strengthen the security of E-commerce and the digital payment industry. On the side of the third-party providers, E-wallet application developers should also work on more secured features of their application to ensure efficient and effective digital transactions. Digital wallet providers must enhance security measures through integration of advanced technology capabilities. On the other hand, the government through Bangko Sentral ng Pilipinas (BSP) should update policies and procedures to promote the use of digital wallet and ensure safety of digital users. To attain a cashless society, BSP and the Philippine government should take comprehensive initiatives and programs that will improve the mobile connectivity and digital infrastructure. hile the private sector should collaborate with the government to improve their financial system's security and services of the country. Future research may also consider a larger sample size and other age group of respondents to understand their preference on the use of digital wallet applications. Lastly, other research methods can be used to determine and analyze the impacts of digital wallets on financial inclusion.

References

1. Balinbin, A. (2021, March 7). Cybercrime to increase further as transactions shift online. Retrieved December 14, 2021, from BusinessWorld Onlinewebsite:https://www.bworldonline.com/cybercrime-to-increase-further-as-transactions-shift-online/
2. Bangkok Sentral ng Pilipinas (February 2009). BSP Circular No. 649 Series 2009. Guidelines governing the issuance of electronic money (emoney) and the operations of electronic money issuers (EMI) in the Philippines.
3. Bangko Sentral ng Pilipinas (2022). 2022 Status of Digital Payments Retrieved fromhttps://www.bsp.gov.ph/PaymentAndSettlement/2022_Report_on_E-payments_Measurement.pdf
4. Ching, M. (2017). Challenges and Opportunities of Electronic Payment Systems in the Philippines. Retrieved from https://www.dlsu.edu.ph/wp-content/uploads/pdf/conferences/research-congress proceedings/2017/HCT/HCT-I-006.pdf
5. GCash Customer Protect Terms and Conditions. Retrieved December 16, 2021, from GCash website: https://www.gcash.com/customer-protect-terms-and-conditions/
6. Rachna, & Singh, P. (2013). Issues and Challenges of Electronic Payment Systems. International Journal for Research in Management and Pharmacy, 2(9), 25–30. Retrieved from http://raijmr.com/wp-content/uploads/2014/02/3_25-30-Rachna-et-al.pdf

Note: All the tables in this chapter were made by the authors.

Advancing Sustainable Science and Technology for a Resilient Future – Sai Kiran Oruganti et al. (eds)
© 2024 Taylor & Francis Group, London, ISBN 978-1-032-79020-6

PUP Microbank: Microbial Collection Repository with Status Monitory and Request Assessment

20

**JaysonJamesMayor*, Sherilyn Usero,
Kimberly D. Francia, Louie B. Gonzales,
Maria Patricia R. Habaan, ReaSophiaL.Pasco**
College of Computer and Information Sciences,
Department of Information Technology Polytechnic
University of the Philippines

**Aleta C. Fabregas, Gary Antonio C. Lirio,
Armin S. Coronado**
Research Institute for Science and Technology, Center for
Computer and Information Sciences Research, Polytechnic
University of thePhilippines

Abstract: Microbial Culture Collections (MCCs) manage enormous amounts of data, specifically information about microorganisms, their taxonomy, and the methods for maintaining, preserving, and accessing these microorganisms. This difficult task necessitates the development of a system that enables MCC administrators and curators to navigate the microbial-related data files quickly, responsively, and efficiently. As a recognized affiliate of the Philippine Network of Microbial Culture Collections (PNMCC), the Polytechnic University of the Philippines (PUP) opted to develop a monitoring system that curates, stores, manages, and documents the processes associated with microbial culture collections, accessioning, and preservation. The PUP Micro Bank: Microbial Collection Repository with Status Monitoring and Request Assessment is a proposed system for storing microbial data. PUP MicroBank is a web-based system that was developed in accordance with ISO 9126, an international standard for software assessment. The system includes a color-coded monitoring alert system that monitors microorganism viability and notifies users of necessary management actions for cultured organisms. Additionally, the system supports microbe orders and deposit requests from clients who wish to have their microorganisms banked and deposited in the culture collection. As for the System Development, Waterfall Model was used because it is the most compatible for this project. Since it is less iterative and adaptable methods as progress currents in largely one way complete the phases of conception. The system was evaluated using the ISO 9126 software evaluation questionnaire accomplished by experts and prospective end users (institute administrators, faculty, and clients). The PUP MicroBank received an overall mean of 4.14, indicating that it is a good system in relations to functionality, portability, reliability, usefulness, usability, satisfaction, efficiency, ease of learning, andmaintainability.

Based on the results of the examiners' validation test, the PUP Micro Bank as an automated system application complied with the ISO 9126standard by simplifying the processes for data retrieval, microbe management, microbe distribution, data culture updates, and preservationrecords.

Keywords: PU PMicro bank web-based system, ISO9126, Data management, Microbial collection

1. Introduction

Idealistic perspective about scientific breakthroughs and future disclosures that can prompt establishing new examinations in microbiology about antibiotics is one of the numerous objectives of the PUP Institute for Science and Technology Research's Center for Life Sciences Research in the Polytechnic University of the Philippines. The research center creates understudies to become their insightful and intellectual aspects to augment the limit of the students'

learning in the field of biology. One of the benefits of Microbe Culturing is the utilization in variety of studies and researches by students in their theses, tests and scientific purposes, if a certain microbe specie has a huge potential and has a unique characteristic that can be used as anti-agent ingredient used for curing pathogen-causing diseases and pharmaceuticals. Pathogenic microbes are based on Biohazard levels 1, 2, 3, 4. Biohazard level 1 and 2 indicates normal pathogens while Bio Hazard 3 and 4 indicates harmful pathogens.

*Corresponding author: sbusero@pup.edu.ph

DOI: 10.1201/9781003490210-20

The research center collects, tests, grows, preserves, stores and cultures different types of bacteria, protists, and fungi, or in general terms known as microbes. Every set of microbes comes with a variety of unique characteristics. The admin of the research center is the one who leads in storing, management, and distributing of microbes.

The PUP ISTR's Center for Life Sciences Research has an extension service which encourages clients to develop their capabilities and production in research culture. Since the research center does not charge for any amount of payment, instead they require the clients to sign a Memorandum of Understanding (MOU). MOU stands for formal agreement between the teams involved outlining the standings and particulars of an accepting, including each gatherings' requirements and responsibilities.

2. Literature Review

According to Yogesh Shouche (2014), Culture gatherings play a vital part in the preservation and viable use of microbial assets. They also stipulate the dependable biological material for high feature study and education in the form of reference strains, components for quality control, etc. The developments in molecular biology have stemmed in the continual innovation of new microbial taxa and straining and there is a need to reservation to make them accessible to other researchers, teaching and for biotechnological utilization. Discrete laboratories are unable to do this due to lack of monetary support and manpower. This part thus participated by gathering a culture. The researchers proposed a web-based system that can be accessible by other researchers, students, professionals, etc.

The computer system has been the primary medium for program of results of bacteriology laboratory purposes to patients' accounts since 1971. The gathering of information linking to tests accomplished on more than one million samples during this period of time has provided exceptional chances for using the computer not only for organizational tasks, but particularly for technical and scientific functions that are clinically related. These contain the examination and control of quality of lab performance, epidemiologic observing and forecast, learning of the vulnerability of organisms to antibiotics and of the development of resistant.

straining, and examination to a diversity of true and probable difficulties. Certain of these submissions are illustrated by reproductions of computer- generated tables, graphs and reports. (Lawrence J. Kunz, 2016)

In relation to the system proposal, the researchers proposed the system to be able to resolve the management of the rapid growth of microbial collections with the problems on shortage of staffs, also the setting up of data bank and retrieval system to be able to easily access the specific microbial collection.

From "Method in Coding Information on the said Microbial Strains for Computers", contributed by Morrison R, Micah L. K, and Rita R. C (2016). Experience with computers in microbial documentation and sorting is now sufficient to warrant planning for the establishment of an international repository for cultures and culture collections. This would be an invaluable resource to be used in many profitable ways. In fact, the present population of computers with the available machine operating and storage capacities and software (i.e. programs and programing versatilities) makes the concept of an international microbiological data bank feasible. The maximum data grid for present information can easily be handled by the models of computers available to major university and research centers in many countries of the world. A microbiological repository, as a centralized repository of information of all types, should include historical, nomenclatural, classificatory, and epidemiological information for groups and clones of such microorganisms as bacteria, rickettsiae, chlamydiae, and viruses.

Literature on taxonomy of microorganisms can be located on computer at an international center for rapid call-up and consultation by any individual in any other country at any time. Identification, classification and nomenclature of microorganisms requires with the same diligence in information search as other scientific doings and the lack of right away access to pertinent literature on microorganisms as is available, for example, on chemical composites, puts a major hurdle in the way of those microbiologists seeking to identify and classify microorganisms, whether for purposes of pure taxonomy or for such applications as medical diagnosis, A computer library with ready call-up to all or specific components of appropriately stored data would meet these needs. Information is never "lost" since all data put into computer storage are retained and can be called up by appropriate query to the computer at any time.

3. Objectives

To develop a PUP Microbank: Microbial Collection Repository with Status Monitory and Request Assessment

3.1 Specific Purposes

The researcher's goal to accomplish the following matters:

1. To save a detailed record for each microbe specimen.

2. To properly monitor the microbe specimens that are stored.

3. To validate the depositor's purpose when requesting for a microbe.

4. To be able to track the current status of all the microbes.

5. To design a system that can easily retrieve data.

6. To provide an efficient system for entire processes and less paperusage.

4. Methodology

The research center collects, tests, grows, preserves, stores and cultures different types of bacteria, protists, and fungi, or in general term known as microbes. Every set of microbes comes with a variety of unique characteristics. The admin of the research center is the one who leads in storing, management, and distributing of microbes.

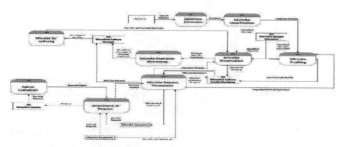

Fig. 20.1 Data flow diagram: microbial collection repository with status monitoring and request assessment

The Micro Bank: Microbial Collection Repository with Status Monitoring and Request Assessment is a proposed system for the well- development of PUP ISTR's Center for Life Sciences and Research's processes and assessments in the scheduling for requesting, depositing and testing of microbes. The system also aims to advance in achieving and keeping the variation of records about microorganisms in a sorted way, and has a feature of color coding for alert levels for the administrator of the system to help them keep track of the microbes' lifespan.

Test Methodology/Procedures

- **Testing Plan:** During the testing process of the system, the team will make sure that the following test plans will be put through.

- **Component Testing:** the testing of an discrete software component of module. Each part of the system is independently tested. All of the functions of each module of the system must be checked using the White Box method in which the tester or the developer must know the inner parts of the systems as it examines the program assembly and derives test data from the program logic or code.

- **Structure testing:** the whole system is verified as a whole per the desires using the Black Box method where it scrutinizes the functionality of an application based on the stipulations. This is from general requirement stipulations and includes all the collective quantities of the system.

- **Functional Testing:** disregards the internal parts and focuses only on the output to checked or verify if the system is behaving according to the specifications. This helps to uncover bugs that might occur frequently.

- **Acceptance Testing:** the client verifies whether the stream of the system is as per the trade necessities or not and if it is as per the needs of the user using the Black Box method. This is usually the last phase of the phase of the testing which the software goes into implementation and is also called the User Acceptance Testing (UAT).

Quality Plan

ISO 9126 was used for the evaluation of the software. This International standard supports in creating a reliable framework for assessing the software. ISO 9126 is the most suitable for our system. As, it is also widely used in the software engineering community and adapted to various domains. By using the Functionality, Reliability, Usability, Efficiency, Maintainability, Portability, Usefulness, Satisfaction and Ease of Learning Criteria it is easy to use and understand.

Overall, Usability got the highest mean with 4.52 and a verbal interpretation of Strongly Agree and the overall mean of the system evaluation is 4.14 and has a verbal interpretation of Agree. The evaluation of the proposed is positive and ready for implementation.

Table 20.1

ISO 9126 – Overall Results Table

DESCRIPTION	MEAN	VERBAL INTERPRETATION
ISO 9126		
Functionality	.31	Agree
Reliability	.01	agree
Usability	.52	Strongly Agree
Efficiency	.24	Agree
Maintainability	.72	Agree
Portability	.09	Agree
Usefulness	.19	Agree
Satisfaction	.25	Agree
Ease Of Learning	.97	Agree
OVERALL MEAN	.14	Agree

Development Tools

The following developmental tools was used while developing the system are the following:

XAMPP,MySQL,PHP,HTML,CSS,Bootstrap,jQuery , JavaScript,Visual Studio Code,CodeIgniter,Git Hub

System Prototype

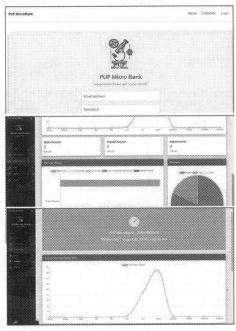

Fig. 20.2   PUP microbial dashboard

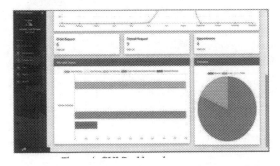

Fig. 20.3   GUI dashboard

5. Results and Discussion

The overall result for the PUP Microbank: Microbial Collection Repository with Status Monitory and Request Assessment is 4.14 with the verbal interpretation of Agree. The Usability got the highest, with a mean of 4.52 and a verbal interpretation of Strongly Agree. With all the nine (9) criteria that was used to decide eight (8) of the criteria are "Agree" and one (1) got the Strongly agree.

6. Conclusion

The following are the proved solutions provided by the system from the problems it encountered:

Inefficient data gathering procedures was given a solution by providing an automated and efficient system to manipulate the data input easier. The inconsistency in monitoring of microbe expiration was solved by creating an automated microbe-monitoring module with color coding legend for warning and status levels. With the help of the color coding legend, keeping track of the mircobe's life expectancy helps the admin to easily monitor if a microbe is nearing its expiration, therefore re-culturing it before it even expires.

Difficulty in monitoring voluminous amount of microbe data was solved by providing an automated system that lets the voluminous amount of data to be stored in the database instead of being piled up when used with papers, making the admin handle the voluminous data with ease. The system was able to provide automatic assigning of accession code per microbe. By putting data queries, sorting different kind of documents such as microbes' profile becomes much easier. By providing an automated system with sub-module of microbe request and deposit assessment, the process of ordering and depositing of microbe is now centralized and efficient for both the user and the administrator.

7. Recommendation

For the future researchers who might seek for a similar study, effective upgrade can be done with the system by some features that did not reach the research development plan due to criticality that can make huge effect to the project plan.

- Data Analytics on Maintenance- by applying data analytics to the system, the user can gather all of the necessary data from the internet to use when filling up data fields.

- Data Analytics on Advance Searching- by applying data analytics to the system when doing advance searching can make the processes faster and more informative compared with manual maintaining of theinformation.

- Data Analytics on Microbe Expiration Monitoring- by applying data analytics to the microbe expiration monitoring can farm voluminous data to predict the expiration that will be based on the appropriate algorithms.

- Emailing system/SMS notification- putting this feature can help the administrator to monitor the microbe expiration by receiving electronic mails or SMS text message. This feature will send notification and information to show the status of microbe in the collection.

With all of the recommendation, the PUP MicroBank: Microbial Collection Repository with Status Monitoring and Request Assessment will surely continue to improve and become better with the help of its users who will attend to their needs with ease.

References

1. Avinash Sharma & Yogesh Shouche (2014), Microbial Culture Collection (MCC) and International Depositary Authority (IDA) at National Centre for Cell Science, Pune. Indian Journal of Microbiology volume 54, pages 129–133

2. Doaa Nabil, Abeer Mosad & Hesham A.Hefny,(2011)Web-Based Applications quality factors: A survey and a proposed conceptual model, Egyptian Informatics Journal, Volume 12, Issue 3, November 2011, Pages211-217

3. Rachid Lahlali, Saroj Kumar, Lipu Wang, Li Forseille, Nicole Sylvain, Malgorzata Korbas, David Muir, George Swerhone, John R. Lawrence, Pierre R. Fobert,, Gary Peng and Chithra Karunakaran (2016), Cell Wall Biomolecular Composition Plays a Potential Role in the Host Type II Resistance to Fusarium Head Blight in Wheat. Frontiers in Microbiology, ORIGINALRESEARCH published: 27June2016 doi: 10.3389/fmicb.2016.00910, pages 2 -12

4. Morrison Rogosa, Micah L. Krichevsky, and Rita R. Colwell (2016), Coding Microbiological Data for Computers , Springer; Softcover reprint of the original 1st ed. 1986 edition, ISBN-10 : 1461293863

5. I.Ghasemzadeh, S.H.Namazi. National Center for Biotechnology Information, U.S. National Library of Medicine8600.Rockville Pike, Bethesda MD, 20894 USA. https://www.ncbi.nlm.nih.gov/pmc/articles/PMC 5319273/

Advancing Sustainable Science and Technology for a Resilient Future – Sai Kiran Oruganti et al. (eds)
© 2024 Taylor & Francis Group, London, ISBN 978-1-032-79020-6

Experimental Study Modeling Straight-bladed Vertical Axis Wind Turbine with Central Panel to Aid Self-Starting Capability

21

Rizal M. Mosquera*

Department of Mechanical Engineering, College of Engineering and Agro-Industrial Technology, University of the Philippines Los Baños, Philippines

Marita Natividad T. De Lumen

Department of Mechanical Engineering, College of Engineering and Agro-Industrial Technology, University of the Philippines Los Baños, Philippines

Abstract: This study aims to assess self-starting capability of straight-bladed vertical axis wind turbine (VAWT) using three different airfoils. Experimental tests were conducted in laboratory setup with model of airfoils at wind speed of 5.5 m/s. Airfoil model, with chord length of 250 mm and height of 300 mm, was tested using blower and duct system. Five attachment locations along the chord line were tested, with incremental distances of 25 mm. Varying pitch angle by 15°, total of 90 tests were performed. The results showed that asymmetric airfoil blade attached at location #2, with 15° toe-out pitch angle, achieved highest efficiency of 21.36%. Incorporation of symmetrical central panel addressed common disadvantage of self-starting in VAWTs. Based on this experimental study, it is recommended to further explore and prototype VAWT with power output sufficient for single-family use in locations with reliable wind source to meet their electrical energy needs.

Keywords: Airfoil, Central panel, Chord length, Fixed pitch angle, Self-start

1. Introduction

It was long ago that wind plays a vital role in the history of mankind specifically for vertical axis wind turbine as Persian windmill which has been in existence since 1,000 B.C. being the oldest windmill design (Zafar & Staubach, 2018). As more concern was put on the environmental effect of traditional sources of energy, the development of wind turbines for generating electricity became more interesting (Daniyan et al., 2018)

Wind power can be known in a particular location by measuring or data logging of wind velocity for whole year. Its availability may produce higher yield from a minimum of 3 m/s wind velocity, (Kaygusuz, 2001) to a tremendous cubical increase of power corresponding to increase of wind velocity.

Vertical axis wind turbines (VAWT) are known for their ability to catch wind from all directions perpendicular to the axis of rotation, and do not need yaw mechanisms, rudders or downwind coning. Their electrical generators can be positioned close to ground, and hence easily accessible. A disadvantage of some VAWT designs are not self-starting (Zhu et al., 2015). A starting mechanism is needed after detecting a sufficient and sustainable wind velocity for wind turbine's operation will be investigated for its elimination making the wind turbine self-starting.

New concepts of vertical axis wind machines are being introduced such as helical Darrieus types wind turbine (Karimian & Abdolahifar, 2020). It is particularly use in urban environment (Siddiqui & Hasan, 2015) where they are considered safer due to their lower rotational speeds avoiding risk of blade ejection.

2. Objective of Study

The general objective of this study is focused on the development of a two-bladed SBVAWT that aids self-start without the help of an external mechanism. Specifically, the study will carry out following sub-objectives to:

1. Design the two-bladed SBVAWT that features drag and lift effect using airfoil blades that contributes to solve self-start capability.

[1]rmmosquera2@up.edu.ph

DOI: 10.1201/9781003490210-21

2. Fabricate two-bladed SBVAWT with central panel model for testing its self-starting capability.

3. Test the self-start capability of two-bladed SBVAWT with and without central panel by varying:

   3.1. Airfoil blade attachment to spoke

   3.2. Pitch angle,

4. Determine optimum pitch angle and airfoil blade attachment location of two-bladed SBVAWT with or without central panel that self-starts the SBVAWT.

With given parameters in its optimum value, the optimum performance has also been taken into account as expected.

3. Scope and Delimitations of Study

The study limits its focus to the use of a SBVAWT with or without central panel attachment, blade attachment location and having pitch angle manipulation settings ranging from -15° to 15° with intervals of 15°.

The drag effect with a drag coefficient value 2.3 for half tube facing concave area means a lot but the other half convex area has a drag coefficient value of 1.2 which yields to 1.1 drag coefficient difference (White, 2011).

VAWT has two types, lift type and drag type and is dependent on turbine blade configuration. The problem was established that Darrieus type wind turbine is not self-starting even though wind is sufficient enough for conversion to other forms as to prime mover of generator to produce electrical energy. The second statement is another problem for sustaining its rotation as it needs a tip speed ratio of about 2 – 4 for its proper operation (Jin et al., 2015).

It was mentioned by Ragni et al., (2015), in one of their airfoil design write-ups that "airfoil geometry can be characterized by coordinates of upper and lower surface. This was done by Eastman Jacobs in early 1930's to create a family of airfoils known as national advisory committee for aeronautics (NACA) Sections."

This study will also be following NACA parameters in development of air foil design for the interest of general public.

In a horizontal axis wind turbine (HAWT), there exists a design with diffuser to augment vortex which is enhancing the pressure differential between incoming and out flowing air (Elsayed, 2021). In the case of VAWT, it can either be a Savonius type or Darrieus type, there are no cases yet of augmentation such that it is omni-directional. Panel augmentation requires directing the wind turbine to wind direction or else it would be a restraining factor in harnessing energy from the wind.

The number of blades to be convenient with central panel is two, plus central panel totaling to three blades. The test result were configured on how blades will be arranged in relation to central panel setting.

4. Methodology

Based on literature of air foil design, NACA blade profile designation was adopted as thickness and other configuration of airfoil was based on percentage of chord. An airfoil with 24% thickness was then developed and produced to symmetric airfoil (NACA 0024), 4% cambered located at 40% of chord from toe of airfoil (NACA 4424), and asymmetric or heavily aft with 7% cambered located at 40% of chord length from toe (NACA 7424) and with circular chord line having radius of curve equal to the radius of blade from the axis of rotation. The three blades are of the same size at 24% thick based on chord length. A metal sheet of gauge #24 thick was used for ease of fabrication, stiffness and durability.

For a self-starting Darrieus VAWT, a two-bladed wind turbine was designed because the other half of cylindrical revolution has a lower power coefficient as in Savonius type. Since one of airfoil blades is going against wind (wind ward), the other air foil blade is going along wind (leeward). This set-up has positive net torque that makes turbine turn starting it at some considerable wind speed based on starting torque load.

Three types of airfoils were produced and tested. The result revealed that, airfoil with optimum performance was selected as candidate and have been produced with different sizes pertaining to self-starting effect at some fixed value of airfoil attachment location and pitch angle.

Airfoil blade that was selected was subjected to energy optimization test by varying its point-mount location along chord line. In every point-mount location, pitch angle was varied being toe-in to toe-out by 15° in every variation to find pitch angle with most energy utilization ranging from -15° to 15° with 0° being circumferential tangent line of swift area.

Figure 21.1, is the family structure of VAWT was line connected and traced coming up to hybrid lift and drag type SBVAWT. Central panel was introduced to help augment harnessing energy from wind. Just like horizontal axis wind turbine that uses airfoil, the angle of attack is one important factor maximizing lift effect with minimum drag effect. In this case pitch angle is varied and tested its effect in every variation finding maximum effect and fixing it at that value after finding optimum airfoil profile at 24 % thickness. With provision of central panel, the distance of airfoil blade is also a matter of study optimizing its performance.

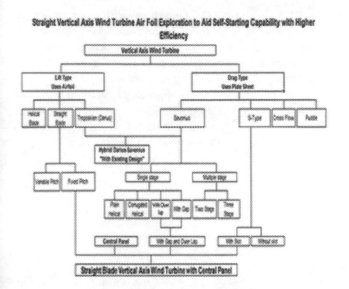

Fig. 21.1 Straight vertical axis wind turbine airfoil exploration

The drawing in Figure 21.2; is the set-up of pair of airfoils in s-form without central panel. The set-up has a big space at the center whereby at times there was a wayward that would let air pass without any effect in rotating wind turbine. The airfoil will not capture energy from air, letting it pass through center or sides

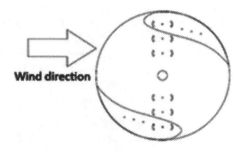

Fig. 21.2 Top view of wind turbine using airfoils without central panel

A central panel as in Figure 21.3 was proposed to be provided making a leeway to capture air passing through center producing torque and start to rotate wind turbine. In this set-up, with provision of central panel allowing bigger volume at entrance and reducing it at a smaller volume at exit. It presses passing wind and produces an imbalance force to the tail part of airfoil producing torque and start turning the wind turbine.

A provision of central panel was developed to be used for the wind turbine model. With the size of airfoil and the capability of fan and duct system to produce air velocity as the source of wind during model testing, a bicycle dynamo was used for the availability of resources because of its small capacity. The blower and duct system with capacity of 100 Watts with

circular discharge diameter of 300 mm can only produce an air velocity of $8\frac{m}{8}$ average. Considering combined efficiency of 25% and an air density of 1.21 $\frac{kg}{m^3}$, using formula:

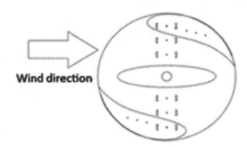

Fig. 21.3 Top view of wind turbine using airfoils with central panel

Wind Power (P_{wind}):

$$P_{wind} = \frac{1}{2}\rho A_s v^3 \tag{1}$$

Where:

ρ – Air density

v – Wind velocity

A_2 – Swift area

Swift area calculation:

$$A_s = L \times W \tag{2}$$

$A = (0.30m) \times (0.30,)$

$A = 0.09m^2$

Therefore:

$$P_{wind} = \frac{1}{2}\left(1.2\frac{kg}{m^3}\right)\left(0.09m^2\right)\left(8\frac{m}{s}\right)^3$$

$P_{wind} = 27.87$ Watts

Wind Turbine Efficiency (η_{wind})

$$\eta_{wind} = \frac{P_{Electrical}}{P_{wind}} \tag{3}$$

Where:

$P_{Electrical}$ - Electrical Power

$P_{Electrical} = (\rho_{wimd}) \times (P_{wind})$

$P_{Electrical} = (0.25) \times (27.87 \text{ watts})$

$P_{Electrical} = 6.96$ watts

In the absence of a dynamometer, torque produced by electric field of dynamo was made to serve as initial load of wind turbine. The following formula prevails,

$$P_{Elect'l.} = 2\pi Tn \qquad (4)$$

Where:

$P_{Elect'l}$ – Power output of the dynamo

T – Electrical magnetic field torque of dynamo

n – angular velocity of the wind turbine in rpm

Torque is dependent of angular velocity produced by wind turbine. Setting the bicycle dynamo as initial load of turbine, the model was developed with provision of mounting and un-mounting airfoil at the same time increasing or decreasing swift area and also with capability of varying pitch angle and airfoil attachment location.

Figure 21.4 is a $\frac{1}{4}$ inch thick metal plate was used as base plate testing panel for stability not to be falling down once hit by wind. The height of four legs that carry the metal panel are designed to house the bicycle dynamo hub and bearing attachment. A cylindrical bearing assembly was designed and aligned in place perpendicular to base metal plate that holds the shaft where the spoke that holds air foils are connected. (See Figure 21.4)

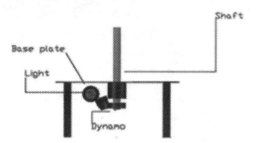

Fig. 21.4 Wind turbine model base plate testing panel assembly front view drawing

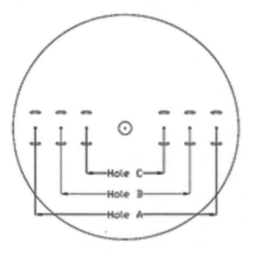

Fig. 21.5 Wind turbine model acrylic panel top view drawing

Figure 21.5, is the CAD drawing of acrylic panel which serves as spoke of wind turbine test model that holds airfoil being tested. Holes A, B and C were provided so as to vary the mount location of airfoils in pairs in reference to central axis in testing the effect of solidity for same airfoil. These holes have circular slots on its upper and lower part which is symmetric in form with 50mm in diameter with its axis same as that of the holes. The slots with size of 3mm form an arc of 30° being 15° and -15° measured along the line perpendicular to spoke line.

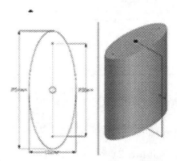

Fig. 21.6 Wind turbine model central panel top and perspective view drawing

Figure 21.6, is the CAD drawing of central panel which is elliptical in form with minor axis length 100 mm and major axis length of 250 mm. The center is provided with 25mm hole for the purpose of mounting it to central shaft that connects to generator as the load. It has also two 3mm diameter hole symmetric to central hole where acrylic panel are intended to be mounted that serves the spoke of airfoils.

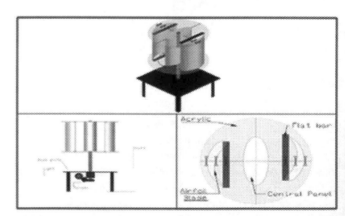

Fig. 21.7 Wind turbine model complete assembly with central panel drawing

Figure 21.7. is the CAD drawing of complete wind turbine model assembly with central panel set-up. It is composed of base panel assembly with tubular bearing assembly and bicycle dynamo as load, central shaft that is connected to tubular bearing assembly and acrylic panel assembly that serves as spoke and hub that holds airfoil in place.

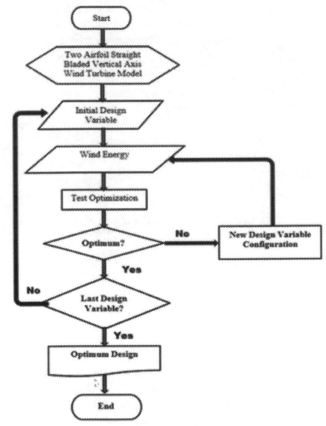

Fig. 21.8 The iteration flow chart finding optimum design parameter

Three varieties of airfoil design have been set for model testing; it has served as input data in MS Excel calculation program. A constant wind speed data was also a factor that was used as input energy in testing wind turbine until self-starting parameters and configuration were identified as output of the study.

The study was all about finding self-starting parametric configuration of SBVAWT and to optimize power coefficient as dependent variable. In doing so, the data were gathered iteratively as in Figure 21.8.

The angular velocity output of test was measured using digital tachometer and used as input in a scattered diagram of power coefficient as ordinate versus airfoil location attachment as abscissa. The scattered diagram was observed to each out of 90 tests subjecting each to a wind speed capacity of blower to find the ideal self-starting configuration for the design using model. The output of iteration test was used as basis for input

data of MS Excel calculation program to produce design parameter values that has an input of power requirement needed in designing SBVAWT of form depending on findings of model testing.

5. Results and Discussions

Testing the effect of central panel to SB-VAWT was done to three airfoils types with same thickness at 24% of chord length (Please see Table 21.1.). The result shows satisfactory performance for SB-VAWT with central panel with all azimuth angle initial condition self-starts at wind velocity fixed to 5.5 m/s to all three airfoils used.

Table 21.1 The effect of central panel to SBVAWT testing data

Effect of Central Panel Testing Data										
Initial Condition:	Wind Turbine Diameter:		0.30m	0.20m		0.20m				
	Wind Turbine Height:		0.30m							
	Initial Load:		50 g							
Profile	Airfoil Size	No. of Blades	Blade Orientation	Central Panel	Wind Vel. Incoming	RPM	Tip Speed	TSR	Solidity	Remarks
Symmetric	250 x 60	2	Circumferential	without	5.5	0	0.00	0.00	12.94	Stops when start
		2	Circumferential	with	5.5	100.4	1.58	0.29	27.70	self-starting
Cambered	250 x 60	2	Circumferential	without	5.5	193.5	3.04	0.55	11.80	Needs starting
		2	Circumferential	with	5.5	134.3	2.11	0.38	26.53	self-starting
Asymmetric	250 x 60	3	Circumferential	without	5.5	148.5	2.33	0.42	19.41	Needs starting
		2	Circumferential	without	5.5	198.4	3.12	0.57	12.94	Needs starting
		2	Circumferential	with	5.5	159.4	2.50	0.46	27.70	self-starting

The test data results of model (Figures 8 and 9.) were made at different five attachment locations varying from 50 mm to 150 mm from toe of airfoil. In every attachment location there were three trials conducted to verify result at that attachment location. With those three trials, average was taken to make result simpler and provide it with graphical presentation to make it easier to analyze. Figures 8 and 9 are the summarized scattered diagram performance result of test based on airfoil used with or without central panel for asymmetric airfoil.

Test performance shows among six set-ups (please see Figure 21.9 and 21.10.) asymmetric airfoil blade with central panel has highest efficiency of 21.36% at airfoil attachment #2 (Refer to Figure 21.11.) toe-out fixed pitch angle. This result shows that central panel help in increasing the performance of wind turbine other than helping wind turbine self-start for a SBVAWT using two-bladed airfoil. It concludes that set parameters can be used in designing prototype wind turbine and be tested in natural wind condition for validation of model in large scale applications. The prototype design should be a two-bladed straight vertical axis wind turbine using asymmetric airfoil with configuration of airfoil attachment #2 with toe-in pitch angle along tangent line of its circumferential attachment perpendicular to radial line.

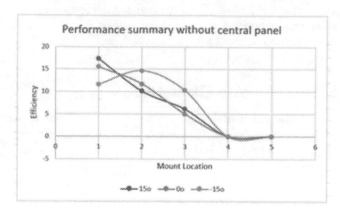

Fig. 21.9 Performance summary chart of asymmetric airfoil without central panel

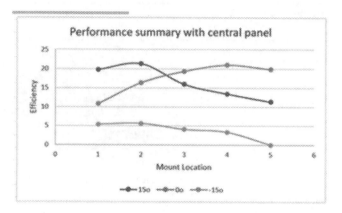

Fig. 21.10 Performance summary chart of asymmetric airfoil with central panel

6. Summary of Findings

This study started from newly combined Savonius and Darrieus wind turbine energy transformation principle, with this, orientation of blades with or without central panel contributed to self-start rotation wind turbine. The researcher conducted several tests to find the highest power efficiency to different pitch angle of airfoil blades with or without central panel.

In Table 21.1., the three airfoil blade designs were tested, testing result shows that set-up with central panel self-start to all three kinds of airfoil blade while set-up without central panel did not self-start.

7. Conclusions

Based on testing conducted on developed model, the following are findings of the study.

Three types of airfoil model were constructed and tested at different variable configuration settings. With series of test of these three types of airfoil profile and configured at varying pitch angle in five different attachment location along chord line, it was found out that central panel help all three thicker airfoils self-start with considerable performance for SBVAWT.

With 15 configurations test set-up per airfoil to with and without central panel, highest achieved efficiency was 21.36% at acrylic attachment A. Airfoil attachment location #2, 15° toe out pitch angle for asymmetric airfoil has highest performance among all model configurations.

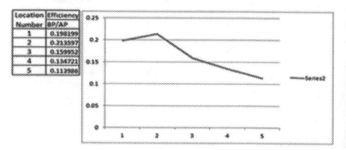

Fig. 21.11 Performance summary graph of asymmetric airfoil blade with central panel wind turbine 15o toe out

This result shows that this type of wind turbine, SBVAWT with inclusion of central panel to its design has potential to solve the self-starting capability of VAWTs without the help of any other external mechanisms. In general, it can be applied in developing a renewable energy transformation from wind resource for power generation when validated by prototyping.

References

1. Daniyan, Daniyan, O. L., Adeodu, Azeez, & Ibekwe, K. S. (2018). Design and Simulation of a Wind Turbine for Electricity Generation. International Journal of Applied Engineering Research, 13(23).
2. Elsayed, A. M. (2021). Design optimization of diffuser augmented wind turbine. In CFD Letters (Vol. 13, Issue 8).
3. Jin, X., Zhao, G., Gao, K., & Ju, W. (2015). Darrieus vertical axis wind turbine: Basic research methods. In Renewable and Sustainable Energy Reviews (Vol. 42, pp. 212–225). Elsevier Ltd.
4. Karimian, S. M. H., & Abdolahifar, A. (2020). Performance investigation of a new Darrieus Vertical Axis Wind Turbine. Energy, 191.
5. Kaygusuz, K. (2001). Renewable energy: Power for a sustainable future. In Energy Exploration and Exploitation (Vol. 19, Issue 6).

6. Ragni, D., Ferreira, C. S., & Correale, G. (2015). Experimental investigation of an optimized airfoil for vertical-axis wind turbines. Wind Energy, 18(9).

7. Siddiqui, M. S., & Hasan, S. M. (2015). Optimized design of a straight blade urban roof top vertical axis wind turbine. 2014 International Conference on Energy Systems and Policies, ICESP 2014.

8. White, F. M. (2011). Fluid Mechanics seventh edition by Frank M. Whitc. In Powcr.

9. Zafar, U., & Staubach, M. S. P. (2018). Literature review of wind turbines. Researchgate.Net.

10. Zhu, J., Huang, H., & Shen, H. (2015). Self-starting aerodynamics analysis of vertical axis wind turbine. Advances in Mechanical Engineering, 7(12).

Note: All the figures and table in this chapter were made by the authors.

Advancing Sustainable Science and Technology for a Resilient Future – Sai Kiran Oruganti et al. (eds)
© 2024 Taylor & Francis Group, London, ISBN 978-1-032-79020-6

The Applications of Artificial Intelligence in Relation to the Marketing Strategy of Fast Food Businesses

22

Airene Resty P. Aprosta*

Rizal Technological University,
Manila, Philippines

Abstract: The research aimed to determine the potential of Artificial Intelligence in the fast-food businesses. The study employed a descriptive design wherein a survey was conducted using the modified SERVQUAL model questionnaire. The said questionnaire then underwent reliability and validity tests with the help of statisticians to gather data from the respondents. Convenience sampling was used in choosing the respondents for the survey and they are composed of 71 employees and 403 customers engaged in QSRs within the selected barangays of Taguig City. The statistical formulas utilized to treat the data included the use of frequency, Percentage age, rank, weighted mean, Pearson Correlation Coefficient (r), and one-way ANOVA. It is hoped that this study would assist other businesses in becoming more knowledgeable and how to best use their resources to keep their operations running using AI applications for them to improve the services that they will offer.

Keywords: Artificial intelligence (AI), Self-service kiosks (SSKs)/self-service technologies (SSTs)

1. Introduction

For more than 250 years, technological advancements have been the primary drivers of economic growth. Key innovations include the steam engine, electricity, and internal combustion engines. Now, Artificial Intelligence and machine learning have emerged as the most promising areas of general-purpose technology, with the potential to inspire further discoveries and revolutionize the economy (Brynjolfsson & McAfee, 2017).

Self-ordering kiosks (SOKs) are a type of self-service technology (SST) that are replacing traditional interactions between service providers and customers to improve service quality and customer experience. SSTs are being adopted by businesses to facilitate efficient and effective service delivery. As a result, it is important to understand the role of kiosks in various marketing initiatives. Quick service restaurants like McDonald's utilize AI-driven voice recognition ordering systems, employing synthetic human voices and natural language processing algorithms to take orders. Artificial intelligence encompasses various capabilities such as voice and image recognition, decision making, machine learning techniques, and semantic search. The use of AI significantly improves operational efficiency, particularly in food production and delivery management. Machine learning, a subset of AI, involves a program or specialized machine with learning capabilities that analyze input data and provide relevant suggestions to help predict and maintain inventory (Gondaliya, S. H., & Sharma, A. K., 2023). Natural Language Processing (NLP) is an emerging field under Artificial Intelligence. AI-powered platforms utilize NLP and machine learning algorithms to process and validate orders quickly, reducing human error and turnaround time. This approach increases efficiency, saves time and money, and ultimately leads to higher customer satisfaction and loyalty. Additionally, AI can identify fraudulent orders, providing fraud alerts to prevent or detect scams.

2. Literature Review

Artificial Intelligence

Artificial intelligence (AI) has been around for a while, with researchers developing computer systems that can

[1]arpaprosta@rtu.edu.ph.

DOI: 10.1201/9781003490210-22

mimic human actions. The food industry implements these technologies to enhance profits and elevate customer service levels professionally, ensuring widespread appeal (Agbai, 2020). The article "Benefits of Artificial Intelligence in the Restaurant Industry" by Deputy Team (2019) explains that AI is a field of computer science that creates intelligent machines. AI is an essential part of the tech industry, emulating cognitive skills like learning and problem-solving.

Applications of Artificial Intelligence in the Fast-Food Industry

Quick service restaurants often employ self-ordering kiosks alongside cashiers. These kiosks increase efficiency during busy periods and reduce the workload of cashiers, cutting labor costs. Additionally, the machines are more accurate than human workers and prompt customers to place larger orders.

Self-Service Technologies and Self-Service Kiosks in Quick Service Restaurant

Self-service technologies (SSTs) are increasingly popular due to technological advancements and businesses' desire to lower costs or increase convenience. Some customers find SSTs faster, cheaper, and easier. In the restaurant industry, modern technology adoption is critical for brands to meet customer needs and remain competitive.

AI-based chatbots in customer service

AI-based chatbot systems are gaining popularity in e-commerce settings for real-time customer service. These systems communicate with customers through live chat interfaces or conversational software agents.

AI-enabled personalization

Digital platforms offer customers greater convenience and personalization options. AI-powered personalization aims to meet implicit customer demands and increase ROI by delivering relevant information through the appropriate channel at the right time (Desai, 2022).

Artificial intelligence and machine learning in food industries

Machine learning enhances traffic control, streamlines orders, optimizes inventory, and elevates customer satisfaction in the food industry. Efficient and precise food deliveries without delays or time-consuming communication are ensured by its application in food delivery. (Khan, 2022). Furthermore, the article "Supervised learning for arrival time estimations in restaurant meal delivery" by Hildebrandt, F. D., & Ulmer, M. W., 2022) precise delivery estimates are crucial for customer satisfaction. Inaccurate delivery times can result in dissatisfaction. Accurate estimates enhance service perception, guide customer selections, and improve delivery speed, ensuring fresh food.

3. Research Method

The study used descriptive research methodology to assess customer behavior and satisfaction in utilizing service technology applications of selected fast-food businesses in Taguig City. For the data gathering, the researcher employed a survey questionnaire.

The researchers utilized a correlation research design to investigate the potential relationship between customer satisfaction and behavior on Self-Service Technology (SST) after collecting and refining the data. As such, a correlation design was deemed appropriate.

The instrument is composed of five (5) parts:

Part 1: contains the respondents' profiles: age, sex, civil status, highest educational attainment, monthly income, and previous visits to the restaurant.

Part 2: contains the respondents' profiles: age, sex, civil status, highest educational attainment, monthly income, years in service, and job position level.

Part 3: contains questions with respect to customer satisfaction with self-servicing kiosks/ self-service technology (SST) in quick- service restaurants.

Part 4: Customers Behavior of Self-Service Kiosk (SSKs) / Self-Service Technology (SSTs) Applications in Quick Service Restaurants.

Part 5: Is there a significant relationship between the respondents' assessment on customer satisfaction and customer behavior of the applications of self-servicing kiosks (SSKs)/self-service technology (SSTs) in quick service restaurant?

Customer's Profile

The profile of the 403 customer respondents in terms of the data gathered can be seen below:

The ages of the customers are 18 to 25 (78.91%), 26 – 35 (14.39%), 36-55 (5.71%), and 56 – 65 (0.99%). They are identified as Male (53.85%) and Female (46.15%). Their status is single (93.30%) and Married (6.70%). Their highest educational attainment was identified as Bachelor's Degree (41.69%), Master's Degree (0.99%), Doctorate Degree (0.25%), and Vocational or Special Courses (Others) (57.07%). The monthly income ranges to Less than Php 15,000 (76.92%), Between Php 15,001 – Php 20,000 (10.92%), Between Php 20,001 – Php 30,000 (4.47%), and Php 30,001 and above (7.69%). While the Previous Visit to a Restaurant result to Never (1.74%), Everyday (20.60%), 1-2 weekly (38.46%),1-2 times per 2 weeks (27.79%), 1-2 times a month (10.67), and 1-2 times a year (0.74%).

The results of the customer's profile show interest in self-service kiosks is increasing among all age groups, with

18-24-year-olds being the most likely to visit restaurants with kiosk options. Based on the data from the National Health and Nutrition Examination Survey (2018) published by the Centers for Disease Control and Prevention, men tend to eat fast food for lunch, while women consume it as a snack. Single individuals are less sensitive to global products at fast-food restaurants. In terms of educational attainment, millions of Filipinos graduated from technical and vocational programs in the last six years. The poverty rate among Filipinos is 23.7%. On average, quick-service restaurants are consumed once a week in the Philippines.

Employees Profile

The profile of the 71 employee respondents in terms of the data gathered can be seen below:

The ages of the employees are 18 to 25 (77.46%), 26 - 35 (19.72%) and, 36 – 55 (2.82%). They are identified as Male (40.85%) and Female (59.15%). Their status is Single (95.77%) and Married (4.23%). Their highest educational attainment Bachelor's Degree (45.07%) and Vocational or Special Courses (54.93%). The monthly income ranges to Less than Php 15,000 (76.06%), Between Php 15,001 – Php 20,000 (14.08%), Between Php 20,001 – Php 30,000 (5.63%) and Php 30,001 and above (4.23%). The years in service of the employees are Less than 1 year (39.44%), 2 years to less than 4 years (46.48%), 4 years to less than 6 years (8.45%), 7 years to less than 9 years (4.23%) and 10 years and up (1.41%). While the Job Position Level was identified as Managerial (22.54%), Supervisory (7.04%), Internal Manager Training (46.48%) and Rank and File (23.94%).

The results of the employee profile are young and willing to work hard. According to the Philippine Statistics Authority's NCR Gender Factsheet, women make up a large percentage of the workforce, and many employees are single. According to the Bureau of Labor Statistics, educational attainment has a significant impact on position, with college graduates more likely to be in management. The monthly income for fast-food chain employees is less than Php15,000.00, while area managers earn over Php30,000.00. The majority of employees who completed the survey are in Internal Manager Training.

How do the respondents assess the customer's satisfaction with the self-service kiosks (SSKs)/self-service technology (SSTs) applications in quick service restaurant?

Table 22.1 shows the respondents' assessment in terms of Customer Satisfaction, where prompt and accurate service leads to increased customer satisfaction and repeat business. Kiosks can be customized to facilitate orders for those with accessibility challenges, including hearing or language barriers. The implementation of Self Ordering Machines in fast-food restaurants enhances customer satisfaction by providing easy-to-use and quality service. Generally,

Table 22.1 Customer's satisfactions with the self-Service Kiosks (SSKs)/self-service technology (SSTs) applications in quick service restaurants

Satisfaction	Weighted Mean	Verbal Interpretation
1. The kiosk and SST provide a satisfying ordering experience.	4.52	Very Satisfied
2. The kiosk and SST can provide what customers want.	4.51	Very Satisfied
3. The kiosk and SST can avoid ordering errors.	4.33	Very Satisfied
4. The kiosk and SST provide complete information, such as meal choices and prices.	4.59	Very Satisfied
5. The kiosk and SST provide clear images of the different menu items.	4.52	Very Satisfied
6. The kiosk and SST can provide good quality service.	4.50	Very Satisfied
7. The kiosk gives freedom to customize the menu according to customer's taste.	4.44	Very Satisfied
8. The kiosk allows the customer to browse the menu conveniently.	4.53	Very Satisfied
9. The kiosk product quality offered by the quick service restaurant meets my expectations.	4.45	Very Satisfied
10. The kiosk provides faster service time.	4.42	Very Satisfied
11. The kiosk is very effective.	4.46	Very Satisfied
12. The kiosk can give customers a happy feeling.	4.47	Very Satisfied
13. The kiosk can give customers a comfortable feeling.	4.46	Very Satisfied
14. The kiosk layout makes it easy for customers to find what they need.	4.49	Very Satisfied
15. Using self-service kiosks and SST is a pleasurable experience.	4.58	Very Satisfied
Total	4.48	Very Satisfied

customers are satisfied with the Self Ordering Machine's performance.

How do the respondents assess the customer's behavior of self-servicing kiosks (SSKs)/ self-service technology (SSTs) applications in quick service restaurants?

Table 22.2 shows the respondents' assessment in terms of Customer Behavior, customers regard the ability to customize orders as a significant benefit. The touchscreen technology is easy to use, which makes placing orders seamless, even for first-time users. Over three-quarters of consumers believe

Table 22.2 Customers behavior of self-service kiosk (SSKs) /self-service technology (SSTs) applications in quick service restaurants

Behavior	Weighted Mean	Verbal Interpretation
1. A kiosk takes orders without social judgment from a waitress or cashier.	4.46	Very Satisfied
2. The kiosk reduces waiting time and frees up sales staff for other tasks or personalized customer service.	4.45	Very Satisfied
3. The kiosk is a line-busting solution that gives customers full control over their purchases and payments.	4.43	Very Satisfied
4. The kiosk avoids the social friction of clients fearing being misunderstood or appearing unsophisticated in front of clerks.	4.55	Very Satisfied
5. The kiosk changes the way customer think and act.	4.38	Very Satisfied
6. The kiosk and SST let clients choose without judgment, making them feel more comfortable.	4.54	Very Satisfied
7. The kiosk and SST allow customers to place orders more easily.	4.54	Very Satisfied
8. The kiosk and SST allow customers to check their own orders for accuracy.	4.55	Very Satisfied
9. The kiosk and SST are convenient and easy to use, so even first-time customers can place orders smoothly.	4.44	Very Satisfied
10. The kiosk and SST allow customers to place orders without assistance.	4.44	Very Satisfied
Total	4.48	Very Satisfied

technology improves convenience, and 70% say it speeds up service and enhances order accuracy (Hoshmand, 2018).

Is there a significant relationship between the respondents' assessment on customer satisfaction and customer behavior of the applications of self-servicing kiosks (SSKs)/ self-service technology (SSTs) in quick service restaurants?

Table 22.3 depicts the significant correlation between satisfaction and behavior has a strong relationship based on the Pearson r value of .799. The result signifies that there is evidence of a relationship between satisfaction and behavior.

Table 22.3 The significant relationship between the respondent's assessments in customer satisfactions and customer behavior of the application of self-service kiosk (SSKs)/self-service technology (SSTs) in quick service restaurants

Aspect		r-value	p-value	Decision	Remarks
Satisfaction	Behavior	0.799	<0.001	Reject Ho	Significant Strong Correlation

Based on the study entitled Consumers' Buying Behavior and Consumer Satisfaction in the Beverage Industry in Taiwan, found a positive correlation between consumer buying behavior and satisfaction in the beverage industry. The correlation analysis showed a statistically significant value of 0.788 (p<0.01), supporting the hypothesis that there is a significant relationship between buying behavior and satisfaction. (Shih, S. et al., 2015).

Table 22.4

Scale Measurement	Scale	Verbal Explanation
4.21 - 5.00	5	Very satisfied
3.41 – 4.20	4	Satisfied
2.61 – 3.40	3	Neither satisfied nor dissatisfied
1.81 – 2.60	2	Dissatisfied
1.00 – 1.80	1	Very dissatisfied

The Researcher used the Likert Scale which all the questions were answerable by a five-point option in ranging respondents' responses.

4. Conclusion

Artificial Intelligence enhances digital capabilities to elevate customer experience and satisfaction in quick-service restaurants. AI can also make it easier for fast-food companies to integrate digital personalization and recommendation technology into their services to increase customer satisfaction and become more analytical. The software acquisition aims to enhance customer experience and provide convenient delivery options.

5. Recommendations and Future Research Directions

1. The fast-food chain offers personalized touchscreens to meet consumer needs. Options include a Get Help button for kiosk issues, an Ask for Assistance button to get help from a nearby employee, and a How it Works option to help customers use the kiosk.

2. The Researcher suggests using AI-Assisted Self-Service Kiosks or Computer Assisted Machines for customers to easily select and customize their orders.

References

1. Agbai, Chidinma Mary. (2020). Application of artificial intelligence (AI) in the food industry. GSC Biological and Pharmaceutical Sciences. 13. 171-178. 10.30574/gscbps.2020.13.1.0320

2. Brynjolfsson, E., & Mcafee, A. (2017). The business of artificial intelligence. Harvard Business Review

3. Deputy Team. (2019, August 21). Benefits of Artificial Intelligence in the Restaurant Industry. https://www.deputy.com/blog/benefits-of-artificial-intelligence-in-the-restaurant-industry

4. Desai, D. (2022). Hyper-personalization: an AI-enabled personalization for customer-centric marketing. In Adoption and Implementation of AI in Customer Relationship Management (pp. 40-53). IGI Global.

5. Gondaliya, S. H., & Sharma, A. K. (2023, May). A Review: Artificial Intelligence in Restaurant Business. In International Conference on Applications of Machine Intelligence and Data Analytics (ICAMIDA 2022) (pp. 397-402). Atlantis Press.

6. Hildebrandt, F. D., & Ulmer, M. W. (2022). Supervised learning for arrival time estimations in restaurant meal delivery. Transportation Science, 56(4), 1058-1084.

7. Khan, R. (2022). Artificial intelligence and machine learning in food industries: A study. J Food Chem Nanotechnol, 7(3), 60-67.

8. Shih, S. P., Yu, S., & Tseng, H. C. (2015). The Study of Consumers' Buying Behavior and Consumer Satisfaction in Beverages Industry in Tainan, Taiwan. Journal of Economics, Business and Management, 3(3), 391–394.

9. Hoshmand, S. (2018, May 14). Here's Why Consumers are Demanding Self-Service Kiosks.

Note: All the tables in this chapter were made by the authors.

Advancing Sustainable Science and Technology for a Resilient Future – Sai Kiran Oruganti et al. (eds)
© 2024 Taylor & Francis Group, London, ISBN 978-1-032-79020-6

Vehicle Rustproofing Practices of Automotive Shops in Municipalities of Surigao Del Norte, Philippines

23

Reymond P. Piedad*

Surigao Del Norte State University, Philippines

Abstract: Corrosion can jeopardize the vehicle's structural integrity however rust-proofing a vehicle can increase the amount of time it can be used and ensure the dependability and functionality of the equipment. Ecological conditions might be taken into consideration to ensure appropriate care and maintenance practices for vehicles. The study utilized a descriptive-survey research design to evaluate the best practices among sixteen (16) automotive shop owners in the province for the services provided to their clients in terms of rust-proofing for car undercarriage. The results revealed a universal application among respondents that very much applied to the fundamental steps of rust proofing. The use of fresh tap water for cleaning, using sandpaper, a steel brush, and portable grinder for rust removal, and used primer paint to protect the metal underneath from various constituents present in air and impurities. The results show that using rubberized undercoating appears to be very common and recommended in terms of the ecological factors in our area due to salty moisture.

Keywords: Automotive, Car undercarriage, Rubberized undercoating, Rust proofing

1. Introduction

One major problem for motorists all over the world is rust. However, the issue has been greatly lessened by contemporary vehicle design and the extensive use of rust-resistant materials and treatments (Roberge, Pierre R. 2019). (Zhang, 2014) All vehicle companies offer rust-proofing warranties, however, their services are only available for a short period of time. Atmospheric gases and ecological variables such as moisture, active ionic substance, and oxidizing agents, airborne salt concentrations contribute to the degradation of metal components in vehicles (Wang et al. 2023). Major problems of corrosion of steel, including metallic parts of vehicles, occur in any place, whether in first or third-world countries, under the influence of the marine climate (Seechurn et al 2022). The province of Surigao del Norte is located at the tip-most parts of Mindanao Island's neighboring marine environment. In addition, the province has the two known Siargao Islands and Bucas Grande, which rest in the Philippine Sea. Based on many research findings, rust and corrosion in the chassis of all automobiles have contributed by the ocean's chemical-physical counterparts.

To prevent rust and corrosion, rustproofing of a car is an essential task to ensure that its chassis is protected during the rainy season with all the salt from the ocean and other harmful elements on the road.

2. Method and Materials

Research Approach

The study employed descriptive research design using survey method and homogenous sampling technique. The design was suitable for this study to describe the assessment of rustproofing practices among automotive shop owners in Surigao del Norte.

Respondent

The respondents of the study were thirteen (13) automotive shop owners in Surigao City, particularly in Brgy. Taft, Brgy. Washington, Brgy. San Juan, Brgy. Luna, Brgy. Rizal, Brgy. Quezon, Brgy. Sukailang, Brgy. Poctoy, Brgy. Lipata and three (3) others automotive shop owners from the Municipality of Dapa, Siargao Island, Surigao del Norte.

*Corresponding author: nreymondpiedad1986@gmail.com

DOI: 10.1201/9781003490210-23

Fig. 23.1 The location of the study area

Research Environment

The study was conducted around automotive shops in Surigao del Norte. The location map is shown in Plate 1.

Research Instruments

The study employed a researcher-made questionnaire that encompassed the profile of the respondents and the vehicle rustproofing practices based on the general procedure in automotive shops in Surigao del Norte. The indicators considered four general steps as follows (1) vehicle cleaning such as the use of detergent bar or powder, shampoo, liquid SOSA, and use of water. (2) rust removals such as the uses of sandpaper, steel brush, portable grinder, perform cutting and repair, sandblasting, and use of pressurized water. (3) chemical applications such as rubberized undercoating, polyurethane undercoating, wax/paraffin-based undercoating, asphalt-based undercoating, cosmoline, oil-based rustproofing, and far harsher chemicals. (4) finishing touches such as primer paint, enamel, acrylic, urethane, water-based paints, and Bondo esque.

Validation. The questionnaire was checked by the experts in Research and Technology for corrections and revisions. Suggestions were carried and the questionnaire was revised and finalized. One (1) validator holds a Master of Arts in English Language degree teaching, with expertise in descriptive-qualitative research. Another, a Master of Arts in Mathematics Education graduate, validator, with expertise in quantitative research.

Data Analysis

The data retrieved from the researcher-made questionnaire was analyzed using descriptive and inferential statistics. The frequency count and percentages were used to describe the profile of the respondents. The mean and standard deviation were used to describe the extent of the rustproofing practices, and analysis of variance (ANOVA) for the determination of significant differences.

3. Results and Findings

Profile of the Respondents

This section showed the profile of the participants. A total of sixteen (16) private firms that practiced vehicle rustproofing in automotive shops in Surigao del Norte responded the survey. Their shop status was classified as a practitioner, rented, or owned. Most of the participants owned the business, accounted 62.5% (10 out of 16), rented the space of their shops comprising 25% (4 out of 16), and practitioners without shops of their own comprise 12.5% (2 out of 16) respondents. Further, Table 23.1 presented the years of shop operation of the respondents. As displayed in the table, most of the participants started their operation 5 years and above, representing a total of 62% (10 out of 16), for 3-4 years shop operations comprise 19% (3 out of 16), and for 1-2 years operation represents 19% (3 out of 16) respondents.

Table 23.1 Frequency and percentage years of shop operation of the respondents.

Years of Shop Operation	Frequency (%)
1-2 years	3 (19%)
3-4 years	3 (19%)
5 years and above	10 (62%)
Total	**16 (100%)**

Extent of Rustproofing Practices in Automotive Shops in Surigao del Norte

The level of cleaning practices of rustproofing was presented in Table 23.2. The result shows that the cleaning practices of rustproofing among automotive shops in Surigao del Norte primarily focused on the use of detergent bar soap, detergent powder soap, shampoo, liquid SOSA, and water. Besides, the fifth indicator was very much practice followed by 2nd indicator using powder detergent. Water was used for the initial cleaning in the shops and oftentimes with powder detergent as confirmed by most practitioners/respondents:

"Basically, using water to remove dust, sometimes we add powder detergent as necessary to easily remove some coagulated soil particles"

These can be attributed to the fact that water plays an integral role in carwash results and system performance (Sun et al, 2023). (Seechurn et. Al, 2022) also added when using water could minimize the effort requirements while removing stains and producing top-notch final wash results.

In addition, indicator 1 was less practiced. The use of detergent bar soap was less practiced in shops for it was no need to use common home cleaners on the paint like hand dish soap for doing could remove the protective wax on car paint. (Consumer Report, 2017). However, the use of shampoo and

Table 23.2 Cleaning practices of rustproofing among automotive shops

Indicators	Mean ± SD	Qualitative Description
1. Usage of Detergent Bar Soap	2.06 ± 0.93	Less Practiced
2. Usage of Detergent Powder Soap	2.94 ± 1.34	Much Practiced
3. Usage of Shampoo	1.56 ± 0.96	Less Practiced
4. Usage of Liquid SOSA	1.25 ± 0.77	Least Practiced
5. Usage of Water	3.88 ± 0.50	Very Much Practiced

Legend: [*(1.00–1.49, Not Practiced); (1.50–2.49, Less Practiced); (2.50–3.49, Much Practiced); (3.50–4.00, Very Much Practiced)]*

liquid SOSA was least practiced in the shops in Surigao del Norte. As some respondents stated:

"Shampoo and Liquid SOSA were not used in automotive shops for they are good only when the paint of the vehicle's body is already done."

Furthermore, as seen in Table 23.3, rust removal practices among automotive shops includes the use of sandpaper, steel brush, portable grinder, cutting and repair rusted part, sandblasting, and pressurized water. However, indicators 1, 2, and 3 were very much practice. The use of sandpaper, steel brush, and portable grinder were utilized for rust removal in the vehicle parts in the shops. While indicator 6 was much practiced, the use of pressurized water. As the respondents attested:

"That depended on the rust type, however it's a kind of procedure that followed right after the other "

The other indicators were found to be the less or least practiced respectively. This pointed out that for mild removal of soft paint coatings or coatings that are not securely adhered to the

Table 23.3 Rust Removal Practices among Automotive Shops

Indicators	Mean ± SD	Qualitative Description
1. Use of Sand Paper	3.44 ± 1.03	Very Much Practiced
2. Use of Steel Brush	3.88 ± 0.34	Very Much Practiced
3. Use of Portable Grinder	3.56 ± 1.03	Very Much Practiced
4. Cutting & repair rusted part	2.06 ± 1.44	Less Practiced
5. Sand Blasting	1.06 ± 0.25	Least Practiced
6. Use of Pressurized water	2.50 ± 1.46	Much Practiced

Legend: [*(1.00 – 1.49, Not Practiced); (1.50 – 2.49, Less Practiced); (2.50 – 3.49, Much Practiced); 3.50 – 4.00, Very Much Practiced)]*

Table 23.4 Chemical application practices among automotive shops.

Indicators	Mean	Qualitative Description
1. Rubberized Undercoating	3.75 ± 0.77	Very much Practiced
2. Polyurethane Undercoating	2.38 ± 1.36	Less Practiced
3. Wax / Paraffin-Based Undercoating	2.13 ± 1.41	Less Practiced
4. Asphalt-Based Undercoating	1.63 ± 0.96	Less Practiced
5. Cosmoline	1.13 ± 0.34	Least Practiced
6. Oil Based Rust Proofing	1.44 ± 1.03	Least Practiced
7. Far Harsher Chemicals	1.06 ± 0.25	Least Practiced

Legend: [*(1.00 – 1.49, Not Practiced); (1.50 – 2.49, Less Practiced); (2.50 – 3.49, Much Practiced);(3.50 – 4.00, Very Much Practiced)]*

metal, such as flaky paint, scaly rust, or even weld spatter, steel or wire brushes and portable grinder are best. Also, the use of sandpaper was found to be very much practiced by the participants. It is believed as such because of its tougher abrasive performance and is the best for rust removal. It also does not deteriorate when the surface is wet.

Furthermore, Table 23.4 showed that the application of chemical practices among automotive shops in Surigao del Norte primarily on the application of harsher chemicals, such as rubberized undercoating, polyurethane undercoating, wax or paraffin-based undercoating, asphalt-based undercoating, Cosmo line and oil-based rustproofing. Among the seven (7) materials, the use of rubberized undercoating was very much practiced for the initial application of chemical practices in the shops. This is because the finest protective coating for a car's undercarriage is typically rubberized undercoating because it is simple to apply. It protects surfaces from rust, dampness, dents, and dings. The use of rubberized undercoating is secure for wheel wells and quarter panels, and it dries with a supple rubbery feel (Zhang et. Al, 2014).

Table 23.5 showed that the finishing touches practices among automotive shops in Surigao del Norte primarily on the use of primer paint, using enamel, using acrylic, using urethane, water-based paint, and bondo esque. Among the six (6) materials, the use of primer paint was very much practiced for the initial finishing touch practices in the shops. Prior to painting, materials are coated with a primer or undercoat. A higher surface adherence, increased paint endurance, and added protection for the material being painted are all benefits of priming. It was mentioned that whether painting fresh drywall, aged wood, bare metal, previously painted brick, or any other surface, a coat of primer is advised. Primers are essentially flat, sticky paint that is intended to cling well and serve as a stable foundation for subsequent layers of paint.

Table 23.5 Finishing touches practices among automotive shops

Indicators	Mean ± SD	Qualitative Description
1. Primer Paint	2.69 ± 1.54	Much Practiced
2. Using of Enamel	2.25 ± 1.39	Less Practiced
3. Using of Acrylic	2.06 ± 1.34	Less Practiced
4. Using of Urethane	2.50 ± 1.55	Much Practiced
5. Water-based Paints	1.19 ± 0.75	Least Practiced
6. Bondo Esque	1.00 ± 0.00	Not Practiced

Legend: [(1.00 – 1.49, Not Practiced); (1.50 – 2.49, Less Practiced); (2.50 – 3.49, Much Practiced); (3.50 – 4.00, Very Much Practiced)]

Differences on the Practices employed in Rustproofing among Automotive Shops when grouped according to Profile Variables were presented on this section.

Table 23.6 ANOVA results in rustproofing practices among automotive shops when grouped according to shop status

Practices	F	p-value	Interpretation
Cleaning	0.28	0.76	Not Significant
Rust Removal	1.98	0.19	Not Significant
Chemical Applications	1.08	0.37	Not Significant
Finishing Touches	0.82	0.46	Not Significant

Significant @ alpha 0.05

The ANOVA results between shop status were presented in Table 23.6. The result revealed that the rustproofing practices employed by participants grouped according to shop status were not statistically different ($p > 0.05$) from each other. This means that regardless of the status of the shop; owned, rented, or the simple practitioner they have similar practices or rendered services towards cleaning, rust removal, chemical application, and finishing touches.

Table 23.7 ANOVA results in rustproofing practices among automotive shops grouped according to the types of vehicle served.

Practices	F	p-value	Interpretation
1. Cleaning	1.73	0.22	Not Significant
2. Rust Removal	0.49	0.80	Not Significant
3. Chemical Applications	0.76	0.62	Not Significant
4. Finishing Touches	3.67	0.04	Significant

Significance @ alpha 0.05

Moreover, Table 23.7 presented the ANOVA results among the automotive shops according to the types of vehicles served in the rustproofing practices

The table showed that the types of vehicles served were not statistically different ($p > 0.05$) in cleaning, rust removal, and chemical application practices. Besides, regardless of the types of vehicles served, the rustproofing practices on cleaning, rust removal, and chemical application were similar. In addition, the practices on finishing touches differ significantly ($p < 0.05$) in chemical application, for example, employing a rubberized undercoating is believed important depending on the car type, uses, and purposes. This is so because the best protective coating is typically provided for the car's underside. In addition, it protects surfaces from rust, moisture, dents, and dings. (Wang et al. 2023).

4. Conclusion

The rustproofing practices in Surigao Del Norte can be broadly divided into four steps: cleaning the vehicle, removing the rust, applying chemicals, and applying finishing touches. For cleaning, water, and detergent powder soap are very much practice. For removing rust, a steel brush, a portable grinder, and sandpaper are commonly used. Further, rubberized undercoating for chemical applications; and primer paint for finishing touches are mainly practiced. Thus, regardless of the vehicle rustproofing practices in automotive shops in Surigao Del Norte, rubberized undercoating is highly recommended for rustproofing used vehicles, and for new vehicles, it is also advised before utilization.

References:

1. Roberge, Pierre R. (2019). Handbook of Corrosion Engineering. 3rd ed. New York: McGraw-Hill Education. https://www.accessengineeringlibrary.com/content/book/9781260116977
2. Zhang, D., Bowden, R. L., Yu, J., Carver, B. F., & Bai, G. (2014). "Association Analysis of Stem Rust Resistance in U.S. Winter Wheat." PLOS ONE, 9(7), e103747. https://doi.org/10.1371/journal.pone.0103747
3. Wang, Z., Liu, J., Wu, L., Han, R., & Sun, Y. (2013). Study of the corrosion behavior of weathering steels in atmospheric environments. Corrosion Science, 67, 1-10. https://doi.org/10.1016/j.corsci.2012.09.020
4. Seechurn, Y., Surnam, B.Y.R., Wharton, J.A.]Marine (2022). Atmospheric corrosion of carbon steel in the tropical microclimate of Port Louis. Materials and Corrosions. https://doi.org/10.1002/maco.202112871

Note: All the figure and tables in this chapter were made by the authors.

Advancing Sustainable Science and Technology for a Resilient Future – Sai Kiran Oruganti et al. (eds)
© 2024 Taylor & Francis Group, London, ISBN 978-1-032-79020-6

Association of Financial Literacy Among Filipino Gen Z College Students: A Structural Equation Modeling Approach

24

Renato E. Apa-ap*

Research Management Cluster Coordinator, Research Management Office-College of Science, Polytechnic University of the Philippines.

Mecmack A. Nartea[1]

Chief, Research and Extension, College of Business Administration- Graduate Studies, Polytechnic University of the Philippines.

Abstract: The primary intention of this article is to explore the association of financial attitude, financial behaviour and financial knowledge towards financial literacy of selected Gen Z college students in Metro Manila, Philippines. The Researchers utilized the data gathered through a questionnaire via an online platform from randomly selected of 439 respondents that satisfies our requirements. The methodology applied to this study was quantitative approach using a survey questionnaire and data gathered were analyzed using Structural Equation Modeling (SEM) with the support of a statistical Package for precision of calculations. Our findings shows that there was an association among financial attitude, financial behavior, and financial knowledge on financial literacy bearing a coefficient of -0.08, 0.98 and 1.69 respectively and statistically significance as manifested on the computed p-values (0.035, 0.000, 0.003; $p < 0.05$). The findings of this research can be a basis to bring in more awareness to Filipino people regarding financial literacy specially on Gen Z College Students. Thus, we recommend that future researchers may investigate the possibility whether demographic profile of the respondents may perhaps affect the three factors of financial literacy and dig up more variables that may influence more on financial literacy.

Keywords: Financial attitude, Financial behavior, Financial knowledge, Financial literacy, Gen Z, Structural equation model

1. Introduction

Understanding your financial capabilities is extremely necessary for you to make financially responsible decisions. Agreeing to the National Economic and Development Authority (NEDA), the Filipino conviction upon acknowledgement of wages, as generally known, is that upon receiving of earnings, expending take place in before conserving. What remains is saved. If there's not a bit left, then, there's nothing more to save. Established on the 2019 Financial Inclusion Survey overseen by the Bangkok Sentral ng Pilipinas, 41% of the adults got one correct answer only out of three financial literacy questions, 27% got two correct answers, and only 8% got all three questions while 24% obtained a zero score. This paper replicates the work of Kamini Rai, Shikha Du and Miklesh Yadav on Financial Literacy with different respondents.

Being financially ignorant becomes challenging for people residing in poverty to shift their financial conditions, which can lead to different factors that affect financial decisions. First, capability to gain more understanding by reflecting on the crucial consequences of financial judgments. For example, if there is an opportunity to either pay rent or feed the household, then an individual's financial judgement may be conflicted.

It is not a "absence of financial literacy." that impacts decisions. As a substitute, low-income families have a monopoly over their financial matters. Second, application of a good attitude in financial management. Such, having a decent financial attitude is very essential for overseeing the firmness and improvement of family's financial with an intent to acquire financial satisfaction.

Lack of budgeting and wasteful lifestyle and consumption patterns causes serious problems. A lot of students face this problem (Wardani, Susilaningsih & Sangka, 2017). This

*Corresponding author: r_apaap@yahoo.com, reapa-ap@pup.edu.ph
[1]mackharvester@gmail.com

DOI: 10.1201/9781003490210-24

poses a serious threat since students must learn the importance of financial decisions as young as possible. It's crucial that students understand the importance of financial literacy because it's lifesaving. The obligation of financial learning is envisioned because of consumers' absence of understanding and predisposition to making incorrect decisions (Kadoya & Rahim Khan, 2020).

Financial attitude implies to ideas and ideals linked to different notions of private finance that can form the financial behavior in making decisions (Priyadharsini,2017). Also, Mendes-Da-Silva (2016) states that financial attitude leads to monetary comportment. Thus, having a positive financial attitude will end up having positive financial behavior. Additionally, the financial knowledge learned contributes to the attitude they possessed in handling financial decisions. Overall, financial learning is not only the comprehension of economic knowledge but also has other elements, namely, financial talents and financial mindset (Dewi et al., 2020).

The integration of the three variables is strongly applicable for the extent of the financial learning of college students. It is necessary to highlight the pressures of economic judgement making and how others are impacted corresponding to the meaning of financial learning.

Statement of the Problem

Generally, this study seeks to answer the question: Does financial attitude, financial behavior and financial knowledge significantly associated towards financial literacy?

Specifically, it seeks to resolve the subsequent problems:

1. Is financial attitude significantly related towards financial literacy?
2. Is financial behavior significantly related towards financial literacy?
3. Is financial knowledge significantly related towards financial literacy?

Significant of the study

This article hopes to promote supplementary evidence and benefit certain groups, organizations, and individuals regarding financial literacy. Furthermore, the study could be of importance to the following:

Students – The research's goal is arranged to help the students improve the power to appreciate and successfully use several financial competences, as well as personal financial administration, accounting, and financing.

Academe – The results of this investigation study can be a learning model in different schools/universities. Through this research, educational institutions can use the findings in this study as their reference for them to create programs that can help their students in making wise financial decisions.

Scope and Limitation

The target respondents are from the Gen Z college students. This study will be composed of 439 randomly selected students in Metro Manila. The method for the survey will be facilitated through an online questionnaire since this is the most effective utilization of available resources to achieve over all intentions of the research.

The estimated time range of the investigation from the survey to members of the cooperatives to the outcome of the research ranges from May – July 2022. The range will vary depending on the availability of the data that will be analyzed quantitatively.

Theoretical Framework

According to OECD (2013), financial literacy suggests the kind of thinking and behavior and developing the knowledge to make efficient decisions to enhance the financial welfare of individuals. Literacy is perceived as an enlarging collection of learning, skills, and approaches, which people build on all through life, instead of as a stable capacity, a position to be opposed, with analphabetism on one edge and learning on the another. Literacy entails more than the propagation of stored knowledge, even though calculating financial knowledge is an essential component in the appraisal. It also includes the utilization of rational and functional competences.

Conceptual Framework

In the study conducted by Nano (2015), it was revealed that financial attitude, behavior, and knowledge are significantly related to each other. Similar result was made by the research of Bhushan & Medury (2014). Having these three dimensions of financial literacy, people can make wise financial decisions to achieve financial satisfaction (Németh & Zsótér, 2017), (Kumar, et al, 2017).

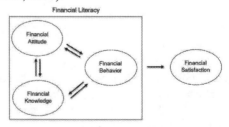

Fig. 24.1 Conceptual Framework

Operational Framework

The definition of OECD reveals that financial attitude, financial behavior, and financial knowledge are the essential attributes of having financial literacy. In the study conducted by Nano (2015), these main dimensions showed a statistically significant linkage towards financial literacy. To achieve the research objective, this study contains financial attitude, financial behavior, and financial knowledge as explanatory variables while financial literacy acts as the response variable.

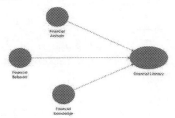

Fig. 24.2 Operational framework

2. Methodology

This study practices the explanatory method which is intended for the researchers to collate evidence about current existing circumstances vital in the chosen field of the study. This approach allows the investigators to construe the hypothetical meaning of the discoveries and supposition improvement for further investigations. To realize this endeavor, the researchers employed an online questionnaire that was distributed randomly to 439 Gen Z college students as our respondents.

The instruments used to gather the necessary data are survey-questionnaires, which are stipulations for descriptive methods. The questionnaires were structured 5-point Likert scale design instrument in such a way that the respondents will be able to understand and answer it easily. Also, the researchers constructed the questionnaires in appropriate form to adapt adequate and relevant information required from the respondents. Additionally, the feedback form is distributed into four (4) parts. The first part focused on the financial attitude of the respondents. Another part concentrated on the financial behavior. These two parts were adapted from the study of Susan Smith Shockey (2002).

The researchers conducted it to 30 respondents and was facilitated through an online platform. A Cronbach Alpha test was utilized to check consistency of the device established on the gathered data from 30 respondents and it obtained a reliability coefficient of 0.878 as a whole and interpreted as "Good" reliability. To analyze the information gathered, Structural Equation Modeling (SEM) was employed, and raw data was managed through STATA software for accuracy of calculations.

3. Results and Discussions

The researchers investigated the relationship of the variables and the indicators associated using Structural Equation Modeling, as presented in the figures.

Table 24.1 Frequency and percentage distribution of the respondents according to their demographics

Variables	Demographics	Frequency	Percentage
Gender	Male	215	48.97
	Female	220	50.11
	Prefer not to say	4	0.98
Year Level	1st year	110	25.06
	2nd year	101	23.01
	3rd year	115	26.20
	4th year	113	25.74
Age	16 years old	15	3.42
	17 years old	24	5.47
	18 years old	10	2.28
	19 years old	31	7.06
	20 years old	221	50.34
	21 years old	80	18.22
	22 years old	20	4.56
	23 years old	10	2.28
	24 years old	21	4.78

The table above indicates the frequency of the samples applied in this study. The sample of this study consists of 48.97% male students and 50.11% female students while 0.98% of the respondents preferred not to say their gender. Based on their frequency and percentage, the respondents are female dominated. In terms of year level, In terms of the respondents year level, most of the respondents are from third year level with 26.20%. On the other hand, half of the respondents also are 20 years old with a percentage of 50.34%.

Table 24.2 Descriptive analysis of the respondents according to profile

Descriptive Measures	Gender	Age	Year Level
Sample Size	439	439	439
Mean	1.72	20.40	2.46
Median	2.00	20.00	2.00
Standard Deviation	0.49	1.16	0.81
Minimum	1.00	16.00	1.00
Maximum	3.00	24.00	4.00
Skewness	-0.52	-0.17	0.01
Kurtosis	-0.62	1.87	-0.48

Above table the showed that the mean age of the respondents is 20.4 (mean = 20, standard deviation =1.16). In terms of year level, Respondents were mostly from third year level (mean = 2.46). Lastly, most respondents were female (mean =2.0). Additionally, the researchers performed a test of normality and data revealed negative skewness (to the left) for both gender and age (Gender=-0.518; Age=-0.171) and 0.0104 for year level. Estimates are less than or inside the range of +1 and -1 which is the normal value. Kurtosis, on the other hand, showed values Gender=-0.623, Age=1.87 and Year Level=-0.476 which are less than and within the normal of +3 to -3. Therefore, values indicate normal symmetry.

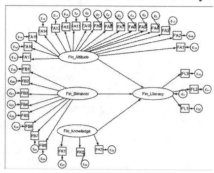

Fig. 24.3 Structural model for financial literacy with inner and outer loadings

Above Figures, denotes the regression of the different variables and their indicators. The inner model illustrates that Financial Behavior, Financial Attitude and Financial Knowledge predicts Financial Literacy. It also shows that most of the outer loadings are acceptable values which are greater than 0.50 thres hold.

Table 24.3 Structural model assessment of financial literacy

Paths	Paths Coefficients	p-values	Remark
Financial Attitude → Financial Literacy	-0.08	0.035	Significant
Financial Behavior → Financial Literacy	0.98	0.000	Significant
Financial Knowledge → Financial Literacy	1.69	0.003	Significant

Path scrutiny was worked out to assess the established hypothesis. Above table shows that, the result indicates that financial attitude, financial behavior, and financial knowledge were significant as manifested on their computed p-values. This indicates that the above variables influence/affect the financial literacy of Gen Z college students. This result confirms the study of Nano in 2015 which indicates that the above three variables have a significant linkage with Financial Literacy.

Table 24.4 Structural model fit selection criteria of financial literacy

	R-Square	RMSEA	CFI	TLI	SRMR
Financial Literacy	0.99	0.04	0.89	0.88	0.03

Legend: RMSEA-Root Mean Squared of Approximation; CFI-Comparative Fit Index; TLI-Tucker-Levis Index SRMR-Standard Root Mean Squared Residual.

Above fit indices indicate an acceptable model, the R-square is close to 1.0 standard and RMSEA less than 0.05. However, both CFI and TLI are less than 0.95 to be considered a good fit but still are close to 1.0, which under acceptable fit. On the case of SRMR the value is close to 0.10 which can be considered also as acceptable fit.

4. Conclusion

Founded on the results and discussion, all latent variables exhibit a positive substantial relationship with the Financial Literacy of the respondents except for financial attitude which is negative with acceptable fit indices on the model.

5. Recommendation

Based on the conclusions, the researchers recommend that a further study be done involving profile of the respondents as a moderating variable.

Additionally, the researchers recommend that a course subject on Financial Literacy be integrated to all programs offered in colleges and universities.

References:

1. Ameliawati, M., & Setiyani, R. (2018). The influence of financial attitude, financial socialization, and financial experience to financial management behavior with financial literacy as the mediation variable. KnE Social Sciences, 811-832.
2. Andarsari, P. R., & Ningtyas, M. N. (2019). The role of financial literacy on financial behavior. Journal of Accounting and Business Education, 4(1), 24-33.
3. Chaulagain, R. P. (2018). Contribution of Financial Literacy on Behaviour: A Nepali Perspective. Journal of Education and Research, 8(2), 75-92.
4. Çoşkun, A., & Dalziel, N. (2020). Mediation effect of financial attitude on financial knowledge and financial behavior: The case of university students. International Journal of Research in Business and Social Science (2147-4478), 9(2), 01-08.
5. Dewi, V., Febrian, E., Effendi, N., & Anwar, M. (2020). Financial Literacy among the Millennial Generation: Relationships between Knowledge, Skills, Attitude, and Behavior. Australasian Accounting, Business and Finance Journal, 14(4), 24-37.

6. Ismail, S., Faique, F. A., Bakri, M. H., Zain, Z. M., Idris, N. H., Yazid, Z. A., ... & Taib, N. M. (2017). The role of financial self-efficacy scale in predicting financial behavior. Advanced Science Letters, 23(5), 4635-4639.

7. Kumar, S., Watung, C., Eunike, J., & Liunata, L. (2017). The Influence of financial literacy towards financial behavior and its implication on financial decisions: A survey of President University students in Cikarang-Bekasi. Firm Journal of Management Studies, 2(1).

8. Németh, E., & Zsótér, B. (2017). Personality, Attitude and Behavioural Components of Financial Literacy: A Comparative Analysis. Journal of Economics and Behavioral Studies, 9(2), 46-57.

9. Susilowati, N., Kardiyem, K., & Latifah, L. (2020). The Mediating Role of Attitude Toward Money on Students' Financial Literacy and Financial Behavior. JABE (JOURNAL OF ACCOUNTING AND BUSINESS EDUCATION), 4(2), 58-68.

10. Wardani, E. W., & Sangka, K. B. (2017). Faktor-Faktor yang Memengaruhi Literasi Keuangan Mahasiswa Program Studi Pendidikan Akuntansi Fakultas Keguruan dan Ilmu Pendidikan Universitas Sebelas Marct. Tata Arta: Jurnal Pendidikan Akuntansi, 3(3).

Note: All the figures and tables in this chapter were made by the authors.

Advancing Sustainable Science and Technology for a Resilient Future – Sai Kiran Oruganti et al. (eds)
© 2024 Taylor & Francis Group, London, ISBN 978-1-032-79020-6

MEDSPEAK: A Linux Based Speaking Prescribed Medicine Reminder Alert Project

25

Engr. Jose Marie B. Dipay, PCpE[1]

Research Management Cluster Coordinator, Research Management Office, Assistant Professor IV, Institute of Technology, Doctor in Engineering Management, Open University System, Polytechnic University of the Philippines, Country.

Engr. Julius S. Cansino, MIT, PCpE[2]

Associate Professor I, Chairperson, College of Computer Engineering, College of Engineering,

Doctor in Engineering Management, Open University System, Polytechnic University of the Philippines

Arvin R. dela Cruz, PCpE[3]

Polytechnic University of the Philippines,

Chairperson, Doctor in Engineering Management, Open University System, Polytechnic University of the Philippines

Ginno L. Andres[4]

Faculty, Doctor in Engineering Management, Open University System, Polytechnic University of the Philippines

Abstract: This project aims to develop Medspeak, a Linux-based talking alarm gadget designed to prompt users to take their prescribed medication as directed. The paper addresses shortcomings in conventional medicine reminder systems, proposes enhancements for such devices, and evaluates the responses of the target audience in terms of usability, comfort, and performance. Data collection employed a combination of questionnaires and scheduled in-person interviews involving a diverse group of participants, including fifteen (15) nurses, thirty (30) caregivers, and eighty (80) medicine users from various countries. Conventional medicine reminder devices often lead to issues such as incorrect dosages, medication duplication, confusion, and loss of track of medications. Linux-based intelligent medication reminder devices, exemplified by Medspeak, prove to be more user-friendly for patients, caregivers, and nurses, potentially expediting device development. The Linux-based evaluations of Medspeak reveal high scores across the board, including functionality (4.62), dependability (4.52), usefulness (4.68), and performance (4.68). Regulatory bodies could potentially recommend this developed device for individuals who require medication management.

Keywords: Medicine users, Linux operating system, Medical drugs, Intelligent device, Caregivers

1. Introduction

Modern science has significantly impacted lives, enhancing various aspects, including healthcare. The application of database technologies in creating drug reminder systems is widely embraced by programmers. This paper introduces an innovative approach to monitor and manage pharmaceutical dosage equipment.

Patients reliant on regular prescription medications have greatly benefited from medication reminder devices for over a decade. These devices have evolved significantly, and the necessity of a reliable medication reminder device became evident, especially in the early days of Linux, to ensure accurate administration of prescription drugs. Addressing prescription medication abuse is a global concern, and the healthcare industry's need for more effective medication management is growing, with Linux-powered devices playing a pivotal role.

A medication reminder device plays a crucial role in ensuring the accurate administration of various drugs. The current design focuses on mitigating the likelihood of accidental drug misuse. This article delves into how technology, including computers and automatic reminder systems, can assist

[1]jmdipay@gmail.com, [2]jscansino@pup.edu.ph, [3]arrdelacruz@pup.edu.ph, [4]glandres@pup.edu.ph

DOI: 10.1201/9781003490210-25

patients in adhering to their prescribed medication schedules. Additionally, the significance of managing devices with respect to medication accessibility and timing is explored. This section also outlines the approach taken in this study to design "Medspeak: A Linux Based Speaking Prescribed Medicine Reminder Alert Project," keeping in mind individuals struggling with medication adherence and management.

2. Methodology

Data for this study will be collected through a mixed-method approach involving both qualitative and quantitative techniques. Qualitative methods aim to describe and comprehend the phenomenon under study by exploring its breadth, depth, and complexity (Juni & Afiah, 2014). To identify the essential features required in a new system that can address the limitations of existing patient monitoring systems, in-depth interviews will be conducted with a diverse participant group, including fifteen (15) nurses, thirty (30) caregivers, and eighty (80) medicine users from various countries. Private conversations will be held with the participants, and a list of potential subjects and questions will be prepared to facilitate the discussions with the researchers.

Quantitative research methods will be employed to establish conceptual connections between variables that can be quantified (Zikmund, 2003). User satisfaction with the distinctive MEDSPEAK will be evaluated using an International Organization for Standardization (ISO 9126) questionnaire. Criteria such as reliability, efficacy, usability, and usefulness will be assessed to determine the system's overall approval. The research will adopt a descriptive research methodology, combining both qualitative and quantitative data collection to explain and evaluate the current situation and condition of the proposed system (Sevilla, 1992). The study's design will be a synthesis of experimental and action research methodologies. Trials conducted by the proponent will demonstrate the system's durability and precision. Any survey research methodology that involves asking respondents about the system's performance and operation qualifies as a measuring technique.

Figure 25.1 shows that block diagram of the MEDSPEAK. The system operates through a user-friendly interface, allowing individuals to input their medication reminders effortlessly. Users utilize a standard keyboard to input crucial details, such as the date, time, and dosage for each medication reminder. These details are securely stored using the Raspberry Pi. At the specified times, as defined by the user, the system retrieves the medication information and employs text-to-speech technology to convert the text into clear, audible reminders. These reminders are played at the scheduled intervals, ensuring patients receive timely and

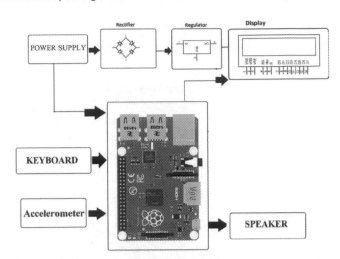

Fig. 25.1 Block diagram of the MEDSPEAK

Source: Author

understandable cues for their medication regimen. This integrated process creates a fully automated medication reminder system, effectively enhancing patient adherence and medication management.

3. Result and Discussion

The difficulties experienced while doing conventional observation on unwell individuals

The interviews with participants highlighted three (3) issues associated with traditional health observation for sick patients:

1. Traditional health observation systems may lead to improper sickness monitoring.
2. Duplication and confusion are prevalent in using traditional observation systems.
3. Maintaining accurate sickness records proves to be challenging.

The opinions of respondents on how to build features that would help solve the issues that have been experienced

Respondents' valuable insights suggest two potential system features to address the identified issues:

1. Leveraging the Internet of Things to expedite and streamline the process of creating a health observation system.
2. Designing an intelligent health observation system based on the Internet of Things that offers a simpler solution for physicians, nurses, and patients compared to conventional methods.

Respondents' level of acceptance toward the developed system

On Functionality

The respondents indicated that Medspeak: A Linux Based Speaking Prescribed Medicine Reminder Alert Project is functionally acceptable, with substantial agreement. The system received high approval ratings regarding functionality, earning a mean overall score of 4.62.

On Reliability

The respondents expressed a high level of satisfaction with the system's dependability, indicating a strong agreement. Medspeak: A Linux Based Speaking Prescribed Medicine Reminder Alert Project gained significant approval in terms of reliability, achieving an average score of 4.52 across all categories.

On Usability

The respondents demonstrated a favorable response to the system's usability, with a very strong level of agreement. Medspeak: A Linux Based Speaking Prescribed Medicine Reminder Alert Project was well- received concerning usability, garnering a mean overall score of 4.68.

On Performance

Respondents indicated a high degree of approval for Medspeak's performance, with substantial agreement. The system received positive feedback on its performance, with a mean score of 4.59 derived from responses across all categories.

4. Conclusion

In conclusion, Medspeak: A Linux Based Speaking Prescribed Medicine Reminder Alert Project demonstrated exceptionally satisfactory results, achieving a total mean score of 4.61. The study's findings underscore the system's potential to address. The MEDSPEAK serves as a versatile and indispensable tool in the healthcare landscape. Its utility extends to a broad spectrum of users, including elderly individuals who may occasionally forget their regular medication schedules. Moreover, the system's simplicity allows it to be easily navigated by illiterate users, enabling them to store medication-related information conveniently. Beyond convenience, the system plays a critical role in ensuring patient safety by mitigating the risks associated with improper medication mix-ups. By offering clear and timely auditory reminders, it stands as a guardian against the adverse effects of medication errors. Overall, this innovative system not only enhances medication adherence but also contributes significantly to the well-being of individuals by promoting the safe and effective management of their healthcare routines.

Reference

1. World Population Ageing: 1950-2050, United Nations Population Division.
2. Slagle, J.M., Gordon, J.S., Harris, C.E., Davison, C.L., Culpepper, D.K., Scott P. and Johnson, K.B., (2011) "MyMediHealth – Designing a next generation system for child- centered medication management", Journal of Biomedical Informatics, Vol. 43, No. 5, pp. 27-31.
3. Becker, E., Metsis, V., Arora, R., Vinjumur, J.K., Xu, Y. and Makedon, F. (2009) "Smart Drawer: RFID- Based smart medicine drawer for assistive environments", Proc. of Pervasive technologies related to assistive environments, June, pp 1-8.
4. Prasad, B., (2013) "Social media, health care, and social networking", Gastrointest Endosc. Vol. 77, pp 492–495.
5. Smart Medication Dispenser Suraj Shinde, Nitin Bange, Monika Kumbhar, Snehal Patil Assistant Professor, SETI, Panhala1 UG Student, SETI, Panhala International Journal of Advanced Research in Electronics and Communication Engineering (IJARECE) Volume6, Issue4, April2017
6. Britsios, J. (2017). Webnauts.net. Retrieved from Why usability is importantto you:http://www.webnauts.net/usability.html.
7. Crossman, A. (2017, March 02). thoughtco. Retrieved from Understanding Purposive Sampling: An Overview of the Method and Its Applications: https://www.thoughtco.com/purposive-sampling-3026727
8. Rouse,M.(2017), searchmicroservices.techtarget. Retrieved from functionality: http://searchmicroservices.techtarget.com/definition/functionali ty.
9. Sevilla, C. G. (1992). Research Methods: Revised Edition. Quezon City: Rex.
10. Thomas M. Cover, Joy A. Thomas (2006). Elements of Information Theory. John Wiley& Sons, New York.

Advancing Sustainable Science and Technology for a Resilient Future – Sai Kiran Oruganti et al. (eds)
© 2024 Taylor & Francis Group, London, ISBN 978-1-032-79020-6

Impact of Microclimate on Noise Reduction

26

Utami Retno Pudjowati*
Bambang Sugiyono Agus Purwono
School of Business and Management, Universitas Ciputra,
Surabaya, Indonesia

Burhamtoro[1]
Politeknik Negeri Malang, Indonesia
School of Business and Management, Universitas Ciputra,
Surabaya, Indonesia

Eko Naryono[2]
Politeknik Negeri Malang, Indonesia
School of Business and Management, Universitas Ciputra, Surabaya, Indonesia

Abstract: The temperature drops and the humidity in the air rises when there is a microclimate. The intensification of sound in settlements along the toll road's edge is caused by the increased volume of traffic, which can be muffled by plants. This study was carried out along the Waru-Sidoarjo toll road at kilometers 23, 27, 31, and 33, each with a different planting system. The research method using multiple linear regression analysis. The results were concluded that the average temperatures decrease in the in the morning was 4.53°C; 3.85°C in the afternoon and 1.63°C in the evening, The air humidity average increase in the morning, afternoon and evening is 4.13%; 3.13% and 1.89%, The increase winds speed's average in the morning, afternoon and evening is 0.53m/s; 0.47m/s and 0.47m/s. Decrease in noise level with changes in microclimate variables in the morning, afternoon and evening by 9.23 dB; 4.19 dB and 10.47 dB. The benefit of this research is to help lower the temperature of the earth's surface, so that global warming can be overcome by having plants around us.

Keywords: Noise reduction, Micro-climate, Settlement, Toll road

1. Introduction

Global Warming is felt by everyone on the surface of this earth. Human activities are very difficult to reduce the temperature of the earth. It is hoped that humans can adapt by using the plants around them. The toll road is used by everyday people as part of the transportation infrastructure to get where they need to go and gives them a sense of security and comfort while driving on the highway. According to Widagdo et al. (2003), the toll road in Indonesia is always congested on a daily basis due to the desire of numerous motorists to pass (Widagdo et al., 2003). People are willing to live in settlements around the highway due to the highway's narrow land and the people's ability to pay for a livable place to stay (Berhitu, 2010). The communities that surround this toll road will be affected by the noise pollution caused by vehicles traveling along the toll road. Last but not least, the objective of creating a toll road for the convenience of all regions was to be stopped. Utilizing noise-reducing vegetation that is planted along the highway's edge is one way to reduce noise. Due to their ability to absorb and dampen sound, plants can aid in noise reduction. In addition, vegetation creates a temperature, humidity, wind speed, and air pressure microclimate (Indriyanto, 2006). The noise level can be greatly reduced by using this microclimate. The alternative to this issue is to create environmentally friendly road conditions. According to Sumaatmadja (2001), this plant's function is to create a microclimate to reduce noise. Other than a wonderful view, vegetation will diminish

*Corresponding author: utami.retno@polinema.ac.id
[1]bambang.sugiyono@ciputra.ac.id, [2]bambang.sugiyono@ciputra.ac.id

DOI: 10.1201/9781003490210-26

the immersion rider mentally. Given this type of toll road, efforts should be made to prevent it and fix it so that users can use it more easily. The foundation of the efforts that will be made is easy and inexpensive environmental conservation. Although vegetation was being planted along the toll road between Waru and Sidoarjo, there were also no plants. This condition will be looked at to see how the microclimate influenced the growth of vegetation to reduce noise in the towns that are located along this highway.

This study aims to find out how the vegetation at the toll road's edge affects microclimate variables and how this affects noise levels. In accordance with Tampubolon (2010), the surface air temperature rises in the afternoon compared to the upper air layers, while the surface air temperature decreases in the evening.so that there will be air expansion during the day. The relationship between temperature and humidity is strong because if the humidity changes, so will the temperature. The lower the humidity, the higher the temperature. This is because the high temperatures would cause water molecules in the air to precipitate (condensate), reducing the amount of water in the air. T various relationships that influence land and water, factors that affect air humidity are closely related to those that affect temperature. The greater the amount of water vapor in the air and on the ground, the higher the humidity. The next factor, whether there is or isn't vegetation, is the effect of altitude, which has the effect that the higher you are, the lower the temperature in the area will be and the higher the humidity will be (Lakitan, 2002). This is true regardless of whether there is or isn't vegetation present. According to climatological data, horizontal wind speeds occur when the grass is planted 2 meters above the ground, directing surface winds in its path. As a driver, the difference in air pressure between the origin and destination and the field resistance in its path determines the wind speed (Tampubolon, 2010).

2. Research Methods

The exploration was done on the Waru-Sidoarjo Expressway at km 23 km 27 km 31 and km 33.At km 23, there is no vegetation; the only ground cover is grass; at km 27, however, there is complete vegetation, including the following: trees, bushes, and shrubs: at km 31, there are bushes and trees, but only trees at km 33.

A: It is located at km 27 with the composition of tree vegetation - shrubs - shrubs

B: is located at km 23 with the composition of the vegetation without vegetation

C: is located at km 33 with the composition of trees

D: is located at km 31 with the composition of trees - shrubs

The first week's data were collected on Fridays from 07:00 to 12:00. Sound Level Meters, Anemometers, Barometers,

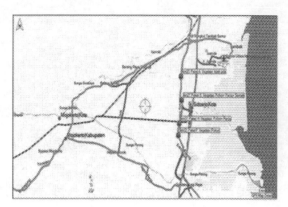

Fig. 26.1 Survey Location

Humidity Meters, Thermometers, GPS, and Roll Meters were utilized at each study site at km 23, km 27, km 31, and km 33. Every one of these areas normally has the similar tree with a dispersing of 10 meters. Temperature, humidity, wind speed and air pressure, and noise level are the required information. Both in front of and behind the tree were measured. Each location was measured five times on a 10-meter-distance tree during the data retrieval process, which was repeated ten times.

Data were collected 1 week are on Fridays at 07:00 until 12:00. At each study site is at km 23 km 27 km 31 and km 33 put the measuring instrument used is a Sound Level Meter, Anemometer, Barometer, Humidity meter, Thermometer, GPS and Roll Meter. Each of these locations typically has the same tree with a spacing of 10 meters. The required data are temperature, humidity, wind speed and air pressure, noise level. Measurements were made in front of the tree and behind a tree. Data retrieval is repeated 10 times, each location was measured 5 times on a tree with a distance of 10 meters.

In the morning and afternoon, there is a decrease in noise, a rise in temperature, an increase in air humidity, and a rise in wind speed. These changes typically take place in places where there is vegetation with complete strata, such as trees, bushes, and bushes.The microclimate is caused by the presence of vegetation that completes these strata.

The tools used for the study are:

Fig. 26.2 Sound level meter (SLM) **Fig. 26.3** Rollmeter

Fig. 26.4 GPS

Fig. 26.5 Anemometer-barometer-humidity meter-thermometer

In the morning and afternoon, there is a decrease in noise, a rise in temperature, an increase in air humidity, and a rise in wind speed. These changes typically take place in places where there is vegetation with complete strata, such as trees, bushes, and bushes. The microclimate is caused by the presence of vegetation that completes these strata.

From the analysis of SPSS 20 (Santoso, 2012) obtained the results of the comparison as follows,

The morning saw the greatest decrease in noise, the morning saw the greatest decrease in temperature, the morning saw the greatest increase in humidity, and the morning saw the greatest increase in wind speed.

The Adjusted R Square value of 0.962 in Table 26.2 indicates that the morning's drop in temperature, rise in air humidity, and increase in wind speed had an effect of 96.2% against a decrease in noise. While the remaining 3.8 percent were influenced by other events that took place close to the study. The results of the day's analysis are shown in Table 26.2.

Table 26.1

Average	Morning	Afternoon	Evening
Decrease Noise	9,23 ± 1,39	4,19 ± 2,35	10,47 ± 2,25
Decrease temperature	4,53 ± 1,10	3,85 ±1,01	1,63 ± 0,38
The increase in air humidity	4,13 ± 0,72	3,13 ± 0,98	1,89 ± 0,72
The increase in wind speed	0,53 ± 0,04	0,47 ± 0,03	0,47 ± 0,02

Table 26.2 Model summary

Model Summary[b]

Model	R	R Square	Adjusted R Square	Std. Error of the Estimate	Change Statistics					Durbin-Watson
					R Square Change	F Change	df1	df2	Sig. F Change	
1	.985[a]	.970	.962	.27187	.970	117.598	3	11	.000	2.493

a. Predictors: (Constant), KENAIKAN KECEPATAN ANGIN, PENURUNAN SUHU, KENAIKAN KELEMBABAN UDARA

b. Dependent Variable: PENURUNAN NOISE

Table 26.3 Model summary

Model Summary[b]

Model	R	R Square	Adjusted R Square	Std. Error of the Estimate	Change Statistics					Durbin-Watson
					R Square Change	F Change	df1	df2	Sig. F Change	
1	.993[a]	.986	.982	.31256	.986	260.146	3	11	.000	1.303

a. Predictors: (Constant), KENAIKAN KECEPATAN ANGIN, PENURUNAN SUHU, KENAIKAN KELEMBABAN

b. Dependent Variable: PENURUNAN NOISE

The Adjusted R Square value of 0.982 in Table 26.3 indicates that the daytime change in temperature, humidity, and wind speed account for 98% of the reduction in noise. The remaining 1.8 percent, on the other hand, was influenced by other events that took place close to the study.

The Adjusted R Square value of 0.993 in Table 26.4 indicates that the effect of a drop in temperature, an increase in air humidity, and an increase in wind speed at dusk is 99.3% greater than the effect of a decrease in noise. While the remaining 0.7% had an impact on other events that took place close to the study.

According to the findings of this analysis, the morning, afternoon, and early evening changes in temperature, air humidity, and wind speed have a negative impact on noise. A microclimate that is characterized by an increase in wind speed, an increase in air humidity, and a decrease in temperature.

Table 26.4 Model summary

Model Summary[b]

Model	R	R Square	Adjusted R Square	Std. Error of the Estimate	Change Statistics					Durbin-Watson
					R Square Change	F Change	df1	df2	Sig. F Change	
1	.997[a]	.995	.993	.18265	.995	702.107	3	11	.000	3.099

a. Predictors: (Constant), KENAIKAN KECEPATAN ANGIN, KENAIKAN KELEMBABAN, PENURUNAN SUHU

b. Dependent Variable: PENURUNAN NOISE

Trees, bushes, and other vegetation with complete strata are to blame for the occurrence of microclimate's.

The ability of air to absorb the force of sound waves that pass through it and propagate through it is influenced by temperature and humidity. In high-temperature air, sound travels more quickly, whereas in low-temperature air, it travels more slowly (Mediastika, 2002).

The water molecules in the air expand and become more tightly packed as a result of the temperature drop and evaporation that follows. Noise will be reduced as a result of the inhibition of sound waves during its propagation (Sun, 2008). Should the developers who will use the land around the highway construct a barrier of this kind, it would assist in reducing the noise that occurs on the highway. This kind of barrier should still pay attention to the surrounding area by employing the function of vegetation to make the surrounding area not only beautiful but also quieter. In the meantime, the type of plants, vegetations and the distance between the motorway and the resettlement site should be taken into consideration when determining the settlement's location on the highway.

3. Conclusion

From these results it can be concluded that:

Changes in the variable micro-climates with plants along the edge of the highway,

(a) The decline in the average temperature on the morning of 4.53 ° C; on the day of 3.85 ° C and in the afternoon at 1.63 ° C

(b) The increase in average air humidity in the morning of 4.13%;   during the day amounted to 3.13% and 1.89% in the afternoon

(c) The increase in average wind speed in the morning at 0.53 m/s; on the day of 0.47 m/s and in the afternoon of 0.47 m/s.

Decrease noise levels with the change of micro climate variables on the morning of 9.23 dB, 4.19 dB for daytime and early evening of 10.47 dB.

Suggestions for the toll road operators are as follows:

(a) Maintain shade trees and reforestation on the highway banks at all times. In order to maximize microclimate and minimize noise in residential areas, shade trees should be planted in a manner that follows the pattern of the vegetation component of trees, shrubs, and bushes.

(b) To ensure that the vegetation's noise reduction process remains safe and effective, held a tree planting so that replacement trees have parents for a certain amount of time.

References

1. Berhitu, P. T., & Matakupan, Y. (2010). The developing Feasibility study in Ambon Water Front City. Jurnal Teknologi. Volume 7 Nomor 1. Pp. 767–781.
2. Widagdo, S., Gunawan, A., Nasrullah, N., & Mugnisjah, W. Q. (2003). Studi Tentang Reduksi Kebisingan Menggunakan Vegetasi dan Kualitas Visual Lansekap Jalan Tol Jagorawi. Forum Pascasarjana Volume 26 Nomor 1 Januari 2003 : 41–50. IPB. Bogor.
3. Indriyanto. (2006). Ekologi Hutan. Jakarta: Penerbit PT Bumi Aksara.
4. Lakitan, B. (2002). Dasar-Dasar Klimatologi. Cetakan Ke-2. Raja Grafindo Persada. Jakarta.
5. Mediastika C.E. (2002). Akustika Bangunan: Prinsip-prinsip dan Penerapannya di Indonesia. Penerbit Erlangga.
6. Santoso, Singgih. (2012). Panduan Lengkap SPSS Versi 20. Jakarta: PT Elex Media Komputindo
7. Sumaatmadja, N. (2001). Metode Pembelajaran Geografi. Bumi Aksara. Jakarta.
8. Sun, G. C., Zuo, S. Liu, M. L., Steven G. Mc Nulty, & J. M.Vose. (2008). Watershed Evapotranspiration Increased Due to Changes in Vegetation Composition and Structure Under a Subtropical Climate. Journal of The
9. American Water Resources Association. Vol. 44 No. 5. October (2008). pp. 1164– 175.
10. Tampubolon, S. (2010). Pengaruh Kecepatan Angin dan Suhu Udara Terhadap Kadar Gas Pencemar Karbon Monoksida (CO) di Udara Sekitar Kawasan Industri Medan (KIM). Departemen Fisika Fakultas Matematika dan Ilmu Pengetahuan Alam Universitas Sumatera Utara. Medan.

Note: All the figures and tables in this chapter were made by the authors.

Advancing Sustainable Science and Technology for a Resilient Future – Sai Kiran Oruganti et al. (eds)
© 2024 Taylor & Francis Group, London, ISBN 978-1-032-79020-6

An Intelligent System on Performance Prediction for Job Placement Using Case-based Reasoning

27

Mark JaysonLay*
School of Graduate Studies AMA Computer University, Philippines

Jenny Lyn Abamo[1]
School of Graduate Studies AMA Computer University, Philippines

Abstract: Finding work is one of the challenges for a new graduate to fulfill,the struggle to market oneself and get hired. The lack of skills, abilities,experience, competition, and specific qualifications are factors that affect graduates in landing a Job. This study aims to conceptualize Case-Based reasoning as a data mining tool in job placement process. The system provides a mechanism for higher education institutions, company, and students to boost their output through prediction utilizing Case-Based Reasoning. The test case has shown how CBR retrieval procedure may be used to compare student's characteristics on local and global similarities and benchmarks based on the standards set by the organization. The research aims to develop a system that will measure the graduate's chance of landing a job and prepare for future opportunities.

Keywords: Case-based reasoning, Intelligent system, Performance prediction, Job placement

1. Introduction

Securing a job is one of the struggles that a fresh graduate faces. From unrealistic expectations of the graduate's to high standard qualifications set by the company nowadays. The truth is, it's not only about the qualifications but there are other factors that affect job placement.Interpersonal and intrapersonal must be developed. Interpersonal, is used when exchanging ideas professionally. While intrapersonal, is about your own attitude, values, and ideas. Both are being assessed when applying for a job. Having the right skill and knowing the correct answer to specific question would dictate if the applicant can carry on to the next part of the application process.Now the question ishow one person would prepare for the job interview and examination. This research focuses on how case based reasoning could be used as a tool to improve job placement for AMA graduates. Using experimental design as a quantitative research approach and deciding on a set of steps to take and observe links between variables to check the hypothesis.

2. Purpose and Description

The research will focus on the Semicolon Connects which is a digital marketing staffing agency for coaches: they offer Strategic, Simple, Cost-effective, and Proactive solutions to their clients.

The system should be able to supplement this process by providing the shortlist of candidates that would be trained and be future professionals in different fields.

A. General Objective

This research aims to test the effectiveness of using CBR as a tool to improve job placement for Information Technology graduates andprovide companies with a shortlist of applicants for their vacant positions.

B. Specific Objectives

1. To build a system that is based on Case-Based Reasoning and improve job placement for AMA Computer College and Semicolon Connects;

*Corresponding author: markjaysonlay@gmail.com
[1]jlvabamo@amaes.edu.ph

DOI: 10.1201/9781003490210-27

2. To create a website that uses latest version of PHP and Bootstrap for dynamic page content, collection of data and to control user-access;

3. To develop a concept of predicting job placement using Case Based Reasoning.

3. Significance of the Study

The AMA Computer College will benefit from this research. Utilizing the system will enable the student to assess their readiness prior to submitting an application. The system will be able to produce a short list of candidates for Semicolon Connects' open positions.

AMA BSIT Graduates – This research will help the student assess their performance and suggest learning paths to improve results when finding a job.

Institution – This research will allow the academic manager to measure the performance of their graduates and be able to adjust the curriculum.

Employees – For their employment openings, Semicolon Connects' staff will be able to access a shortlist of graduates.

4. Scope and Limitation

This study focuses on the following scope:

1. Application of CBR for job placement prediction.
2. Creating a system for Semicolon Connects and AMACC.
3. Create a data mining tool for job placement process.
4. Create a system that will determine the likelihood of IT graduates to land a Job.

Limitations for this study are as follow:

1. The system is exclusively designed for Semicolon Connects and AMACC.
2. The specimens for this research are Information Technology students from AMACC.
3. Test questions that will be used for this research will be provided by the beneficiary

5. Review of Related Literature and Studies

Performance prediction is a significant examination feature of instructive information mining. Most models extricate understudy conduct highlights from grounds card information for expectation. (Chen, et al. 2023)

Case-based reasoning (CBR), a technique for analogical thinking which is normal and critical in human discernment, has as of late arisen as a significant thinking procedure. CBRincludes tackling new issues by recognizing and adjusting answers for comparative issues put away in a library of previous encounters. The significant stages in the deduction pattern of CBR are to recover cases from the library which are generally pertinent to the central issue and adjust the recovered cases to the ongoing info. (Bonissone et al. 1998)

A plan for education that takes artificial intelligence into account: Higher Education Institutions (HEI's) should keep simulated intelligence in mind while building courses to demonstrate its main notion and uses. It will be advantageous to set up classes, roundtable discussions, and studios that focus on simulated intelligence. (Galvez et al. 2019)

While moving toward modernization and commonality among task, expectations for graduates with great programming abilities is on the rise. In equal, there have been a rising number of understudies who find it hard to accomplish the abilities vital to land the IT position they want. (Kuma et al. 2022)

Using an altered version of Case Based Reasoning (CBR), an AI viewpoint that provides a mechanism to nurture information based frameworks, an information-based framework using Case Based Reasoning (CBR) offered a solution to identify the qualified students. It was selected for two convincing reasons: first, it is obvious that significant human master talent is demonstrated using this methodology; second, when compared to rule-based approaches, it requires less information designing and simpler execution. CBR is more flexible when used with unique SQL scripts than static rules embedded in code(Canlas, 2020).

Every educational institutionrecognizes the importance of campus placement in assistingstudents in reaching their objectives. The big student datasetcan be used as a resource for obtaining the associated datavia data mining categorization. Data mining techniques arewidely employed in the field of education, where there aremany different approaches and methods for learning. Apredictive model is created to determine the placementcategory for which students are qualified based on theirprior academic and extracurricular achievements. Thealgorithm will also recommend additional abilities neededfor future hiring, which could aid students in preparing forplacement.(Rao et al. 2019).

6. Synthesis

The proposed system primarily focuses on creating an intelligent system with the contribution of the pertinent gathered articles and studies. The review of related literatures and studies helped conceptualize and supply information in the development of the research.

Performance prediction is important feature in which plays a vital role in Improvement of education. (Chen et al 2023).

Intelligent system is being used by multiple industries under various fields of discipline (Rao et al. 2019)

Artificial Intelligence (AI) and Machine Learning (ML) are widely utilized in connection to problem solving and training (Bonissone et al. 1998), (Galvez et al. 2019).

One reason for the low employment rate in the Philippines is education. (Galvez et al. 2019).

Case Based Reasoning may be applied to fix these problems (Kuma et al. 2022).

7. Technical Background

The current manual workflow of the chosen beneficiary is too tedious, time-consuming, and prone to mistakes that can lead to problems. To solve this, the researchers develop a web-based application tailored to the problems of the beneficiary as stated in the past chapter. In this section, the researchers discuss the technical aspects used in the development of the proposed study.

8. Software Design

This section discusses the software used in the development of the system. It is essential for the researchers to have tools that would help in developing the application. Several tools are utilized in the development process that will deliver the application in a cost-efficient manner. The proposed system is developed using a PHP framework called bootstrap. To give additional dynamic functionalities to the components structured by HTML, CSS, and JavaScript, JQuery and AJAX are utilized. MySQL is a Relational Database Management System (RDBMS), that is backed-up by Oracle and based on Structured Query Language. The application uses an open-source cross-platform web server called XAMPP to serve the application and the database for the management and interaction with the data.

9. Conceptual Design

Conceptual Framework is a versatile analytical tool with several applications. It can be used in numerous fields of employment where a comprehensive picture is required. It is used to organize and categorize ideas.

Figure 27.1 represents the illustration of the proposed system. There are three mainmodules, first is Semicolon Module this is where a personnel from the establishment will post all the job listing and questions for the prediction exam. Second the student module this answering question as well as providing data will be the role of the graduates. And lastly third is the CBR module this is where all the data are stored and all the computations are made.

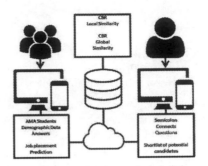

Fig. 27.1 Conceptual framework

10. Machine Learning

Machine Learning (ML) is a form of artificial intelligence (AI), software programs that can predict outcomes more accurately without having to be explicitly instructed to do so. Machine learning algorithms forecast new output values using historical data as input. This may process may involved some or all of this steps data collection, data penetration, selection of a model, training, evaluation, tuning and prediction.

11. Cased Based Reasoning

A paradigm of the artificial intelligence (AI) and cognitive science known as case-based reasoning (CBR) which depicts the reasoning process as being predominantly memory-based. (Canlas, 2020)

Its strategy primarily focuses on recovering and reusing solutions that have worked on a similar issue before. (Canlas, 2020)

The four main steps of case-based reasoning are: retrieve, reuse, revise, and retain. These actions form the backbone of the system's learning cycle as shown in Fig. 27.2 (Bonissone et al. 1998)

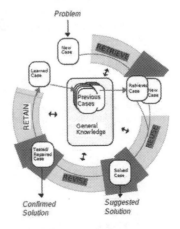

Fig. 27.2 CBR cycle

12. Conclusion

The test cases using the prototype illustrated how Case-Based Reasoning could be used to determine and choose students who were qualified for job placement.

The suggested model can be visualized as a multi-level implementation with a client and server architecture that will enable the different devices to communicate with CBR system. The knowledge base is stored on a MySQL server, and while retrieval process is being done using PHP scripts. Organizations can draw conclusions utilizing quantitative and empirical evidence, as well as with some degree of flexibility, rather than only relying on a static rule base and fixed database, thanks to the percentage format used to display similarity results. After the algorithm for local which are sim(a,b)=1-(a-b)/range for numeric and $if a$ = b then 1 or if a ≠ b then 0 for discrete, for global sim(a,b)=1/(sum of weights) x sum of weights x (a,b) similarity where applied the output below was produced.

Shown in Fig. 27.3 is a prediction output from a job with the following requirements shown in Fig. 27.4 the prediction result is a 98% chance to get hired using the given data.

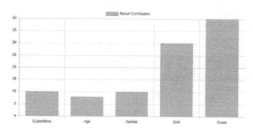

Fig. 27.3 Sample prediction

Shown in Fig. 27.4 are sample Job requirement and the results of graduates after taking the test. Requirement is the parameters for prediction along with the weights that are set by the employer. The applicant's data following the applicant row all three data was used to generate a prediction.

	Experience	Age	Gender	Skills	Score
Requirement	1	25	Male	3	20
Applicant	1	20	Male	3	20
Weight	10	10	10	30	40

Fig. 27.4 Sample data

References

1. Chen, Y., Wei, G., Liu, J., Chen, Y., Zheng, Q., Tian, F., Zhu, H., Wang, Q., & Wu, Y. (2023). A prediction model of student performance based on self-attention mechanism. Knowledge & Information Systems, 65(2), pp. 733–758. doi.org/10.1007/s10115-022-01774-6
2. Ruspini, E., Bonissone, P., &Pedrycz, W. (Eds.). (1998). Handbook of Fuzzy Computation (1st ed.). CRC Press. doi.org/10.1201/9780429142741ImprintCRC Press eBook ISBN9780429142741
3. Artificial intelligence: policy paper Reagan L. Galvez*, Alvin B. Culaba, Elmer P. Dadios, and Argel A. Bandala Journal of Computational Innovations and Engineering Applications 4(1) 2019: pp. 1–5
4. Kuma, M. K., Pranav, K. S., Gowtham, D., & Abhishek, S. (2022). Students performance analysis system using cumulative predictor algorithm. International Journal for Research in Applied Science and Engineering Technology, 10(5), pp. 3277–3285. doi.org/10.22214/ijraset.2022.43073
5. Ferddie Quiroz Canlas. (2020). Student Internship Placement Using Modified Case-Based Reasoning: An Implementation Model. International Journal of Advanced Science and Technology, 29(8s), pp. 4279-4289. Retrieved from http://sersc.org/journals/index.php/IJAST/article/view/25459
6. Abhishek S. Rao, Aruna Kumar S V, Pranav Jogi, Chinthan Bhat K, Kuladeep Kumar B, Prashanth Gouda, "Student Placement Prediction Model: A Data Mining Perspective for Outcome-Based Education System," International Journal of Recent Technology and Engineering (IJRTE) ISSN: pp. 2277-3878, Volume-8 Issue-3, September 2019, DOI: 10.35940/ijrte.C4710.098319.

Note: All the figures in this chapter were made by the authors.

Advancing Sustainable Science and Technology for a Resilient Future – Sai Kiran Oruganti et al. (eds)
© 2024 Taylor & Francis Group, London, ISBN 978-1-032-79020-6

28

Ticketing Platform for QCU Computing Department with Serverless Video Communication

Gerard Nathaniel Ngo*
School of Graduate Studies AMA
Computer University, Philippines

Jenny Lyn Abamo
School of Graduate Studies AMA
Computer University, Philippines

Abstract: Education industry reassessed its customer approach engagement in light of pandemic's impact. They used to handle inquiries, concerns, and issues face-face inside the campus. When social distancing was implemented, previous approach was unfeasible. Ticketing platforms boomed, during pre and post pandemic. The Quezon City University has large population of students and was also affected by pandemic. They are using multitude of platform to resolve student and parent inquiries. Current method is working but inefficient and slow. The research objective is to improve the method of resolving inquiries and concerns of constituents by developing a dedicated platform for directing all concerns. Accessible to departmental employees and enable direct resolution of issues in ticket forms. Accounts will be tagged to respective departments and effectively facilitate tickets routing. Rapid Application Development is applied, this methodology is selected to deliver working product as soon as possible to make changes and improvements as needed.

Keywords: Customer support, Helpdesk platform, Ticketing system, Process improvement, Rapid appication development

1. Introduction

Good customer service is essential for educational institutions to remain competitive. It helps to satisfy students and stakeholders, build relationships, and increase enrollment. The pandemic has limited face-to-face interactions, but made the world open and adapt to online transactions. This research aims to develop a single platform to direct all concerns in one application, helping to improve customer service and track the progress of each concern.

2. Purpose and Description

Main Purpose

The main purpose of the research is to develop a ticketing platform for the beneficiary to improve the overall handling of students concerns or inquiries. To have an effective and efficient tool to provide solution of its stakeholders.

Objective of the Study

The main goal of the research is to increase the effectiveness of the Department in resolving issues from its stakeholders by decreasing the resolution time in resolving inquiries and concerns in the current method being used by the QCU Computing Department.

Specifically, the research aims:

1. To implement a ticketing system to consolidate all inquiries and concerns into a single platform;

2. To improve the resolution time of resolving inquiries and concerns in the form of tickets by setting a specific timeframe allowed to answer each ticket; and

3. To improve the quality of resolving tickets by forwarding the ticket to the correct department for proper resolution of issues, concerns, or inquiries.

[1]gncngo@amaes.edu.ph. (jlvabamo@amaes.edu.ph)

DOI: 10.1201/9781003490210-28

3. Significance of the Study

The research will provide a platform which would improve the service being rendered by the department to its student, specifically pertaining to resolving issues.

To employees, they can monitor and give solution that concerns their department. They can reply to the incoming inquiries as quickly as possible and efficiently.

To students, they can easily communicate to their professors, school staff, and school administration. Concerns can easily be raised in a form of ticket and solution will be provided via a reply.

To parents, school inquiries can be submitted as tickets, and they will receive feedback and information. This platform improves collaboration and transparent communication to build trust.

4. Scope and Limitation

The scope of the research is to develop a ticketing platform with user management, access levels, department management, ticket monitoring, ticket tagging, ticket forwarding, serverless video communication, knowledge base, and reporting and analytics.

The system is not integrated with the enrollment platform, so the administrator must upload the enrolled student list and mark students who are no longer enrolled. Video communication is only between students and administrators, and cannot be forwarded or include additional agents. Video communication is subject to administrator availability, and closed tickets cannot be reopened. Only enrolled students can file tickets, and visitors can only file inquiries. Anything not stated in the scope of the study is a system limitation.

5. Related Literature

As technology continues to evolve, IT companies have become increasingly popular. However, manual tracking of day-to-day transactions such as incidents and requests from clients or internal operations can be challenging. To address this, IT companies utilize ticketing systems to efficiently track and log all records they collect. (Aglibar, K.D.M. et. al., 2022)

Agung Security Podomoro Group faces frequent inventory issues with its heavy reliance on electronic inventory. Reports of damaged inventory are submitted by the location user or admin to Asset Management, which is then passed onto technical support. However, technical support has a hard time at handling the large volume of inventory reports, leading to the need for a benchmark system that can record damage reports and their history for easier future handling of similar issues. (Khasanah, S. N. et. al., 2020)

PT. Bank Mega Tbk requires a computerized helpdesk ticketing system to enhance their complaint reporting process and improve customer service. Currently, the high volume of complaints coming into the IT Support section is not being optimally managed, as other units are reporting through various channels without clear lines, resulting in a lack of recording and frequent data redundancies. To address these challenges, the authors are conducting an analysis and design of an information system (IS) for helpdesk (HD) ticketing to support IT Support staff. (Pt, P., & Mega, B. et. al., 2020)

To improve complaint handling processes, a web-based integrated Helpdesk application that is easily accessible to the employees and the EDP/IT division is necessary. The application is designed using tools such as Flow map and UML and follows the Waterfall method for system development. Xampp and MySQL are used as supporting software and for program documentation. With the web platform connected to the MySQL database, complaints can be delivered and handled more efficiently and effectively. (Wati, E. F., & Maryadi, D. et. al., 2019)

6. Synthesis

Overall, the available literature and studies suggest that the use of a ticketing system can bring significant benefits to an organization, including improved performance and efficiency in resolving issues, higher levels of customer satisfaction, and better tracking and organization of inquiries. Therefore, it is recommended that companies consider implementing a ticketing system to improve their overall operations and customer service.

The research would provide a platform which would improve the service being rendered by the department to its student, specifically pertaining to resolving issues.

7. Technical Background

The beneficiary of this research currently uses Google Forms and social media to handle student requests. This process is time-consuming and inefficient, taking 1 hour to 3 days to forward a request to the right department and resolve the issue.

To improve the current flow and process of handling student concerns and inquiries, the study will develop a ticketing platform with multiple department support, escalation procedures, ticket status tagging, time tracking, serverless video communication, and a dashboard for administrators to monitor employee responsiveness, performance, and assertiveness.

8. Software Design

The researchers used Rapid Application Development methodology to develop the platform to aid in improving the resolution time of inquiries and concerns of the clients of the institution. RAD is an iterative and incremental approach to software development, which emphasizes on rapid prototyping, user involvement and evolutionary delivery.

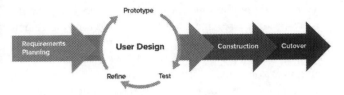

Fig. 28.1 Methodology

The RAD methodology is composed of various stages, such as requirements planning, user design, rapid construction, and cutover. During the requirements phase, the researcher is with constant communication with the beneficiary collecting the required features of the system. Within the User Design phase, prototypes of modules are developed, tested and refined to provide quality assurance before constructing the final version and including it to the main system.

9. Conceptual Framework

The researchers conceptualizes the development of the proposed study to provide a single platform to address all concerns and issues of the beneficiaries' constituents. The platform will be deployed on a hosting site which will contain all the web pages and database of the platform. The platform will be accessible thru the internet of its specific domain.

Fig. 28.2 Ticketing platform for QCU computing department with serverless video communication

With performance as the focus during the development phase, the platform is developed using HTML, CSS, bootstrap framework, featherjs, fontaswsome for its front end to have the most minimal impact in loading the platform while maintaining the proper UI/UX design for user friendliness.

The backend is being handled by PHP, jQuery to have synchronous backend, and MySQL database to store all the relevant information needed to run the platform.

10. Database Schema

Figure 28.3 displays different tables and its connections that stores the user's data on the ticketing platform.

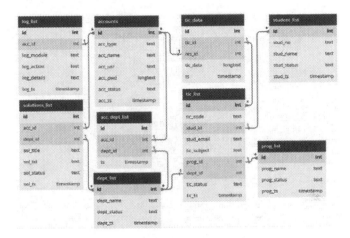

Fig. 28.3 Database schema

11. System Flowchart

Figure 28.4 displays flow of the system from the client side.

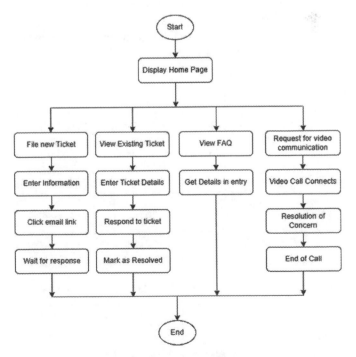

Fig. 28.4 System flowchart

The first module is for filling new tickets, this is the primary module of the client side as its the main purpose of developing the platform. The second module is used to view and send updates to an existing ticket and track its history. The third module is for viewing frequently asked questions to educate the clients regarding the policies and common procedures of the institution. The last module is to allow the client to request a video communication with an available staff assigned to take calls.

12. System Evaluation

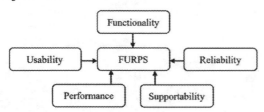

Fig. 28.5 FURPS

FURPS is a software evaluation model that stands for Functionality, Usability, Reliability, Performance, and Supportability. Each of these dimensions provides a different perspective to evaluate software and help developers identify areas that need improvement.

FURPS is an effective tool for software evaluation that provides a comprehensive evaluation of software products. It measures functionality, usability, reliability, performance, and supportability, allowing the researcher to identify areas for improvement and create the system better. By evaluating these dimensions, the researcher can improve software quality, increase user satisfaction, and reduce the total cost of ownership.

Table 28.1 FURPS evaluation results

Category	Aspect	Mean Score	Acceptability Level
Functionality	Compliance	4.55	Highly Acceptable
Functionality	Security	4.90	Highly Acceptable
Usability	Human Factors	4.59	Highly Acceptable
Usability	Overall Aesthetics	4.41	Highly Acceptable
Reliability	Accuracy	4.71	Highly Acceptable
Reliability	Predictability	4.86	Highly Acceptable
Performance	Speed of Processing	4.53	Highly Acceptable
Performance	Efficiency	4.75	Highly Acceptable
Supportability	Adaptability	4.55	Highly Acceptable
Supportability	Configuration	4.57	Highly Acceptable

The Table 28.1 shows a highly acceptable evaluation summary across all categories and aspects, including functionality, usability, reliability, performance, and supportability. This highlights the system's strong performance and adherence to FURPS criteria, instilling confidence in its ability to meet the needs of users.

13. Conclusion

By using a single platform for directing all concerns would improve customer service by allowing tickets to be routed to the correct department and resolved efficiently. Serverless video communication would also improve the handling and resolution of concerns. Improved customer satisfaction would lead to a positive reputation and increased student retention, helping the institution achieve its goal of providing the best assistance to its constituents.

References

1. Aglibar, K. D. M., Alegre, G. C. T., Del Mundo, G. I., Duro, K. F. O., & Rodelas, N. C.(2022). Ticketing system: A descriptive research on the use of ticketing system for project management and issue tracking in IT companies. International Journal of Computing Sciences Research. Advance online publication. doi:10.25147/ijcsr.2017.001.1.90

2. Khasanah, S. N., Kuryanti, S. J., Hermanto, & Adiwihardja, C. (2020). IT-Helpdesk System Design With Waterfall Model (Case Study : Agung Podomoro Group). Jurnal Mantik, 4(1), 56–60.

3. Pt, P., & Mega, B. (2020). Sistem Informasi Helpdesk Ticketing. SISTEM INFORMASI HELPDESK TICKETING PADA PT. BANK MEGA Tbk, 22(2), 201–207.

4. Wati, E. F., & Maryadi, D. (2019). Helpdesk System At PT Himalaya Everest Jaya Jakarta. SinkrOn, 3(2), 229. https://doi.org/10.33395/ sinkron.v3i2.10053

5. Cassandra, C., Hartono, S., & Karsen, M. (2019). Online Helpdesk Support System for Handling Complaints and Service. In Proceedings of 2019 ICIMTech 2019 (pp.314–319).Institute of Electrical and Electronics Engineers Inc. Doi. org/10.1109/ICI

Note: All the figures and table in this chapter were made by the authors.

Advancing Sustainable Science and Technology for a Resilient Future – Sai Kiran Oruganti et al. (eds)
© 2024 Taylor & Francis Group, London, ISBN 978-1-032-79020-6

Usability of Human Resource Information System of Apayao State College

29

Mc-ornoc T. Lamaoa*

Faculty, Apayao State College, Conner, Apayao, Philippines.

Thelma D. Palaoag[1]

Professor, University of the Cordilleras, Baguio City, Philippines.

Abstract: An institution's most dynamic and valuable asset is its people. Thus, any organization must implement data and service management systems to maximize its human resources. The researchers evaluate HRIS usability. The developed HRIS's usability was tested using the USE questionnaire and the 5-point Likert scale level of agreement. According to the study, all groups of respondents rated the developed HRIS as useful with a mean of 4.27 (strongly agree), ease of use with 3.91 (agree), and satisfaction with 3.89 (agree). The study's findings support the system's full implementation to boost efficiency and improve customer service. It was further concluded that the developed system solves and reduces the manual system's problems in Recruitment and Selection, Personal Data Sheet Management, Attendance Monitoring, Employee Evaluation, and Leave Ledger Management of the Human Resource Management and Development Unit and employees. Potential users are pleased with the system's overall functionality due to its efficiency, usefulness, and ease of use.

Keywords: Human resource, HRIS, Human resource information system, Assessment, Usability

1. Introduction

In the 21st century, human resources are crucial for an organization's success, driving business plans and goals. Maximizing human resource potential and efficiently managing data and services is vital. Information systems, including Human Resource Information Systems (HRIS), have become indispensable in today's globalized and technologically advanced business environment.

The Apayao State College is a lone higher education in Apayao with two (2) campuses namely: Conner Campus and Luna Campus. At present, the College has 95 employees for both campuses. Human Resource Management operations and services in Apayao State College are totally manual-based. There is no existing system or ICT-developed software being used to integrate HRM operations and services for faster and easier transactions. The Human Resources Management Department (HRMD) Unit needs to have a Human Resource Information System that would greatly improve the unit's operations and delivery of its services, contribute to the effectiveness of manpower activities, and manage the administration of HR processes and procedures. (Khaled, R., 2022). Because of these scenarios, the researchers were inspired to conduct a study and develop an HRIS for the HRMD of Apayao State College to help HR personnel Execute HR tasks and provide prompt and pertinent HR services with efficiency and effectiveness, adapt to today's changes and technological advances, and cope with market and social pressures. The researchers primarily aimed to assess the system's usability level when being used by the end user.

2. Literature Review

HRIS systems have transformed HR functions, boosting efficiency and communication skills among HR staff (Johnson & Gueutal, 2020). Integrating effective HR management with technology can significantly enhance productivity and profitability. In today's globalized and technology-driven business landscape, information systems are integral to various HR tasks (Zhou et al., 2019). HRIS systems have revolutionized HR service delivery, promoting better communication, knowledge sharing, and efficiency

*Corresponding author: amackylamao@gmail.com
[1]btpalaoag@gmail.com

DOI: 10.1201/9781003490210-29

among HR professionals, thus demonstrating their value to organizations. Key HRIS benefits include improved planning, cost control, enhanced decision-making, and increased overall productivity (Zhou et al., 2019). These systems store extensive employee data, offer easy access, and maintain comprehensive workforce records.

3. Methodology

The study used a descriptive-quantitative approach to assess the HRIS's usability at Apayao State College's HRMD. Usability was evaluated through the USE questionnaire.

The study employed an interview guide and a survey questionnaire to gather accurate data on system usability from end users. Data analysis utilized the weighted mean, interpreting respondents' assessments using a 5-point Likert scale.

Usability Evaluation

Usability involves improving usability during initial design (Prastyo, D., et al., 2019), while usability testing assesses user accessibility and perspective (Othman, M. K., et al., 2022). This study evaluates the Human Resource Information System's (HRIS) usability and development using the USE questionnaire (Zhou et al., 2019). Usability is crucial for HRIS adoption and user satisfaction, determined by factors like usefulness, ease of use, and satisfaction (Lubis et al., 2020).

This usability assessment gauges interaction and information delivery (Lubis et al., 2020) via a 21-question USE questionnaire using a five-point Likert scale in Table 29.2 (Harna, T. G., et al., 2022). The 66 respondents, comprising 56 employees, 6 HRD staff, and 4 IT experts, are listed in Table 29.1. Total enumeration was employed for the HRMD personnel and IT experts, while employee respondents were selected using stratified random sampling.

Table 29.1 Respondents of the study

Respondents Type	Number
HRMD Personnel	6
IT Experts	4
Employees	56

Qualitative data interpretation is presented in Table 29.2 to explain the results that match the study's category averages and overall usability averages.

4. Discussion and Findings

This section covers the developed system and the assessment findings of its usability and features by respondents.

Table 29.2 Data interpretation

Scale	Mean Range	Descriptive Interpretation
5	4.20 – 5.00	Strongly Agree
4	3.40 – 4.19	Agree
3	3.00 – 3.39	Just Right
2	1.80 – 2.59	Disagree
1	1.00 – 1.79	Strongly Disagree

The HRIS for Apayao State College's HRMD consists of five modules to automate HRMD personnel's major functions which are as follows:

Recruitment and Selection Module will rank the applicants by discipline and keep records of applicants for reference.

Personal Data Sheet (PDS) Management Module automatically archives employee PDS, handles service records, and generates necessary reports, including gender ratio and educational degree percentages during specific periods.

Attendance Monitoring Module generates CS Form 48, excluding calculations for tardiness, undertime, and absences due to blank time entries and attachments on specific dates.

Leave Ledger Management Module will facilitate the updating and computing of the used and unused leave/service credits.

The *Performance Evaluation module* will facilitate employees' performance each semester for NBC 461 promotion evaluations and generate a summary report.

Physical Design of the Developed System

The following screenshots illustrate the system's interface and its usage.

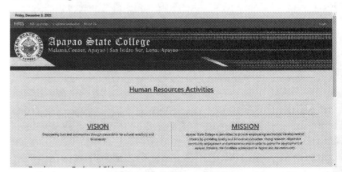

Fig. 29.1 Main page

This page will display HRM&D announcements and activities and the ASCs Vision, Mission, and Development Goals and Objectives.

Login requires a username and password. Users and admins can log in using their Employee IDs as usernames and default Employee ID numbers as passwords, which must be changed for security after login.

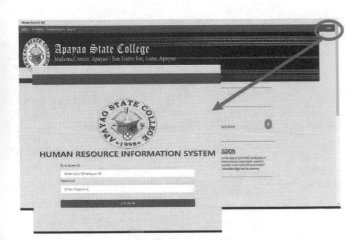

Fig. 29.2 Log in page

Administrator Page

Fig. 29.3 Administrator page

The Administrator Page will be managed by the Human Resource Officer and Staff to add users, update, delete, and set the settings, and generate reports such as PDS, Leave Ledger, and others to be used by other entities.

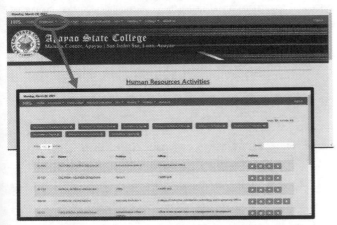

Fig. 29.4 Employee Tab

The employee tab shows reports on education, rank, and other personal information, leave records, PDS, DTR, and evaluations. It also includes a search engine and screen entry count.

The Timelog tab shows the attendance monitoring page with a time-in and time-out section on the left side of the screen. It includes an ID number box for entering employee IDs and buttons for time-in and time-out.

The Jobs tab manages job openings, with an application form for basic information and document attachments. It also lists applicants for HR review and displays position timelines.

Fig. 29.5 Timelog tab

Fig. 29.6 Jobs tab

Employee Page

Fig. 29.7 Employee page

The employee page consists of six (6) tabs namely: profile tab which displays the pre-inputted PDS, Leave Ledger which displays the used and unused leaves, Employees Evaluation which displays the evaluation sheet, daily time record which displays the printable DTR, Job Opening which displays the job vacancies and other.

The developed system's usability and its features were evaluated by the respondents using the USE tool.

In Fig. 29.8, researchers assessed the system's usability. The highest mean score (4.68) suggests it mostly meets needs, but the lowest (4.03) shows room for improvement to meet all expectations (Balcita, A. P., et al., 2019; Maligat, D. E., et al., 2020). Respondents strongly agreed the system is useful regarding time, effectiveness, and user expectations (Balcita, A. P., et al., 2019).

In Fig. 29.9, user-friendliness excels with a high weighted mean of 4.50, but error recovery and simplicity received the lowest rating at 3.56. This indicates a need for a simpler design and improved error recovery for a better user experience (Balcita et al., 2019).

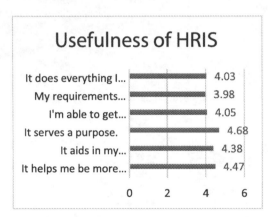

Fig. 29.8 Usefulness of HRIS of ASC

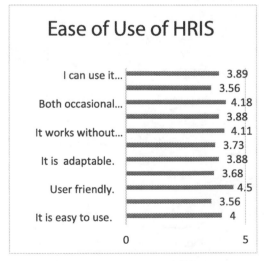

Fig. 29.9 Ease of use of HRIS

Figure 29.10 shows HRIS satisfaction for ASC. Functionality scored high at 4.15, but overall satisfaction was lower at 3.68, echoing Balcita et al. (2019) findings, indicating user satisfaction with room for improvement in the user experience.

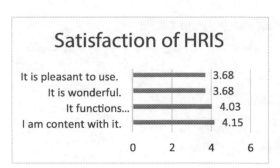

Fig. 29.10 Satisfaction of HRIS

Table 29.3 summarizes usability findings. "Usefulness" received the highest average rating (4.27), indicating strong user satisfaction. "Ease of use" scored an average of 3.91, signifying agreement on its user-friendliness. "Satisfaction" averaged 3.89, indicating alignment with user contentment. Overall, usability averaged 4.02, suggesting general acceptance. These results support the need for full system implementation to enhance HR services in line with modern demands and technology (Khaled, R., 2022).

Table 29.3 Usability of HRIS of ASC

Indicator	Mean	Descriptive Interpretation
Usefulness	4.27	Strongly Agree
Ease of Use	3.91	Agree
Satisfaction	3.89	Agree
Overall Weighted Mean	4.02	Agree

5. Conclusion

The system's full implementation resolves manual HR tasks, benefiting both HR and employees. Users appreciate its efficiency but have noted usability issues. Researchers recommend management's technical support to optimize the system's performance. Future researchers could consider adding menu items for increased efficiency and improved customer service in the HRIS.

References

1. Balcita, A. P., Kupsch, S., & Palaoag, T. D. (2019). KITIKIT: A Mobile App for Searching Hand-carved Wood Products. In Proceedings of the 2019 4th International Conference on Intelligent Information Technology (pp. 57-61).
2. Harna, T. G., & Syaifurrahman, E. (2022). User Experience Evaluation of Building Materials Application Using Usability Testing Based on ISO 9241–11 Standard and The USE Questionnaire. In 2022 10th International Conference on Cyber and IT Service Management (CITSM) (pp. 1-6). IEEE.
3. Johnson, R., & Gueutal, R. (2020). Transforming HR through technology. The use of E-HR and HRIS in organizations.

4. Khaled, R. (2022). Impact of the Human Resources Information System on the Effectiveness of Strategic workforce planning case study British Petroleum Company in Algeria.

5. Lubis, B. O., Salim, A., & Jefi, J. (2020). Evaluasi Usability Sistem Aplikasi Mobile JKN Menggunakan Use Questionnaire. Jurnal Saintekom, 10(1), 65-76.

5. Maligat, D. E., Torio, J. O., Bigueras, R. T., Arispe, M. C., & Palaoag, T. D. (2020). Web-based knowledge management system for camarines norte state college. In IOP Conference Series: Materials Science and Engineering. IOP Publishing

6. Othman, M. K., Nogoibaeva, A., Leong, L. S., & Barawi, M. H. (2022). Usability evaluation of a virtual reality smartphone app for a living museum. Universal Access in the Information Society, 21(4), 995-1012.

7. Prastyo, D., & Bakhtiar, M. Y. (2019). Development of the Human Resources Information System (HRIS) Based on Usability Analysis with Usefulness, Satisfaction, Ease to Use (USE) Questionaire and Cognitive Walkthrough Methods. bit-Tech, 1(3), 148-160.

8. Zhou, L., Bao, J., Setiawan, I. M. A., Saptono, A., & Parmanto, B. (2019). The mHealth App Usability Questionnaire (MAUQ): development and validation study. JMIR mHealth and uHealth, 7(4), e11500.

Note: All the figures and tables in this chapter were made by the authors.

Advancing Sustainable Science and Technology for a Resilient Future – Sai Kiran Oruganti et al. (eds)
© 2024 Taylor & Francis Group, London, ISBN 978-1-032-79020-6

Thematic Analysis of the Learning Management System of Samar State University: The User Perspectives

30

Sweet Mercy Pacolor*
Samar State University, Philippines.

Thelma Palaoag[1]
University of the Cordilleras, Philippines

Abstract: The use of learning management system has become more popular in higher education institutions through the advancement of delivering quality educational services. Access to LMS offers a platform in modern education thus integrating information and communication technologies has changed how students are taught and how they learn. This study aims to find out the perceptions of the faculty and students on the use of an LMS in higher education. Strategy in the collection and analysis of data, the researchers used qualitative research design. The main source of data was gathered through an open-ended questionnaire for thematic analysis through NVivo 12 software. The results highlight the four categories based on thematic analysis of data towards using LMS. These were advantages, disadvantages, features, and problems users of the LMS encountered. The implementation of LMS has both positive and negative implications. Continual improvement of the developed LMS addressing the problems must be undertaken.

Keywords: Higher education, e-learning, Information and communications technology, Moodle, NVivo 12

1. Introduction

The majority of higher education institutions in the Philippines currently have Learning Management Systems (LMS) in place. These web-based LMSs are designed to assist in educational endeavors. According to Unwin et al. (2010), the most extensively used LMS in the area is Blackboard, Sakai, KEWL, and Moodle. They include a variety of capabilities that allow faculty members to exchange learning materials and provide synchronous and asynchronous interaction with their students (Vovides et.al. 2007). LMS is used by institutions to enhance the conventional face-to-face delivery model where faculty members create and distribute digital resources online. Here, the LMS serves as a digital library for learning resources (Vovides et. al 2007). To reach more students across diverse geographic borders, some institutions, notably those that provide distance education, have started fusing LMS with conventional face-to-face delivery (Andersson & Gronlund 2009). These factors have led to a rise in LMS implementation among Filipino higher education institutions in the past few years. Today LMS is adopted by almost all higher learning institutions in the country. This adoption has been caused by the need to

continue the learning process despite the threat posed by the pandemic. Supported by a number of agencies, including the Office of the President, the Commission on Higher Education (CHED), the Department of Budget and Management (DBM), and other public and private educational institutions. These organizations have been allocating a range of resources to help universities adopt and implement various LMS. Many institutions in developed nations have implemented LMS with success. Students' academic performance has improved, dropout rates have decreased, and successful in raising student satisfaction with the course given (Naveh et al. 2012). In an effort to reap the same advantages as their counterparts elsewhere, Philippine institutions have been implementing them. In contrast to institutions in affluent nations, HEI in the Philippines is situated in a diverse environment and faces unique problems. Because of this, implementing and adopting these methods does not ensure that institutions will receive the same advantages as those in wealthy nations. The goal of this study is to establish whether the use of LMS is satisfying its capability and identify its advantages, challenges, and problems encountered while utilizing the LMS, and determine solutions to improve the utilization of LMS.

*Corresponding author: sweetmercy.pacolor@ssu.edu.ph
[1]tpalaoag@gmail.com

DOI: 10.1201/9781003490210-30

2. Methodology

Strategy in the collection and analysis of data, the researchers used the qualitative research design. The main source of data was gathered through an open-ended questionnaire for thematic analysis. The questions asked were thematic with advantages, disadvantages, features, and problems in utilizing the LMS. In the selection of faculty and student respondents, random sampling was utilized by the researchers. A total of 30 faculty respondents from the College of Arts and Sciences and 60 fourth-year student respondents taking Bachelor of Science in Information Technology at Samar State University-College of Arts and Sciences utilize Moodle as their LMS. NVivo 12 software was used for the context analysis of the result. The word frequency query was used, and the most frequently used words in the discussions were examined using the word cloud tab. The word cloud analysis uses different font sizes, with bigger fonts used for frequently occurring terms. The research respondents were assured of the confidentiality of their responses. Furthermore, the respondents were informed about the way in which the data was processed and the use of a computer and had the choice to voluntarily consent or decline to participate in the study. The data collection phase stated 100 percent of the respondents answered the open-ended questionnaire through google forms. Thematic analysis was done through NVivo 12 software to the obtained data, organized, analyzed, and interpreted.

3. Results and Discussion

Figure 1 shows the analysis of the advantages of LMS as perceived by the faculty and students as respondents. As gleaned from the word cloud analysis, the words that are blown up are educational, LMS, easy, students, and organization. Figure 30.1 denotes that they consider the LMS can save the organization time and money by allowing the easy administration of tracking students' progress and generating reports in a user-friendly web-based environment; easy to access the necessary data and information, easy to access to the online class as well as easy to access to instructional materials and assessment without being in the classroom.

Fig. 30.1 Word cloud on advantages of LMS

Furthermore, the respondents mentioned that the LMS organizes the subject contents and learning management which is helpful in conducting flexible learning. Students can download their learning materials, study, read and learn anywhere and anytime. LMS improves the educational approaches facilitated through face-to-face learning as well as online learning.

Fig. 30.2 Word cloud on disadvantages of LMS.

Figure 30.2 shows the analysis, through word cloud, of the disadvantages of LMS as perceived by the respondents. As gleaned from the word cloud analysis, the words that are blown up are connection, system, LMS, require, and caused. It denotes the need for a good internet connection to access and utilizes the LMS. Poor internet connection has challenges in logging into the accounts of the respondents, difficult data traffic to upload and download files, and submission of the requirements was challenging as well. It is also difficult for some students caused for those who are living the remote areas need to find a place with a good signal to have access to the internet. Furthermore, it requires a gadget like a laptop, computer, tablet, or smartphone to fully utilize and access the LMS. The system requires self-discipline for online learning hence self-study and self-learning, not all students can easily understand and need interaction and discussion to attain learning. Online learning may create a sense of isolation and hard to comprehend and sometimes lessons need to be further discussed to fully understand.

Fig. 30.3 Word cloud on features of LMS.

Figure 30.3 shows an analysis of the features of LMS as perceived by the faculty and student as respondents. As gleaned or the word cloud analysis, the words that are blown up are upload, transfer, activities, LMS, and content. Figure 3 denotes that LMS has a centralized utilization of the learning materials where you can upload and download files and transfer the same coursework to another class with the same content and activities. LMS supports multiple learning modes for faculty it can post lessons in a video, audio, eBooks, presentation, and simulation while students can also take online assessments, upload their requirements, and communicate with their instructors. Furthermore, the university has developed a customized system of LMS that makes it easier to administer, upload and evaluate the university's online learning initiatives. It provides instructional resources to the students including online courses and real-time teaching. Both desktop computers, smartphones, and other devices running the Android operating system can access and use LMS. As long as they have an internet connection, it is accessible to all faculty and students, wherever they may be.

Fig. 30.4 Word cloud on problems of LMS.

Figure 30.4 shows the analysis through word cloud, the problems of LMS as perceived by the respondents. As gleaned from the word cloud analysis, the words that are blown up are connection, communication, system, LMS, and personalization. It denotes the biggest problem the faculty and student respondents encountered in utilization of the LMS is the poor intermittent internet connection. It requires a good signal from a communication network to have a stable internet connection. Without internet connectivity, you cannot access the system of LMS with its content learning materials and course management system. Furthermore, the student respondents mentioned that they have come across the complexity of downloading the resources, uploading submission requirements, and during online assessment especially if the quiz/exam is set to take it once. A few systems' functionalities of the LMS are difficult to understand, with personalization it has a restriction of large memory files that cannot be uploaded due to the limit of storage capacity.

4. Conclusion

The respondents shared similar perceptions of the positive attitude toward the utilization of LMS. Both faculty and students can benefit greatly from LMS. The use of LMS improves the management of learning resources and facilitates data and information access for both faculty and students. Enhances adaptability for effective time management. The disadvantage of LMS using Moodle involved the software is not user-friendly for the non-technical user, and some functionality is difficult to understand. As it requires a stable internet connection to function, not all students can access LMS in remote areas. The LMS is for both faculty and students; faculty may upload online lessons, assignments, and quizzes and enter assignment grades. It not only saves time but also enhances student performance and learning experiences. The problem faced by faculty while using the LMS involves the limited upload size from the activity or resources. The university has provided facilities to facilitate the utilization of LMS, training was provided for faculty, and orientation to the students. SSU has enhanced and upgraded its network system and internet connectivity and established a fiber-optics network on the campus. The implementation and application of LMS have both positive and negative outcomes. Continual improvement of the developed LMS addressing the problems must be undertaken. The conduct of intensive training programs for the faculty and students should be comprehensive for a better understanding, experience, and interaction of Moodle as the model of technology in designing the content and learning goals.

References

1. Andersson, A., & Gronlund, A. (2009). A Conceptual Framework for E-learning in Developing Countries: A Critical Review of Research Challenges. Electronic Journal of Information Systems in Developing Countries (EJISDC).1 16.
2. Naveh, G., Tubin, D., & Pliskin, N. (2012). Student Satisfaction with learning management systems: a lens of critical success factors. Technology, Pedagogy and Education. 21(3): 337 350.
3. Thuseethan, S., & Achchuthan, S. (2018). Usability Evaluation of Learning Management System in Sri Lankan Universities.
4. Unwin, T., & et., a. (2010). Digital learning management systems in Africa: myths and realities. Open Learning: The Journal of Open and Distance Learning. 25(1):5 23.
5. Vovides, Y., Sanchez-Alonso, S., Mitropoulou, V., & Nickmans, G. (2007). The use of e-learning course management systems to support learning strategies and to improve self-regulated learning. Educational Research Review. 2(1): 64 74.
6. Yousaf, F., Shehzadi, K., & Aali, A. (2021). Learning Management System (LMS): The Perspectives of Teachers. Global Social Sciences Review. VI (I):183 196.

Note: All the figures in this chapter were made by the authors.

Advancing Sustainable Science and Technology for a Resilient Future – Sai Kiran Oruganti et al. (eds)
© 2024 Taylor & Francis Group, London, ISBN 978-1-032-79020-6

Integrating Augmented Reality to Omani HEIs: A Covid19 Inspired Case Study-Based Model

31

Khalfan Al Masruri*
Muscat College, Oman

Ferddie Quiroz Canlas[1]
Muscat College, Oman

Hameetha Begum[2]
Muscat College, Oman

Reshmy Krishnan[3]
Muscat College, Oman

Abstract: The COVID-19 pandemic forced Governments to impose lockdowns and temporary closures of establishments to curb the spread of the virus affecting the economy in general but all sectors of society - education in particular. A recent study conducted on higher education institutions in Oman revealed that students become passive learners, teacher-dependent and lack motivation for independent learning during online classes due to an impersonal approach to teaching. Lecturers had difficulty to conduct and assess practical sessions. Despite many reported perceived practical implications of augmented reality, there is a scarcity of studies within higher education institutions in Oman. This project aims to explore the potential of AR and develop an augmented reality-based learning system for selected theoretical and practical courses in Muscat College as models by developing a prototype that will provide an immersive experience and immediate learning feedback for the students and facilitate independent learning through an interactive system.

Keywords: Augment reality, ICT, Teaching and learning, Education technology

1. Introduction

The Novel Coronavirus broke out by the end of December 2019. To prevent the spread of the virus, government agencies had imposed lockdowns. This action has repercussions not only on the economy in general but on almost all sectors of society and education. This negatively affected around 978 million learners in 131 countries (World Health Organization, 2020).

As early as the 15th of March 2020, the Supreme Committee of the Sultanate of Oman suspended all schools and universities, followed by a month-long lockdown. In the ensuing months, the Ministry of Higher Education (MOHE) decreed that all higher education institutions (HEIs) shift from traditional face-to-face classes to online mode.

Even in the pre-pandemic semesters, students are attentive within 10 to 15 minutes of the class and eventually drop out. This manifestation of the loss of interest varies from frequently going out of the room and the use of mobile phones in class (Bradbury, 2020). During online class, it is reported that 48.1% of Omani students do not cooperate during online classes due to various reasons such as internet connection issues, inexperience in learning online, etc. Furthermore, most students are dependent learners and require more assistance from teachers. Due to the perceived distance barrier and impersonal approach to teaching, students tend to become passive learners thus resulting in a loss of interest (Slimi, 2020). Moreover, assessing, and guiding students in doing practical activities becomes cumbersome to teachers.

Bradbury (2020) also emphasizes the role of educators in designing learning-teaching materials and environments that make students engaged and maintain their interest throughout the session. This endeavour encompasses various pedagogical principles and teaching methodologies underpinning learning

*Corresponding author: ferddie@muscatcollege.edu.om.

DOI: 10.1201/9781003490210-31

outcomes. However, shifting to augmenting online teaching takes more effort. This is evident when it comes to the assessment of practical and laboratory exercises and activities students do online.

The proponents believe that augmented reality will help in addressing the issues mentioned based on convincing applications of it to various fields. Besides, augmented reality supplements a real-world environment with digital views using assistive tools like smartphones and tablets. Long before the pandemic, simulations using either or both virtual and augmented reality applications were in place to replace environments that pose risks to students or are too expensive. For instance, flying aeroplanes, performing surgeries, and underground mining are dangerous and expensive for students and universities.

This project aims to explore the potential of AR and develop an augmented reality-based learning system for selected theoretical and practical courses in Muscat College as models. The said prototype will provide an immersive experience for the students and facilitate independent learning through an interactive system. Also, it will augment the lecturers' delivery of theoretical and practical sessions, wherein students can obtain assistance and immediate feedback on their learning.

2. Review of Related Studies

Augmented Reality (AR) and Virtual Reality (VR) have become more affordable, allowing teachers easy access to AR learning activities. With the growing trend of online education, the use of digital technologies has become a critical issue.

There are many applications of AR applications in education, i.e. engineering and computing (Nesenbergs, Abolins, Ormanis, & Mednis, 2021). AR brings immersion experience for learning and generates new learning (Mystakidis, Berki, & Valtanen, 2021).

Another stream of research discussed how AR does affect student learning. As an example, augmenting the e-learning experience for students is the subject of (Bednarz, Caris, & Dranga, 2019) in their proposed immersive virtual reality prototype. Using a four-meter hemispherical dome projection screen that simulates underground mining and an iPhone, users can interact directly with an application. This virtualization simulates a risky environment for students to explore in real life. With the same goal, researchers from Chung Yuan Christian University, (Chan, Wu, Jong, & Lin, 2022) developed a web-based infrastructure melding both virtual and physical Physics laboratories. Before going to actual experiments, students can practice with the simulated environment. This setup avoids dangers for students who

will do the experiments for the first time and without prior knowledge of the use of laboratory equipment.

Despite this evidence, limited research studies have examined the role of image-based and content-based AR on student acceptance behavior in the context of Oman. (Alsqria & Al Salmi, 2020) explored the impact of augmented reality on 10th-grade female students in an Islamic education in Oman. The study revealed that the AR control group had higher imaginative thinking scales. Similarly, (Al Shuaili, Al Musawi, & Muznah, 2020) found that 10th-grade male Omani students using mobile AR applications scored higher in their achievement and attitudinal tests as compared to those who used the normal teaching approach in social studies. (AlBuraiki, Adbullah, & Khambari, 2022) observed that Omani students are starting to lose interest in studying STEM-related disciplines. They proposed a model based on their study that explored the students' readiness and acceptance of AR and how it captures their interest.

3. Methodology

Data Collection and Sample

The study will involve BCS Software Engineering and BCS Computer Systems and Networking freshmen students taking ICT Competency Workshops as respondents of the study. A perception-based questionnaire will be given to solicit their experience with the use of augmented reality applications. Furthermore, empirical analysis and comparisons of the student's assessment results will be conducted to underpin the use of the app in other modules.

Proposed AR-Based System Architecture

The proposed AR-based learning application (Figure 1) is a multi-tier approach system that is comprised of entities such as front-end applications for students' mobile/tablets interacting via the internet with the application server that houses the database, AR algorithm engine, and learning/ course materials.

Students access the learning materials and interact with the AR system using their mobile phones or tablets. The application server coordinates and manages transactions made to the system. The database holds pertinent information related to but not limited to students' information, course information, learning progress, and feedback. The AR algorithm engine processes and facilitates students' interaction with the system and melds the learning contents and real environment. The learning institution (using an interface) regularly updates learning contents and materials.

To make a robust application, the application server will be using various AR engines as no single engine can address all the requirements. Unity Game Engine provides a native cross-platform AR API for various platforms. It will work with ARToolKit to target Windows, Linux, and OS X, ARKit for iOS, and ARCore for Android. Unity has gained a good reputation in the AR realm. Wikitude, another AR engine will be used to support students' activities that involve tracking, geolocation, cloud recognition, and distance-based scaling features

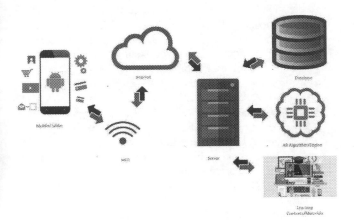

Fig. 31.1   Proposed AR-Based System Architecture

As per the User Interface, Flutter will be used to develop a cross-platform application. Laravel, NodeJS, and CSS to design the interface for the web version of the application.

Excerpt of the Visual Presentation of the Application's Prototype

Initially, the students access the application using a link in a QR code as depicted in Fig. 31.2. This can be in a printed format or stored electronically.

Fig. 31.2   QR code to launch App

Once the app is launched as shown in Fig. 31.3, the students have options either to load the learning materials or proceed to the assessment part.

Fig. 31.3   Dashboard

Figure 31.4 shows an example of the prompt when a student is in the learning mode.

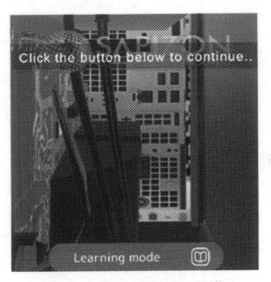

Fig. 31.4   Learning mode

During the learning mode, the student can manipulate the AR model in a 360-degree mode. Zooming and panning are also supported. Each component can be magnified. Links were provided to obtain the description and discussions. External references were also embedded to supplement students' learning. There is also a dropdown menu to select the chapters and topics. Figure 31.5 shows the AR model of the internal components of a computer system. Students can detach and reattach the computer part while reading the topic related to the part under scrutiny. Similarly, during assessment, students will be given the same model with parts disassembled. The app will notify the students of mistakes committed. Scores will be provided at the end of the activity.

Fig. 31.5   AR model environment

Proposed Statistical Treatment of Data

With the assumption that the distribution is not normal, the study will employ Chi-Square, Kruskall-Wallis, Scheffe Post Hoc, and Spearman Rho Tests to treat the data. Using PSPP, an open-source data science tool, the following hypotheses will be tested:

1. There is no significant difference between the courses and experiential perception of the integration of AR into the module between the students and academic staff.

2. There is no significant difference in the results of the assessment of students taking the AR-integrated module and students taking the traditional module.

3. There is no significant correlation between the use of AR in the module in the student's assessment results.

3. Conclusion

AR has been proven as an effective tool to augment the traditional way of teaching as proven by various studies. Despite all these, Oman has few studies exploiting the benefits of AR. This study initially highlighted the ability to realize such prototypes to augment classroom teaching (as inspired by COVID-19), especially the practical sessions and prepare students prior to their hands-on examination. The proponents intend to conduct an empirical analysis of the survey results as the continuation of this study.

Acknowledgements: This project was funded by the Ministry of Higher Education, Research, and Innovation – Oman for 2021-2023 through Musat College

References

1. Al Shuaili, K., Al Musawi, A., & Muznah, R. (2020). The Effectiveness of Using Augmented Reality in Teaching Geography Curriculum on the Achievement and Attitudes of Omani 10th Grade Students. *Multidisciplinary Journal for Education, Social and Technological Sciences, VII*(2), 20-29.

2. AlBuraiki, A. A., Adbullah, S. I., & Khambari, M. N. (2022). Enhancing Students' Interest in STEMRelated Subjects at Omani Post-Basic Schools through Application of Augmented Reality. *Proceedings of the 30th International Conference on Computers in Education. Asia-Pacific Society for Computers in Education*. Kuala Lumpur.

3. Alsqria, R. M., & Al Salmi, M. N. (2020). The impact of using augmented reality technology on imaginative thinking improvement of 10th grade female students in Islamic education in Sultanate of Oman. *International Journal of Educational and Psychological Studies, VIII*(2), 463-474.

4. Bednarz, T. P., Caris, C., & Dranga, O. (2019). Human-computer interaction experiments in an immersive virtual reality environment for e-learning applications. *20th Australasian Association for Engineering Education Conference*. Adelaide.

5. Bradbury, N. A. (2020, June 10). *The Science of Attention: How To Capture And Hold The Attention of Easily Distracted Students*. Retrieved from InformED: https://www.opencolleges.edu.au/informed/features/30-tricks-for-capturing-students-attention/

5. Chan, T., Wu, Y., Jong, B., & Lin, T. (2022). A Web-Based Virtual Reality Physics Laboratory. *Advanced Learning Technologies, IEEE International Conference*. Athens.

6. Mystakidis, S., Berki, E., & Valtanen, J. (2021). Deep and Meaningful E-Learning with Social Virtual Reality Environments in Higher Education: A Systematic Literature Review. *Applied Sciences, XI*(5).

7. Nesenbergs, K., Abolins, V., Ormanis, J., & Mednis, A. (2021). Use of Augmented and Virtual Reality in Remote Higher Education: A Systematic Umbrella Review. *Education Sciences, 11*(8).

8. Slimi, S. (2020). Online learning and teaching during COVID-19: A case study from Oman. *International Journal of Information Technology and Language Studies, IV*(2), 45-56.

9. World Health Organization. (2020, April 23). *Corona Virus Disease 2019 (COVID-19) Situation Report - 94*. Retrieved from WHO: https://www.who.int/docs/default-source/coronaviruse/situation-reports/20200423-sitrep-94-covid-19.pdf

Note: All the figures in this chapter were made by the authors.

Advancing Sustainable Science and Technology for a Resilient Future – Sai Kiran Oruganti et al. (eds)
© 2024 Taylor & Francis Group, London, ISBN 978-1-032-79020-6

CatBoosting Advanced Ensemble Machine Learning Techniques to Predict Heart Disease

32

S. Hameetha Begum*
Muscat College, Sultanate of Oman.

S. N. Nisha Rani[1]
Fatima Michael College of Engineering and Technology, Madurai, India.

S. M. H. Sithi Shameem Fathima[2]
Syed Ammal Engineering College, India

T. SivaKumar[3]
M. Kumarasamy College of Engineering, India

Abstract: Heart disease is a life-threatening disease that is rapidly increasing in both developed and developing countries and, as a result, causes death. Cardiovascular disease kills roughly one in each four people, and it affects all racial and ethnic groups. Researchers are developing smart systems that can diagnose this disease accurately with the aid of machine learning algorithms. In this study, few Classification methods were demonstrated in this paper including Decision Tree (DT), K Nearest Neighbour (KNN), Naïve Bayes (NB), Random Forest (RF), Gradient Boosting followed by Cat Boosting to improve prediction accuracy of heart disease. The models were assessed based on their accuracy, precision, recall, and F1-score. Cat Boosting was utilised in the proposed model to detect signs of having a heart illness in a specific individual. The model's effectiveness is measured and performed best with 100% accuracy and AUC score of 1. It is expected that the proposed system will help doctors diagnose heart problems more effectively.

Keywords: AUC, CatBoosting, KNN, RF, ROC

1. Introduction

Heart disease can potentially cause death without showing any outward signs, earning "dumb thug". Machine learning methods can be quite helpful in this regard. Although there are many different ways that heart disease can present itself, there is a common a group of important risk factors that determine whether or not someone has a heart disease risk. Machine learning is attracting the attention of many researchers, because it shortens the diagnostic process while proving accuracy and efficacy. In order to anticipate diseases, machine learning (ML) is crucial (Obasi & Omair Shafiq, 2019). It determines whether a patient has a certain illness type based on an effective learning technique (Sagheer et al., 2019; Obasi & Omair Shafiq, 2019). Machine learning techniques can be used to diagnose a variety of diseases, this paper is primarily concerned with the process of diagnosing

heart disease because it is now the top reason of mortality and detection in advance of heart disease can save many lives.

2. Related Work

Numerous papers have been published in the literature on the utilisation of machine learning methods to diagnose heart disease (Chandra et al., 2019) with multiple feature selections to classify heart disease diagnosis as healthy or unhealthy (Thomas & Princy, 2016). This study examines various classification methods for estimating each person's risk level based on factors like blood pressure, gender, age, pulse rate, and cholesterol. Data mining classification techniques such as Decision Tree Algorithm, Naive Bayes, KNN, Neural Network etc. are used to categorise the patient risk level. In the study (Yaqoob et al., 2020), recommended that the predictive accuracy of k-nearest neighbors (k-NN) and support vector

*Corresponding author: hameetha@muscatcollege.edu.om

DOI: 10.1201/9781003490210-32

machines (SVM) for ischemic heart disease (IHD) based on important risk factors be compared. The accuracy rate, sensitivity, specificity, and area under the receiver operating characteristics (ROC) curve (AUC) of both models were evaluated on the training and testing datasets. On the testing dataset, the support vector machine (SVM) outperformed the k-nearest neighbors (k-NN) model with higher values for accuracy (86.67%), sensitivity (80%), specificity (90%), and AUC (94.1%). On the test dataset, the accuracy, sensitivity, specificity, and AUC values for the support vector machine (SVM) were greater than those for the k-nearest neighbors (k-NN) model. Researchers are therefore becoming more and more interested in using machine learning approaches to forecast heart disease. Proposed CatBoost stands out as the superior choice for heart disease prediction compared to other algorithms due to several limitations of the latter methods.

3. Machine learning models

This section covers the existing machine learning models utilised in the study in detail before introducing the suggested Cat boosting classifier. The proposed model was examined with the baseline machine learning models already in use. A Cat Boosting classifier is used in this part to categorise heart disease patients according to their symptoms. Prior to data collection, the methodology involves preprocessing. The chosen classifiers are then trained and assessed once more on the benchmark dataset using a conventional 5-fold cross validation technique. Its primary goal is to aid in diagnosing heart disease and predicting the likelihood of heart-related conditions

In this study, various ML techniques are used to build the models. Briefly described the techniques as follows: Decision trees are binary trees that divide the dataset until only pure leaf nodes, or data with a single type of data class, are left (Fan et al., 2017). KNN algorithm is part of a group of models known as 'lazy learning' algorithms. This means that it solely relies on a stored training dataset instead of undergoing a distinct training phase (Gárate-Escamila et al., 2020; Hameetha Begum & Nisha Rani, 2021; Barun et al., 2022;) This also implies that when a classification or prediction is made, all computation occurs. Navie Bayes Model is employed in a variety of classification problems. According to Bayes' theorem , the probability of one event happening given the occurrence of another is measured by conditional probability. Although they are quick and easy to implement, their main drawback is the requirement for independent predictors. The Random Forest model operates on the assumption that a collection of independent models, also known as decision trees, can yield precise outcome than a standalone decision tree. Each tree contributes a classification when Random Forest is employed for classification, and forest chooses

classification with the most "votes" The ability of random forest to automatically adjust for decision trees' propensity to over fit to their training set is its most practical benefit. The bagging method and random feature selection virtually minimise the over fitting issue when applying this algorithm, which is great because over fitting produces wrong results. Furthermore, Random Forest typically maintains its accuracy even when some data is absent (Yadav & Pal, 2020). Gradient boosting is a training method that learns several models in a highly progressive, cumulative, and sequential way. Learning in gradient boosting occurs through minimising the loss.

The Cat Boost algorithm introduced a new method for regularization in boosting models known as Minimal Variance Sampling. This approach is modified variant of the Stochastic Gradient Boosting method, which involves weighted sampling. Cat Boost handles categorical features using one-hot encoding. This boosting technique is designed to prioritize the processing of categorical data and boost trees using an ordering approach. Mean target encoding replaces category with mean target value. The average target is calculated by using formula.

$$Average\ Target = \frac{Current\ count + prior}{Maximum\ count + 1} \tag{1}$$

In the context of heart disease prediction, different machine learning algorithms offer distinct strengths and applicability based on the dataset's characteristics and the desired outcome. K-Nearest Neighbors (KNN) may be well-suited for small datasets with straightforward feature relationships, while Random Forest proves effective in handling both numerical and categorical features, capturing complex non-linear interactions among multiple risk factors. Decision Trees are valuable for initial exploratory analysis, providing interpretable rules for feature importance in certain scenarios with limited features and simple relationships. Support Vector Machine (SVM) can be beneficial when the dataset is relatively small and well-structured. On the other hand, CatBoost stands out as a robust choice for heart disease prediction, effectively handling categorical features and mitigating biases without extensive preprocessing. CatBoost's powerful gradient boosting capabilities, a well-rounded approach can be adopted to predict heart disease accurately. The final algorithm selection should consider the dataset characteristics, interpretability requirements, and computational resources available, ensuring optimal predictive performance.

The dataset comprises 303 entries with 14 columns, representing various patient attributes. The attributes include age, sex, chest pain type, resting blood pressure, serum cholesterol level, fasting blood sugar, resting electrocardiographic results, maximum heart rate achieved, exercise-induced angina, ST depression induced by exercise relative to rest, the slope of the peak exercise ST segment, the number of major vessels colored by fluoroscopy, and thalassemia status. The target variable is binary, indicating the presence (1) or absence (0) of heart disease.

4. Results and Discussion

Various assessment criteria, notably Recall, Precision, Accuracy and F1-score are employed in this study to assess the effectiveness of ML classifiers using components from the confusion matrix. Researchers can use a confusion matrix to gauge a classification problem's performance rate based on four key variables. False Positive (FP), True Positive (TP), True Negative (TN), and False Negative (FN). The values of the parameters used in the calculation of an accuracy score are displayed in a classification report, which is made up of numerous constituent variable parameters.

Table 32.1 Classification report summary of various ML binary classifier

ML Models	Precision	Recall	F-Score	Accuracy (%)
Decision Tree	0.88	0.88	0.88	89.2
K Nearest Neighbor	0.82	0.84	0.82	82.3
Navies Baye	0.87	0.86	0.86	86.89
Random Forest	0.88	0.89	0.88	89.3
Gradient Boosting	0.95	0.92	0.93	96.67
CAT Boosting	1.0	1.0	1.0	100

Table 32.1 presents a comparison of the performance of various machine learning models for a classification task. The evaluation is based on four key metrics: Precision, Recall, F-Score, and Accuracy. The models, Gradient Boosting and CAT Boosting, achieved the highest performance across all metrics. Based on the evaluation of the ML models, it can be concluded that CAT Boosting are the most effective models for the classification task, achieving perfect Precision, Recall, F-Score, and Accuracy.

Figure 32.1 shows the analysis of F-Score, Precision, and Recall alone, it is evident that Gradient Boosting and CAT Boosting are the top-performing models, excelling in achieving a balance between Precision and Recall, with CAT Boosting achieving perfect scores. Selecting the most suitable model for a specific application should consider the trade-off between Precision and Recall based on the specific requirements and cost of false positives and false negatives.

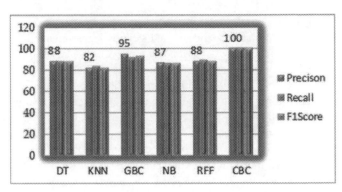

Fig. 32.1 Classification summary of ML classifiers

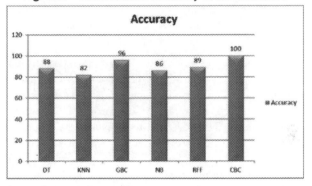

Fig. 32.2 Classification accuracy of various ML models

Figure 32.2 shows the analysis in terms of accuracy alone, CAT Boosting stands out as the top-performing model, achieving a perfect accuracy of 100%. This indicates that the model made no mistakes and accurately classified all instances in the dataset. Gradient Boosting follows closely behind with a high accuracy of 96.67%, demonstrating strong overall performance. Based on the provided evaluation metrics, the Decision Tree model shows good precision, recall, and F-score indicating a balanced performance in identifying positive cases. The accuracy of 89.2% suggests it performs reasonably well on the given dataset. However Decision Trees tend to be prone to overfitting, especially with deeper trees. This might limit its generalization to unseen data. It may not be the best choice for datasets with complex relationships that require a more sophisticated model. KNN demonstrates reasonable precision, recall, and F-score. The accuracy of 82.3% suggests that KNN's performance is not as strong as some other models. It is computationally expensive during prediction as it needs to calculate distances to all neighbors. It may not work well with high-dimensional or imbalanced datasets. Naive Bayes shows good precision, recall, and F-score. The accuracy of 86.89% indicates decent overall performance, but it might not capture complex relationships in the data, making it less suitable for datasets with intricate interactions between features. Random Forest demonstrates good precision, recall, and F-score (around

0.88), and its accuracy of 89.3% is commendable. It reduces overfitting compared to individual Decision Trees and handles both numerical and categorical features effectively. Despite mitigating overfitting, Random Forest may still struggle with capturing very complex non-linear relationships as it relies on ensembling relatively simple trees. Gradient Boosting shows excellent precision, recall, and F-score (around 0.93), and its accuracy of 96.67% is the highest among the models evaluated. Gradient Boosting models, including CatBoost, may require more computational resources during training compared to simpler models. They might also be more sensitive to hyperparameter tuning.CAT Boosting exhibits perfect precision, recall, and F-score of 1.0, indicating it correctly predicts all positive cases. Its accuracy of 100% is impressive and suggests the model fits the data well.

The ROC curve is compressed into a single floating-point number using the numerical metric known as Area Under the ROC Curve (AUC). The performance across all categorization criteria is summed up by the AUC metric. AUC values are between 0 and 1. The AUC Score of several ML classifiers is shown in Figure 32.3. A positive sample chosen at random has a 100% chance of being given a greater probability than a negative sample chosen at random, according to the ROC-AUC score of 1.0. Concluding that performance of Cat Boosting is best compared to other models.

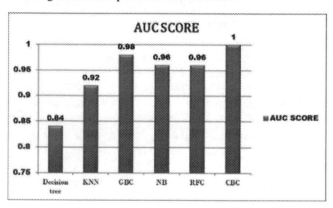

Fig. 32.3 AUC score of different ML binary classifiers

5. Conclusion

In this study, various machine learning classifiers were used to assess the performance of the heart disease prediction model. Cat Boosting surpassed the base classifiers in terms of accuracy, precision, recall, F-score, and AUC ROC. Cat Boosting classifier outperforms with highest accuracy of 100%. ML models predict a patient's risk of developing cardiovascular diseases by considering medical history, lifestyle factors, genetic markers, and clinical data. This risk stratification helps cardiologists identify high-risk patients who may benefit from more frequent monitoring and personalized preventive measures, optimizing disease management. It is essential to recognize that ML models are not intended to replace human expertise but rather act as valuable tools to augment cardiologists' capabilities. They offer additional insights and support in complex decision-making processes, enhancing the efficiency of healthcare delivery. Collaborations between ML experts, clinicians, and healthcare professionals are essential to maximize the potential benefits to safeguard patient privacy.

Acknowledgments: This work was supported by Muscat College, Sultanate of Oman

References

1. Sagheer, A., Zidan, M., & Abdelsamea, M. M. (2019). A novel autonomous perceptron model for pattern classification applications. *Entropy, 21*(8), 763.
2. Obasi, T., & Omair Shafiq, M. (2019). Towards comparing and using Machine Learning techniques for detecting and predicting Heart Attack and Diseases. *2019 IEEE International Conference on Big Data*, 2393–2402.
3. Gárate-Escamila, A. K., Hajjam El Hassani, A., & Andrès, E. (2020). Classification models for heart disease prediction using feature selection and PCA. *Informatics in Medicine Unlocked, 19*, 100330.
4. Barun, A., Nanda, M., Jarwal, M. K., & Sipal, D. (2022). Various disease predictions using machine learning. *2022 IEEE Region 10 Symposium (TENSYMP)*, 1–6.
5. Yaqoob.M, Iqbal.F, & S. Zahir , (2020). Comparing predictive performance of k-nearest neighbors and support vector machine for predicting ischemic heart disease. *Journal of Advanced Scientific Research, 1* (1).
6. Fan, Q., Wang, Z., Li, D., Gao, D., & Zha, H. (2017). Entropy-based fuzzy support vector machine for imbalanced datasets. *Knowledge-Based Systems, 115*, 87–99.
7. Yadav. D. C. & Pal.S. (2020), Prediction of heart disease using feature selection and random forest ensemble method. *International Journal of Pharmaceutical Research, 12*(04).
8. Hameetha Begum, S., & Nisha Rani, S. N. (2021). Model evaluation of various supervised machine learning algorithm for heart disease prediction. *2021 (ICSECS-ICOCSIM)*,119–123.
9. Thomas, J., & Princy, R. (2016). Human heart disease prediction system using data mining techniques. 2016 International Conference on Circuit, Power and Computing Technologies (ICCPCT), 1-5.
10. Chandra Reddy, N. S., Shue Nee, S., Zhi Min, L., & Xin Ying, C. (2019). Classification and feature selection approaches by machine learning techniques: Heart disease prediction. *International Journal of Innovative Computing, 9*(1).

Note: All the figures and table in this chapter were made by the authors.

Advancing Sustainable Science and Technology for a Resilient Future – Sai Kiran Oruganti et al. (eds)
© 2024 Taylor & Francis Group, London, ISBN 978-1-032-79020-6

Development of Electrochemical Oxygen Microsensor for the Measurement of Dissolved Oxygen in Natural Waters

33

Denson Liday*

Quirino State University, Philippines. PCIEERD-DOST Complex, Gen. Santos Ave., Bicutan 1631, Sibol St, Taguig, Metro Manila, Philippines

Nobelyn Agapito[1]

Quirino State University, Philippines. PCIEERD-DOST Complex, Gen. Santos Ave., Bicutan 1631, Sibol St, Taguig, Metro Manila, Philippines

Samuel Dulay[2]

Quirino State University, Philippines. PCIEERD-DOST Complex, Gen. Santos Ave., Bicutan 1631, Sibol St, Taguig, Metro Manila, Philippines

Abstract: Oxygen is vital for mammals, and notable changes in its levels can act as an important clinical sign for different metabolic dysfunctions in both natural surroundings and healthcare settings. This article presents the creation and utilization of an electrochemical microsensor, utilizing a platinum wire, designed to measure dissolved oxygen (DO) concentration in milligrams per liter (mg/L) within static natural water at room temperature. In laboratory settings, researchers conducted experiments using a Phosphate buffered saline solution (PBS) and manipulated the fraction of inspired oxygen (O_2). These experiments yielded a sensitivity of 1.7 microamperes per milligram per liter (uA/mg.L-). The developed sensor was able to correlate estimated concentration of the water sample in an in-situ measurement and its direct correlation with the standard solution were extrapolated assuming 8:3 mg/L as DO at ambient temperature 25°C during the day. The developed microsensor response time at t_{95} gave a rapid and reached a stable signal in just less than 10 seconds.

Keywords: Electrochemistry, Amperometry, Oxygen sensors, Dissolved oxygen

1. Introduction

The O_2 levels is very important for aquaculture as well as to natural waters that holds wild aquatic life. Significant efforts have been directed towards improving methods and technology for measuring dissolved oxygen (DO) in water due to its crucial role in environmental monitoring, including water quality assessment, agriculture, and medical diagnostics (Fenghong Chu, et al., 2008). Hence, it's vital to create suitable in-situ monitoring tech and calibration methods to tackle instability problems linked to these sensing technologies. Despite significant efforts, the recent development of dissolved oxygen (DO) monitoring sensors has seen limited success in meeting the necessary criteria (Kim et al., 2007). Electrochemical sensors, known for their

sensitivity, selectivity, miniaturization potential, and cost-effectiveness, have long been favored for diverse analytical applications. Electrochemical DO sensors have limitations: oxygen consumption affects accuracy, frequent calibration required, interference vulnerability, slower response times vs. optical sensors, temperature sensitivity needing compensation, and possible drift needing recalibration. Recognizing these limitations is crucial for accurate measurements via proper maintenance and calibration. When it comes to oxygen (O2) sensing, noble metals like platinum (Pt) and gold (Au) have traditionally served as the most commonly used transducer platforms. Alternatively, carbon-based substances have been utilized to accomplish the same objective and have displayed promising capabilities in monitoring oxygen levels in brain tissue in real-time (Bolger, 2010). In this paper, a

*Corresponding author: densonmarianoliday@gmail.com
[1]nobelyn.agapito@qsu.edu.ph, [2]dulaysamuel@yahoo.com

DOI: 10.1201/9781003490210-33

cost-effective electrochemical O_2 microsensor based on Pt and silver (Ag) wire were developed for monitoring DO of aquaculture natural water at ambient temperature to evaluate amount of DO.

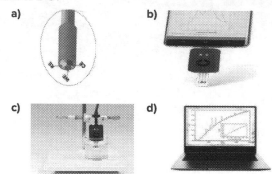

Fig. 33.1 Schematic representation of the newly developed electrochemical microsensor for oxygen, which was evaluated using a standard solution: (a) wire sensors, b) carbon screen printed sensors, (c) android mobile phone system measurement, and (d) conventional laptop measurement

This microsensor, a standout feature of this study, seamlessly connects to smartphones and computers via dedicated apps. This groundbreaking compatibility boosts convenience and accessibility, addressing constraints noted in previous literature concerning traditional DO monitoring devices.

2. Materials

Necessary materials, including KCl, 5 wt. % Nafion® perfluorinated resin solution, and PBS tablets, were sourced from Sigma-Aldrich (Spain). Electrodes consisted of insulated platinum (Pt) and silver (Ag) wires with 99.9% purity and 0.125 mm diameter (Pt wire with coating resulting in approx. 0.150 mm diameter) from Advent Research Materials in the UK. Electrochemical measurements were conducted using a custom-made sensit device by PalmSens, a Dutch manufacturer. For in-vitro testing, the dissolved oxygen (DO) concentrations were assumed to be 8.3 mg/L under ambient atmospheric conditions (Bozorg-Haddad et al., 2021).

3. Oxygen sensor Preparation

The preparation of the oxygen sensor involved employing a three-electrode configuration. This system comprised of a constructed Ag/AgCl wire serving as the reference electrode (RE), a Pt wire as the counter electrode (CE), and a Pt microwire that was modified with a membrane functioning as the working electrode (WE). To prepare the reference electrode, the Ag wire underwent anodization by being immersed in a saturated solution of KCl (3M) for a duration of 2 minutes. This process was facilitated by a 9V

battery, after which the wire was left to dry overnight. To protect the modified Ag wire, it was submerged in a 5% w/w Nafion® solution for 2 minutes. Subsequently, the wire was cured at 100 °C for 1 hour. The insulating layer on one end of the Pt wire was delicately removed to establish electrical connections with the portable potentiostat, specifically the sensit smart device from PalmSens. The exposed tip of the Pt wire served as the working electrode (WE) and was coated with 2 μL of 5% w/w Nafion®. The coating was allowed to dry overnight. Before utilization, the Nafion®-modified sensors were soaked in a PBS pH 7.4 buffer solution for 20 minutes to allow for swelling.

4. In-vitro Measurements

Cyclic voltammetry (CV) was performed using a scan range spanning from +0.2 to -1.0 V, with a scan rate of 100 mV/s. Additionally, chronoamperometry (CA) experiments were conducted by maintaining a constant potential of -0.7 V for a duration of 300 seconds. In all experiments, multiple measurements (n≥3) were performed using different sensors to ensure reliability. In-vitro experiments were performed using a PBS buffer to calibrate the O2 concentration, utilizing a two-point calibration approach. The calibration process involved exposing the PBS solution to the surrounding conditions at room temperature (25°C) to establish a dissolved oxygen (DO) content of 8.3 mg/L, serving as one calibration point. For the second calibration point, a zero DO solution was prepared by dissolving at least 1 g of sodium sulfite (Na_2SO_3), an oxygen scavenger, in 1 L of distilled or deionized (DI) water. Additionally, a catalyst and indicator in the form of 1 mg of cobalt salt (such as $CoCl_2 \cdot 6H_2O$) were added. The solution was immediately sealed to prevent re-absorption of O_2 and allowed to react for a suitable period of time. Prior to taking measurements on samples, the calibration process was always performed.

5. In-situ Measurements

The developed oxygen sensor was then directly tested in in-situ samples where the three-electrode system were directly immersed on water samples in the QSU-farm pond irrigation system at 25-28°C. The QSU-farm pond irrigation system is a crucial resource for agricultural and aquatic activities within the university. Located at the heart of the campus, it not only enhances the landscape but also serves practical purposes. It draws water from multiple sources, including groundwater, rainfall, and higher area canals. With an estimated area of 15,000 m2 and a potential volume of 50,000 m3, this pond is designed to collect, distribute, and store water for various purposes such as crop irrigation, livestock watering, and aquaculture support. Additionally, it doubles as a water source for nearby structures in case of fire emergencies.

Fig. 33.2 Shows the QSU-farm pond water system (a) and map of the sampling stations (b)

6. Electrochemical Characterization

The characterization of the reference electrode (RE) involved the use of cyclic voltammetry on an anodized silver wire coated with Nafion®. For the working electrode (WE), a graphite lead pencil was utilized, while a copper wire was employed as the counter electrode (CE). The characterization was performed in a 50 mM solution of mixed ferrocyanide/ferricyanide. The cyclic voltammogram displayed distinct reversible peaks in the voltage range of approximately -0.1 to 0.7 V, as shown in Fig. 33.3 (R). A Pt wire coated with Nafion® was utilized to develop an oxygen (O2) sensor. Nafion® is a biocompatible polymer made of perfluorinated polysulfonate that is commonly employed in biosensors. Its main purpose in this sensor is to act as a barrier against interfering anionic species. Additionally, Nafion® possesses the unique ability to allow gas to permeate through the polymer, enabling it to reach the electrode surface for subsequent reduction (Zimmerman, 1991; Heither-Wirguin, 19960. The reduction mechanism of O2 depends on the electrode material and can occur through either a two-electron or four-electron process (Wu et al., 1996 ; Xiao et al.,2017) . The four-electron process, observed in materials such as carbon graphite, enables complete reduction of oxygen to water without the formation of intermediate products.

Cyclic voltammetry (CV) was carried out to evaluate the reduction potential of the Pt wire-Nafion® sensor. The experimental findings revealed a clear peak for O2 reduction, observed between -405 and -305 mV in the voltage range.. However, this peak vanished when the solution containing a 0.0 mg/L induced dissolved oxygen was introduced, as illustrated in the circled area in the figure.

7. Sample Measurement

Dissolved oxygen (DO) represents the oxygen available to fish, invertebrates, and all organisms in aquatic environments. It is crucial for the survival of most aquatic plants and animals. For instance, fish cannot thrive for extended periods in water with DO levels below 5 mg/L. The concentration of DO in water, measured in milligrams per liter (mg/L), varies with temperature. As temperature rises, water holds less oxygen, resulting in lower DO levels. In contrast, cold

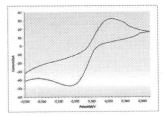
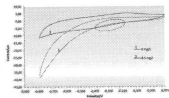

Fig. 33.3 Cyclic Voltammogram characterization of microwire Pt wire as WE in: (L) pH 7.4 PBS solution – atmospheric adsorbed O2 (8.3 mg/L) and Sodium Sulfite (NA2SO3) 0 mg/L oxygen induced solution and (R) 50 mM of ferro-/ferricyanide solution.

water typically has higher DO levels compared to warm water. At sea level and a temperature of 25°C, the DO value in freshwater is approximately 8.3 mg/L. Furthermore, at a constant temperature, as elevation increases, the DO content decreases. The developed oxygen sensor was utilized to measure DO levels at different stations within the QSU fishpond. This evaluation aimed to assess the performance of the developed sensor device. Figure 33.4 shows the two-point calibration plot for the developed sensor. Due to limited resources, it has been assumed that the DO at atmospheric ambient condition, i.e 25°C in 8.3 mg/L DO from air in the atmosphere. The sensor sensitivity holds 1.7 uA/mg.L-. The unknown DO concentration of the sample was then measured amperometrically (Fig. 33.4b) and the obtained signal (n=3) was calculated using the two-point calibration. The recorded values of current signal were then extracted from the average signal from 5-9 seconds running time as this is where the region where stability of the sensor's response time starts (t95%). The calculated value was then plotted against the calibrators (Fig. 33.4c) and it showed that station 1 holds 7.5 mg/mL higher than the rest of sampling areas stations 2, 3, and 4 containing from 6.5, 6.3 and 6.4 mg/L respectively. This signifies that it doesn't hold a saturated absorbed O2(8.3 mg/L) of a normal freshwater level at 25°C possibly due to having been contaminated with other organic materials that are present in the water itself preventing oxygen aeration in water. Additionally, one potential reason for the low levels of dissolved oxygen (DO) in the pond could be the breakdown of organic matter, including decomposing plants, animals, and waste from animals or humans, that are present in the water. When DO levels fall below approximately 6 mg/L, it can have harmful consequences for the pond's ecosystem (Swistock et al., 2017). In ponds, insufficient dissolved oxygen (DO) often leads to fish mortality. This typically happens when aquatic plants and algae rapidly die off, especially during the summer or after treatment with aquatic herbicides. Fish die-offs due to low oxygen levels are most common during hot and dry periods when algae experience rapid growth and subsequent decay. The organisms that decompose the dead algae consume large amounts of oxygen, leaving insufficient oxygen for the survival of fish. In deeper ponds, the lower regions are particularly prone to having low DO concentrations due to insufficient aeration (Kibria, 2016). Consequently, the low

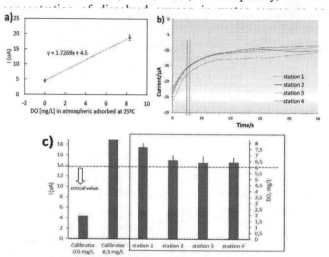

Fig. 33.4 (a) Two-point calibration curve of the developed O2 sensor based Pt-Nafion coated wire WE, (b) Amperometric measurement of water samples in-situ, and (c) Correlation of calculated water samples from the calibrators. To enhance the dissolved oxygen (DO) levels in water, re-aeration methods can be employed. This involves the introduction of oxygen from the air into the water, primarily through processes involving turbulence. Examples of such processes include the movement of water over rocks (such as rapids, waterfalls, and riffles) or the generation of wave action using mechanical devices. During daylight hours, photosynthesis by aquatic plants also contributes to the production of oxygen. Consequently, DO levels tend to be highest in the late afternoon and lowest in the early morning hours before sunrise.

8. Conclusion

The DO concentration fishpond of QSU is still within the limit for a surviving cultured fish that are present in the pond. But then, the results obtained must be confirmed with a commercial standard device with known high accuracy to validate the results from the developed microsensors. The in-situ measurements and assessment of the DO level of the pond and irrigation system will be utilized as basis for future intervention. Furthermore, this preliminary study on the DO content of the fishpond is substantial enough to obtain water quality in this season to understand the total yield of harvest for cultured fish. Additionally, the developed DO microsensor device is an alternative analytical method to easily track DO on natural waters giving enough information in real-time

owing to its sensor attributes such as being a rapid and cost-efficient developed device when compared to cumbersome and expensive DO devices available in the market. This study's oxygen sensor is notable for smartphone compatibility via a dedicated app, removing the need for additional equipment and promoting user-friendliness. This innovation simplifies DO level tracking, benefiting researchers and environmental agencies in data collection. It aligns with prior studies stressing continuous DO monitoring for ecosystem health. In summary, this sensor surpasses existing limitations and supports previous literature goals, improving aquaculture, productivity, and water quality monitoring.

References

1. B.R. Swistock, T. Sharpe, W. E., and McCarty (2017). Management of Fish Ponds in Pennsylvania.
2. C. Heitner-Wirguin (1996). Recent advances in perfluorinated ionomer membranes: structure, properties and applications, J. Memb. Sci. 120 1–33.
3. C.-S. Kim, C.-H. Lee, J.O. Fiering, S. Ufer, C.W. Scarantino, H.T. Nagle (2007) Manipulation of Microenvironment With a Built-In Electrochemical Actuator in Proximity of a Dissolved Oxygen Microsensor, IEEE Sens. J.4568– 575.https://doi.org/10.1109/JSEN.2004.832857.measurements, Sensors Actuators B Chem. 128 179–185.
4. F.B. Bolger, S.B. McHugh, R. Bennett, J. Li, K. Ishiwari, J. Francois, M.W. Conway, G. Gilmour, D.M. Bannerman, M. Fillenz, M. Tricklebank, J.P. Lowry. (2011). Characterisation of carbon paste electrodes for real-time amperometric monitoring of brain tissue oxygen, J. Neurosci. Methods. 195 135–142.https://doi.org/10.1016/j.jneumeth.2010.11.013
5. G. Kibria (2016). Blue-green algal toxins/cyanobacterial toxins (BGA), climate change and BGA impacts on water quality, fish kills, crops, seafood, wild animals and humans. https://doi.org/10.13140/RG. 2.1. 1306.9765/1.
6. J.B. Zimmerman & R.M. Wightman (1991). Simultaneous electrochemical measurements of oxygen and dopamine in vivo., Anal. Chem. 63 24–8. https://doi.org/10.1021/ac00001a005.
7. J. Wu & H. Yang (2013) Platinum-Based Oxygen Reduction Electrocatalysts, Acc. Chem. Res. 46 (2013) 1848–1857.
8. O. Bozorg-Haddad, M. Delpasand, H.A. Loáiciga (2021). Water quality, hygiene, and health, in: Econ. Polit. Soc. Issues Water Resour., Elsevier: pp. 217–257. https://doi.org/10.1016/B978-0-323-90567-1.00008-5.
9. T. Xiao, F. Wu, J. Hao, M. Zhang, P. Yu, L. Mao (2017). In Vivo Analysis with Electrochemical Sensors and Biosensors, Anal. Chem. 89 300–313. https://doi.org/10.1021/acs.analchem.6b04308.
10. Z.F. Fenghong Chu, Haiwen Cai, Ronghui Qu (2008). Dissolved oxygen sensor by using Ru-f luorescence indicator and a U-shaped plastic optical fiber, CHINESE Opt. Lett. 6 401–404.

Note: All the figures in this chapter were made by the authors.

Advancing Sustainable Science and Technology for a Resilient Future – Sai Kiran Oruganti et al. (eds)
© 2024 Taylor & Francis Group, London, ISBN 978-1-032-79020-6

Comparative Analysis of Water Potability Prediction Based on Classical Machine Learning Algorithms and Artificial Neural Networks

34

Rob Christian M. Caduyac*

Department of Electrical Engineering,
University of the Philippines, Los Baños, Philippines

John Francis R. Chan[1]

Department of Electrical Engineering,
University of the Philippines, Los Baños, Philippines

Abstract: For the past years, consumption of contaminated water became a norm due to water scarcity which caused several human health problems. Hence, water potability prediction is vital in the prevention of water-related health risk. This study aims to generate a water potability predictor using classical machine learning algorithms and artificial neural networks (ANN). Logistic regression (LR), support vector machine (SVM), and random forest (RF) were used for the former while two-layer and four-layer were used for the latter. ANN-based sensitivity analysis was used to identify most influential water potability parameter. A cross-validation was performed and the performance metrics - F1-Score, Area under the Curve (AUC), True Positive Rate (TPR), True Negative Rate (TNR), positive predictive value (PPV), and negative predictive value (NPV) were averaged. Results showed that ANN2 outperformed all models aside from TPR and NPV. Lastly, hardness turned out to be the most influential parameter for water potability.

Keywords: Water potability, Water potability prediction, Machine learning, Artificial neural networks, Sensitivity analysis

1. Introduction

Approximately 829,000 people die each year from illnesses caused by drinking contaminated water (Lin L, et al., 2022). Hence, there is an increase in the number of research studies that focus on the evaluation of water potability (Rozynek, P, et al., 2021).

Machine learning is one of the most accurate methods when it comes to data analysis (Sewak, M., et al., (2018); Sokolov, A et al., (2018)). Recent machine learning studies for water potability prediction include using K-Nearest Neighbours and ANN (Rozynek, P. et al., 2021) and using active SVM (Zhao, R, 2021). While these methods show algorithms for water potability prediction, there is a need for optimized architecture for better prediction.

The main objective of the study is to compare the effectiveness of classical machine learning algorithms (LR, SVM, and RF) and ANN in water potability prediction. Also, most influential parameter for water potability will be determined. Through this study, an accurate water potability prediction can be achieved.

2. Methodology

Figure 34.1 shows the overall methodology workflow.

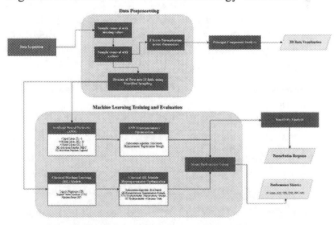

Fig. 34.1 Overview of the classical ML and ANN methods for Water potability prediction.

Data Acquisition

The main source of data in this study is Kaggle, which has a Google Scholar rating of five stars that proves its data quality

*Corresponding author: rmcaduyac1@up.edu.ph
[1]jrchan2@up.edu.ph

DOI: 10.1201/9781003490210-34

credibility. There are total of ten parameters, with potability serving as an output that indicates whether the water is safe to drink. The link to the data source is https://www.kaggle.com/adityakadiwal/water-potability.

Data Preprocessing

This was done to compensate for the missing values and the presence of outliers. Samples with missing values were removed while the interquartile range (IQR) method was used to remove outliers. The output is then normalized using z-score normalization for visualization.

Ten percent of the preprocessed data will serve as test set, while the remaining will be divided into 10-fold using stratified sampling for machine learning training and evaluation.

Data Visualization

Principal component analysis (PCA) is an unsupervised dimensionality reduction technique that is dependent on the variability of the data. In this study, the two largest eigenvectors were used to reduce the high-dimensional preprocessed data for 2D visualization plot.

Machine Learning Training & Evaluation

Two ANN of different layer sizes were used for comparison with classical machine learning models, namely LR, SVM and RF. The next subsections discuss the design of each model.

1. Artificial Neural Networks (ANN): 2-layer and 4-layer ANN architectures were designed in the study. For both architectures, its input, hidden, and output layers are composed of 9, 50, and 1 neuron respectively. Rectified linear unit and sigmoid were used as activation for hidden and output layer respectively.

2. Logistic Regression (LR): This model uses an L2 regularization term and a binary cross-entropy cost function, which is optimized using the limited-memory Broyden Fletcher Goldfarb Shanno (LBFGS) algorithm.

3. Support Vector Machines (SVM): This model uses a radial basis function as the kernel for the SVM. The kernel will then serve as an input for the cost function of the model.

4. Random Forest (RF): The design is a combination of different decision trees made from random features and sample selection. Here, each tree has at least 10 nodes, and the hyperparameter is the number of trees.

A grid search was performed on all models for hyperparameter optimization. All models aside from RF used regulation strength. F1-Score was selected as the metric for selecting the best hyperparameter per model due to the unbalanced class distribution of the data.

ANN Sensitivity Analysis

Sensitivity measure utilizes the chain rule of differentiation to solve the partial derivative of the output layer with respect to the input layer (Pizarroso, J. et al., 2021). Calculation of sensitivity measure can be solved from the partial derivatives of the inner neural network layer. Upon solving all the sensitivity for each sample, the mean sensitivity was used as a sensitivity measure.

Performance Metrics

The performance metrics to be considered are F1-score, AUC, TPR, TNR, PPV, and NPV. Since the study is a multi-classification type, one-versus-rest was applied wherein the metrics are obtained by getting the average of metrics values per classes.

1. F1-Score – this metric is commonly used when the dataset is unbalanced. It is computed by taking the harmonic mean of TPR and PPV which penalizes extreme values from either one of TPR and PPV (Hicks, S. et al, 2021).

2. Area under the curve (AUC) – this represents the overall performance of the classifier (Hossin, M. et al., 2015). It was proven empirically by (Huang J. et al., 2005) that AUC is a better metric for classification purposes.

3. True Positive Rate (TPR) – it is defined as the ratio of the correctly classified positive patterns relative to the total correct predictions. This metric is commonly used for studies that desire to only miss a few positive instances (Hicks, S. et al, 2021).

4. True Negative Rate (TNR) – TNR is calculated by taking the ratio of the correctly classified negative instances () versus all the negative samples .

5. Positive Predictive Value (PPV) – PPV measures how accurate is the classifier given that the classifier outputs a positive prediction (Monaghan, T. et al, 2021).

6. Negative Predictive Value (NPV) – NPV measures how accurate the classifier is given that it outputs a negative prediction.

3. Results and Discussion

Preprocessing

Figure 34.2 shows the summary of the whole preprocessing steps done on the raw data of 3276 original number of samples of dataset, 1265 samples were removed due to missing data. Furthermore, 182 outliers, shown in Fig. 34.3, were removed since they fell outside the boxplot. In summary, only 1829 samples remain. For visualization using principal component analysis, z-score normalization was applied across the input parameters and its output served as an input to PCA.

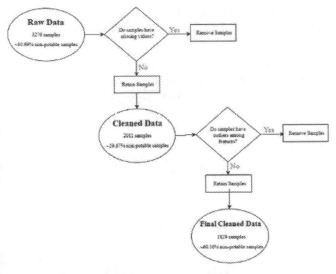

Fig. 34.2 Data Preprocessing

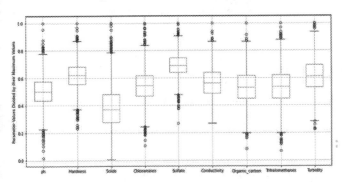

Fig. 34.3 Box and Whisker's Plot of Input Parameters

Principal Component Analysis of Water Potability

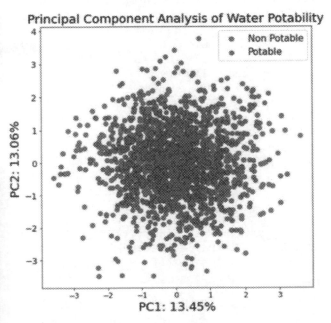

Fig. 34.4 2D plot of first 2 PCA components

Principal Component Analysis

The variation between potable and non-potable samples using 2D PCA plots is shown in Fig. 34.4. However, only 26.51 % of the total variance was represented by the top two PCA values combined. This shows that more components were necessary to capture the relevant information from the preprocessed data. Since there are nine (9) input parameters, the largest number of eigenvectors that can be obtained through PCA is also nine.

Shown in Fig. 34.5 is the explained variance values for each eigenvector obtained using PCA. The input parameters are highly uncorrelated and nonlinear based on the explained variance values; therefore, a nonlinear model must be used for water potability prediction.

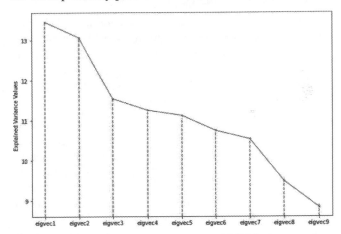

Fig. 34.5 Explained Variance values of 9 PCA eigenvectors.

Model Performance

The performance of ANN2 and ANN4 architectures was compared to the classical machine learning models (LR, SVM, and RF) using the F1-score, AUC, TPR, TNR, PPV, and NPV metrics. For hyperparameter tuning, F1-score was used due to the unbalanced distribution of data (~40% vs. ~60% potable and non-potable samples).

Table 34.1 summarizes the metric values for each model as well as their optimal hyperparameter values. ANN2 outperforms all models in terms of the F1-score, AUC, TNR, and PPV while SVM and LR are the best in terms of TPR and NPV respectively. The observed higher model performance of ANN2 versus ANN4 is due to the model complexity of ANN4, which makes inaccurate tuning of learning rate. Overall, ANN2 is the best as it has a good F1-score and AUC. When dealing with classifier's judgment on whether water sample is potable, ANN2 is also the best as it has the highest PPV. If the water potability application desires to have less missed positive instances, then SVM might be used. Lastly, LR can be used if it is desired to test whether water sample is non-potable.

Table 34.1 Summary of ANN2 sensitivity analysis

Inputs	(Change in Output Neuron vs Inputs)
PH Level	0.566%
Hardness	1.061%
Solids	0.170%
Chloramines	0.448%
Sulfate	-0.512%
Conductivity	-0.790%
Organic Carbon	-0.126%
Trihalomethanes	0.221%
Turbidity	0.514%

D. ANN2 Sensitivity Analysis

The sensitivity analysis was conducted from ANN2 being the best model. The training weights were used when computing layer derivatives. Since the model training and evaluation involves stratified 10-fold for stability of metrics, mean sensitivity was computed. Table 34.2 shows the results of the perturbation response per input parameters.

Results showed that "Hardness" is the most influential parameter for every output change. This implies that "Hardness" affects water potability prediction the most. Low metrics can also be associated with the low perturbation response. This implies that there may be other better input parameters that may be used to improve the water potability prediction.

4. Conclusion

In this study, the performance of water potability prediction based on classical ML and ANN was compared. It was found that ANN2 has the best F1-score among all the machine learning algorithms used in this paper, as it outscored the other algorithms in four out of six performance metrics. One of the main reasons for this is the balanced complexity of the ANN2 method that is suitable for non-complex and non-linear data in terms of the number of input parameters. It

was also determined that hardness is the input parameter that water potability is most sensitive to, based on the sensitivity analysis that was performed.

The data used for this study only comprises nine input parameters; however, there are other water parameters that have not been considered. It is recommended to increase the number of input parameters to be considered for improvement of the predictor. Moreover, the deployment of the water potability prediction in embedded systems should also be considered for the expansion of the study for the sake of having a real-time test on water samples from large-scale communities.

References

1. Lin L, Yang H and Xu X (2022) Effects of Water Pollution on Human Health and Disease Heterogeneity: A Review. Front. Environ. Sci. 10:880246. doi: 10.3389/fenvs.2022.880246

2. Rozynek, P. & Rozynek M. (2021). Water Potability Classification using Neural Network. 2021 International Conference of Yearly Reports on Informatics Mathematics and Engineering (ICYRIME). Retrieved from http://ceur-ws.org/Vol-3118/p05.pdf.

3. Zhao, R. (2021). The Water Potability Prediction Based on Active Support Vector Machine and Artificial Neural Network. 2021 International Conference on Big Data, Artificial Intelligence and Risk Management (ICBAR), pp. 110-114, doi: 10.1109/ICBAR55169. 2021.00032.

4. Sewak, M., Sahay, S. K., & Rathore, H. (2018). Comparison of Deep Learning and the Classical Machine Learning Algorithm for the Malware Detection. 2018 19th IEEE/ACIS International Conference on Software Engineering, Artificial Intelligence, Networking and Parallel/Distributed Computing (SNPD). doi:10.1109/snpd.2018. 8441123.

5. Sokolov, A. N., Pyatnitsky, I. A., & Alabugin, S. K. (2018). Research of Classical Machine Learning Methods and Deep Learning Models Effectiveness in Detecting Anomalies of Industrial Control System. 2018 Global Smart Industry Conference (GloSIC). doi:10.1109/glosic.2018.8570073.

6. Pizarroso, J., Portela, J., Munoz, Antonio., (2021). NeuralSens: Sensitivity Analysis of Neural Networks. Journal of Statistical Software. doi.org/10.18637/jss.v000.i00.

Table 34.2 Summary of ANN2 sensitivity analysis

Performance Metrics	ANN2	ANN4	LR	SVM	RF
F1-Score	66.14±0.86%	59.60±1.67%	37.65±0.41%	65.59±1.49%	59.47±1.61%
AUC	65.89±0.75%	59.54±1.65%	49.98±0.17%	65.42±1.36%	60.71±1.26%
TPR	64.63±3.52%	52.06±2.25%	0.00±0.00%	69.49±2.81%	69.27±3.48%
TNR	71.25±1.03%	67.46±1.29%	60.01±0.08%	69.98±0.92%	66.34±0.79%
PPV	50.41±4.87%	49.18±2.70%	0.14+0.43%	43.56±2.31%	30.41±2.65%
NPV	81.36±3.89%	69.91±2.25%	99.82±0.38%	87.27±1.60%	91.00±1.57%
Best Hyperparameter	C=0.0003	C=0.0001	C=1	C=2.5	estimators=100

7. Hicks, S. et al, (2021). On evaluation metrics for medical applications of artificial intelligence. MedRxiv: The Preprint Server for Health Science. doi.org/10.1101/2021.04.07.21254 975.

8. Hossin, M. Sulaiman, M. (2015). A Review on Evaluation Metrics for Data Classification Evaluations. International Journal of Data Mining & Knowledge Management Process. doi.org/10.5121/ijdkp.2015.5201.

9. J. Huang and C. X. Ling, (2005) "Using AUC and accuracy in evaluating learning algorithms", IEEE Transactions on Knowledge Data Engineering. doi.org/10.1109/ TKDE.2005.50

10. Monaghan, T. et al, (2021). Foundational Statistical Principles in Medical Research: Sensitivity, Specificity, Positive Predictive Value, and Negative Predictive Value. Medicina 2021. doi.org/10.3390/medicina57050503.

Note: All the figures and tables in this chapter were made by the authors.

Advancing Sustainable Science and Technology for a Resilient Future – Sai Kiran Oruganti et al. (eds)
© *2024 Taylor & Francis Group, London, ISBN 978-1-032-79020-6*

Predicting CKD Stages and Urgency Level with K-Means Clustering via Data Mining

35

August Anthony N. Balute*

Philippine Christian University, Philippines

Marjorie Cuyes Solomon[1]

AMA University Maximina St., Philippines

Abstract: Data mining holds an unprecedented potential for the clinical benefits industry to enable prosperity structures to methodically use data and assessment to perceive inadequacies and best practices that work on the individual fulfilment. It is transcendently being utilized for disease expectation and achieves possibly helpful and justifiable examples in information gathered including its grouping. In this review, K-implies grouping calculation, an unaided learning calculation, was utilized to recognize the bunches and number of patients as per age, orientation, direness level for dialysis and chronic kidney disease (CKD) stages, where the important outcomes act as a supportive device for clinic managers and specialists to all the more likely distinguish the earnestness of dialysis with the comparing CKD stage and the situation with the general number of kidney patients. This will support dynamic cycles using AI procedures.

Keywords: Chronic kidney disease (CKD), Cross-industry process for data mining (CRISP-DM), Electronic medical record (EMR), K-means clustering algorithm, Knowledge discovery in databases (KDD), R programming

1. Introduction

AI in the medical care industry can furnish specialists with electronic data upgraded by the force of examination and AI. Progressed investigation can likewise show a patient's gamble for stroke, coronary course disease and kidney disappointment in view of regularly pulse readings, research facility test results, socioeconomics and most recent clinical preliminary information. Gigantic datasets past the level of human limit can be continually unique over into evaluation of clinical experiences helping experts in coordinating and giving idea, inciting unrivaled results, lower costs, and expanded patient fulfillment and individual satisfaction. AI can be prepared to take a gander at electronic clinical records and recognize irregularities in this manner working on the exactness and expectation of diseases provoking finding and preventive screening of sicknesses or diseases. Doctors representing things to come will regulate and survey readings that have been at first perused by a machine, utilizing AI like a cooperative accomplice that recognizes explicit areas of spotlight on high likelihood areas of concern considering the morals included (Corbett E. 2017).

The overall target of this study is to urge a model to be utilized in social event the truthfulness level and CKD time of dialysis patients. The gather likewise desires to:

1. To decide the huge characteristics that can assist with grouping the level of criticalness and CKD phases of dialysis.

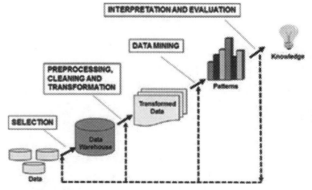

Fig. 35.1 Knowledge discovery of databases

*Corresponding author: j.doe@gmail.com
[1]marjorie.cuyos@gmail.com

DOI: 10.1201/9781003490210-35

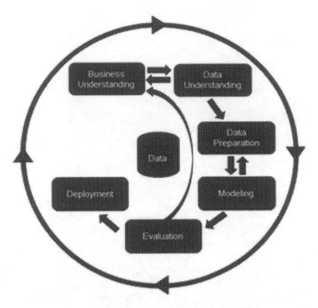

Fig. 35.2 CRISP-DM process

2. To use K-implies Bunching Calculation in anticipating the level of desperation and CKD Phases of dialysis.

3. To make a model that can assist with deciding the level of earnestness and CKD phases of dialysis.

This stage entails comprehending and defining the end users' goals, as well as the location of the knowledge discovery process and any pertinent prior information.

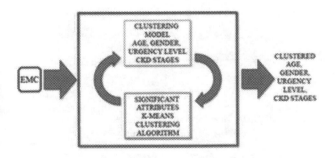

Fig. 35.3 Conceptual framework of the study

To model the K-means Clustering Algorithm, the system will use the training dataset using the patient records or medical records. These will cluster the CKD stages of urgency level of dialysis of a patient.

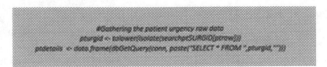

Fig. 35.4 Data transformation

The database stores uploaded patient records from a CSV file format and extracted in R Programming Language.

R Composing PC programs was used in this survey since it gives a wide variety of verifiable straight and nonlinear illustrating, quantifiable tests, gathering tree, and graphical methodologies, and is significantly extensible. R gives an Open Source course to collaboration in making a model.

Review structures are ready for the adequacy test and overview assessment structures are an additional element in the framework dashboard. This is to affirm that the computer programming nature of the review is ISO/IEC 25010 Frameworks and programming Quality Prerequisites and Assessment Agreeable. Questions were gathered into eight (8) organized set qualities. These are Down to earth Sensibility, Execution Efficiency, Likeness, Usability, Trustworthiness, Security, Practicality and Flexibility

Table 35.1 Likert scale

Range	Interpretation
4.51 – 5.00	Highly Acceptable
3.51 – 4.50	Acceptable
2.51 – 3.50	Uncertain
1.51 – 2.50	Unacceptable
1.00 – 1.50	Highly Unacceptable

Each of the sections is represented by one (1) person. The total number of respondents is eleven (11). The comments, suggestions, and recommendation will be noted and may be recommended if found to be valuable in the enhancement of the system.

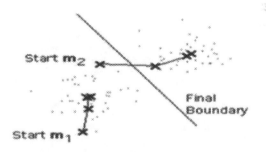

Fig. 35.5 Illustrates the means m1 and m2 move into the centers of two clusters.

K-infers Gathering Computation head capacity is to pack the amount of patients as demonstrated by age, direction, genuineness level and CKD stages. The outcomes will be a critical consider decision making of the Informed authorities, Emergency focus Chiefs and future Specialists.

The outcome demonstrates unequivocally that the majority of individuals had stage 5 CKD. These patients are advised to have a kidney transplant and require three times daily dialysis.

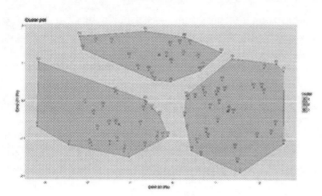

Fig. 35.6 Illustrates how the K-means clustering algorithm clusters the electronic medical records specifically by clustering the patient's age

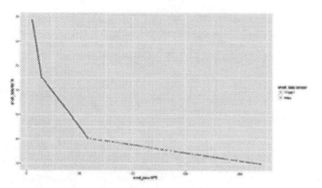

Fig. 35.7 illustrates how the K-means clustering algorithm clusters the electronic medical records by patient's gender. Male patients represent blue color while female patients represent red color

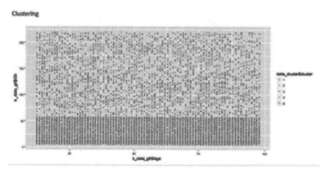

Fig. 35.8 Presents the cluster of patients according to CKD stages through K-means Clustering Algorithm. Color Pink represent Stage 1, the color green represents Stage2, color beige represents Stage 3, blue represents Stage 4 and lavender represents Stage 5.

When patients are identified, tracked, and given the right interventions and treatment regimens, they can obtain better, more inexpensive healthcare services, especially if they have chronic or renal failure. This research can raise

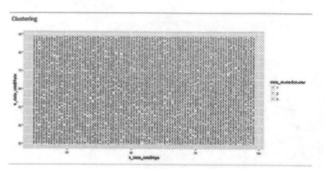

Fig. 35.9 Shows K-means clustering algorithm clusters the patients according to its urgency level. The color red represents patients under emergency Care, blue represents Moderate level and green represents urgent level

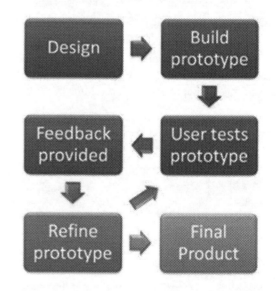

Fig. 35.10 Urgency level code

Fig. 35.11 Shows the course of Transformative Prototyping Model; (1) Plan, (2) Form Model, (3) Client test model, (4) Input gave, (5) Refine model, (6) End result [10]

patient satisfaction, deliver more patient-centered care, save expenses, and boost the dialysis center's operational efficiency while maintaining high-quality care.

The goal of this study is to create a model that might be used to group the CKD Stages and desperation levels of dialysis patients.

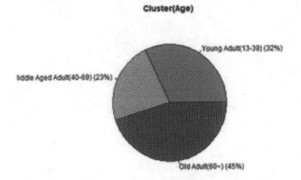

Fig. 35.12 Cluster Age - Delineates that young adults (ages 13 to 39) show 32 percent (32%) bunched, moderately aged adults (ages 40 to 69) show 23% (23%) bunched, and older adults (ages 60 or more) show 45% (45%) bunched, which is the most notable in positioning. This clearly demonstrates that older patients who are adults are in a more serious state while receiving critical

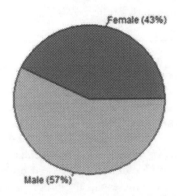

Fig. 35.13 According to bunch orientation, male patients have the most conceded in both CKD and ARF, at 57% (57%), while female patients have the lowest surrendered in both conditions at 43% (43%). This demonstrates that male patients are more commonly diagnosed with CKD and ARF.

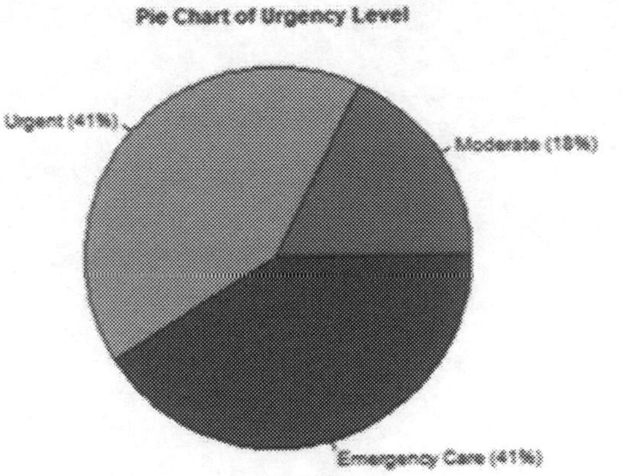

Fig. 35.14 The clustered urgency level for a patient is shown by Cluster Urgency Level. The level of moderate care has the lowest percentage (18%), whereas levels of urgent and emergency care have the highest number (41%). This demonstrates that the majority of people with CKD and ARF do not take their medications as directed, which pushes up both the urgency level

References

1. "The health catalyst". Retrievefromhttps://www.healthcatalyst.com/clinical- applications-of-machine-learning-in- healthcare 2017Morino, K., Hirata, Y., Tomioka, R., Kashima, H., Yamanishi, K., Hayashi, N.,Aihara, K. (2015). Predicting disease progression from short biomarker series using expert advice algorithm. Scientific Reports. https://doi.org/10.1038/srep08953

2. Khanmohammadi, S., Adibeig, N., & Shanehbandy, S. (2017). An improved overlapping k-means clustering methodfor medical applications. Expert Systems with Applications.https://doi.org/10.1016/j.eswa.2016.09.025

3. Chauhan, P., & Shukla, M. (2015). A review on outlier detection techniques on data stream by using different approaches of K-Means algorithm. In 2015 International Conference on Advances in Computer Engineering and Applications.https://doi.org/10.1109/ICACEA.2015.7 164758

Note: All the figures and table in this chapter were made by the authors.

Advancing Sustainable Science and Technology for a Resilient Future – Sai Kiran Oruganti et al. (eds)
© 2024 Taylor & Francis Group, London, ISBN 978-1-032-79020-6

36

A Smart Healthcare System Using an Artificial Intelligence for Detecting Heart Disease

Girish Kalele*

Chitkara University Institute of Engineering and Technology, Chitkara University, Punjab, India

Abstract: Smart healthcare has developed from the standard health service. The advent of artificial intelligence (AI) will broaden several opportunities for the healthcare field. For heart patients to be treated before a heart incident comes, it is crucial to forecast the development of heart disease. A number of HD prognostic and diagnostic methods have been established. The need for early HD recognition may be further increased by the absence of trustworthy data sets. In order to effectively detecting HDs, this research introduces the crow search optimized attentive long/short term memory (CSO-ALSTM) approach. Accuracy, precision, recall, and f-measure metrics are used to evaluate the effectiveness of the suggested CSO-ALSTM. The experimental findings show that, when compared to other approaches for detecting HDs, the suggested method achieves the highest efficiency.

Keywords: Healthcare, Heart disease (HD), Artificial intelligence (AI), Crow search optimized attentive long/short term memory (CSO-ALSTM)

1. Introduction

The most serious health problem is heart disease (HD), which has affected many individuals all over the globe. Breathing difficulties, muscular weakness, and swollen ankles are prominent signs of HD. When cutting-edge diagnostic tools and skilled physicians are unavailable, HD is a major health concern. Many lives may be saved via accurate detection and timely treatment (Li et al. 2020). Hence, the automated detection of cardiac problems is one of the most crucial and challenging medical issues in the actual world. Heart illness impairs the functioning of blood arteries and produces coronary illnesses, which damage the person's body, particularly in adults and the elderly. The World Health Organization estimates that cardiovascular diseases cause more than 18 million deaths annually worldwide (Koch et al. 2018). HD is often diagnosed by a doctor after reviewing the patient's medical records, the results of their physical exam, and any concerning symptoms. Moreover, the US spends $1 billion per day on HD therapies. Heart conditions include hypertensive, heart attacks and stroke are the leading causes of mortality in America. In order to properly cure

cardiovascular patients before such a heart attack or stroke may happen, early detection of HD is essential (Ali et al. 2020). Wearable technologies and medical examinations help with the identification of cardiovascular disease. It may be challenging for doctors to rapidly and properly diagnose patients using electronic medical checks, especially if such tests provide information that might be used to assess the individual's risk for cardiovascular disease. These EMRs are not organized, and their size is continually growing as a result of the many medical tests performed on a regular basis (Mansour et al.2021). In order to diagnose cardiac problems, detection systems are now also used to constantly evaluate the patient's body both inside and outside. First and foremost, tracking cardiac patients requires a large and challenging effort that combines the use of wearable sensors and EMRs. Second, it might be difficult to identify pertinent and useful aspects from the data for forecasting HD. In order to discover the covert signs of heart issues and forecast heart illness before a cardiac attack happens, an operating system that can automatically combine the collected information from various data and EMRs is necessary (Elayan et al. 2021). Many approaches for predicting and diagnosing HD using

*Corresponding author: girish.kalele.orp@chitkara.edu.in

DOI: 10.1201/9781003490210-36

data mining methods and hybrid models have been presented recently. A data mining approach derives risk variables from unstructured textual input. In contrast, a hybrid approach consists of two distinct strategies that function together more effectively than each way alone. Artificial intelligence will play a crucial part in the delivery of smart healthcare. (Subahi et al. 2022). This research proposes a new, intelligent healthcare monitoring system that makes use of ensemble AI to anticipate cardiac conditions.

2. Related Works

Nashif et al. (2018) suggested a prototype for a cloud-related system to use ML to anticipate the onset of cardiac disease. An effective ML approach should be employed for the accurate identification of heart illness; that approach was produced through a unique study among multiple ML algorithms in a Javafx Freely Accessible Data Mining Engine, WEKA. The present fog algorithms have a lot of drawbacks and concentrate on either lowered reaction times or reliability of outcomes, but not both. To integrate collective DL with Edge operating systems, (Tuli et al. (2020)) created a unique framework HealthFog and implemented it for a practical implementation of autonomous HD analysis. Using IoT technologies, HealthFog provides health insurance as a cloud resource and effectively maintains user-requested cardiac patient data. A machine learning-based cardiac disease diagnostic method that is efficient and accurate. The system uses classification techniques like SVM, ANN, NB, etc to remove features that are not relevant or redundant. (Ahmed (2019)) suggested a fast-conditioned similarity feature selection approach. Feature ranking methods improve accuracy rate and classifier model execution time. Using a machine learning method, (Natarajan et al. 2023) suggested a foundation for the Internet of Things-based wireless sensor networks (ML). Wearables such as thermometers, glucometers, heart-rate monitors, and chest-mounted monitors all send data to one centralized cloud storage system through an Internet of Things device. ML was employed to filter out irrelevant information in the collected information and to assess the control voltage for use in patient diagnostics.

3. Methodology

The suggested smart healthcare monitoring system (SHMS) architecture is dissected in this chapter. Technology is used in a SHMS to diagnose, treat, and monitor patients in many areas of care. The SHMS uses the suggested crow search optimized attentive LSTM ensemble AI model to provide cardiac illness predictions. Improving diagnostic precision is one of the key focuses of AI applications in the healthcare industry.

Data collection: Two HD datasets evaluated the model, which is considered to identify patients with HD, is done with a number 1 (present) or 0 (absent). Our research included 14–16 variables to determine the patient's health. Filtering handles incorrect values in both databases. To assess the model's accuracy, we pooled this dataset. 597 instances with 14 characteristics comprise the dataset. Nominal datasets cannot utilize AI models.

Disease detection model using crow search optimized attentive LSTM (CSO-ALSTM)

The bias and weighted parameters of the CALSTM model are optimized using CSO. In the avian world, crows have a higher reputation for intelligence than any other creature. Possibilities are tremendous because of its huge brain and short frame. The human mind, according to the body-to-brain idea, is relatively small. A large data set proves that crows are intelligent. A survey found that crows are both self-aware and adept at toolmaking. A crow's ability to recognize a familiar face and provide a threat signal to its flock is evidence of its intelligence. Furthermore, as the bird exits the nest, it watches other birds and chases them to locate the hidden spot of food and seize it. When stealing food, crows hide it from the real birds so they don't get caught. To protect its food, it uses the thief's expertise to predict the thief's next move. There are just a few crow norms provided here,

- It is located in the category
- It may recall the position of food kept in a concealed spot.

Then, there are stations in N dimensions made up of huge crows, where C is the total number of crows and u is the time-varying location of a crow in a search space. Below is an evaluation performed using the function shown here.

$$X^{w,iter}(o = 1,2, ..., D; iter = 1,2, ...) \, iter_{max} \quad (1)$$

Where $X^{w,iter} = [X_1^{w,iter}, X_2^{w,iter}, ..., X_D^{w,iter}]$ are like the repetitions with the larger count. A crow might be used to help recall a hidden location. In this context, $T^{w,iter}$ is meant to be taken as shorthand for the position of crow u's hidden treasure. The crows w has successfully moved to a more advantageous spot. Let's say we're iterating over a scenario in which a crow v has to be hidden somewhere ($T^{x,iter}$). Crow w now intends to follow Crow x to the hidden location.

$$X^{u,iter+1} = X^{u,iter+1} + l_i \times X^{u,iter+1} \times (T^{u,iter} - X^{u,iter}) \quad (2)$$

Where l_i is assumed to be uniformly distributed among 0 and 1, and $X^{u,iter}$ is the crow's flight time in seconds. Search engine predominates when X^u is small, but massive search dominates when X^u is large.

$$X^{u,iter+1} = \begin{cases} X^{u,iter+1} + l_i \times ekk^{u,iter+1} \times (T^{u,iter} - X^{u,iter}) \, l_i \geq BUO^{u,iter} \\ \text{a random location other wise} \end{cases} \quad (3)$$

Where $BUO^{u,iter}$ means that crow x is known at each iteration.

The term "attentive neural network" refers to a network of neurons that makes use of the attention mechanism. To improve our ability to anticipate both short and lengthy sequences of code, we combined the attention mechanism with LSTM. The suggested LSTM-AttM model opens up a new field of use for itself. Now, the system uses the attn layer among both and concealed states. The following equations tell you how we set up our attention-based LSTM model.

$$B_s = N_s.g_s \tag{4}$$

$$\alpha_s = softmax(B_s) \tag{5}$$

$$d_s = N_s\alpha_s^S \tag{6}$$

The computation for determining the very next phrase at time step s is determined by the currently hidden layers g_s and the set is designed d_s. The ultimate probability, B_s is derived from the vocabulary spaces generated by the softmax function. A vector of results, H_s.

$$H_s = \tanh(w^h(g_s) + w^n(d_s)) \tag{7}$$

$$z_s = softmax(w^h H_s + a^x) \tag{8}$$

Where $w^h \in \mathbb{R}^x$ and $w^h \in \mathbb{R}^x$ are a pair of learnable projecting vectors, $a^x \in \mathbb{R}^x$ is a bias vector that may be adjusted during training, and x is the word amount. The LSTM-AttM network improves the suggested model's efficiency since the attention mechanism allows for the precise collection of characteristics from input data.

4. Results and Discussion

In this part, we will discuss the results of the studies conducted to demonstrate the efficacy of the suggested CSO-ALSTM in predicting cardiovascular illness. We evaluated the suggested AI model against the support vector machine (SVM), random forest, and naive Bayes classifiers to determine its efficacy.

Having accurate data sets is essential for categorization. The proportion of accurate replies, both favorable and adverse, should be determined. The accuracy attained is measured by the proportion of accurate predictions. Figure 36.1 shows a comparison of the accuracy and precision of currently used and suggested approaches The recall value characterizes the databases that are pertinent to the other demand search. The F1 measure, which is the modulation index of the weight vectors of accuracy and recall, is used to determine how well an analysis performs relative to the gold standard. In Fig. 36.2, we see a comparison between the recall and f1 measures of the existing and suggested techniques.

Our suggested model improves upon the accuracy of the SVM, the RF, and the NB as compared to the outcomes of the generalized feature representation technique. Table 36.1 depicts the comparison result. Furthermore, the results demonstrate that the characteristic weight must be class-dependent when used for prediction.

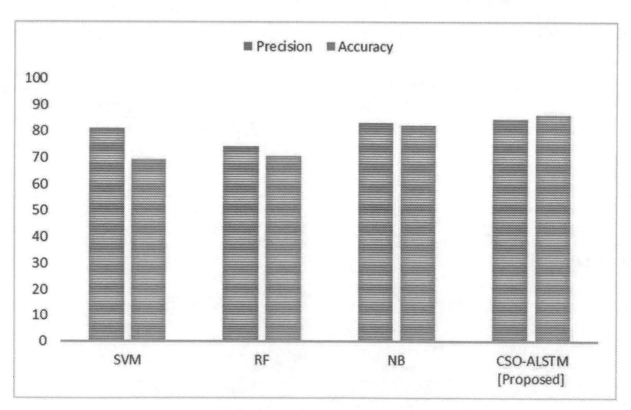

Fig. 36.1 Findings of accuracy and precision

5. Conclusion

We developed an SHM system employing an aggregation AI model and feature fusion approaches to enhance cardiovascular disease detection and enable clinicians promptly and correctly identify heart patients. The suggested technology discovers and thoroughly analyses the most relevant risk indicators in rising healthcare information to effectively forecast cardiovascular disease before a stroke or heart attack occurs. This novel system finds the best characteristics and quantifies their dataset importance, improving prediction performance. The AI system should remove any personally identifiable data from the patient data before using it for analysis in order to protect patient privacy.

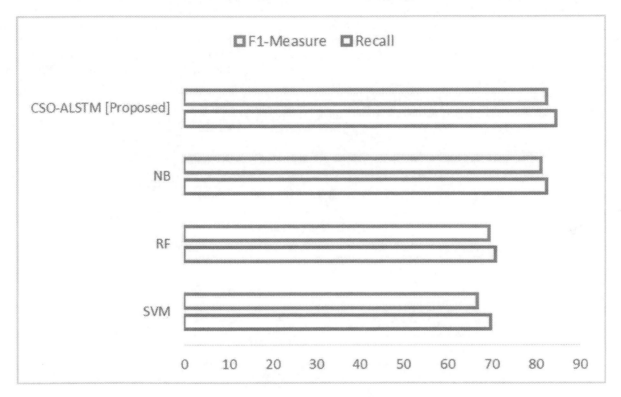

Fig. 36.2 Findings of recall and f1 measure

Table 36.1 Outcomes of the existing and proposed methods

	Precision	Accuracy	Recall	F1-Measure
SVM	81.1	69.7	69.8	66.7
RF	74.5	70.7	70.8	69.4
NB	83.2	82.4	82.4	81.2
CSO-ALSTM [Proposed]	84.9	86.4	84.5	82.4

References

1. Li, J.P., Haq, A.U., Din, S.U., Khan, J., Khan, A. and Saboor, A., (2020). Heart disease identification method using machine learning classification in e-healthcare. IEEE Access, 8, pp.107562-107582.
2. Koch, M., (2018). Artificial intelligence is becoming natural. Cell, 173(3), p.531.
3. Ali, F., El-Sappagh, S., Islam, S.R., Kwak, D., Ali, A., Imran, M. and Kwak, K.S., (2020). A smart healthcare monitoring system for heart disease prediction based on ensemble deep learning and feature fusion. Information Fusion, 63, pp.208-222.
4. Mansour, R.F., El Amraoui, A., Nouaouri, I., Díaz, V.G., Gupta, D. and Kumar, S., (2021). Artificial intelligence and the internet of things enabled disease diagnosis model for smart healthcare systems. IEEE Access, 9, pp.45137-45146.

5. Elayan, H., Aloqaily, M. and Guizani, M., (2021). Digital twin for intelligent context-aware IoT healthcare systems. IEEE Internet of Things Journal, 8(23), pp.16749-16757.

6. Subahi, Ahmad F., Osamah Ibrahim Khalaf, Youseef Alotaibi, Rajesh Natarajan, Natesh Mahadev, and Timmarasu Ramesh. (2022). "Modified Self-Adaptive Bayesian Algorithm for Smart Heart Disease Prediction in IoT System" Sustainability 14, no. 21: 14208. https://doi.org/10.3390/su142114208

7. Nashif, S., Raihan, M.R., Islam, M.R. and Imam, M.H., (2018). Heart disease detection by using machine learning algorithms and a real-time cardiovascular health monitoring system. World Journal of Engineering and Technology, 6(4), pp.854-873.

8. Tuli, S., Basumatary, N., Gill, S.S., Kahani, M., Arya, R.C., Wander, G.S. and Buyya, R., (2020). HealthFog: An ensemble deep learning-based Smart Healthcare System for Automatic Diagnosis of Heart Diseases in integrated IoT and fog computing environments. Future Generation Computer Systems, 104, pp.187-200.

9. Ahmed, G., (2019). Management of artificial intelligence-enabled smart wearable devices for early diagnosis and continuous monitoring of CVDS. International Journal of Innovative Technology and Exploring Engineering, 9(1), pp.1211-1215.

10. Natarajan, R.; Lokesh, G.H.; Flammini, F.; Premkumar, A.; Venkatesan, V.K.; Gupta, S.K. A Novel Framework on Security and Energy Enhancement Based on Internet of Medical Things for Healthcare 5.0. Infrastructures (2023), 8, 22. https://doi.org/10.3390/infrastructures8020022

Note: All the figures and table in this chapter were made by the authors.

Advancing Sustainable Science and Technology for a Resilient Future – Sai Kiran Oruganti et al. (eds)
© 2024 Taylor & Francis Group, London, ISBN 978-1-032-79020-6

37

Artificial Intelligence Assisted Intelligent COVID-19 Monitoring Framework

Sanjay Bhatnagar*

Chitkara University Institute of Engineering and Technology, Chitkara University, Punjab, India

Abstract: COVID-19 has a high contagiousness rate, making it an extremely dangerous virus. To accurately and promptly detect an infection, clinical help is necessary. COVID-19 patient data may be automated and quantified remotely with the use of artificial intelligence (AI). A researcher has developed various deep learning and IoT-based methods to categorize COVID-19 patients. The lack of reliable data sets may make early patient detection even more important. This study offers a unique hybrid swarm-intelligent augmented recurrent neural network to efficiently monitor COVID-19 patients' (HS-ERNN) approach. The success of the proposed HS-ERNN technique is assessed using accuracy, specificity, sensitivity, and f-measure criteria. The experimental results demonstrate that the recommended HS-ERNN technique achieves the maximum efficiency when compared to alternative strategies for intelligent COVID-19 monitoring.

Keywords: Intelligent COVID-19 monitoring, Clinical support, artificial intelligence (AI), Hybrid swarm-intelligent augmented recurrent neural network (HS-ERNN)

1. Introduction

The biggest difficulty facing every government in terms of health care is safeguarding its citizens against sudden illness epidemics when diseases may spread from one person to another. The COVID-19 pandemic is a current worry around the world. Most people with COVID-19 have a fever, trouble breathing, coughing, chest tightness, runny nose, etc. It is getting worse all over the world. The virus spreads when two people are close to each other. If someone shows symptoms, they should stay away from other people for their safety. People are scared by this case, and they want to know more and more about what's going on. As COVID-19 has been deemed a hazard to the public's health, the government and other organizations are making a lot of effort to limit its spread. (Singh, et al., (2020)). The "World Health Organization" declared the coronavirus to be a pandemic on March 11, 2020. Unfortunately, there is still no effective medication or vaccination. It will likely take more than a year to make a vaccine that works, particularly since the nature of the virus has not yet been fully understood. (Otoom, et al., (2020)). At the moment, slowing the spread of the Coronavirus is the only way the globe can fight it. (i.e. "flatten the curve") by taking steps like cleaning your hands, avoiding people, and using face masks. Yet, by identifying (or predicting future) such cases in advance and keeping a watch on them, technology might help curb their spread. The average time for COVID-19 to mature is 5 days, although it may take anywhere from 1 to 14 days. Studies from South Africa and the UK indicate that the novel COVID-19 versions of Delta and Omicron have an incubation time of 4 days and 2 days, respectively. However, with the correct knowledge and care, COVID-19 patients may be healed. The emphasis shifts from the disease to anticipating and preventing the spread of the virus by continuously monitoring the pandemic. AI-powered COVID-19 monitoring mechanisms raise privacy concerns.

---

*Corresponding author: sanjay.bhatnagar.orp@chitkara.edu.in

DOI: 10.1201/9781003490210-37

2. Related Works

The authors demonstrate that most AI implementations employed Convolutional Neural Networks (CNN) on X-ray and CT images, whereas the bulk of mathematical modeling used Susceptible Exposed Infected Removed (SEIR) and Susceptible Infected Recovered (SIR) models (Mohamadou, et al., (2020)). For identifying and taking care of COVID-19 patients, the use of RFID and body-wearable sensor technologies supported by the fog-cloud platform is advised, as well as the possible applications of robotic and AI-based healthcare technologies in the battle against the COVID-19 pandemic. The Preferred Reporting Items for Systematic Reviews and Meta-Analyses (PRISMA) technique is used to conduct a systematic search for this literature, and it is then thoroughly examined (Sarker, et al., (2021)). COVID-19, an infectious virus with different genes, has a big effect on the health of people all over the world. COVID-19 is an infection caused by Severe Acute Respiratory Syndrome CoronaVirus-2 (SARS-CoV-2) (Rajesh, et al., (2021)). While ML algorithms can uncover complex patterns in large datasets, AI-based models may still be helpful (Irudayasamy, et al., (2021)). "Smart Mobile Technology," or SMT is described to be "Smart" since it can instantly deliver information if you request it and this knowledge may be put to good use. Nowadays, smartphones come with capabilities like cameras, capturing videos, using a GPS, playing games, sending and receiving emails, and doing web search apps for a variety of uses (Vaishya, et al., (2020)).

3. Proposed Methodology

Data Collection

The COVID-19 Open Research Dataset collection had 14251 validated COVID-19 cases. For each instance, the data contains a wide range of various forms of information. For machine learning, the data was pre-processed and arranged.

Symptom Data Collection and Uploading

The goal of this component is to collect real-time symptom data by connecting a variety of wearable sensors to the user's body. The main symptoms were determined using a dataset of actual COVID-19 patients. The symptoms were coughing, fever, sore throat, tiredness, and shortness of breath.

Quarantine/Isolation Center

Patients who have been isolated or quarantined in a hospital are surveyed in this section. These records include information that is both non-medical and medical (or technical). In every report for health data, time-series data for the following symptoms are provided.

Data Analysis Center

The data center contains algorithms for machine learning and data analysis. These methods may be used to create a model for COVID-19, and the processed data can be seen in real-time on a screen. Based on user-uploaded the model can then be used to rapidly spot or anticipate possible COVID-19 problems using real-time data. The patient's response to treatment may also be predicted using the model.

4. Hybrid Swarm-intelligent Enhanced Recurrent Neural Network (ERNN)

Enhanced Recurrent Neural Network

The output for Enhanced Recurrent neural networks (ERNNs) has feedback loops $z_1(t)$, ..., $z_{nj}(t)$ to the network inputs $x(t)$. These loops' presence significantly affects the network's capacity for learning. The output for a new data is calculated and supplied again to set the changed input. That process is maintained until the output stabilizes. The following are the stages in this process:

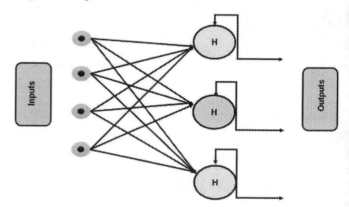

Fig. 37.1 Enhanced Recurrent neural network architecture

$$a_m(t) = \sum \varphi_{nm} X_m(t) + \sum \sigma_{nm} h_m(t-1), n = 1, \ldots \quad (1)$$

$$h_m(t) = F(a_m(t)), n = 1 \ldots, o_H \quad (2)$$

$$b_n(t) = \sum a_{nm} h_n(t), n = 1, \ldots, n_J \quad (3)$$

$$z_n(t) = G(b_n(t)), n = 1, \ldots, o_N \quad (4)$$

Where φ_{nm} σ_{nm} and *anm* are the processing units, that make up the network. The hidden neurons that receive input during the learning process calculate modifications to the network's weights the input vector x(t). The non-linear change between the neurons is represented by the letter F. After trial and error, the hidden neuron number o_H is identified by trial and error. A simple ERNN construction is provided in Fig. 37.1 demonstrates how the hidden neurons as well as the input neurons provide input values to the middle-layer neurons. Enhanced Recurrent neural networks (ERNNs) are made up of three layers in this study: one input, two hidden, and one output layer. O such that V1 has the largest possible variance. HS-ERNN may utilize COVID-19 infection, hospitalization, and vaccination distribution data to generate hypotheses and make choices.

Particle Swarm Optimization Phase

The original purpose of this algorithm was to replicate the social cooperation of flocks of birds and schools of fish. In this hybrid technique, PSO is employed to calculate the value of σ. The optimum value is assigned to a particular pattern selected from the training dataset and is started with a swarm value of σ. To arrive at a final global, the same process may then be used for every pattern in the training dataset; as a result, this optimization model creates the PSO technique's structure. There are two components to this method: particles and swarm. After every movement, each particle modifies and updates its location until it achieves the ideal position. The following example reduces the PSO structure and its components, including the regional and huge variances in the initial search space, to more clearly demonstrate how optimization works in our classification phase. A d-dimensional search space will be discussed $S \subset S_{train}$ and a swarm including p particles; every particle ith is in the form of the vector $a_i = (a_{i1}, a_{i2}, ..., a_{id}) \in S$. For this element, the particle's speed is $v_i = (v_{i1}, v_{i2}, ..., v_{id}) \in S$ the best earlier position achieved by the ith particle in S is denoted by $\hat{a}_i = (\hat{a}_{i1}, \hat{a}_{i2}, ..., \hat{a}_{id}) \in S$, where gi is the index of the particle, among all the particles in the near area of ith particles, attained the best prior position; t is the iteration timer. To update the location of each particle, apply the equation below:

$$a_i(t+1) = a_i(t) + v_i(t+1) \tag{5}$$

The following equation is used to increase each particle's area:

$$v_i(t+1) = wv_i(t) + c_1 r_1 (\hat{a}_i(t) - a(t)) + c_2 r_2 (\hat{a}_{gi}(t) - a_i(t)) \tag{6}$$

In the equations above, i stands for the particle index. ω is the inertial (or constriction) coefficient, c_1 and c_2 are the coefficients of acceleration ($0 \leq c_1, c_2 \leq 2$), and r_1 and r_2 are the random variables ($0 \leq r_1$ and $r_2 \leq 1$) is regenerated in every area update.

$$v_i(t+1) = \chi(t) + c_1 r_1 (\hat{a}_i(t) - a_i(t)) + c_2 r_2 (\hat{a}_{gi}(t) - a_i(t)) \tag{7}$$

Alternatively, here the parameter χ the restriction factor is what decides which PSO version to use. Equations (7)

and (8) are mathematically similar, although choosing the appropriate parameters results in significant changes. The following formula is used to derive the constriction factor.

$$\chi = \frac{2k}{|2 - \phi - \sqrt{\phi^2 - 4\phi}|} \tag{8}$$

5. Result and Discussion

In this study, we evaluate the sensitivity, specificity, accuracy, and f-measure of the, "Multiple Linear Regression" (MLR), Random Forest (RF), and Support Vector Machines (SVM) to our suggested methodology (Maheswari, et al., (2021)). We compared to other approaches for intelligent COVID-19 monitoring, the suggested HS-ERNN method achieves the highest efficiency in healthcare

Specificity: To evaluate the validity of trial results, sensitivity analyses look at how many changes to the methodology, models, values of unclear variables, or assumptions impact the results. Figure 37.2 illustrated by specificity compared to our methodology.

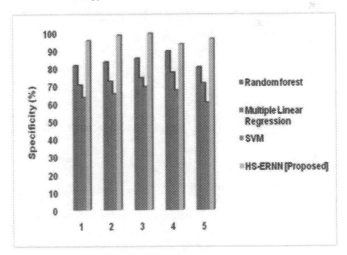

Fig. 37.2 specificity compared to the proposed

Sensitivity: An analysis of sensitivity and specificity is used to assess a test's effectiveness. It may be used in medicine to assess the effectiveness of a test used to identify an illness or in quality control to find a flaw in a produced product. Figure 37.3 illustrated by sensitivity compared to our methodology.

Accuracy: A test's accuracy is determined by its ability to differentiate between sick and healthy individuals with accuracy. Figure 37.4 illustrated by accuracy compared to our methodology.

F-measure: The recall and accuracy of a system are represented by the weighted harmonic mean, or F-measure. As compared to the other methodologies, our proposed methodology has the best results for the medical industry.

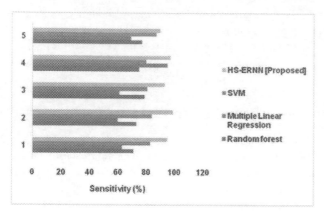

Fig. 37.3 sensitivity compared to the proposed method

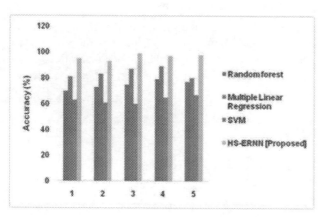

Fig. 37.4 Accuracy compared to the proposed method

6. Conclusion

This framework proposes using IoT to reduce the impact of dangerous illnesses. To create an artificial intelligence (AI) based prediction model for illness and to optimize the treatment response, the proposed framework was used to analyze health data of proven COVID-19 patients and information about possible COVID-19 cases. The framework also conveys these findings to medical professionals so that they may act quickly in response by doing any additional clinical testing required to confirm the situation, to suspected cases identified by the prediction model. In the future makes it possible to isolate confirmed patients and provide them with the required medical treatment. The recommended HS-ERNN technique eventually provides the maximum level of healthcare efficiency when compared to other methods for intelligent COVID-19 monitoring framework. AI methods

that protect privacy should be employed as possible for compliance monitoring.

References

1. Singh, P. and Kaur, R., 2020. An integrated fog and Artificial Intelligence smart health framework to predict and prevent COVID-19. Global transitions, 2, pp.283-292.
2. Otoom, M., Otoum, N., Alzubaidi, M.A., Etoom, Y. and Banihani, R., 2020. An IoT-based framework for early identification and monitoring of COVID-19 cases. Biomedical signal processing and control, 62, p.102149.
3. Tellez Gaytan, J.C., Ateeq, K., Rafiuddin, A., Alzoubi, H.M., Ghazal, T.M., Ahanger, T.A., Chaudhary, S. and Viju, G.K., 2022. AI-Based Prediction of Capital Structure: Performance Comparison of ANN SVM and LR Models. Computational Intelligence & Neuroscience.
4. Bhatia, M., Manocha, A., Ahanger, T.A. and Alqahtani, A., 2022. An artificial intelligence-inspired comprehensive framework for Covid-19 outbreak control. Artificial Intelligence in Medicine, 127, p.102288.
5. Mohamadou, Y., Halidou, A. and Kapen, P.T., 2020. A review of mathematical modeling, artificial intelligence, and datasets used in the study, prediction, and management of COVID-19. Applied Intelligence, 50(11), pp.3913-3925.
6. Sarker, S., Jamal, L., Ahmed, S.F. and Irtisam, N., 2021. Robotics and artificial intelligence in healthcare during COVID-19 pandemic: A systematic review. Robotics and autonomous systems, 146, p.103902.
7. Rajesh, N. and Christodoss, P.R., 2021. Analysis of origin, risk factors influencing COVID-19 cases in India, and its prediction using ensemble learning. International Journal of System Assurance Engineering and Management, pp.1-8.
8. Irudayasamy, A., Ganesh, D., Natesh, M., Rajesh, N. and Salma, U., 2022. Big data analytics on the impact of OMICRON and its influence on the unvaccinated community through advanced machine learning concepts. International Journal of System Assurance Engineering and Management, pp.1-10.
9. Vaishya, R., Javaid, M., Khan, I.H. and Haleem, A., 2020. Artificial Intelligence (AI) applications for the COVID-19 pandemic. Diabetes & Metabolic Syndrome: Clinical Research & Reviews, 14(4), pp.337-339.
10. Wang, L.L., Lo, K., Chandrasekhar, Y., Reas, R., Yang, J., Eide, D., Funk, K., Kinney, R., Liu, Z., Merrill, W. and Mooney, P., 2020. Cord-19: The covid-19 open research dataset. ArXiv.
11. Maheswari, K.G., Nalinipriya, G., Siva, C. and Raj, A.T., 2021. Real-Time Health Monitoring Using IoT With Integration of Machine Learning Approach. Machine Learning for Healthcare Applications, pp.249-259.

Note: All the figures in this chapter were made by the authors.

Advancing Sustainable Science and Technology for a Resilient Future – Sai Kiran Oruganti et al. (eds)
© 2024 Taylor & Francis Group, London, ISBN 978-1-032-79020-6

38

Hybrid Artificial Intelligence Algorithm for Medical Data Classification

Raman Verma*

Chitkara University Institute of Engineering and Technology,
Chitkara University, Punjab, India

Abstract: Computational intelligence is being increasingly used in medical diagnostics, and there are many different medical uses for this. It is possible to classify many medical diagnosis processes as smart data classification. Several investigators have tried to apply various techniques in the medical field to increase the precision of the classification process. Techniques with higher accuracy rates will offer more data that is sufficient for recognizing possible sufferers and enhancing diagnostic performance. Thus, a unique hybrid Artificial Intelligence (AI) approach for classifying medical data is proposed in this study. Stochastic gradient descent optimization (SGDO) and probabilistic neural network (PNN) make up the suggested technique. The best features are chosen for the PNN technique's input in this case using the SGDO algorithm. Comparing existing methods for classifying medical data, the suggested technique performs favorably.

Keywords: Computational intelligence, Medical diagnostics, Medical data classification, Artificial intelligence (AI), Stochastic gradient descent optimization (SGDO), Probabilistic neural network (PNN)

1. Introduction

Worldwide healthcare systems create enormous volumes of data from several sources. Determining the trends and tiny changes in genetic, radiological, laboratory, or clinical data that consistently discriminate morphologies or enable accurate predictions in wellness activities is crucial, despite the information's intricacy for a natural having soul great. The massive growth in data produced by different diagnostic equipment and nodes within the healthcare systems is what propels the field of clinical treatment (Saravi, et al., (2022)). A patient's health and medical history are both included by the term medical data that also includes demographic data, medical records, laboratory test results, imaging data, prescription records, and clinical notes. Medical information is important for patients, doctors, and researchers because it gives a complete picture of a patient's medical history and allows them to make educated choices regarding their diagnosis, course of treatment, and continuing care.

The best ways to stop illnesses like cancer, brain tumors, and heart disorders from killing people are early diagnosis and treatment. Information mining and artificial intelligence techniques have shown to be effective in this respect for supplying personal information for an early diagnosis. High-dimensional data, however, may be expensive to measure and keep, complex to study, and almost impossible to show (Prakash, et al., (2022)).Its general term in this area of research is artificial intelligence or AI. It educates digital learning of human skills including teaching, judgment, and using machines to imitate human judgment reasoning. A learning system called AI's main applications just like its focus assembles information studies, investigates various kinds of data storage, then employs those techniques to create mimic intellect. AI was vital to current societal growth or has resulted in a revolutionary effect on labor efficiency and expense savings people management architecture, and work overload (Zhang, et al., (2021), Raja Gopal, et al., (2021)).

We routinely use AI software as well as its help within their everyday routines. The COVID-19 epidemic has highlighted

*Corresponding author: haraman.verma.orp@chitkara.edu.in

DOI: 10.1201/9781003490210-38

our dependency on AI systems and the need for support. Across a variety of industries, including medicine, economics, telecommunications, travel, farming, and much more, Ai technologies are becoming more and more influential. Living in a contemporary world means constantly interacting with programs that use artificial intelligence. (Akgun, et al., (2021), and Irudayasamy, et al., (2022)). They analyze the hybrid artificial intelligence algorithm for medical data classification.

2. Related works

Kavitha, et al., (2021) mentions that the suggested study used datasets and the Detroit stroke database. Including regression and categorization were utilized. Random Forest and Selection tree artificial intelligence algorithms are used. This computational figure's innovative method is formed. Its solution uses three machine learning techniques: Random Forest, Decision Tree, and Hybrid Model (Hybrid of random forest and decision tree).

Khalaf, et al., (2022) proposed a Blinder Oaxaca-based Shapiro Wilk Neutrosophic Fuzzy (BO-SWNF) data analytics approach created for use in remote medicine. The Cultural phenomenon database is used to gather data. The Blinder Oaxaca Linear Regression-based preprocessing methodology eliminates duplicating information. The Blinder Oaxaca function is used to significantly boost effectiveness. To provide a reliable research study, the Shapiro Wilk Neutrosophic Fuzzy algorithm is used.

Barik, et al., (2021) presented two different machine learning sets of methods for diabetes assumption one of them is a hybrid method, while the other is a classification-based approach. They choose the random forest method for categorization. They have selected the Boost algorithm for the hybrid method

3. Proposed Methodology

Dataset

Using the UCI dataset, which has received a lot of citation in the research [29], three typical clinical records were chosen to test the effectiveness of the suggested approach. One of them, the mammary database, has 699 samples of breast cancer in reported trials along with 9 numerical variables and 1 classification variable. The classification of measurements and deleterious the aim factor. With 395 samples and 6 numerical variables and 1 target variable, the hepatic information is separate. This is a limited choice collection made up of 768 samples of American Indian ancestors over the age of 21. Every example has 8 numerical variables and 1 category factor.

Also, all the data were pre-processed, including noise reduction, abnormality elimination, and normalisation treatment, to provide clean and immediately accessible information, that assisted in improving the categorization performance of the following models and enable more accurate comparison of the testing results.

Probabilistic Neural Network

Consider that we're presented with a weight matrix $x \in \mathbb{R}^n$ that originally belonged to particular standard framework classes $g = 1, 2, ...$ Here G denotes metric possibility also matrix x belongs to the category g, cg denotes the price for classification of the column onto g, while pdf denotes the softback. $y_1(x)$, $y_2(x)$, ..., $y_G(x)$ .Recognized over all courses. The column x is again assigned to that same category g by the Conditional probability whenever g if $p_g c_h y_h(x) > phchyh(x)$. Usually $p_g = p_h$ and $c_g = c_h$, Therefore as result, if $y_g(x) > y_h(x)$, any column x must form a class g.

Since distributions of a sample group are often uncertain in real-world data classifiers, hence it is necessary to develop an estimate of the probability density function $y_g(x)$. The probability density function for several factors may be explained by using the Parzen approach for that purpose.

$$y(x) = \frac{1}{l} \sum_{i=1}^{l} W_i(x, x_i) \qquad (1)$$

Where $W_i(x, x_i) = \sigma_1^{-1} ... \sigma_n^{-1} F(\sigma_1^{-1}(x_{i1} - x_i), ..., \sigma_1^{-1}(x_{in} - x_n)$, $F(.)$ is the weighting its amount of training sets would be a variable that must be used carefully, while $\sigma_1, ..., \sigma_n$ signify all variance calculated using the average of the n factors $x_1, ..., x_n$, typically, a popular option for weighing is indeed the Linear distribution.

Theequation (1) explains the PNN's architecture and workings. Gaussian functionfollows a transform into account as the activating again for probability density function and suppose that this value is produced again for cases of category g.

$$y_g(x) = \frac{1}{l_g (2\pi)^{\frac{n}{2}} (\det \sum_g)^{1/2}}$$

$$\sum_{i=1}^{l_g} exp\left(-\frac{1}{2}\left(X_{g,i} - X\right)^T \sum_g^{-1}\left(X_{g,i} - x\right)\right) \qquad (2)$$

While g is the number of instances of category g, (g, j) stands for the scaling factor linked only with j^{th} variable and the gth class, and x_{g-1}, I am the learning array $(i = 1, ..., l_g)$ by the category g. A single PNN bridge's $g = 1 ..., G$ summing cells is provided by the equation in (2). The patterning cells that make up the components of the level above supply the contribution to the total that is calculated across all of the gth class samples. Hence, g nodes represent the gth summing photoreceptor data. Ultimately, depending just on the results of all the cells in the summing level, the hidden layers calculate the column x's result in line with Bayes' selection method.

$$G^*(x) = \arg \max_g \{y_g(x)\} \qquad (3)$$

Where $G^*(x)$ represents the shirt's anticipated category for x. The design level thus needs vertices $l = l_1 + ... + l_G$. For each characteristic and subclass, a unique smoothness factor is used in this study. According to equation (2), the choosing of this version of picking forces the inevitable storage of a $G(x)$ matrix of σ's. As a result, the gth summing cell delivers the signals (2) to the judgment layers with $\sigma_{g,j}$ as the intrinsic parameter. This leads to the computation of the density function for the jth variable of each class g using this strategy, it is possible to highlight the similarities between the matrices within a single category. The steepest descent technique is used to calculate the values of s. PNN may be used to examine patient information, including medical history, laboratory test results, and vital signs, in order to identify illnesses and forecast that treatments will go. PNN is a tool that can be used to analyze massive collections of medical data to find new relationships and correlations between various medical factors that might help clinical practice and medical research.

Stochastic Gradient Descent Optimization

The stochastic gradient descent (SGD) method has been made simpler. SGD has a number of advantages over other optimization strategies in terms of improving the accuracy and efficacy of medical data analysis. By adding randomness to the model updates, SGD optimization may enhance the generalization performance of machine learning models using medical data. Instead of precisely computing the slope, every cycle guesses the gradienton the basis of a single randomly picked example

$$u_{d+1} = u_d - \gamma_t \nabla u O(h_d, u_d) \qquad (4)$$

The instances chosen at random at each iteration determine the outcome of the stochastic process (u_d, $d = 1...$). The noise created by this streamlined approach is believed to not affect how acts under its expectation.

Given it isn't essential to recall which instances occurred in the early weeks, the stochastic technique may process examples on the deployed system. In this case, the stochastic gradient descent immediately reduces the expected risk since the examples are randomly selected mostly from test dataset distributions.

4. Result and Discussion

In this section, we discuss in detail about the findings of the hybrid artificial intelligence algorithm for medical data classification. Accuracy and precision are the parameters used to identify the effectiveness of the proposed methodology. Data categorization has various applications and is crucial to the processing of medical data. Medical data categorization issues range in complexity and variety, are quickly impacted by noise and equipment flaws during data processing, and have a significant impact on the physician's assessment of the lesion site. As can be observed, the SGDO-PNN algorithm's parameters are quite high; thus, the following stage will concentrate on increasing the SGDO-PNN algorithm's operational effectiveness while lowering its parameters. Other patient physiological information may also be considered (Wang, et al.,2022)

Accuracy

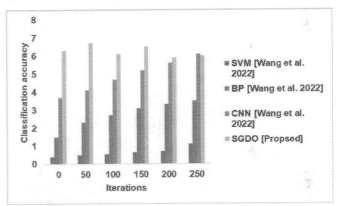

Fig. 38.1 Accuracy

The definition of accuracy, which is "the level that the outcome of a study corresponds to the proper value or a standard," basically relates to how closely a measurement matches its set point. The accuracy of medical data is also dependent on data security.

Precision

The quantity of data that a statistic can convey via its digits is called precision, and it shows how near one or many measurements are separated apart. It is independent of accuracy. The degree of agreement between measurements or observations is referred to as the precision of medical data. Efforts may be taken to eliminate errors and ensure that the data is as accurate as possible in order to guarantee the correctness of medical data.

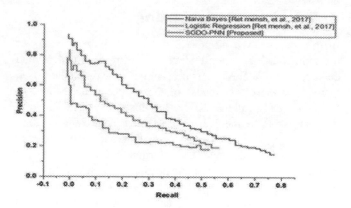

Fig. 38.2 Accuracy

5. Conclusion

The categorization of medical data discusses the current state of classification technology and presents pertinent knowledge points. Based on this, the SGDO-PNN model-based medical data classification approach is suggested in this study, and the classification outcome is promising. About illness data and the approach suggested in this research produced the best classification accuracy (97.5% respectively) and quickest running time. The SGDO-PNN algorithm's parameters are also evidently rather high, thus the next phase will concentrate on enhancing SGDO-PNN's operational effectiveness while lowering parameters. Other patient physiological information may also be considered. Medical data is delicate and vulnerable to issues with data privacy and security. Healthcare providers may overcome this by putting in place safeguards like data encryption and access restrictions to safeguard patient data from unwanted access. In future, there remain a number of additional areas that need to be investigated with regard to different elements of healthcare data, such as safety, security, reliability, etc. The above part offers a more comprehensive overview of big data analytics in health systems, various topics like health data features (like complexities and wide scale), data analytics tasks (like latitudinal display and analysis), and goals (like real-time, personal privacy, and expert collaboration). The exponential growth of digital health data will be advantageous for health informatics in the future.

References

1. Saravi, B., Hassel, F., Ülkümen, S., Zink, A., Shavlokhova, V., Couillard-Despres, S., Boeker, M., Obid, P. and Lang, G.M., 2022. Artificial intelligence-driven prediction modeling and decision making in spine surgery using hybrid machine learning models. Journal of Personalized Medicine, 12(4), p.509.
2. Prakash, P.S. and Rajkumar, N., 2022. HSVNN: an efficient medical data classification using dimensionality reduction combined with a hybrid support vector neural network. The Journal of Supercomputing, 78(13), pp.15439-15462
3. Zhang, C. and Lu, Y., 2021. Study on artificial intelligence: The state of the art and prospects. Journal of Industrial Information Integration, 23, p.100224.
4. Akgun, S. and Greenhow, C., 2021. Artificial intelligence in education: Addressing ethical challenges in K-12 settings. AI and Ethics, pp.1-10.
5. Irudayasamy, A., Ganesh, D., Natesh, M., Rajesh, N. and Salma, U., 2022. Big data analytics on the impact of OMICRON and its influence on the unvaccinated community through advanced machine learning concepts. International Journal of System Assurance Engineering and Management, pp.1-10.
6. Kavitha, M., Gnaneswar, G., Dinesh, R., Sai, Y.R. and Suraj, R.S., 2021, January. Heart disease prediction using hybrid machine learning model. In 2021 6th international conference on inventive computation technologies (ICICT) (pp. 1329-1333). IEEE.
7. Khalaf, O.I., Natarajan, R., Mahadev, N., Christodoss, P.R., Nainan, T., Romero, C.A.T. and Abdulsahib, G.M., 2022. Blinder Oaxaca and Wilk Neutrosophic Fuzzy Set-based IoT Sensor Communication for Remote Healthcare Analysis. IEEE Access.
8. Barik, S., Mohanty, S., Mohanty, S. and Singh, D., 2021. Analysis of prediction accuracy of diabetes using a classifier and hybrid machine learning techniques. In Intelligent and Cloud Computing: Proceedings of ICICC 2019, Volume 2 (pp. 399-409). Springer Singapore.
9. Wang, L. and Zuo, K., 2022. Medical Data Classification Assisted by Machine Learning Strategy. Computational & Mathematical Methods in Medicine.
10. Rotmensch, M., Halpern, Y., Tlimat, A., Horng, S. and Sontag, D., 2017. Learning a health knowledge graph from electronic medical records. Scientific reports, 7(1), pp.1-11.
11. Lashari, S.A., Ibrahim, R., Senan, N. and Taujuddin, N.S.A.M., 2018. Application of data mining techniques for medical data classification: a review. In MATEC Web of conferences (Vol. 150, p. 06003). EDP Sciences.

Note: All the figures in this chapter were made by the authors.

Advancing Sustainable Science and Technology for a Resilient Future – Sai Kiran Oruganti et al. (eds)
© 2024 Taylor & Francis Group, London, ISBN 978-1-032-79020-6

Artificial Intelligence-based Framework for Predicting Academic Performance in Online Learning

39

Tannmay Gupta*

Chitkara University Institute of Engineering and Technology,
Chitkara University, Punjab, India.

Abstract: Artificial intelligence (AI) is a developing technology that is attracting interest for its ability to analyze student activities and assess student progress. One kind of breakthrough in information technology is the increasing usage of online learning. Every learning technology program attempts to enhance the standard of education while simultaneously increasing student engagement online. Teachers may raise their student's involvement in the teaching-learning process by selecting instructional materials that are appropriate for each students learning preferences. This study's goal was to develop an AI strategy for the prediction of academic performance in online learning using a Dwarf Mongoose Optimized with Artificial Neural Network (DMO-ANN). Educational datasets for participants in the study were gathered from the university. Results of the study demonstrate that DMO-ANN is an effective strategy when compared to existing approaches in terms of accuracy, prediction rate, RMSE, and MAE. It also covers several aspects of internet-based learning.

Keywords: Artificial intelligence (AI), Online learning, Dwarf mongoose optimized with artificial neural network (DMO-ANN), Academic performance, Information technologygradient descent optimization (SGDO), Probabilistic neural network (PNN)

1. Introduction

The students and professors engage personally to obtain information in conventional learning management systems (LMSs). The Internet has changed a wide range of industries, especially education. Considering its flexibility, e-learning has become popular in most educational institutions. (Stadlman, et al., 2022). Institutions all across the globe were compelled to switch from traditional classroom settings to online learning because of the COVID-19 pandemic's unique conditions to guarantee that classes could continue (Ting, et al., 2022). While this online mode gives the student freedom, there is no social contact between both the learners and the lecturers. AI makes it possible to adopt data-driven judgments for practical purposes to solve issues with teaching and learning (Doe, et al., 2020, Kauri, et al., 2021). When used by educational professionals to analyze information regarding student achievement, AI provides a methodical approach for gathering and modeling such data to create smart tools that

quickly and effectively evaluate students' learning (Begonia, et al., 2022, and Mahesh, et al., 2022).

2. Related Works

The methods used for evaluation may be improved by taking into account some factors that affect the education programs and adults versus kids. These elements, which include the amount of sleep a person gets, the time of day, the amount of noise, the surroundings, and their mood, are directly related to their cognitive abilities and emotional states, which might impact how well they do on an online exam. The aim is to determine how architecture students' efficiency in an online test in India is affected by their higher energy, mood, sleep habits, and time of day (Arum Kumar, et al., 2022). Estimating a value, grouping, efficiency forecasting and coaching are all parts of the artificially intelligent tool for evaluation and evaluation. According to pupil participation and test results, each person's ratings are calculated by making use

*Corresponding author: tannmay.gupta.orp@chitkara.edu.in

DOI: 10.1201/9781003490210-39

of recurrence neural network (RNN). Cao, et al., 2022. is to understand how the epidemic of viruses and the expansion of virtual classrooms have an impact on college student's mental health. We use a range of statistical and machine-learning algorithms to analyze data collected by the Faculty of Public Governance at the University of London, Austria, in association with a worldwide consortium of universities, other educational institutions, and student unions. Students of all abilities find it difficult to keep up with a traditional educational system, and teachers have historically been overworked. Education is one of the many sectors being transformed by AI Digitalization has already resulted in indicates and goal setting for education Evaluation of the happiness of kids with using the advancement our school via training is crucial establishments in higher learning make use of the advantages of e-learning. Personality, curricula, student-based ideas happiness, and noticed utility are used to model the motivation to utilize e-learning (Devi, et., 2022).

3. Proposed Methodology

Data set

Online distance learning (ODL) is a kind of using digital instruction among professors and instructors communicating via online meetings and electronic channels [1, 2]. The smallest swabs were collected and produced 85 data were used by -Power (edition 3.1.9.4). with an alpha score of 0.05, a high consumption of 0.80, as well as an effect size of 0.15. The last example of 207 achieved was more than the necessary minimum. This article's data survey file, which included 207 rows and 24 columns, was saved as a Microsoft Excel spreadsheet. According to Table 1, a code was given to each object. Numeric, cardinal, or scale measurements were made for the items. Profiles of responders are shown in Table 2. 49.8% of the 207 answers were students who identified as male, and 50.2% were women. Most of the pupils were above the age of 21. The majority of the pupils seem to be from architectural and engineering programs, plus managerial and business programs. 42.5% of responders do not have any previous ODL training, whereas 57.5% of participants have (Shi-Hui, et al., 2022)

Artificial Neural Network

Artificial neural networks (ANN) are utilized both for the classification of data based on input observations as well as for the development of neural predictions regarding the grade point averages of students. Both of these strategies rely significantly on supervised machine learning as their primary method of data analysis. It is common practice to write the ANN model in the form of a straightforward mathematical function, such as:

$$Z\sim = (V, U) \tag{1}$$

$Z\sim$ *output vector* and V, – *input factors*. *U weight parameters*. Within an artificial neural network, a connection is denoted by a scalar value denoted by, U, which is a vector of weight parameters. The following equation is determine the values of z_i that are produced by the zth neuron in vector Z' of elements U and X_j:

$$z_i = \theta(\sum U_{ji}X_j) \tag{2}$$

The value of the weighted sum of inputs is communicated to the output layer through the utilization of the hyperbolic tangent activation function θ. The node vi is going to be the one that is activated in the succeeding input layer:

$$vi = \theta(zi).vi = \theta(yi) \tag{3}$$

To achieve the objectives that have been specified, this rule makes adjustments to the neuron weights U_{ji} based on the mistakes that have been determined. We can determine how far off the computed BP-based ANN is from the outputs that were desired by using the formula that is presented here:

$$F = 0.5\sum\sum Mi(yi - si)^2 \tag{4}$$

where si target neuron value in output and is the total output neurons The steepest descent is also known as the gradient descent. When it comes to the resolution of non-linear issues, it beats other training algorithms because of the accelerated convergence to the optimal solution that it offers. The following algorithm demonstrates a new method for approximating the Gaussian matrix that is analogous to the Gauss-Newton approach:

$$uji + 1 = uji - [\Gamma I + \zeta J] - 1\Gamma f^t \tag{5}$$

where J Jacobian matrix, ek network error,

wij updated weight, wij current weight and ζ damping factor

The Gauss-Newton method for solving small issues ζ and the gradient descent algorithm for solving large problems. The formula for determining the mean squared error is as follows:

$$MSE = \sum_{Mi}\sum_{j=1}\left(zji - sji\right)^2 \tag{6}$$

Dwarf mongoose optimized (DMO)

The Protest Learning Mechanism is accompanied by the Dwarf Mongoose Optimization Algorithm (DMOA).

$$V_{j+1} = V_j + phi \times peep \tag{7}$$

Where as V_j is the imaginary choice produced by adding zeroes and then one. *phi* and *peep* define the boundaries of the search engine. carers, the alpha cluster, and scouts are three of the organizations that form the Destination management cluster. Each team presented differently to get the meal, and each group's specifics are as follows: provide as

Alpha Group

If the populace was formed, the adequacy of each approach was assessed. The algorithm evaluates the potential values for each health species, and a dominant female (α) is specific and reliant on this likelihood

$$\alpha = \frac{fit_j}{\sum_{j=1}^{m} fit_j} \tag{8}$$

M we have to do with the number of stoats in the alpha group. *peep* is the voice of the dominating female, who maintains order in the home. Mongooses slumber during the whole main that. *Our tn$_j$* is fixed ϕ. The DMO that was used to create a prospective fast-food chain

$$V_{j+1} = V_j + \text{phi} \times \text{pee} \tag{9}$$

was offered in Equation (10) then all the repetitions, whereas phi signifies the uniformly distributed arbitrary value in −1 and 1

$$tn_j \frac{fit_{j+1} - fit_j}{\max\left\{\left|fit_{j+1} - fit_j\right|\right\}} \tag{10}$$

Equation (10) comprises the average value of tn_j.

$$\varphi = \frac{\sum_{j=1}^{m} tn_j}{m} \tag{11}$$

This technology evolves to the scouts' phase, where the next fruit or vegetable or rest mound is presumed if the childcare changes requirement is satisfied.

Scout Group

Although it is known that mongooses won't return to previous sleep mounds, the scout looks for the next lying piles, taking care to search with this strategy, foraging and scouting were done simultaneously. A successful or failed search for a brand-new sleeping mound followed this journey. Mouse migration, in particular, depends on their overall effectiveness.

$$V_{j+1} = \begin{cases} V_j - DE * phi * rand * \left[V_j - \vec{N}\right] \text{if } \varphi_{j+!} > \\ V_j + DE * phi * rand * \left[V_j - \vec{N}\right] \end{cases} \tag{6}$$

In *rand*, I use an arbitrary integer between 0 and 1 as an example.

$$DE = \left(1 - \frac{iter}{Max_{iter}}\right)^{\left(2 - \frac{iter}{Max_{iter}}\right)} \tag{12}$$

whereas is the factor controlling the viper band.

$$\vec{N} = \sum_{j=1}^{m} \frac{V_i \times tn_j}{V_i} \tag{13}$$

whereby vector determined the mongoose's journey to the innovative nesting heap.

Babysitters Group: The alpha female (mother) led the remainder of the party on daily foraging forays while the babysitters were often inferior group members who continued to care for the young and cycled frequently. Refreshing the usage of the sitter exchange parameter replaces the scout and food source data that were previously specified by the family members.

4. Result and Discussion

In this section, we discuss in detail the findings of the AI-based framework for predicting academic performance in online learning. Accuracy and precision are the parameters used to identify the effectiveness of the proposed methodology. They conducted research and worked together using digital sources. Digital displays and screens were essential for classroom pupils because they shared learning outcomes and promoted engagement. To promote the expansion of student knowledge in intelligent settings, it focuses on contextual, customized, and simplified learning. Teaching theory, the development of educational technology, teacher intelligence artificial leadership, social and cultural structures, the rise of technologies, and contemporary society are just a few of the challenges that smart teaching must overcome (Cao, et al., 2020). The potential of an instrument to measure an exact number is known as accuracy. In other words, it refers to how closely the measured value resembles a reference or genuine value. Little readings may be taken to determine accuracy. Figure 39.1 depicts the accuracy comparison. The modest result decreases the calculation's inaccuracy.

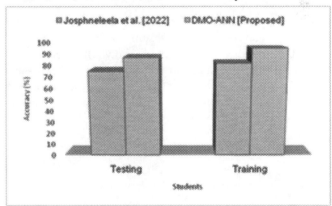

Fig. 39.1 Accuracy

The world has recently felt the throb of AI. One measure of a machine learning model's effectiveness is precision or the standard of a model correctly predicts the outcome. Using the ratio of the total number of predicted positives to the actual positives. Figure 39.2 depicts the comparison of precision.

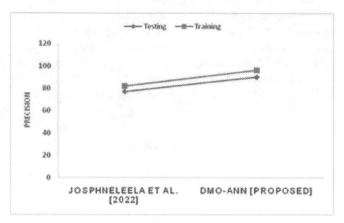

Fig. 39.2 Precision

5. Conclusion

To meet the difficulties of a smart learning environment, this research provides an artificial intelligence-based methodology for forecasting academic achievement in online learning. Each student in the class should have access to the Internet and wireless devices since this will improve their learning experience. They conducted research and worked together using digital resources. Smart displays and screens were essential for classroom pupils because they shared learning outcomes and promoted engagement. It emphasizes contextual, individualized, and simplified learning to support student knowledge acquisition in intelligent settings. Smart teaching must overcome some challenges, including pedagogical theory, the development of educational technology, artificial intelligence in teachers, social and cultural structures, the expansion of innovations, and challenges found in contemporary society. The experimental findings demonstrate that, when compared to other widely used approaches, the suggested AI strategy improves the predictions and accomplishment ratio in terms of students' online learning behavior and interaction.

References

1. Stadlman, M., Salili, S.M., Borgaonkar, A.D. and Miri, A.K., 2022. Artificial Intelligence Based Model for Prediction of Students' Performance: A Case Study of Synchronous Online Courses During the COVID-19 Pandemic. Journal of STEM Education: Innovations and Research, 23(2), pp.39-46.
2. Ting, T.T., Teh, S.L. and Wee, M.C., 2022. Kaur, P., Kumar, H. and Kaushal, S., 2021. Affective state and learning environment based analysis of students' performance in online assessment. International Journal of Cognitive Computing in Engineering, 2, pp.12-20.
3. Deo, R.C., Yaseen, Z.M., Al-Ansari, N., Nguyen-Huy, T., Langlands, T.A.M. and Galligan, L., 2020. Modern artificial intelligence model development for undergraduate student performance prediction: An investigation on engineering mathematics courses. IEEE Access, 8, pp.136697-136724.
4. Bagunaid, W., Chilamkurti, N. and Veeraraghavan, P., 2022. AISAR: Artificial Intelligence-Based Student Assessment and Recommendation System for E-Learning in Big Data. Sustainability, 14(17), p.10551
5. Kaur, P., Kumar, H. and Kaushal, S., 2021. Affective state and learning environment based analysis of students' performance in online assessment. International Journal of Cognitive Computing in Engineering, 2, pp.12-20.
6. T. R. Mahesh, Dilip Kumar, V. Vinoth Kumar, Junaid Asghar, Banchigize Mekcha Bazezew, Rajesh Natarajan, V. Vivek, " Blended Ensemble Learning Prediction Model for StrengtheningDiagnosis and Treatment of Chronic Diabetes Disease", Computational IntelligenceandNeuroscience, vol. 2022, ArticleID 4451792, 9 pages, 2022. https://doi.org/10.1155/2022/4451792
7. Arun Kumar, U., Mahendran, G. and Gobhinath, S., 2022. A Review on Artificial Intelligence Based E-Learning System. Pervasive Computing and Social Networking: Proceedings of ICPCSN 2022, pp.659-671.
8. Cao, W., Wang, Q., Sbeih, A. and Shibly, F.H.A., 2020. Artificial intelligence-based efficient smart learning framework for an education platform. Inteligencia Artificial, 23(66), pp.112-123.
9. Devi, J.S., Sreedhar, M.B., Arulprakash, P., Kazi, K. and Radhakrishnan, R., 2022. A path towards child-centric Artificial Intelligence based Education. International Journal of Early Childhood, 14(03), p.2022.
10. Shi-Hui, S., Chaw, L.Y., Aw, E.C.X. and Sham, R., 2022. Dataset of international students' acceptance of online distance learning during COVID-19 pandemic: A preliminary investigation. Data in Brief, 42, p.108232.

Note: All the figures in this chapter were made by the authors.

Advancing Sustainable Science and Technology for a Resilient Future – Sai Kiran Oruganti et al. (eds)
© 2024 Taylor & Francis Group, London, ISBN 978-1-032-79020-6

40

Swarm Optimized Robust AI Algorithm for Biomedical Data Classification

Sachin Mittal*

Chitkara University Institute of Engineering and Technology,
Chitkara University, Punjab, India

Abstract: Applications of artificial intelligence (AI) or computerised systems in medical diagnosis are growing in popularity lately. A medical doctor's ability to diagnose diseases is frequently aided by decision systems that use AI. The vast expansion of data in the field of biomedical engineering has created significant challenges for data processing and analysis. A significant difficulty results from the enormous volume of biological data leverage. Swarm optimised k means cluster-adaptive Bayesian (SO-KB) for biomedical data is presented in this research as a remedy for this issue. To remove noise from the gathered data, we first pre-process it using min-max normalisation. After that, we use our suggested approach to classify biological data. Particle swarm optimization (PSO) is used in this instance to optimise the KB technique's characteristics. According to the experimental findings, the suggested SO-KB method for classifying biomedical data has the maximum accuracy of existing techniques.

Keywords: Biomedical, Artificial intelligence (AI), Data processing and analysis, Swarm optimised k means cluster-adaptive Bayesian (SO-KB)

1. Introduction

Machine learning methods have been effectively used in some biological fields, including the diagnosis and prognosis of cancer and other difficult illnesses, as well as the detection of tumors. The so-called "curse of dimensionality," which is caused by the fact that biomedical data are often defined by a small number of occurrences and are presented in a high-dimensional higher dimensional space, is one of the main problems with biomedical data analysis and mining. Insufficient classification accuracy is caused by irrelevant characteristics, and locating potentially beneficial information is made more difficult as a response (Wood, et al., 2019). As excluding pointless information makes it easier to visualize data and better comprehend computer models, selection has emerged as one of the key sub-fields in biomedical data categorization. Also, a wise choice may lower the cost of database administration and storage by reducing the need for measurement and storage (Anagaw, et al., 2019). Medical data has grown quickly in the healthcare sector in recent years. The United States produced a zettabyte of healthcare data in 2018. This accumulation of medical data, particularly photographs, has necessitated the development of new techniques based on artificial intelligence (AI), & machine learning (ML). It is crucial to do research in this area since the use of information technology in the healthcare industry creates prospects for the creation of novel diagnostics and therapies. The practice of arranging and categorizing various sorts of medical and healthcare data according to their traits and features is known as biomedical data classification (Tchito, et al., 2020).

2. Related Works

Alomari, et al., 2021 suggested a novel hybrid filter-wrapper method for selecting the top-ranked genes. According to the findings, the suggested technique performs admirably on the remaining seven datasets and produces the best results on four of the nine datasets. The experimental result showed how well the suggested strategy searched the gene search

*Corresponding author: sachinm84824@gmail.com

DOI: 10.1201/9781003490210-40

space and was able to identify the most advantageous gene pairings. Pashaei, et al., 2022 proposed a novel wrapper feature selection technique for biological data categorization based on the chimp optimization algorithm (ChOA). The results of the experiments show that the suggested methods may successfully eliminate the least important elements and increase classification precision. Li, et al., 2016 addressed the issue of unbalanced datasets, which is prevalent in biomedical applications, by using a unique class-balancing strategy called the adaptive swarm cluster-based dynamic multi-objective synthetic minority oversampling technique. Baliarsingh, et al., 2019 examined the first filters Fisher score and minimal redundancy maximum relevance (mRMR). The experimental findings demonstrate that, as compared to the current approaches, the suggested framework delivers superior CA with less NSG. Baliarsingh, et al., 2021 evaluated a hybrid method for choosing the best gene subset, and categorizing cancer is suggested utilizing simulated annealing (SA) and Rao algorithm (RA). Experiment results have shown that the method we've suggested picks discriminating genes with excellent classification accuracy. Moeini, et al., 2020 created a hybrid algorithm that combines the simulated annealing (SA) and grey wolf optimizer (GWO) techniques for feature selection in biomedical data. The results of the evaluations, which were determined by many difficult biological benchmarks, demonstrated that the offered methodologies surpass those of their competitors.

3. Proposed Work

Min-max normalisation

While utilizing the Min-max approach to normalize the data, the original data are linearly transformed. The letters y_{min} and y_{max} stand for a variable's lowest and highest values in the samples. The equation (1) below describes how the Min-max method transfers a value of this variable, to a value:

$$u' = \frac{u - y_{min}}{y_{max} - y_{min}} + y_{min} \tag{1}$$

The Min-max normalization uses a linear mapping to scale a variable in the training samples in the range $[y_{min}, y_{max}]$ to $[-1, 1]$ (or $[0, 1]$). However, when the testing samples fall outside the training data range of the variable, the scaled values will be beyond the confines of the interval $[-1, 1]$, which might lead to problems in certain applications.

Swarm optimised k means cluster-adaptive Bayesian

The PSO algorithm views the process of problem optimization as the act of feeding. The two parameters that characterize each particle in the PSO algorithm its location and velocity

are abstracted as solutions to the optimization problem. The following equation (2) and equation (3) are used to update the particle's location and velocity after each iteration:

$$z_{id}(s+1) = uz_{id}(s) + d_1 q_1 \left[pbest_{jc}(s) - y_{jc}(s) \right] +$$
$$d_2 q_2 \left[gbest_c(s) - y_{id}(s) \right] \tag{2}$$

$$y_{jc}(s+1) = y_{jc}(s) + z_{jc}(s+1) \tag{3}$$

where s for the number of iterations, i stands for the ith particle, d for the dimension, y_{jc} for the ith particle's location in the d dimension, confined to the range [*popmin*, *popmax*], and $z_{id}(t)$ for the particle's velocity; Fig. 40.1 illustrates the two-dimensional vector representation of the above formulae. The velocity formula is shown in Fig. 40.1(a), and the variables $uz(s)$ and z produce the velocity $z(s + 1)$. The position formula is shown in Fig. 40.1(b), where the location of the original particle $y(s)$ and the new particle $y(s + 1)$ are added. PSO has been shown to be successful in finding the best solutions for challenging issues including feature selection and parameter tweaking in the categorization of biological data.

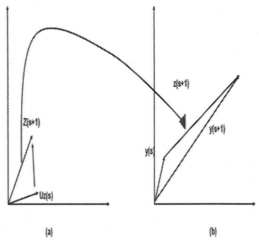

Fig. 40.1 Updating of a PSO algorithm with vector representation in two dimensions. (a) Velocity updates schematic. (b) position update schematic

K-means

The K-means technique was the most straightforward unsupervised learning algorithm that could address the most pressing clustering problems. With its simplicity, usefulness, and modest cost, K-means continues to be a popular option because of its consistent presentation across a variety of difficulties. K-means had, on the one hand, been largely accepted in a variety of applications. On the other hand, constant efforts have been made to advance the execution k-means. Conventional k-means is still not notably effective to handle a web-scale of data, although the time complexity was linear to the data size. The method utilizes a simple and direct approach to characterize a given data set using a preset number of previously generated groups.

One starting idea is to describe each of the "k" centroids, one for each group. The centroid has to be cleverly placed in a different region with a new idea to get a different result. The next step is to take each point that is located inside a particular data collection and subordinate it to the nearest centroid. The first phase is complete when there is no point still open. Now, we must re-determine 'k' new centroid as the focal points of the clusters arising as a result of the previous progress. Later, when we had this "k" new centroid, additional coupling between similar informative set points and the closest new centroid was required to create a loop. Due to this loop, we may see that the k-alters centroid its territory as desired until the point at which no more modifications are made. As a result of this loop, we could see the k-centroid changed its area stage by stage until all alterations were made. In other words, the centroid is no longer moving. Finally, this kind of method limits the objective function; the squared error function for this kind of case is as follows in equation (4).

$$I = \sum_{i-1}^{l} \sum_{j-1}^{m} y_j^{(i)} - d_i^2 \rightarrow \quad (4)$$

Where $\|y_j^{(i)} - \|d_i^2$ is the measurement of the distance between information point and a "cluster center" d_i^2, which denotes the distance of m information points from their respective cluster centers.

The algorithm was organized as follows:

1. Identify the "K" point in the area where a group of clearly concentrated objects was. Also, "these points" denote the main group of the centroid.

2. Every item should be assigned to a set that includes its nearest centroid.

3. The placements of the "K"-centroid should be recalculated when the complete object has been allocated.

4. Repeat steps two & three until the centroid could not move due to elongation. So, a collection of an item that has to have a metric reduced and computed is harvested in this manner.

4. Bayesian

There are several behaviors in intrusion cases that are comparable to both normal behavior and other intrusion instances. Several algorithms, like K-Means, are also unable to discriminate between incursion occurrences and regular instances with accuracy. We merged the Bayesian classifier and K-Means approach to address this classification flaw. One of the best learning algorithms nowadays is Bayesian. Bayesian models have a very straightforward formulation and rely heavily on the independence assumption. A conditional probability is calculated for each connection by analyzing the relationship between the independent and dependent variables. Applying the Bayesian Theorem, we formulate equation (5):

$$O(G \mid Y) = O(Y \mid G)O(G)/O(Y) \quad (5)$$

Let Y be the record of data. Let G explain the data record Y, which belongs to the specified class C. We need to know O(G|Y), or the probability that a hypothesis Y is correct given an observed data record Y, to categorize data. The posterior probability of G given Y is denoted by the symbol O (G|Y). O (G), which is independent of Y, is based on less information than the posterior probability, O (G|Y), which is based on more information, such as background knowledge. The posterior probability of Y conditioned on G is O (G|Y), similarly. Since the Bayes theorem offers methods for calculating the posterior probability O (G|Y) from O (G), O(Y), and O (Y|G), it is helpful. Equation (6) displays the Bayes rule.

$$O(D_1|Y) = \frac{O(Y|D_1).O(D_1)}{O(Y)} \quad (6)$$

Where D_1 stands for the class category and Y for the data record. A given instance of Y might be broken up into parts called $y_1, y_2 ... y_N$ each of which would be associated with an attribute of $Y_1, Y_2 ... Y_N$.

Equation (7) shows the probability that was attained.

$$O(D_1|Y) = \frac{O(Y_1|D_1), O(Y_2|D_1).....O(Y_m|D_1)O(D_1)}{O(Y)} \quad (7)$$

O(Y) has a constant denominator across all classes. Thus, it may be disregarded, just as in Equation (8).

$$O(D_1|Y) = O(Y_1|D_1), O(Y_2|D_1).....O(Y_m|D_1)O(D_1) \quad (8)$$

5. Results and Discussion

The dominant class is often designated as the negative class, whereas the minority class is identified as the positive class. Considering that the overall accuracy in data with imbalances is often skewed toward the majority class, additional assessment measures including precision (Pre), and area under the curve (AUC) are utilized as backup measures.

Precision: One of the most crucial standards for accuracy is precision, which is well-defined as the ratio of correctly classified cases to all instances of predictively positive data. The precision is calculated using equation (9).

$$precision = \frac{TP}{TP + FP} \times 100 \quad (9)$$

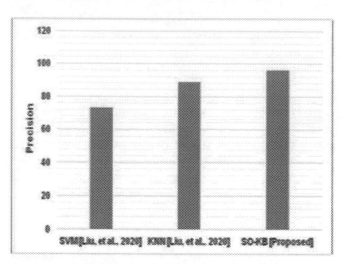

Fig. 40.2 Result of precision

Comparable results for the precision measures are shown in Fig. 40.2. With a 96% precision, the suggested method outperforms the currently used ones, which include KNN in precision (89%), and SVM (73%).

AUC: AUC is increasingly considered when assessing models since it has the potential to rank models based on overall performance and is insensitive to the ratio between the majority & minority classes. AUC is an asset for assessing classification models in biomedical data analysis because it is agnostic to the decision threshold that is used to generate predictions and is more resistant to data imbalances than other metrics such as accuracy or sensitivity.

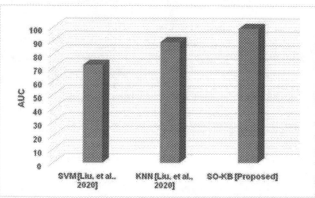

Fig. 40.3 Result of AUC

In Fig. 40.3, SO-KB scored (99%) on the F-Measure, followed by KNN scored (89%), and SVM (at 72%).

References

1. Wood, A., Shpilrain, V., Najarian, K. and Kahrobaei, D., 2019. Private naive bayes classification of personal biomedical data: Application in cancer data analysis. Computers in biology and medicine, 105, pp.144-150.
2. Anagaw, A. and Chang, Y.L., 2019. A new complement naïve Bayesian approach for biomedical data classification. Journal of Ambient Intelligence and Humanized Computing, 10, pp.3889-3897.
3. Tchito Tchapga, C., Mih, T.A., Tchagna Kouanou, A., Fozin Fonzin, T., Kuetche Fogang, P., Mezatio, B.A. and Tchiotsop, D., 2021. Biomedical image classification in a big data architecture using machine learning algorithms. Journal of Healthcare Engineering, 2021, pp.1-11.
4. Alomari, O.A., Makhadmeh, S.N., Al-Betar, M.A., Alyasseri, Z.A.A., Doush, I.A., Abasi, A.K., Awadallah, M.A. and Zitar, R.A., 2021. Gene selection for microarray data classification based on Gray Wolf Optimizer enhanced with TRIZ-inspired operators. Knowledge-Based Systems, 223, p.107034.
5. Pashaei, E. and Pashaei, E., 2022. An efficient binary chimp optimization algorithm for feature selection in biomedical data classification. Neural Computing and Applications, 34(8), pp.6427-6451.
6. Li, J., Fong, S., Sung, Y., Cho, K., Wong, R. and Wong, K.K., 2016. Adaptive swarm cluster-based dynamic multi-objective synthetic minority oversampling technique algorithm for tackling binary imbalanced datasets in biomedical data classification. BioData Mining, 9(1), pp.1-15.
7. Baliarsingh, S.K., Vipsita, S., Muhammad, K. and Bakshi, S., 2019. Analysis of high-dimensional biomedical data using an evolutionary multi-objective emperor penguin optimizer. Swarm and Evolutionary Computation, 48, pp.262-273.
8. Baliarsingh, S.K., Muhammad, K. and Bakshi, S., 2021. SARA: a memetic algorithm for high-dimensional biomedical data. Applied Soft Computing, 101, p.107009.
9. Moeini, F. and Mousavirad, S.J., 2020, October. An Evolutionary Hybrid Feature Selection Approach for Biomedical Data Classification. In 2020 10th International Conference on Computer and Knowledge Engineering (ICCKE) (pp. 623-628). IEEE.
10. Liu, S., Zhang, J., Xiang, Y., Zhou, W., and Xiang, D., 2020. A study of data pre-processing techniques for imbalanced biomedical data classification. International Journal of Bioinformatics Research and Applications, 16(3), pp.290-318.

Note: All the figures and tables in this chapter were made by the authors.

Advancing Sustainable Science and Technology for a Resilient Future – Sai Kiran Oruganti et al. (eds)
© 2024 Taylor & Francis Group, London, ISBN 978-1-032-79020-6

41

Artificial Intelligence-Based Categorization of Art Paintings

Abhiraj Malhotra*

Chitkara University Institute of Engineering and Technology,
Chitkara University, Punjab, India

Abstract: The long history and culture of China have shaped the art form of painting, and many paintings depict the way of life in China at various times. This has enormous significance for the advancement of Chinese culture. In the era of rapid information technology development, image classification has emerged as a core study topic, and the area of art painting image classification has also seen rapid growth. Artificial intelligence (AI) technology has advanced quickly in recent years, and this might speed up the classification of paintings. The painting image classification method based on saliency is then used to categorize the picture meanings, and this research provides a revolutionary classification technique named Chaotic Remora Optimized Long Short-Term Memory (CROLSTM) technology to evaluate the kinds of different creative efforts. According to the findings, the recommended approach may considerably increase the categorization impact of paintings with a better level of accuracy.

Keywords: China, Artificial intelligence (AI), Image classification, Chaotic Remora optimized long short-term memory (CROLSTM)

1. Introduction

The ability of technology to emulate or enhance human cognition, such as logical reasoning and experience-based learning, is known as Artificial Intelligence (AI). Since many years ago, computer programs have employed AI; however, the technology is currently used in a wide range of different goods and services. A new set of enabling conditions for artistic production is provided by AI technology. AI painting has grown into a more famous type of art, and it has a greater link to everyday life for individuals. AI painting uses computer mathematical algorithms to convert two- or three-dimensional visuals into painted artwork that is presented on a screen. AI painting has grown into a more famous type of art, and it has a greater link to everyday life for individuals. AI painting uses computer mathematical algorithms to convert 2or3-dimensional visuals into painted artwork that is presented on a screen (Liu, (2020)). AI is getting extremely intelligent, and it is now being auctioned at major auction marketplaces and in the alternative art market. These data demonstrate that AI-produced artworks are already widely available on the artwork. (Wang, et al., 2019). Art paintings are a form of nature imitation and resemble a miniature version of the actual thing. In addition to reproducing the natural world, photography also could make several clones of artwork. Since it has the potential to replace the artist, AI is a technical advancement of an entirely other order (Sun, et al., 2022). In this regard, the art world suggests a system for automatically categorizing effect paintings based on performance approaches to address the issue of design team studying to describe mature design works, giving artists in landscape art layout and other industries a conceptualized reference surfing method. The impact of AI artworks on the art market has raised several problems concerning AI artworks, including whether they possess artistic innovation or elegance or even if they qualify as works of art (Li, (2021)). By analyzing the issues in Art Paintings, we propose a chaotic remora optimized long short-term memory approach to resolving it.

*Corresponding author: abhiraj.malhotra.orp@chitkara.edu.in

DOI: 10.1201/9781003490210-41

2. Related Work

The article first categorizes the channels of experimental art representation and AI technology development on a timeline based on historical development, then analyses the deconstructive relationship between the two from a macro perspective of the evolution of both technology and art (Shen, et al., 2021). The Kong, (2020) goal was to describe the issues with these implementations after reviewing the present application state of AI in fine arts. Strengthen the environment and aesthetic experience of AI-based art education while also strengthening the intelligent teaching method of art education. The study suggests the AI-Assisted Effective Art Teaching Framework (AIEATF) improves the art learning and environment for AI-oriented artwork learning, develops intelligent teaching methods, and increases the ability to adapt to AI-oriented fine arts. An evaluation model has been created to take the improving effects into account of basis (Rajagopal, et al., 2022). Gong, (2021) talks about the AI may be used to improve the art design program systems at universities. The mechanical nature of AI painting's expressive approaches, however, restricts the variety of paintings. The basic purpose of AI media in the development of interactive art is first covered in the article. The subjectivity and interactions between artists, robots, and other process actors are then examined. The basic structure, stylistic traits, and design phases of Russian contemporary oil painting are first covered in the dissertation. After that, the use of artificial intelligence and the Internet of Things is explained, and the innovation and development of Russian contemporary oil painting are then investigated using a survey strategy (David, (2022)).

3. Proposed Methodology

Dataset

The research investigates the AI-based categorization of art painting pictures and proposes several AI-based techniques to extract art painting. Here, an Art image is constructed to examine the impact of art painting pictures on categorization. This dataset has 60,000 pictures in total, which are broken up into 100 types. There are 600 pictures in each category. The relationship network model is trained to carry out data classification tasks using this data source, which is a dataset of Chinese Art Paintings. Birds and blossoms, rivers and mountains, and humans make up the 3 primary categories of Chinese art painting picture data gathering. After analyzing the image data collection of Chinese fine art painting, 5 training photographs from three various kinds of bloom, bird, landscape, and individual have been extracted, and 15 training images in all have been collected.

Chaotic Remora Optimized Long Short-Term Memory (CROLSTM)

The design of LSTM consists of four gates, for example, Gates for input, output, control, and forget. One definition of input is:

$$j_s = \sigma\left(U_j * \left[g_{s-1}, v_s\right] + a_j\right) \qquad (1)$$

The field may receive the data that was retrieved from the formula previously. The following formula is used by the forget gate to determine the information from the entry of the preceding layer that will be disregarded:

$$e_s = \sigma\left(U_j * \left[g_{s-1}, v_s\right] + a_j\right) \qquad (2)$$

The continuity equation allows the management gate to regulate the input from the whole memory module:

$$e_s = \sigma\left(U_j * \left[g_{s-1}, v_s\right] + a_j\right) \qquad (3)$$

$$\tilde{D}_s = e_s * \tilde{D}_{s-1} + j_s * \tilde{D}_s \qquad (4)$$

The hidden layer g_{s-1}, and the result are updated as follows:

$$O_s = \sigma\left(U_j * \left[g_{s-1}, v_s\right] + a_p\right) \qquad (5)$$

$$g_s = P_p * tanh\left(\tilde{D}_s\right) \qquad (6)$$

The range [-1 to 1] is standardized by using *tanh*; here, U represents the weight matrices, and σ represents the activation process, which is assumed to be sigmoid. The BA receives for automatic hyperparameter change, the learning rate, momentum rate, and dropout rate from each of the LSTM dropout layers. Each hidden layer's g_{s-1}, hyperparameters for s = {1, 2, 3…N} are enhanced by offering a worldwide solution. The improved LSTM's output layer may be understood as:

$$P_s = \sigma\left(U_p\left[g_{s-1}\left[\begin{cases} \overline{v}_{jH}^s & if\left(\overline{v}_{jH}^s > \overline{v}_{jH}^s\right) \\ \overline{v}_{jH}^s - \overline{O}_{jH}^s & otherwise \end{cases}\right\}, v_s\right]\right] + a_p\right) \qquad (7)$$

The only specific parameter optimization that has a minor effect on the effectiveness of the suggested LSTM is learning rate optimization, according to our analysis of the influence of single parameter optimization on the proposed method. The Chaotic Remora completes the optimization process by utilizing the bioactivities of the remora. Remora can cling to swordfish, whales, or other creatures. The integer input H (0 or 1). In certain ways, while addressing optimization issues, the Remora Optimized (RO) will benefit from using both optimization strategies.

Chaotic Remora

Free travel (Exploration): The RO uses the SFO approach to carry out the international, this is based on the sophisticated technique applied to swordfish. The following is an expression for the things stated equation:

$$X_j(s+1)$$

$$= V_{best}(s) - \left(rand \times \left(\frac{V_{best}(s) + V_{round}(s)}{2} \right) - V_{rand}(s) \right) \quad (8)$$

Where $X_j(s + 1)$ is a potential contender for the remora. $V_{best}(S)$ is now in the greatest position. $V_{round}(S)$ is a randomly chosen remora position. s signifies the number of iterations currently in use. Remora may also alter the host based on its experiences. A new applicant role can be created in this scenario by:

$$X_j'(t+1) = X_j(s+1) + randn \times \left(X_j(s+1) - V_j(s) \right) \quad (9)$$

Feed thoughtfully (Exploitation): To feed, remora can also connect to humpbacks. The movement of remora will therefore resemble that of humpback whales. To conduct the local search, RO uses the whale optimization algorithm approach (WOA). In particular, the WOA foam assaulting strategy is deployed. The following is the modified position updating equations:

$$X_j(s+1) = C \times f^b \times cos(2\pi b) + V_{best}(s) \quad (10)$$

$$C = \left| V_{best}(s) - V_j(s) \right| \quad (11)$$

$$b = rand \times (a-1) + 1 \quad (12)$$

$$a = -\left(1 + \frac{s}{S} \right) \quad (13)$$

Where D stands for the spacing between remora. Also, the remora can create a tiny step by utilizing the WOA surrounding prey mechanism, which will further enhance the solution's overall quality. Which is shown in the following way:

$$V_j(s+1) = V_j(s+1) + B \times C' \quad (14)$$

$$B = 2 \times A \times rand - A \quad (15)$$

$$B = 2 \times \left(1 + \frac{s}{S} \right) \quad (16)$$

$$C' = V_j(s+1) - D \times V_{best}(s) \quad (17)$$

Where $V_j(S + 1)$ is the freshly created location of the ith remora. D Refers to the remora factor. That is set to 0.1 in remora optimization.

4. Result and Discussion

The suggested methodology is compared to current methods such as deep neural networks (DNN), ALEX NET, and

VGG-16. Analysis is done using metrics such as Accuracy, Precision, and Recall.

Accuracy

A difference between the outcome and the true number is caused by inadequate precision. The percentage of actual outcomes reveals how balanced the data is overall. Accuracy is assessed using an equation (18).

$$Accuracy = \frac{TP + TN}{TP + TN + FP + FN} \quad (18)$$

Figure 41.1 shows the comparable values for the accuracy measures. When compared to existing methods like DNN, which has an accuracy rate of 59.5%, ALEX NET, which has an accuracy rate of 62.62%, and VGG-16, which has an accuracy rate of 63.01%, the recommended method's CROLSTM value is 90%. The suggested CROLSTM performs well in classifying art paintings and has higher accuracy than other methods.

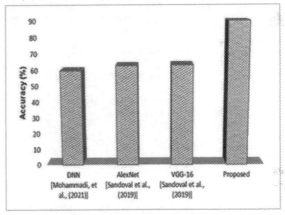

Fig. 41.1 Comparison of accuracy

Precision

The most crucial standard for accuracy is precision, it is clearly defined as the percentage of properly categorized cases in all instances of predictively positive data. Equation (19) is used to compute the precision.

$$Precision = \frac{TP}{TP + FP} \quad (19)$$

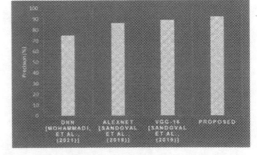

Fig. 41.2 Comparison of precision

Figure 41.2 shows the comparable values for the accuracy measures. When compared to existing methods like DNN, which has an accuracy rate of 75%, ALEX NET, which has an accuracy rate of 86.2%, and VGG-16, which has an accuracy rate of 88.9%, the recommended method's CROLSTM value is 92%. The suggested CROLSTM performs well in classifying art paintings and has higher accuracy than other methods.

Recall

The potential of a model to identify each important sample within a data collection is known as recall. The percentage of TPs divided by the sum of True Positive and False Negative is how it is statistically defined. The recall is calculated using equation (20).

$$Recall = \frac{TP}{TP + FN} \qquad (20)$$

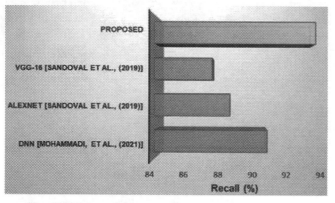

Fig. 41.3 Comparison of recall

Figure 41.3 shows the comparative data for the recall metrics. Recall rates for DNN91%, ALEX NET 88.7%, VGG-16 87.6%, and CROLSTM 94%. The proposed method performed better than the current results with a recall of 94%.

5. Conclusion

The most challenging aspect of picture categorization is defining painting art. It requires a lot of information processing since there are many different types of painting pictures and significant variances between oriental and western painting. It provides a revolutionary classification approach named chaotic remora optimized long short-term memory (CROLSTM) technology to determine the styles of different creative works. Thus, as compared to other current methods, the performance of our AI-based classification of art paintings will be improved. The testing results (accuracy, precision, recall) demonstrate the suggested optimization's greater efficacy in comparison to other well-liked optimization techniques. The results show that the suggested method may significantly improve the classification impact of artworks with a higher level of accuracy.

References

1. Liu, X., 2020, October. An artistic reflection on artificial intelligence digital painting. In Journal of Physics: Conference Series (Vol. 1648, No. 3, p. 032125). IOP Publishing.
2. Wang, Y. and Ma, H., 2019, December. The value evaluation of artificial intelligence works of art. In 2019 International Joint Conference on Information, Media, and Engineering (IJCIME) (pp. 445-449). IEEE.
3. Sun, Y., Yang, C.H., Lyu, Y. and Lin, R., 2022. From Pigments to Pixels: A Comparison of Human and AI Painting. Applied Sciences, 12(8), p.3724.
4. Li, Y., 2021. Intelligent environmental art design combining big data and artificial intelligence. Complexity, 2021, pp.1-11.
5. Shen, Y. and Yu, F., 2021. The influence of artificial intelligence on art design in the digital age. Scientific Programming, 2021, pp.1-10.
6. Kong, F., 2020. Application of artificial intelligence in modern art teaching. International Journal of Emerging Technologies in Learning (iJET), 15(13), pp.238-251.
7. Rajagopal, N.K., Qureshi, N.I., Durga, S., Ramirez Asis, E.H., Huerta Soto, R.M., Gupta, S.K. and Deepak, S., 2022. Future of business culture: an artificial intelligence-driven digital framework for the organization decision-making process. Complexity, 2022.
8. David, L.G., Patra, R.K., Falkowski-Gilski, P., Divakarachari, P.B. and Antony Marcilin, L.J., 2022. Tool Wear Monitoring Using Improved Dragonfly Optimization Algorithm and Deep Belief Network. *Applied Sciences*, 12(16), p.8130.
9. Mohammadi, M.R. and Rustaee, F., 2021. Hierarchical classification of fine-art paintings using deep neural networks. Iran Journal of Computer Science, 4, pp.59-66.
10. Sandoval, C., Pirogova, E. and Lech, M., 2019. Two-stage deep learning approach to the classification of fine-art paintings. IEEE Access, 7, pp.41770-41781

Note: All the figures in this chapter were made by the authors.

Advancing Sustainable Science and Technology for a Resilient Future – Sai Kiran Oruganti et al. (eds)
© 2024 Taylor & Francis Group, London, ISBN 978-1-032-79020-6

42

Deep learning-based Evaluation of Cybersecurity Threats in Smart City Buildings

Rahul Mishra*

Chitkara University Institute of Engineering and Technology,
Chitkara University, Punjab, India

Abstract: Smart cities have risen in popularity lately because of the increasing adoption of Internet of Things (IoT) technologies. The possibility of cyber incidents and threats has risen with the expansion of smart city systems, though. A smart city platform's IoT devices are susceptible to unlawful assaults and threats since they are linked to detectors that are interconnected to big cloud systems. The development of strategies to thwart such assaults and shield IoT devices from malfunction is crucial. The goal of this research is to protect against and minimize IoT cyber threats in a smart city using the firefly-optimized hybrid deep convolutional recurrent neural network (FFO-HDCRNN) method. Findings from experiments show that the suggested technology can detect cyberattacks more successfully than current approaches.

Keywords: Smart city, Internet of things (IoT), Cyber threats, Firefly optimized hybrid deep convolutional recurrent neural network (FFO-HDCRNN)

1. Introduction

Major infrastructure development and execution may benefit from innovations in IoT, distributed energy systems, cloud computing, fog computing, modern communications, and application. Even though these solutions may have numerous advantages, privacy concerns are a significant barrier to achieving these advantages. As the majority of technologies for smart cities are interconnected apps, they are similarly vulnerable to possible intrusions as any other distributed software (Mohamed, et al., 2020, and Kaushal et al. 2022). Ma. C. (2021) is utilized as data proves. Supervised learning, a sort of computer vision, has numerous advantages in the smart city since it closely resembles the process the conscious beings employ to study a specific topic. Deep learning and machine learning both gains from the use of transfer learning, a subfield of intelligent machines. Hyman, et al., 2019 recognition programmers have recently significantly improved classic feature extraction techniques in several different fields, including automation, learning languages, and many more. Structures presently consist of more than just windows, slabs, and chandeliers. Designing

things autonomously gather data from IoT devices, tenant behaviour, and the surroundings using artificial intelligence, IoT, and advanced technologies assess materials. Through combining radar systems, electrical components, and programmed systems that are linked via networks with incredibly sophisticated AI innovation. Social systems include data, monitoring systems, and algorithmic decision-making tools. As a result, cybersecurity is set up, organised, and updated utilising the recently noticed Sengan et al., 2020. Raveendran, et al., 2021, and Saxena et al., (2021) have enabled IoT smart buildings to work largely by physically connecting all of its components and using internet-powered sensing devices. These smart objects gather data on their surroundings and also provided the judgement call centres additional characteristics.

2. Related Work

Sarker, et al., 2022 convey a wide overview of the concept of smart urban biostatistics that can be used as a reference for academics, experts in the business, and decision-makers in a

---

*Corresponding author: rahulmishra84824@gmail.com

DOI: 10.1201/9781003490210-42

nation, especially from a technical standpoint. To calculate the security indicators for ego circuits, we are required by quantitative risk assessment techniques to gather factual analysis, and the correctness of this approach relies on how many computing repetitions are performed. Kalinin, et al., 2021 have put out a fresh strategy for managing cybersecurity threats for the foundation of smart cities founded on the object type, data mining, and quantitative vulnerability assessments. To make the optimal choice, data is processed using advanced analysis, modelling, optimizer, and visualisation services. Many stakeholders do not have a similar comprehension of crucial data protection. The study evaluated smart cities' precautions for cyber security, paying special attention to the legal and technological context. Chen, et al., 2021 Combining dynamic circuits using parametric designs, constrained State models, deep residual nets, perceptron, & classification. The understanding and application of principles linked to Deep Learning, Cyber Security, Smart Cities, and IoT in Smart Cities. Gives a thorough analysis of a variety of current AR and cyber security applications for smart cities, highlighting any possible advantages, restrictions, and unresolved problems that could founding principles issues later on. Since communities employ modern technology, As a result, personal data and cyber warfare have become more crucial. Possible cyber-attack scenarios in the service establishment and the system of the company are discovered and assessed, as well as security risks against the urban planning design (Lee, et al., 2019). Ullah, et al., 2020 malicious attacks call for a comprehensive response. The article aims to provide a comprehensive analysis of DL algorithms that might be used to enhance protection. Capabilities are examined, as well as the potential for DL techniques to be used in IoT platforms to preserve a permissible level of functionality, protection, and privacy concerns, to enhance the overall experience of smart cities.

3. Proposed Methodology

ICS Cyber Attack Dataset: The author Pan et al.,(2019) evaluates Data set collection has five separate datasets they are Power System Datasets, Gas Pipeline Datasets, Energy Management System Data, New Gas Pipeline, and Gas Pipeline and Water Storage Tank. The 37 instances in the Power System dataset are broken down into 28 assault events, 1 no event, and 8 natural forces. There are three kinds of assaults: material input, remote tripping command injection, and transmit setup alteration. Such datasets may be utilized by smart buildings to identify cyber security intrusions.

Convolution Neural Network (CNN): Deep learning (DL) programmes understand images, videos, texts, and sounds.

Most DL algorithms use CNNs. The CNN learns from data and recognises patterns without human background reduction. CNNs are preprocessed and classified. Top layers may be connected. Several sensor-based human activity recognition studies have used CNN. Influences of a CNN for local feature extraction on public datasets and some basic statistical features that preserved time series structure. Going up and down stairs, sitting, standing, and lying activity identification were not 100% accurate. Smartphones and watches abstracted features for 18 side and semi-tasks. A CNN model yielded considerable results. These results are promising, but there is always opportunity for improvement, especially in complex human conduct.

Long Short-Term Memory (LSTM): RNNs handle minute events. RNNs' diminishing gradient issue makes it hard to train them well. They are frequently used in natural language processing, especially in human activity detection, word prediction, language translation, etc. Input-behavior relationships Memory-building DL model modules comprise internal and external repetition. Bidirectional LSTM expands LSTM network storage. These are promising outcomes, but complex human behaviour always has potential for improvement. This framework simultaneously learns from previous and subsequent sequences. The continuous LSTM outperformed the nonlinear method in object recognition.

Gated Recurrent Unit (GRU): LSTM's memory cells increase memory utilization, but it solves RNNs' disappearance gradient problem. GRUs are LSTMs without memory cells. A GRU network's update and reset gate controls each concealed value's shift. To improve sensor-based physical action detection, the GRU network-based DL model improves data. One of the main drawbacks of such a network is that its output at a given time step depends completely on the training data. In some cases, looking back and forward might help make predictions. The Bidirectional long-short-term memory model was also successful in detecting human behaviours using sensor data.

Recurrent Neural Network: The CNN is good at finding qualities while annotating. The CNN model may also examine the input comment thread and successively send it to the RNN model for deeper understanding. Despite the RNN element handling successive estimations to extract features from input data, the CNN-RNN model is called a hybridization. Before specifying specific factors, the CNN-RNN model analyses the fundamental components from each chunk to understand the comment thread formed from the main sequence in blocks.

A dynamic model performed human activity recognition and yielded impressive results. Infancy Simulation design requires

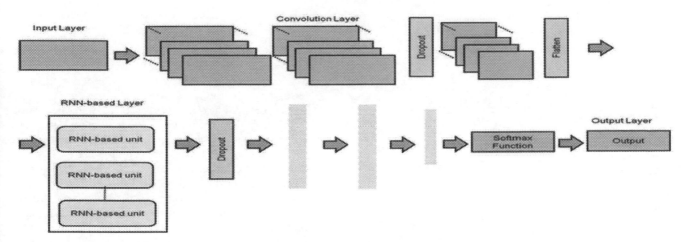

Fig. 42.1 RNN architecture

several learning algorithms and feature subset modifications. Since challenging behaviour is frequent, it's harder to spot. developing wearable and continually recognising human activity. The framework was built to improve the current human activity recognition model and study the effects of numerous elements on complicated that can be managed in light of the wide range of human movements. This is the first study to show that sliding mode control widths, convolution layer combinations, and the bidirectional approach affect DL approaches. We recommend the four baseline DL models to compare composite DL model classification outcomes.

Firefly Algorithm: Firefly is the latest optimization metaheuristic algorithm. Firefly flashing will guide the programme. Algorithms treat spontaneous solutions as dragonflies and assign intensity accordingly on objective function performance. The Optimization Algorithm mimics fireflies using their light to migrate towards each other in the dark. Fireflies are attracted to each other regardless of gender. The brightness also relies on the distance between the target firefly and other fireflies, which affects attraction. Separation decreases attractiveness. Finally, the optimization method's value defines a firefly's luminance. The model equation determines the FA algorithm. Equation (1) provides the firefly's light output.

$$I(r) = I_0 exp\left(-\gamma . r_{ij}\right) \quad (1)$$

Where r = 0 and I_0 are the beginning values, and y is the absorption coefficient. Equation (2), assuming β_0 is the starting value at (r = 0), expresses attraction as follows.

$$\beta = \beta_0 \exp\left(-\gamma . r_{ij}^m\right), m \geq 1 \quad (2)$$

Equation (3) calculates the distance between two fireflies at coordinates r_i and r_j and is known as a Geometrical radius. Where D stands for the problem's complexity and r_{ik} is the kth element of the spatial coordinate of r_j the ith fireflies.

$$r_{ij} = \left|r_i - r_j\right| = \sqrt{\sum_{l=1}^{D}\left(x_{ik} - x_{jk}\right)^2} \quad (3)$$

Equation (4) determines the movement algorithm from the ith firefly to the jth one.

$$x_i(t+1) = x_i(t) + \beta(x_j(t) - x_i(t) + \alpha(rand - 0.5) \quad (4)$$

$x_i(t+1)$ represents the migration of firefly i at iteration $(t+1)$. It is evident that the location of the firefly i at iteration t is represented by the first half of the right side of Equation (4), while the second component represents a measure of attraction. The last step is diversification, where α is the arbitrary shuffle parameter $\alpha \notin [0, 1]$. Algorithm 1 depicts Firefly algorithm.

Algorithm 1: Firefly Algorithm

intro FA parameters (no of people, α, β_0, γ, no of iterations).
The Light intensity denotes the cost function
$f(x_i)$ where $x_i(i = 1,..., n)$.)
While (iter < Max Generation).
  for i = 1:n (n fireflies)
   for j = 1: n (n fireflies)
   if $f(x_j) < f(x_i)$ move firefly i against j,
   end if.
    Upgrade value β with distance r.
   Analysis and modifying the possible idea $f(x_i)$ in quite similarly (4).
  end for j
 end for i
Get the best worldwide perfect by filtering replies.
 end while.
 Show results.

---

4. Result and Discussion

Unsatisfactory accuracy results in a discrepancy between some of the outcome and the true number. The proportion of actual results demonstrates as evenly distributed the data is overall in equation (5).

$$\text{Accuracy} = \frac{TP + TN}{TP + TN + FP + FN} \qquad (5)$$

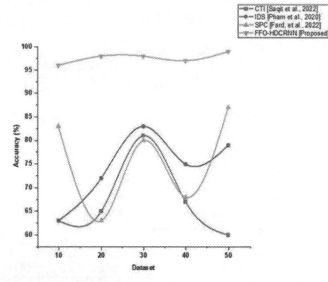

Fig. 42.2 Comparison of accuracy

Fig. 42.2 shows the comparable values for the accuracy measures. When compared to existing methods like CTI, which has an accuracy rate of 60%, IDS, which has an accuracy rate of 79%, and SPC, which has an accuracy rate of 87%, the recommended method's FFO-HDCRNN value is 99%. The suggested FFO-HDCRNN performs well in evaluation of cybersecurity threats and has higher accuracy than other methods.

5. Conclusion

The possibility of cyber incidents and threats has risen with the expansion of smart city systems, though. A smart city platform's IoT devices are susceptible to unlawful assaults and threats since they are linked to detectors that are interconnected to big cloud systems. The study uses the FFO-HDCRNN approach to mitigate and defend against IoT cyber risks in a smart city. Research has shown that the proposed system can identify cyberattacks more effectively than existing methods. The FFO-HDCRNN model will be improved in the future and it will be tested with a variety of hyper parameters, including learning rate, number of iterations, optimizer, and many more. In order to combat alternative DL models, we also seek to introduce our model to more complex tasks.

References

1. Mohamed, N., Al-Jaroodi, J. and Jawhar, I., 2020, July. Opportunities and challenges of data-driven cybersecurity for smart cities. In 2020 IEEE systems security symposium (SSS) (pp. 1-7). IEEE.
2. Hyman, B.T., Alisha, Z. and Gordon, S., 2019. Secure controls for smart cities; applications in intelligent transportation systems and smart buildings. International Journal of Science and Engineering Applications, 8(6), pp.167-171.
3. Sengan, S., Subramaniyaswamy, V., Nair, S.K., Indragandhi, V., Manikandan, J. and Ravi, L., 2020. Enhancing cyber–physical systems with hybrid smart city cyber security architecture for the secure public data-smart network. Future generation computer systems, 112, pp.724-737.
4. Raveendran, R. and Tabet Aoul, K.A., 2021. A meta-integrative qualitative study on the hidden threats of smart buildings/cities and their associated impacts on humans and the environment. Buildings, 11(6), p.251.
5. Sarker, I.H., 2022. Smart City Data Science: Towards data-driven smart cities with open research issues. Internet of Things, 19, p.100528.
6. Kalinin, M., Krundyshev, V. and Zegzhda, P., 2021. Cybersecurity risk assessment in smart city infrastructures. Machines, 9(4), p.78.
7. Chen, D., Wawrzynski, P. and Lv, Z., 2021. Cyber security in smart cities: a review of deep learning-based applications and case studies. Sustainable Cities and Society, 66, p.102655.
8. Lee, J., Kim, J. and Seo, J., 2019, January. Cyber attack scenarios on smart city and their ripple effects. In 2019 international conference on platform technology and service (PlatCon) (pp. 1-5). IEEE.
9. Ullah, Z., Al-Turjman, F., Mostarda, L. and Gagliardi, R., 2020. Applications of artificial intelligence and machine learning in smart cities. Computer Communications, 154, pp.313-323.
10. Kaushal, R.K., Bhardwaj, R., Kumar, N., Aljohani, A.A., Gupta, S.K., Singh, P. and Purohit, N., 2022. Using Mobile Computing to Provide a Smart and Secure Internet of Things (IoT) Framework for Medical Applications. Wireless Communications and Mobile Computing, 2022.
11. Saxena, S., Yagyasen, D., Saranya, C.N., Boddu, R.S.K., Sharma, A.K. and Gupta, S.K., 2021, October. Hybrid Cloud Computing for Data Security System. In 2021 International Conference on Advancements in Electrical, Electronics, Communication, Computing and Automation (ICAECA) (pp. 1-8).

Note: All the figures in this chapter were made by the authors.

Advancing Sustainable Science and Technology for a Resilient Future – Sai Kiran Oruganti et al. (eds)
© 2024 Taylor & Francis Group, London, ISBN 978-1-032-79020-6

43

Analyzing the Data Security Concerns in Cybersecurity Risks for Social Media Sites

Dhiraj Singh*

Chitkara University Institute of Engineering
and Technology, Chitkara University, Punjab, India

Abstract: Although Data security needs are widely utilized throughout a range of academic disciplines and incorporated the fundamental concepts of data security, integrity, and accessibility. When we talk about security, we're talking about restrictions regarding the utilization and storage of certain types of data. Integrity refers to the assurance that data isn't been altered, while availability refers to the assurance that approved individuals may utilize data and related resources when needed. Social media data may be wrongly shared if its confidentiality is compromised, modified inappropriately if its integrity is at risk, and discarded if its accessibility is in danger. As a result, the investigation for this study concentrated on examining the relationship between the requirements for data security and the accessibility, privacy, and accuracy of social media. To determine their association, the outcomes were evaluated and statistically examined.

Keywords: Social media, Data security, Accessibility, Academic disciplines

1. Introduction

In the present day, social networking accounts for a significant portion of internet use, making it one of the main avenues for communication. With the rise in social media and internet use over the last several years, the majority of nations today see cybersecurity as among the most important challenges. This may be because frequent use of social media has emerged as a new habit that quickly spreads to a broad spectrum of individuals (Herath, et al., 2022). Protecting against assaults has become a major and severe problem in the current world, and this includes shielding the computer system from possible threats. Due to the rising cyber threats brought on by technology advancements, new preventive measures must be developed. The rise in cyberattacks on businesses is mostly attributable to an increased reliance on digital technologies, which results in the storage of more personal as well as financial data (Alhayani, et al., 2021). One of the major hurdles to the implementation of digital transformation initiatives is cybercrime, which is accountable for exposing

the weaknesses of technologies (Nifakos, et al., 2021). When technology business began to take shape in the late 1970s, cybercrime arose The aggregate of data from many sources now poses a serious threat to the commonly employed data sensitization technique, and it is quite likely to fail (Sun, et al., 2021). Confidential and data security are the two main security procedures that any firm is continually worried about. We decide to square measure in the very electronic or malware environment in which all the data are stored. While social networking sites provide users a secure environment to communicate with family and close friends, hackers also use these tools to gather personal information (Rajasekharaiah, et al., 2020).

2. Related Works

Alazab, et al., 2021 intend to offer a comprehensive analysis of Federated Learning (FL) that may be used to improve security and prevent threats in real-time. Bhatnagar et al., 2020 investigate a novel idea of a management framework

---

*Corresponding author: dhirajsingh84824@gmail.com

DOI: 10.1201/9781003490210-43

for social media risk education based on several degrees of complexity, from the basic settings menu to sophisticated ideas of personal product marketing. Kalinin, et al., 2021 proposed a novel approach based on abstract class, data mining, and quantitative risk evaluation for addressing cybersecurity risks for the architecture of smart cities. Saxena, et al., 2021 explain Cloud hosting is deployed utilizing the cloud service, cloud provider, cloud hybrids, and collaborative cloud deployment paradigms. A portion of the user organization believes that the network owner may abuse the data security. Refaee, et al., 2022 present a safe and efficient architecture for transmitting health information via IoT using an efficient routing mechanism. Originally, medical data is gathered via different Internet of Things (IoT) devices, such as smart objects and sensors. The original information is preprocessed using methods for data cleansing and data extraction.

3. Proposed Methodology

The association between secrecy, authenticity, and accessibility of cybersecurity in Virtual Social Media was determined via an empirical investigation. Using a five-point likert scale, the intensity of social networking site cybersecurity concerns was measured. Questionnaire forms were used as the main data-collecting instrument again for research. The adoption of relevant statistical procedures, including such total mean assessment and the Multivariate regression Pearson Correlation test, facilitated the study of actual information.

4. Research Model

A research model emphasizing the key elements of secrecy, authenticity, accessibility that have resulted in managing cyber risks in social networking sites online was created and is shown in Fig. 43.1. Based on the study model shown in

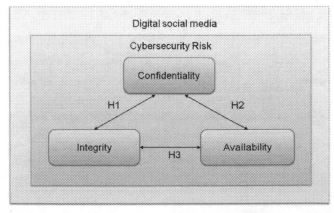

Fig. 43.1 Study model on the effects of cybersecurity risk on privacy, availability, and integrity in social networking sites online

Fig. 43.1, the following hypothesis has been created to evaluate the link between each basic feature of information security needs, namely confidentiality, authenticity, and accessibility. Cybersecurity risk has large and complex consequences on online social networking site privacy, accessibility, and integrity.

H1: Favorable correlation between secrecy and honesty for social networking site cybersecurity risk.

H2: Favorable correlation between secrecy and accessibility for social media sites that pose security risks.

H3: Integrity and availability are positively correlated with cyber risks in Virtual Social Media.

5. Result and Analysis

Demographic Profiles and Analysis

The analysis reveals that the majority of the participants were regarded as appropriate personnel with enough experience to provide reliable answers to all of the questions. Table 43.1 lists the demographic details of the initial research participants. The demographic factors of age, gender, education, employment, income, and location are particularly important in the context of cybersecurity.

Information Security Requirement Mean Score Analysis

As stated in Table 43.2, the average score on the accessibility of data in Digital Social Networks suggests that the majority of the cybersecurity community is concerned about the accessibility element of the severity degree of cyber risks. The highest mean for availability shows that the respondents are very concerned with availability factors. The second-highest average score on data confidentiality demonstrates the cyber society's knowledge of confidentiality challenges in social networking sites. Thus, the cyber community's understanding of confidentiality concerns might potentially contribute significantly to the severity degree of cybersecurity risk. The findings demonstrate that cybercommunities pay concern to the authenticity of information saved, analyzed, and transmitted via Digital Social Media sites. A quantifiable measurement of an organization's cybersecurity posture in regard to predetermined security requirements may be obtained via the use of the Information Security Requirement Mean Score Analysis. This determines an organization's compliance with a variety of security criteria, including access control, data protection, and incident response, and then generates a mean score for that compliance.

Relationship Study of Digital Social Media Confidentiality, Integrity, and Availability

The relevance, strength, and direction of the relationship between these three crucial criteria for data security were then evaluated using the Bivariate Pearson Correlation test on the

Table 43.1 Demographic assessment

Gender of respondent	Percentage (%)
Male	54.5
Female	42.4
Gender of respondent	Percentage (%)
>50 years	9.1
46-50 Years	6.1
41-45 Years	18.1
36-40 Years	17.2
31-35 Years	23.2
26-30 Years	21.2
Years of working experience	Percentage (%)
15-20 Years	32.3
>20 years	17.1
11-14 years	5.1
<=5 years	27.1
6-10 years	15.2
Years of ICT security experience	Percentage (%)
4-6 years	15.2
>6 years	21.2
<1 year	21.2
1-3 years	42.4
Organizational sectors	Percentage (%)
Private company	6.1
Government-link-company (GLC)	9.1
Government agencies	84.8
Industrial cluster	Percentage (%)
Healthcare Government	3.0
Healthcare private	6.0
Higher education	69.7
Creative technology	6.0
Information communication technology	6.0
Services	9.2

Table 43.2 Average cybersecurity risk in terms of availability, confidentiality, and integrity

Information security requirement	Social media cybersecurity risk severity		
	Mean	Std. Dev	N
Confidentiality	3.6093	0.75712	32
Integrity	3.6141	0.63700	32
Availability	3.6410	0.78237	32

6. Discussion

The average score suggests that while using digital social networks, cyber organizations are moderately concerned about the integrity, confidentiality, accessibility of information. In order to decrease the occurrence of cyberattacks and more efficiently manage the risks associated with them, an awareness of information security system should be created to educate online forums on how to utilize, communicate with, and interact through digital social networking sites. Considering the results the demographic study, a key target demographic and generational difference for the cybersecurity awareness campaign has been determined to improve the program's effectiveness. The outcomes of the hypothesis show that the three hypotheses have favorable correlations, as shown in Table 43.3. H1, H2, and H3 were approved as well as the blank was denied with a substantial p-value < 0.001. H1 had a correlation coefficient (r) of 0.627, H2 had one of 0.591, and H3 had one of 0.631.

Table 43.3 Results of the substantial correlation between cyber risks and hypotheses

Hyp.	Sig. (p-value)	Correlation coefficient (r)	Decision	Results
H1	0.000*	0.635	Significant	Moderate +ve
H2	0.000*	0.572	Significant	Moderate +ve
H3	0.000*	0.621	Significant	Moderate +ve

provided research hypotheses. To characterize the strength of their link, the correlation results (r) were produced

A discovery with a p-value just under 0.01 is regarded significant. A (r) value from less than 0.4 denotes a tenuous association, one between 0.4 to 0.7 indicates a promising connection, and one more over 0.7 denotes a strong connection. The examination of mean score findings reveals empirical proof of how cyber societies are worried about the essential component of data security required in Digital Social Media, which may generate cybersecurity threats.

Information integrity and accessibility at the strongest somewhat positive connection level is going to be the source of cyber risks in social media on the internet. As a result, there are some mild interactions between access to information and authenticity on social media sites on the internet. Information privacy and accessibility for cyber risks in social media platforms online, however, had the lowest moderate link. Others have a somewhat good association between secrecy and honesty for cyber risks in social media on the internet, but it is not the only one. The storage, transport, and sharing of the three essential components of data security in social networking sites still add to the cybersecurity hazards event that may have devastating effects on both people and the whole cyberspace infrastructure.

7. Conclusion

This research has experimentally shown the important connection between access to information, privacy, and business and governmental institutions using digital social networks. Nonetheless, the cybersecurity hazards occurrence in social networking sites was caused by the modest amount of connection effects between each data security need. Additional research looks at the degree of cyber community knowledge of danger. For digital societies, a thorough program of risk awareness for cybersecurity is required, according to empirical research. Understanding the connections between these fundamental information security requirements can help clinicians, cybersecurity experts, and technologists reduce the potential effect of risk. Also, the digital social networking site will fully benefit the internet community. Experts in cybersecurity will need to create new methods and tools in order to detect and prevent attacks that make use of artificial intelligence and machine learning, since the usage of these technologies is expected to continue to expand in the near future.

References

1. Herath, T.B., Khanna, P. and Ahmed, M., 2022. Cybersecurity practices for social media users: a systematic literature review. Journal of Cybersecurity and Privacy, 2(1), pp.1-18.
2. Alhayani, B., Mohammed, H.J., Chaloob, I.Z. and Ahmed, J.S., 2021. The effectiveness of artificial intelligence techniques against cyber security risks applies to the IT industry. Materials Today: Proceedings, 531.
3. Nifakos, S., Chandramouli, K., Nikolaou, C.K., Papachristou, P., Koch, S., Panaousis, E. and Bonacina, S., 2021. Influence of human factors on cyber security within healthcare organizations: A systematic review. Sensors, 21(15), p.5119
4. Sun, L., Zhang, H. and Fang, C., 2021. Data security governance in the era of big data: status, challenges, and prospects. Data Science and Management, 2, pp.41-44.
5. Rajasekharaiah, K.M., Dule, C.S. and Sudarshan, E., 2020, December. Cyber security challenges and its emerging trends on the latest technologies. In IOP Conference Series: Materials Science and Engineering (Vol. 981, No. 2, p. 022062). IOP Publishing.
6. Alazab, M., RM, S.P., Parimala, M., Maddikunta, P.K.R., Gadekallu, T.R. and Pham, Q.V., 2021. Federated learning for cybersecurity: concepts, challenges, and future directions. IEEE Transactions on Industrial Informatics, 18(5), pp.3501-3509.
7. Bhatnagar, N. and Pry, M., 2020. Student Attitudes, Awareness, and Perceptions of Personal Privacy and Cybersecurity in the Use of Social Media: An Initial Study. Information Systems Education Journal, 18(1), pp.48-58.
8. Kalinin, M., Krundyshev, V. and Zegzhda, P., 2021. Cybersecurity risk assessment in smart city infrastructures. Machines, 9(4), p.78.
9. Saxena, S., Yagyasen, D., Saranya, C.N., Boddu, R.S.K., Sharma, A.K. and Gupta, S.K., 2021, October. Hybrid Cloud Computing for Data Security System. In 2021 International Conference on Advancements in Electrical, Electronics, Communication, Computing and Automation (ICAECA) (pp. 1-8). IEEE.
10. Refaee, E., Parveen, S., Begum, K.M.J., Parveen, F., Raja, M.C., Gupta, S.K. and Krishnan, S., 2022. Secure and scalable healthcare data transmission in IoT based on optimized routing protocols for mobile computing applications. Wireless Communications and Mobile Computing, 2022.

Note: All the figure and tables in this chapter were made by the authors.

Advancing Sustainable Science and Technology for a Resilient Future – Sai Kiran Oruganti et al. (eds)
© 2024 Taylor & Francis Group, London, ISBN 978-1-032-79020-6

44

A Novel Machine Learning Approach for Detecting Cybersecurity Risks in Industries

Saurabh Lahoti*

Chitkara University Institute of Engineering and Technology,
Chitkara University, Punjab, India

Abstract: In recent years, the political and military aspects of cyber security dangers to critical infrastructure and industry have become more and more significant. In contrast to conventional residential and commercial IT systems, industrial IT, or OT solutions, influence the actual world. The operating lifetime of industrial equipment is also quite long, sometimes several decades. Modifications and updates are time-consuming and frequently unavailable. Both the dangers to industry and the enduring requirements of industrial settings drive the need for an efficient cyber security threats detection system that may be included in an existing model. The network data from industrial operations are evaluated in this research using a machine learning (ML)-a based method called the genetically optimized fuzzy-embedded Bayesian network (GO-FBN) to identify cyber security threats that have been made against the data. Based on the results of the investigation, we conclude that the proposed technology identifies cybersecurity risks more efficiently than conventional methods.

Keywords: Industrial and infrastructure, Cyber security risks, Machine learning (ML), Genetic optimized fuzzy embedded Bayesian network (GO-FBN)

1. Introduction

The process of defending against illegal cyber-attacks on Cybersecurity refers to the protection of electrical devices, networks, servers, computers, smartphones, and data. Industrial Control Systems (ICS) have historically been used in isolated places. The traditional ICS primarily focuses on system operations. Information and network security were not taken into consideration when it was created. Unfortunately, the installation, maintenance, and remote operation of this concept have all grown very costly (Ullah, et al., 2019). In addition to increasing the efficiency of the current system, these new advances have also raised concerns about untested industrial cyber security at all levels in the following of cyber attacks that interrupt routine operations. Operational cyber security necessitates a different approach than the conventional information technology (IT) strategy because of the interconnectivity and interaction between the cyber and physical components in chemical processes (Parker, et al., 2023). Recent cyber attacks have increased

the necessity for developing and using unique cyber security solutions in the OT sector, despite these variations that present difficulties. Once the number of cyber-attacks increased, the sector revealed the flaws in emerging cyber security technologies. Between 2000 and 2019, there were reportedly 77 cybersecurity-related events in critical infrastructure, including the process industry, with the majority of attacks occurring in the energy and oil production sectors (Sarker, et al., 2021).To mention a few, The development of threat-detection systems have used a variety of techniques, including Naive Bayes, Bayesian classifiers, support vector machines (SVM), neural network classifiers, and self-organization maps and references inside (Martínez Torres, et al., 2019). Services' important safety features are found, examined, and carried out by using a security evaluation. It is also stressed to prevent application security problems and vulnerability. By doing a risk assessment, an organization may get a comprehensive understanding of the application portfolio—from the perspective of an attacker. Key risk indicator metrics allow security and business management

*Corresponding author: saurabhlahoti254@gmail.com

DOI: 10.1201/9781003490210-44

to track the evolution of the risk profile and characterize the level of risk that an organization is exposed to. In cyber security operations, for instance, metrics that assess the risks and vulnerabilities offered by various instruments may be employed.

2. Related Works

Genetic algorithms are used to randomly create and equally distribute the beginning population across all viable alternativesa new genetic algorithm for CNN architecture optimization for a specific picture classification issue. (Johnson, et al., 2020).Use The Factor Analysis of Information Risk FAIR methodology to evaluate a G2C e-service that makes use of a third-party supplier for an Amazon Alexa skill that has been compromised by malware (Dreyling, et al., 2021).Using test data, the Median Absolute Deviation (MAD) algorithm model is trained, and the first predictions are generated (Prabhakar, et al., 2022). To present a summary of several methods for identifying malicious URLs, including blacklist-based, approaches based on rules, machine learning, and deep learning (Madhubala, et al., 2022). Many different security methods are available for the cloud to secure Cloud Client (CC) data. The Third-Party Auditor (TPA) Genetic algorithms are used to randomly create and equally distribute the beginning population across all options to confirm that Content Security Policy CSP and the CC are in line with these safety requirements (Hazela, et al., 2022).

3. Proposed Methodology

Data collection: Although process data were collected using LabVIEW aWindows Performance Monitor was used to collect host system data, network traffic data, and a sample rate of one observation every eight seconds.Data on behavior was collected to analyze data-driven models when the system was functioning normally, clear from every cyberattack, and data on abnormal behavior was gathered under theoretical cyberattack scenarios.To maintain system setpoints when heater power rises between 100% and 50% during typical transient operation, the process control system takes action. The engineering workstation and the cDAQ communicate consistently thanks to these operating transients. Although the precise amount of features gathered varies based on the counters used in the Windows Performance Monitor, 47,000 features are typically collected. The method for pre-processing data entails eliminating characteristics with a large proportion of not-a-number (NaN) values and practically constant values, since these characteristics often do not communicate useful information. (Zhang, et al., 2017). The main goal of this stage is to locate and identify possible security issues. To spot unusual activity or symptoms of penetration, it involves monitoring network traffic, examining system logs, deploying intrusion detection systems, and employing security monitoring tools. This data may be targeted by cybercriminals for identity theft or other illegal purposes. Strong data privacy safeguards must be in place, including encryption, access restrictions, and frequent data audits.

4. Genetic Optimized Fuzzy Embedded Bayesian Network

Genetic optimization algorithm: Darwinian evolutionary theories of natural selection and genetics were the inspiration for the creation of genetic algorithms (GA), which are organized but random search engines. To solve the challenge, stimulate everyone's fitness over all succeeding generations. In every generation, there are a certain number of people. Each person represents a point in the search space and a potential solution. Different genetic processes, such as crossover, mutation, etc., must be applied to each person. The primary function of a genetic algorithm is as follows: *1.* Initialization: The starting population is generated randomly and dispersed evenly over all practical alternatives using genetic algorithms. *2.* Selection: The fitness function is used by the selection operator to identify the strongest persons. *3.* Variation: After selecting a promising solution using the fitness function, changes are made to produce new solutions using the crossover (Recombination) and variate operations. *4.* By using the Crossover procedure, people may copulate. The mutation process produces random changes in solutions. *5.* Replacement: The original solution, or a portion of it, is replaced by the population of new candidate solutions after applying crossover and mutation to the group of possible solutions. If the conditions for termination are not filled, the next loop is carried out, starting with selection.

Fuzzy logic: FL is a widely used theory in computer science and refers to variables that have linguistic values, that is, variables values are words or sentences in a natural or artificial language, rather than integers. A linguistic variable is represented by the a quintuple (S, T(S), U, G, and O), where S denotes the variable's name and T(S) the term-set or set of linguistic values associated with S; G provides the grammatical rule that creates the words in T(S), U defines the discourse universe, and Ode defines the semantic rule that assigns meaning to each linguistic value A, a fuzzy part of the discourse world An attribute of U is its membership function μ_A, Where each element of U in the interval corresponds to [0, 1], with $\mu_A(u)$represent the grade of membership. Classical set $A = \{s | s \in U\}$only true or false conclusions are permitted. While fuzzy set $A_{fuzzy} = \{(s, \mu_A) | s \in U, \mu_A \in [0, 1]\}$ is described by the membership function, which indicates how closely an element in U resembles a fuzzy subset.As a result, fuzzy sets provide a way to roughly characterize events such that they may be described using traditional quantitative language. Figure 44.1 illustratesa fuzzy logic system.

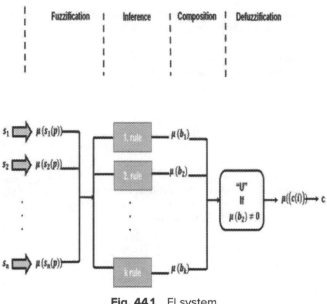

Fig. 44.1 FI system

Bayesian network: BBayesian networks are a popular method for risk analysis and safety management (BNs). One feature of BNs is their ability to manage unclear and unknown variables while still having a pictorial dimension and the capacity to represent highly complex situations. BNs are a type of mathematical probabilistic graphical model,as a pair defined Δ = {F(M, N), P} where F(M, N) P is the model's probabilistic component, while G is its graphical component. F(M, N) is like a directed acyclic graph, When the variables are reflected by the nodes M = {M_1, ..., N_1} and the arcs (N)Show the interconnections between them that are conditional. The variable P is a set of conditional probability tables P(M_i | Pa(M_i)) for each variable M_i, i = 1, ..., l in the network. Pa(M_i) is used to denote the collection of parents of M_i in F M_i = {Y $\in$ M|(Y, M_i)}. ABNs factors the joint probability distribution across all of the variables in M.

$$P(M) = \prod_{J=1}^{n} P(M_i \mid Pa(M_i)) \qquad (1)$$

The marginal probability of M_i is:

$$P(M_i) = \sum_{exceptM_i} P(U) \qquad (2)$$

5. Result and Discussion

In this section,we discuss in detail a novel machine-learning approach for the detection of cyber security risks in industries.

Accuracy precision and sensitivity are the parameters used to identify the effectiveness of the proposed methodology. It is essential to preserve vital infrastructure and guarantee the dependable functioning of energy systems by defending the energy industry from cyberattacks. The existing method ofData Encryption Standard algorithm (DES), Triple data encryption standard (3DES), and The Advanced Encryption Standard algorithm (AES)is compared to the highest efficiency of our proposed method.

Accuracy: The most critical performance indicator for assessing the capability of a classification model (classifier) is accuracy. It refers to how well the algorithm can predict unknown data as well as how well it can learn the data patterns in the dataset. Figure 44.2 shows the highest efficiency of our proposed method when compared to the existing method.

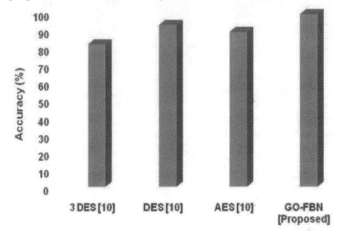

Fig. 44.2 Accuracy compared to the proposed method

Precision: It is important to consider precision as a performance measure. The ratio of accurately observed positive outcomes to all observed positive findings is what determines this. Figure 44.3 shows the highest efficiency of our proposed method when compared to the existing method.

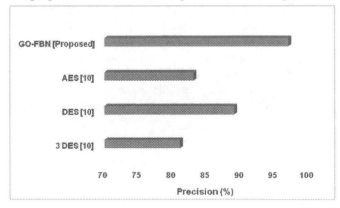

Fig. 44.3 precision compared to the proposed method

Sensitivity: Given a certain set of assumptions, how a particular dependent variable is impacted by different independent variable values Fig. 44.4 shows the highest efficiency of our proposed method when compared to the existing method.

As compared to the other methodologies, our proposed methodology has the highest efficiency in cyber securities in industries.

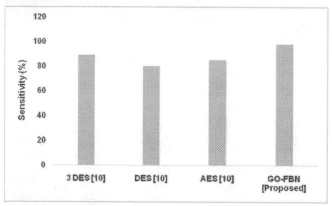

Fig. 44.4 sensitivity compared to the proposed method

6. Conclusion

Enhancing security measures to recognize and respond to cyber-attacks has made cyber security a concern on a global scale. The previously used traditional security solutions are no longer adequate since they are ineffective at identifying previously undiscovered and polymorphic attacks. Techniques for machine learning are essential and play a crucial part in many applications of cybersecurity systems. The unique machine learning approach for cyber security is briefly presented in this research. Even the most splitting ML model finds it challenging to deal with these intrusions because of the unique characteristics of each cyber threat. It is hard to provide a single suggestion based on a single model for all attacks. In a variety of applications, machine learning methods are essential in cyber security systems. The investigation's findings led us to the conclusion that the suggested technique detects cyber security issues more effectively compared to conventional approaches. As cyber security develops and cyber threats evolve, cybersecurity specialists will need to consistently adapt their knowledge, stay up with the most recent trends, and develop innovative techniques to protect data and systems from new threats.

References

1. Ullah, F., Naeem, H., Jabbar, S., Khalid, S., Latif, M.A., Al-Turjman, F. and Mostarda, L., 2019. Cyber security threats detection in the internet of things using deep learning approach. IEEE Access, 7, pp.124379-124389.
2. Parker, S., Wu, Z. and Christofides, P.D., 2023. Cybersecurity in process control, operations, and supply chain. Computers & Chemical Engineering, p.108169.
3. Sarker, I.H., Kayes, A.S.M., Badsha, S., Alqahtani, H., Watters, P. and Ng, A., 2020. Cybersecurity data science: an overview from a machine learning perspective. Journal of Big data, 7, pp.1-29.
4. Martínez Torres, J., Iglesias Comesaña, C. and García-Nieto, P.J., 2019. Machine learning techniques applied to cybersecurity. International Journal of Machine Learning and Cybernetics, 10, pp.2823-2836.
5. Dreyling, R., Jackson, E. and Pappel, I., 2021, July. Cyber security risk analysis for a virtual assistant G2C digital service using a FAIR model. In 2021 Eighth International Conference on eDemocracy& eGovernment (ICEDEG) (pp. 33-40). IEEE.
6. Johnson, F., Valderrama, A., Valle, C., Crawford, B., Soto, R. and Nanculef, R., 2020. Automating configuration of convolutional neural network hyperparameters using genetic algorithm. IEEE Access, 8, pp.156139-156152.
7. Prabhakar, P., Arora, S., Khosla, A., Beniwal, R.K., Arthur, M.N., Arias-Gonzáles, J.L. and Areche, F.O., 2022. Cyber Security of Smart Metering Infrastructure Using Median Absolute Deviation Methodology. Security and Communication Networks, 2022.
8. Madhubala, R., Rajesh, N., Shaheetha, L. and Arulkumar, N., 2022, April. Survey on Malicious URL Detection Techniques. In 2022 6th International Conference on Trends in Electronics and Informatics (ICOEI) (pp. 778-781). IEEE.
9. Hazela, B., Gupta, S.K., Soni, N. and Saranya, C.N., 2022. Securing the Confidentiality and Integrity of Cloud Computing Data. ECS Transactions, 107(1), p.2651.
10. Zhang, M., Xu, B., Bai, S., Lu, S. and Lin, Z., 2017. A deep learning method to detect web attacks using a specially designed CNN. In Neural Information Processing: 24th International Conference, ICONIP 2017, Guangzhou, China, November 14–18, 2017, Proceedings, Part V 24 (pp. 828-836). Springer International Publishing.

Note: All the figures in this chapter were made by the authors.

Advancing Sustainable Science and Technology for a Resilient Future – Sai Kiran Oruganti et al. (eds)
© 2024 Taylor & Francis Group, London, ISBN 978-1-032-79020-6

45

Spoofing Attack Detection in WSN Based on Deep Learning for Cyber Security Enhancement

Aman Mittal*

Chitkara University Institute of Engineering and Technology,
Chitkara University, Punjab, India

Abstract: Several low-cost radio technologies are now being used to enable the wireless communication in a number of key prospective applications, including as smart grids, smart cities, and the Internet of Things (IoT). Nevertheless, since it is so simple for malicious users to passively scan wireless signals on these networks and exploit that information to conduct identity-based attacks, cheap access to these new radio technologies creates a security issue. Deep learning is one of the greatest methods for creating cyber-attack detection systems (DL).To identifying spoofing attacks in WSNs; this study introduces a unique Fuzzy Swarm Optimized Multi-Layer RNN (FSO-MLRNN). We evaluated the effectiveness of the suggested methodology in comparison to the three classical methods. In the performance study, many metrics such as accuracy, false alarm rate, detection probability, misdetection probability, model size, processing time, and average prediction time per sample were evaluated. The simulation produced successful outcomes for our proposed methodology.

Keywords: Internet of things (IoT), Wireless sensor network (WSN), Spoofing attack, Cyber security, Cyber attack

1. Introduction

The Internet demands a higher degree of cyber-security in every working organization since cyber-attacks are always evolving. The technological prowess of the organization is one factor that affects an organization's cyber security. The human resources used in the organizations have an equal impact on it (Kumar, et al., 2021). Regarding technological capabilities that may be held by an organization or linked to the networks of that other organization, cyber security relates to maintaining Integrity, Confidentiality, and Availability (ICA). Several scholars advocated for teaching the next generation about the ideas of cyber-security because of the development and growth of cyber risks (Kaur, et al., 2022). The confidentiality, accessibility, and reliability of data must be guaranteed through a solid and secure computer system. The collection of security measures known as cyber security may be used to safeguard user assets and cyberspace against intrusion and assault. The basic goal of a cyber defense system is that data should be private, essential, and accessible (Shaukat, et al.,

2020). Wireless sensor networks (WSN) have a significant impact on the Internet of mission-critical objects today. Location spoofing attacks, which pretend to be a valid place to access networks, are one important concern. To counter location spoofing threats, several location spoofing detection techniques as well as strong localization algorithms have been created (Gao, et al., 2020). Due to significant cyber-security issues, that have made headlines in several media sources due to significant cyber-attack incidents, the issue has now gained widespread attention in the current digital age. Transportation systems from all sectors of the economy have been one of the main targets of cyber assaults in recent decades (Chowdhury, et al., 2021). We analyze the problems with spoofing attacks and provide a Fuzzy Swarm Optimized Multi-Layer RNN (FSO-MLRNN) method to fix them.

2. Related Works

Augusto-Gonzalez, et al., 2019 presented the GHOST cyber security strategy for IoT-based home automation. It seeks

*amanmittal89212@gmail.com

DOI: 10.1201/9781003490210-45

to solve the security issues brought on by several forms of attacks, including networking, hardware, and application ones. Nejabatkhah, et al., 2020 provided indications for FDI attack creation, detection, and mitigation in smart microgrids. Instances of current cyber-security initiatives throughout the globe are provided, along with crucial smart grid cyber-security standards. The Technology-Organization-Environment (TOE) paradigm is used in this research to evaluate a wide range of variables impacting an organization's preparedness for cyber security as well as how these elements affect both financial and non-financial organizational performance (Hasan, et al., 2021). Rajagopal, et al., 2022 developed a novel quantitative forecasting technique for HR demand prediction using recurrent neural networks (RNNs) and grey wolf optimization (GWO). Refaee, et al., 2022 offered a better routing protocol-based solution for safe and extensible IoT-based care data transmission.

3. Proposed Method

Dataset

Access to the public UCI Machine Learning Repository provided the dataset for this study. The phishing websites dataset was assembled using the Phishtank archive, Miller Smiles archive and Google's search operators. 11,055 different websites make up this dataset. The website belonging to the phishing category has a value of -1, while the non-phishing category has a value of 1. From this dataset, 4898 phishing websites and 6157 non-phishing websites with various website attributes were retrieved.

4. Fuzzy Swarm Optimized Multi-Layer RNN (FSO-MLRNN)

The performance of canonical Fuzzy Swarm Optimized (FSO) has been enhanced by a two-input, one-output fuzzy logic controller (FLC). The current best performance evaluation (CBPE) and the present inertia weight are the two input variables described here; the sole output variable is the change in inertia weight. CBPE is normalized in the manner described below to enable its use in a variety of optimization:

$$\text{NCBPE} = \frac{\text{CBPE} - CBPE_{min}}{CBPE_{max} - CBPE_{min}} \qquad (1)$$

$CBPE_{max}$ Stands for the non-optimal CBPE, where $CBPE_{min}$ is the actual minimum. Be aware that the non-optimal here signifies that the minimization issue cannot be solved by any solution that has higher than or equal to $CBPE_{max}$ (assume minimization problems). The various fuzzy variables in the fuzzy logic system are stated to have numerous fuzzy values and rules. All of the data input that relates to both the past

and the future during a particular time frame may be used to train an MLRNN.

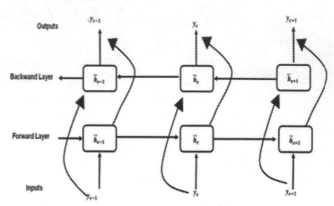

Fig. 45.1 Multi-layer RNN

The idea is to separate the state neurons of a normal MLRNN into 2 segments: one that controls the effective time direction and another that controls the negative time direction. The sources of backward states cannot be related to the outcomes of the forward states, and conversely. It should be noted that this structure becomes a typical continuous forward MLRNN without the backward states, as seen in Fig.1. The output (x), forward hidden layer $\vec{h}_e$, and backward hidden layer $\overleftarrow{h}_e$ are computed from Fig. 45.1.

$$\vec{h}_e = H\left(V_{y\vec{h}}y_e + V_{\vec{h}\vec{h}}\vec{h}_{e-1} + a_{\vec{h}}\right) \qquad (2)$$

$$\overleftarrow{h}_e = H\left(V_{y\overleftarrow{h}}y_e + V_{\overleftarrow{h}\overleftarrow{h}}\overleftarrow{h}_{e+1} + a_{\overleftarrow{h}}\right) \qquad (3)$$

$$x_e = V_{\vec{h}x}\vec{h}_u + V_{\overleftarrow{h}x}\overleftarrow{h}_u + a_x \qquad (4)$$

In the absence of the forward states, a standard MLRNN with a reversed time axis emerges. The normal MLRNN stated earlier requires delays to integrate future data, but in a network that supports all times direction, required information from the present and the future of the period being evaluated may be utilized immediately to reduce the objective function. Since there are no connections between the two kinds of state neurons, the RNN may be expanded into a generalized feed-forward network and trained using the same procedures as a typical unidirectional MLRNN. If, for example, any kind of back-propagation through time (BPTT) is used, the simultaneous forward and backward pass approach become substantially more challenging because value and throughput layers can no longer be modified one at a time. While using BPTT, the forward and backward passes over unfolded RNN over time are carried out very identically to when using a standard MLP. Specific treatment is only necessary at the start and the end of the training data. With forward state values and the backward state inputs at $(q = 1)$ arc unknown. Set as a part of the learning process, but in this instance, individuals are fixed arbitrarily at a value (0.6). The local value variations at ($q = 1$) of the forward value and ($q = 1$) for the backward states are also undetermined and set to zero in this example, assuming that the information beyond a certain point is not essential for the current update, that is the case for the boundaries. Spoofing assaults, that involve an attacker pretending to be a valid network node, may be extremely challenging to detect and prevent using conventional cyber security approaches. A strong and effective spoofing attack detection system may be created to improve cyber security by using the capabilities of FSO-MLRNN. The system may also be vulnerable to adversarial assaults, in which a perpetrator deliberately alters the sensor data to avoid detection. For the purpose of detecting real-time spoofing attacks in massive WSNs, it is effective and scalable.

5. Result and Discussion

Analysis of the FSO-MLRNN methodology in comparison to other methods, including the k-nearest neighbor (KNN), Logistic regression (LR), and Linear discriminant analysis (LDA) are discussed in this section. The findings showed that the FSO-MLRNN technique correctly identified every attack. The FSO-MLRNN system was seen to have achieved an average accuracy of 99%, a precision of 97%, and a recall of 98% (Table 45.1).

Table 45.1 Comparison study of FSO-MLRNN with the existing method

Method	Accuracy (%)	Precision (%)	Recall (%)
LDA	72	79	89
KNN	81	70	75
LR	89	83	81
FSO-MLRNN [Proposed]	99	97	98

Accuracy

The FSO-MLRNN model's detection approach is being compared with other models using an accuracy indicator. The percentage of properly categorized samples in the whole sample is represented by the accuracy. The detection and prevention effectiveness for spoofing attacks. The results signified that the LR, KNN, and LDA models reached a minimum value of 72%, 81%, and 89%, respectively (Fig. 45.2).

$$Accuracy = \frac{TP + TN}{TP + TNFP + FN} \quad (5)$$

Precision

The precision metric to compare the detection approach of the FSO-MLRNN model with existing models. The fraction

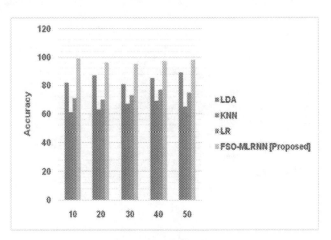

Fig. 45.2 Accuracy

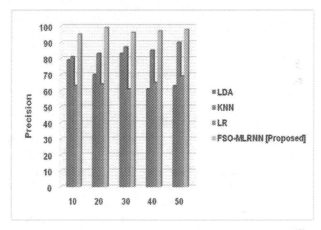

Fig. 45.3 Precision

of appropriately identified attack samples among all assault samples is known as precision. Effectiveness of the approach used to identify spoofing attacks.

$$Precision = \frac{TP}{TP + FP} \quad (6)$$

The findings showed that the minimum values for the LR, KNN and LDA models were 79%, 70%, and 83%, respectively (Fig. 45.3).

Recall

The recall indicator compares the detection method of the FSO-MLRNN model with other existing models. Recall measures the percentage of all attack samples that were successfully categorized. The intrusion detection performance for spoofing attacks. The results demonstrated that the LR, KNN, and LDA models' respective minimum values were 89%, 75%, and 81% (Fig. 45.4).

$$Recall = \frac{TP}{TP + FN} \quad (7)$$

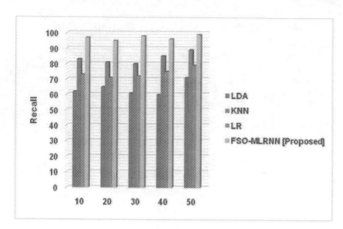

Fig. 45.4 Recall

6. Discussion

The LDA approach fails if the distribution's mean values are shared by both classes. In this situation, a distinct regression problem axis cannot be found using linear discriminant analysis. KNN will not succeed if the data is disorganized into all various classes since it will attempt to locate k's closest neighbor but all the points are random. Logistic regression is the machine learning technique most often employed to solve classification problems. If there are only two classes to predict, it is useful for binary classification issues. The proposed approach is shown to be more efficient when compared to the available technology. The cyber security of WSNs may be improved with the use of deep learning-based spoofing attack detection systems. WSNs can be secured from spoofing attacks and maintain the integrity and reliability of the data passed on in the network by utilizing RNNs to learn the usual activity patterns of the legitimate nodes and detect any anomalies that could indicate a spoofing attack.

7. Conclusion

Smart grids, smart cities, and the web of things are just a few of the key emerging applications that are already using a variety of low-cost radio technologies to offer wireless connectivity. Yet, it's so easy for malicious individuals to passively monitor wireless signals on these networks and utilize that data to carry out identity-based attacks. This study offers a way of detecting spoofing attacks in WSNs and developing cyber attack detection (FSO-MLRNN). A two-input, one-output model has improved the performance. The two input variables discussed here are the current inertia weights, and the only output variable is the change in inertia weight. The other is in charge of the negative time direction, whereas the former is in control of the positive time direction. The inputs of backward states and the outputs of forward states cannot be correlated with one another. For attacks like request spoofing, a middleman attack, and denial of service, an extension is necessary. The cooperative PHY-layer authentication mechanism in mobile WSN is the way of the future.

References

1. Kumar, S., Biswas, B., Bhatia, M.S. and Dora, M., 2021. Antecedents for an enhanced level of cyber-security in organizations. Journal of Enterprise Information Management, 34(6), pp.1597-1629.
2. Kaur, J. and Ramkumar, K.R., 2022. The recent trends in cyber security: A review. Journal of King Saud University-Computer and Information Sciences, 34(8), pp.5766-5781.
3. Shaukat, K., Luo, S., Varadharajan, V., Hameed, I.A. and Xu, M., 2020. A survey on machine learning techniques for cyber security in the last decade. IEEE Access, 8, pp.222310-222354.
4. Gao, N., Ni, Q., Feng, D., Jing, X. and Cao, Y., 2020. Physical layer authentication under intelligent spoofing in wireless sensor networks. Signal Processing, 166, p.107272.
5. Chowdhury, N. and Gkioulos, V., 2021. Cyber security training for critical infrastructure protection: A literature review. Computer Science Review, 40, p.100361.
6. Augusto-Gonzalez, J., Collen, A., Evangelatos, S., Anagnostopoulos, M., Spathoulas, G., Giannoutakis, K.M., Votis, K., Tzovaras, D., Genge, B., Gelenbe, E. and Nijdam, N.A., 2019, September.
7. Nejabatkhah, F., Li, Y.W., Liang, H. and Reza Ahrabi, R., 2020. Cyber-security of smart microgrids: A survey. Energies, 14(1), p.27.
8. Hasan, S., Ali, M., Kurnia, S. and Thurasamy, R., 2021. Evaluating the cyber security readiness of organizations and its influence on performance. Journal of Information Security and Applications, 58, p.102726.
9. Rajagopal, N.K., Saini, M., Huerta-Soto, R., Vílchez-Vásquez, R., Kumar, J.N.V.R., Gupta, S.K. and Perumal, S., 2022. Human resource demand prediction and configuration model based on grey wolf optimization and recurrent neural network. Computational Intelligence and Neuroscience, 2022.
10. Refaee, E., Parveen, S., Begum, K.M.J., Parveen, F., Raja, M.C., Gupta, S.K. and Krishnan, S., 2022. Secure and scalable healthcare data transmission in IoT based on optimized routing protocols for mobile computing applications. Wireless Communications and Mobile Computing, 2022

Note: All the figures and table in this chapter were made by the authors.

Advancing Sustainable Science and Technology for a Resilient Future – Sai Kiran Oruganti et al. (eds)
© 2024 Taylor & Francis Group, London, ISBN 978-1-032-79020-6

46

Malicious Prediction in Wireless Body Area Networks for Enhancing Cyber Security

Anubhav Bhalla*

Chitkara University Institute of Engineering and Technology,
Chitkara University, Punjab, India

Abstract: The Wireless Body Area Network (WBAN), which tracks and gathers biological information from the human body, is a crucial developing technology in the healthcare sector because of the advancement of tiny sensors, embedded systems, and wireless networking. As the compact physical sensor nodes are incompatible with the conventional security techniques, protecting this network against hacked sensor devices is a difficult problem. To categorize a sensor node as trusted or malicious node, we have suggested a trust management framework in this research that is based on the Enhanced Bilateral Naive Bayes (EBNB) classification. During the information transfer, the trustor node will select a trustee node based on the suggested classification algorithm. The developed framework is trained using MATLAB simulation tool and the results are evaluated. The suggested model can correctly identify a sensor node as malicious or trustworthy, according to empirical observations.

Keywords: Wireless body area network (WBAN), Security, Trust, Enhanced bilateral naïve bayes (EBNB)

1. Introduction

A crucial piece of technologies which evaluates several dangerous diseases and perform continual health screenings on patients. A WBAN offers a variety of therapeutic and non-restorative uses while operating near, on, or within a patient's psyche. With their available budgets, wireless sensor networks (WSNs) have the vital but difficult job of safeguarding themselves against cybersecurity risks. Because according to its open architecture and dispersed design, WSNs are susceptible to assaults. A base station may be taken over by a hacker, who can also abuse nodes, spy on communications, insert untrue signals, and edit. Notably hazardous categories it might endanger the safety of WSNs is the denial-of-service (DoS) assault (Alsubaie, et al., 2020).

The deployment of a body system for a variety of purposes, including brain waves, heart rate monitoring, ECG, physiological thermometer, place monitoring, etc. It consists of a unique configuration of sensors with the ability to collect information about biology and physiology and transmit it to the coordinating networks. When long once coordination software obtains such knowledge, software uploads it till evaluation and application in urgent situations on a personal server or in a clinical system. With the aid of WBAN technology, it is feasible to smoothly enable movement while also securely transferring customer records to acute ambulance stations (Remu, et al., 2020 and Khalaf, et al., 2021).

These were two types of cyberattacks carried out by malicious users: inside attacks and outside attacks. Insider attacks are cyberattacks launched by those with authorized access or insider information inside the targeted system or organization. These attackers take use of their lawful access to do harmful acts such as stealing valuable information, destroying systems, or undermining network security from inside. Outside attacks, on the other hand, are cyberattacks carried out by people or organizations without authorized access or insider knowledge. These attackers often attack the system or organization from a distant location, such as the internet or another network. Their purpose is to penetrate

*Corresponding author: bhalla.orp@chitkara.edu.in

DOI: 10.1201/9781003490210-46

network defenses, acquire unauthorized access, and engage in a variety of destructive behaviors. Internal attacks occur when a node inside the network region misbehaves and performs erratically in an attempt to take over other legitimate nodes and use them as pawns for their evil intentions. For instance, a rogue node may utilize information from its packets, such as the transmission key or other data, to attack the whole network. In a malicious attacks from the outside, the attacker is not within the network's perimeter. The outside attacker lacks the identifying tools that sensor nodes have (such as node ID, transmission key, etc.), making this assault simpler for the network to identify than an inside attack (Ismail, et al., 2021 and Rajesh, et al., 2021). The networks assaults using various ML techniques without fully knowing their complete properties. Conventional machine learning algorithms, moreover, cannot offer distinguishing feature descriptions to define the problem of attack detection due to their limitations in model complexity. The fact that sensors are often made to be small and inexpensive makes the installation of IDSs for WSNs more challenging than the development of other systems. The IDS must also be inexpensive to have little impact on the WSN's architecture and have high accuracy in identifying an attacker, even unknown assaults. (AlShahrani, et al., 2021). While the study focuses on detecting rogue nodes, it does not address more complex assaults that may seek to circumvent the suggested trust management architecture. Future research should look at possible responses to advanced attack techniques such as node compromise, masquerade assaults, or collusion among numerous hostile nodes.

2. Related Works

Liu, et al., 2021 mentions that the suggested a few WBAN signal processing topics, as well as networking dependability, WBAN interaction between several systems, bandwidth allocation, and safety of extremely effective professional health care applications. This basis is built on the latest expansion of electronics industry and technological advancements, which facilitate device accessibility, miniaturization, and interaction takes into account a systematic examination carried out with the use of surveys, normative papers, and participant studies. Based on the complete examination, they found WBANs has various operational, standards, and concerns, hurting efficiency and preservation of user safety and privacy. Due to improvements in web browser, they see a growing reliance of future healthcare on WBAN for both clinical and semi uses (Liu, et al., 2021).

Gunduz, et al., 2020 mentions that the suggested the risks and possible remedies associated with an IoT-based smart grid are examined. They provide a thorough analysis of the intelligent based load cyber-security condition while concentrating on different sorts of cyberattacks. They specifically focus on the investigation and debate of attacks and ethical hacking defenses, and protection. In addition to providing a roadmap for future work on internet in smart grid applications, our goal is to provide a thorough knowledge of cyber-security vulnerabilities and remedies.

Azees, et al., 2021 mentions that The recommended encrypted solution helps to ensure the validity of the individuals involved, even if it compromises either the victim's or the doctor's privacy. Whereas there are many conventional encryption systems, like AES and DES, to ensure secrecy, the cryptosystem and lot of great issues are what prevent us from offering a sufficient degree of protection. The study proposes an effective linear cryptosystem encrypting approach to provide a high degree of secrecy with a reduced block cipher than modern cryptographic strategies. It demonstrates the planned work's resistance to different damaging security threats to guarantee that it offers improved security. The data capture component of the proposed method includes a detailed demonstration of the costs, transmission time, and computing cost.

3. Proposed Methodology

Dataset

The Multimodal Intelligent Monitoring in Intensive Care (MIMIC-I and II) collection, that includes precise biological documents taken from over 90 ICU patients known as participants, was utilized in this study. Significantly, the majority of researchers tested the feasibility of the presented methods using the MIMIC sample as a standard. 7 elements make up the sample. Heart rate (HR), systolic arterial blood pressure (ABPsys), diastolic arterial blood pressure (ABPdias), mean arterial blood pressure (ABP-mean), pulse, temperature, respiration rate (RESP), and oxygen saturation (SPO2) with timesteps and date are the features which depict the medical state of the patient (Albattah, et al., 2022).

Enhanced Bilateral Naive Bayes

Enhanced Bilateral Naive Bayes method was inductive theory includes the EBNB theory as a key component. EBNB decision-making involves estimating the personal chance of certain unknown states with limited knowledge, moderating the chance of them happening using the EBNB formula, and then selecting the best course of action based on the expected value and modified probability. Ω is consists of a whole set $C_1, C_2\cdots, C_n \in \Omega$, C_i some main categories cannot coexist, and the ith category is denoted by $P(C_i) > 0$, $i = 1, 2, \ldots, n$ any two categories are mutually exclusive, because and $U_{i=1}^n C_i = \Omega$. For set X, $if P(X) > 0$, so

$$P(C_i) = \frac{P(X|C_i)P(C_i)}{\sum_{i=1}^{n} P(X|C_i)P(C_i)} \quad (1)$$

To categorize the data into the greatest probable group using EBNB the maximum likelihood estimation approach is used that is

$$P(C_i \mid X) = Max(P(C_1 \mid X), P(C_2 \mid X), \ldots P(C_n \mid X)) \quad (2)$$

Only A_j and $X = (A_1, A_2 \ldots, A_k)$. The characteristics are thought to be independent of one another according to Enhanced Bilateral Naive Bayes classification, therefore

$$P(X|C_i) = \prod_{j=1}^{k} P(A_j = x_i \mid C_i) \quad (3)$$

Substituting formula (3) into formula (1) that is:

$$P(C_i \mid X) = \alpha \frac{\prod_{j=1}^{k} P(A_j = x_i | C_i) P(C_i)}{P(X)} \quad (4)$$

Let equation (1) be $\frac{1}{P(X)}$ α (> 0) that is

$$P(C_i|X) = \alpha \prod_{j=1}^{k} P(A_j = x_i \mid C_i) P(C_i) \quad (5)$$

Where N(D) represents the total number of samples in sample set D, , N(C_i), the amount of samples of C i, C_i, $N(C = C_j$, $A_j = x_j)$ the number of samples where attribute A_j is x_j in C_j respectively

$$P(C_i) = \frac{N(C_i)}{N(D)} \quad (6)$$

$$P(A_j = x_j | C = C_j) = \frac{N(C = C_j, A_j = x_j)}{N(C_j)} \quad (7)$$

Substituting formula (6) and formula (7) into formula (5), then

$$(C_i \mid X) = \alpha \prod_{j=1}^{k} \frac{N(C = C_j, A_j = x_j)}{N(C_i)} \cdot \frac{N(C_i)}{N(D)} \quad (8)$$

4. Result and Discussion

In this section, we discuss in detail about the findings of the malicious prediction in body area networks for telecommunications enhancing cyber security. Accuracy and Malicious node detection are the parameters used to identify the effectiveness of the proposed methodology. This article covered the issues faced by the WBAN, as well as how many elements impact its success. The human civilization may gain by increasing the patients' success and effectiveness. The greatest difficulty for such devices is safety. In conclusion, safety needs to be handled at all levels. Given the restricted resources, the security system must be lightweight. Several attack types may be conducted against e-health networks.

Accuracy

Both system accuracy and log loss is useful for figuring out how many samples in the testing set have been correctly categorized as well as if the models has been developed correctly to categorize various attacks classes. The total number of errors for each training sample is represented by the loss. It is clear from the loss plot in the analysis of the log reduction categorical cross entropy for the DL models that the validation set contains very few false forecasting. Figure 46.1(a) shows the accuracy and Fig. 46.1(b) loss of the proposed system.

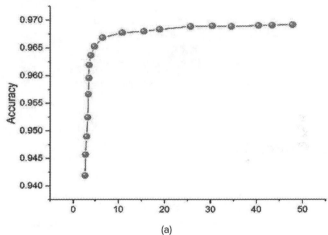

(a)

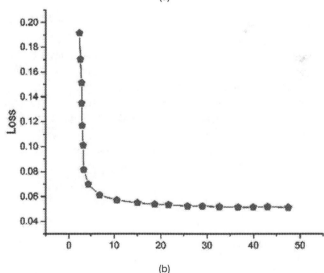

(b)

Fig. 46.1 (a) Accuracy and (b) Loss of proposed system

Malicious Node Detection

The malicious node delays the exchange of knowledge among the transmitter and recipient before replaying the out-of-date data. To use the old information, the unknown nodes compute erroneous locations, the malicious node clogs the knowledge flow between the transmitter and receiver before replaying the stale data. These unknown nodes determine erroneous placements by using out-of-date material. Attacks, as opposed to certain types, may take down an entire system with only single site.

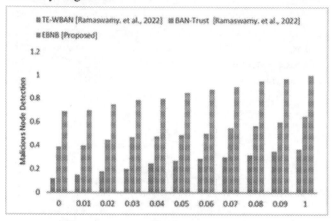

Fig. 46.2 Malicious node detection

5. Conclusion

To identify malicious nodes in the WBAN network, we have suggested EBNB In our model, we've outlined a few progressive phases such node categorization, harmful node detection, and malicious node elimination. The network's coordinator node always retains the authority to choose or reject any minor nodes. As we have said, 99.05% accuracy was achieved in both our training and forecasted models, indicating that WBAN privacy is completely seamless. Future efforts will focus on strengthening the coordinating datatype safety and enhancing decision-making capabilities like the Prime Lending Rate (PLR). Scaling the trust management methodology in bigger WBAN networks. This would require investigating the framework's performance and scalability as sensor nodes expand and exploring distributed trust management or hierarchical trust models.

References

1. Alsubaie, F., Al-Akhras, M. and Alzahrani, H.A., 2020, November. Using machine learning for intrusion detection system in wireless body area network. In 2020 First International Conference of Smart Systems and Emerging Technologies (SMARTTECH) (pp. 100-104). IEEE.
2. Remu, S.R.H., Faruque, M.O., Ferdous, R., Arifeen, M.M. and Reza, S.S., 2020, January. Naive Bayes based Trust management model for wireless body area networks. In Proceedings of the international conference on computing advancements (pp. 1-4).
3. Ismail, S., Khoei, T.T., Marsh, R. and Kaabouch, N., 2021, December. A comparative study of machine learning models for cyber-attacks detection in wireless sensor networks. In 2021 IEEE 12th Annual Ubiquitous Computing, Electronics & Mobile Communication Conference (UEMCON) (pp. 0313-0318). IEEE.
4. AlShahrani, B.M.M., 2021. Classification of cyber-attack using Adaboost regression classifier and securing the network. Turkish Journal of Computer and Mathematics Education (TURCOMAT), 12(10), pp.1215-1223.
5. Khalaf, O.I., Natarajan, R., Mahadev, N., Christodoss, P.R., Nainan, T., Romero, C.A.T. and Abdulsahib, G.M., 2022. Blinder Oaxaca and Wilk Neutrosophic Fuzzy Set-based IoT Sensor Communication for Remote Healthcare Analysis. IEEE Access.
6. Rajesh, N. and Christodoss, P.R., 2021. Analysis of origin, risk factors influencing COVID-19 cases in India and its prediction using ensemble learning. International Journal of System Assurance Engineering and Management, pp.1-8.
7. Liu, Q., Mkongwa, K.G. and Zhang, C., 2021. Performance issues in wireless body area networks for the healthcare application: A survey and future prospects. SN Applied Sciences, 3, pp.1-19.
8. Gunduz, M.Z. and Das, R., 2020. Cyber-security on smart grid: Threats and potential solutions. Computer networks, 169, p.107094.
9. Albattah, A. and Rassam, M.A., 2022. A correlation-based anomaly detection model for wireless body area networks using convolutional long short-term memory neural network. Sensors, 22(5), p.1951.
10. Asam, M., Jamal, T., Adeel, M., Hassan, A., Butt, S.A., Ajaz, A. and Gulzar, M., 2019. Challenges in wireless body area network. International Journal of Advanced Computer Science and Applications, 10(11).

Note: All the figures in this chapter were made by the authors.

Advancing Sustainable Science and Technology for a Resilient Future – Sai Kiran Oruganti et al. (eds)
© 2024 Taylor & Francis Group, London, ISBN 978-1-032-79020-6

47

Enhancement of Cybersecurity: Ensemble Machine Learning Method for Detecting Credit Card Fraud in Banking

Saket Mishra*

Chitkara University Institute of Engineering and Technology,
Chitkara University, Punjab, India

Abstract: A hard and crucial business issue in the financial sector is the early identification of unauthorized credit card transactions. We should primarily deal with the dataset's extremely skewed character, which is that there is relatively little fraud relative to typical transactions. In this work, we propose a novel ensemble machine learning (EML) strategy as a potential remedy for this issue. This paper proposes a binary tree-adaptive support vector machine (BT-SVM) for efficiently identifying credit card fraud in banks. After collecting data samples, noisy data is removed using z-scaling normalization. Then, independent component analysis (ICA) is utilized to extract the best characteristics. To detect credit card fraud, we finally use our suggested BT-SVM approach. Accuracy, recall, precision, and f-measure are all used to evaluate the performance of the suggested strategy. The testing results show that the suggested strategy is more effective than current methods at identifying credit card fraud in banks.

Keywords: Banking, Credit card transaction, Fraud detection, Ensemble machine learning (EML), Binary tree-adaptive support vector machine (BT-SVM)

1. Introduction

The number of credit card transactions has significantly expanded in recent years because of the increasing use of credit and debit cards and the quick growth of e-services like e-commerce, e-finance, and mobile banking. Credit card fraud will undoubtedly result in millions of dollars in losses owing to widespread credit card use and different transaction situations without sufficient verification and oversight (Zheng, et al., 2021). Personal information might be used by thieves to commit credit card fraud. People's private details, the number of their credit card, and one-time passcode (OTP) may all get into the hands of criminals when it came to credit cards via several different channels. Cybersecurity is a practice designed to stop unauthorized access, computer, servers, network, and digital information loss or attack in cyberspace. Organizations must prioritize protecting their financial information, intellectual property, and reputation as an essential part of their business plans. Enterprises and governments use the cybersecurity element to protect their data and ensure its integrity and accessibility. As cyber hazards are considered to be a serious issue in the banking industry, banks must be kept up to date on new technological developments in data security (Ghauri, et al., 2021).

By using the capacity of machine learning methods that may reveal hidden patterns, the financial sector has benefited from the period of technological growth by improving the identification of these financial crimes. The science of teaching machines to learn without explicit instructions is known as machine learning. It's been extensively employed in a variety of fields, including finance, biology, production, bioinformatics, and the medical field. To help companies and financial institutions more reliably identify fraudulent actions, machine learning is primarily employed in fraud detection (Malik, et al., 2020). Machine learning techniques may be divided into two key categories: supervised learning and unsupervised learning. Fraud surveillance is also possible and only limited by the dataset in terms of how to utilize it. Outliers are constantly detected during supervised training, the same as previously. During the last several years, many

*Corresponding author: saketmishra84824@gmail.com

DOI: 10.1201/9781003490210-47

supervised techniques have been used to detect credit card fraud (Trivedi, et al., 2020). Credit card fraud can have a significant impact on customers, leading to financial losses and compromised personal information. Collaboration among banks allows for a more coordinated response to protect customers.

2. Related Works

Chen, et al., 2021 suggest using a deep learning algorithm to propose the Deep Convolution Neural Network (DCNN)-based fraudulent financial detection method. Suggests improving detection accuracy when dealing with vast amounts of data. They offer a mixed strategy that combines supervised and unsupervised approaches to enhance the reliability of fraud detection. Evaluation and testing of unsupervised outliers rates produced at different levels of accuracy are conducted using a real, annotation credit card fraud detection collection. More, et al., 2021 suggest a Random Forest-based fraud detection system that may assist in resolving this practical issue. They suggested technique improves the accuracy of identifying fraud in transactions made with credit cards. Saxena, et al., 2021, Refaee et al., 2022. Large-scale data breaches at merchants, financial institutions, or other entities can expose sensitive credit card information. Fraudsters often exploit this stolen data for unauthorized transactions. Banks need to monitor and react swiftly to such breaches, including reissuing cards and implementing fraud detection measures.

3. Proposed Methodology

Data Set

The dataset contains the purchases that cardholders made using their cards. The period of consolidation is included in the dataset. The transformation turns the input variable-containing dataset into a numeric value(Sailusha, et al., 2020).

Z-Scaling Normalization

Using z-scaling normalization, the data are scaled to have a mean of 0 and a variance of 1. To flexibly adjust the data in the area approximately between 1 and -1 is the factor in the equation. The researcher's selected factor value throughout the research is 4 since this is the value that puts all stock data in the right range.

Extracting Transaction data Features Using ICA

Many Transfer data segments are subjected to the ICA technique to obtain separate vectors from credit card information. A transaction data segment x is employed to train an ICA network to extract companies contributing U, and the recursive feature coefficients U are extracted from x using the learned weight matrix W. A linear combination of the different components U is assumed to be the observation by ICA. The column of A reflects the basis vectors of observation x if A is the inversion matrix of W.

$$u = W - x, x = A.u$$

The training of matrix mix W must be either the de-noising matrix A or the mixing matrix A to obtain the basis functions.

$$\Delta U \alpha \frac{\partial J(z,v)}{\partial U} = \frac{\partial G(z)}{\partial U} \tag{1}$$

$$\Delta U \alpha \frac{\partial J(z,v)}{\partial U} = \frac{\partial G(z)}{\partial U} \tag{2}$$

Moreover, a natural gradient is included to increase convergence speed. In particular, this technique offers the following rule and does not need the inversion of matrix U:

$$\Delta U \alpha \frac{\partial G(z)}{\partial U} u^S u = \left[J \sim \varphi(w) w^S \right] u, \tag{3}$$

U is repeatedly updated in a gradient ascending manner until completion that uses the learning procedure in Eq. (3). The N by N matrix A ($A = U^{-1}$) of the ICA network produces N basis vectors from its N inputs and N outputs.

Support Vector Machines Utilizing a Binary Tree

In this article, they suggest a binary decision tree design that makes use of SVMs to effectively detect credit card fraud in financial institutions binary tree-adaptive support vector machine (BT-SVM) a proposed classifier design, benefits from both the fast computing provided by the binary tree design and the accuracy rate of SVMs. Figure 47.1 illustrates an instance of BT-SVM that uses a binary tree to solve a seven-class pattern recognition issue, with each node employing an SVM to make a binary decision. Before training an SVM classifier, the structure of binary decision sub-tasks should be properly established. At the tree's base, each data is first recognized. The input sequence is either transferred to the left or the right sub-tree at each node of the binary tree, depending on which of the two potential groups it should be assigned to.

Several classes might be found within each of these groupings. Once the recent survey of a node symbolizes the class to which it has been allocated, this is replicated recursively descending the tree. The BT-SVM approach they provide is based on learning an SVM that determines one of the two groups the entering unknown compound should be allocated to after iteratively splitting the categories into two distinct groups at each node of the tree structure. Bank investigators often begin by reviewing transaction data to search for any fraud red flags. It is possible to demonstrate

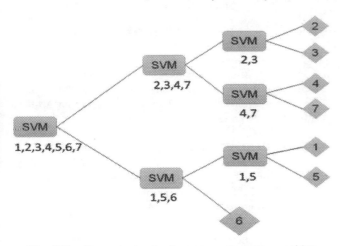

Fig. 47.1 Binary tree adapting support vector machine

whether the cardholder participated in the transaction or not using time stamps, location information, IP addresses, and other components.

Take a look at the following collection of samples: c1, c2,..., cn, each labeled by yi: " x1, x2,..., xm. The procedure continues until there is just one subclass group that defines a node inside the binary tree. This is done by separating all of the groupings into two subgroups using the above-described approach. The categories c2 through c5 are determined to be the most distant from one another after computing the gravitational centers for all classes, taking into account their Euclidean distance, and are subsequently allocated to groups g1 and g2. The subclasses of the left and right subtrees of the root are grouped separately again, resulting in the grouping of classes c7, c4 in g1,1 and c2 and c3 in g1,2 in the left tree node, and classes c1 and c5 in g2,1 and c6 throughout g2,2 in the right node in the tree. Each SVM related to a node in the taxonomy reiterates the idea.

4. Result and Analysis

In this paper, the results of the proposed methodology BT-SVM are explained in detail. It discusses the outcomes of the BT-SVM method for enhancing the detection of credit card fraud. Skimming your credit card, such as at a gas station pump. Hacking your computer. Calling about fake prizes or wire transfers. Phishing attempts, such as fake emails. The well-known performance metrics recall F-measures, accuracy, and precision are used to demonstrate these findings. The efficiency of fraud detection is increased by the BT-SVT approaches, which also makes it a more stable classifier model since it maintains excellent accuracy performance regardless of the amount of data.

Accuracy

The test set estimates are compared against the right labels to determine the classification's accuracy rate. Figure 47.2 accuracy compares the proposed methodology with existing methodologies.

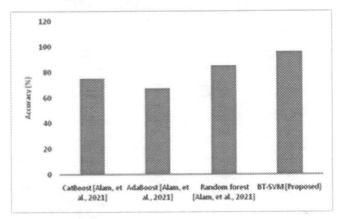

Fig 47.2 Accuracy comparison

F-measures

The weighted harmonic average of the sensitivities and the precision is determined by the accuracy metric known as the F-measure. Figure 47.3 shows the comparison of the proposed methodology with existing methodologies.

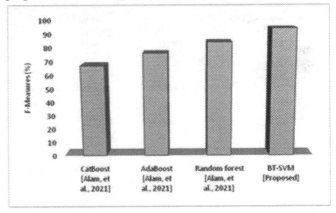

Fig 47.3 F-Measures comparison of proposed with existing

Precision

A measure of precision counts the number of accurately detected true positives. Figure 47.4 shows the comparison of the proposed methodology with existing methodologies.

Recall

The number of positive situations that were properly foreseen as positive is measured by the recall. Another name for it is

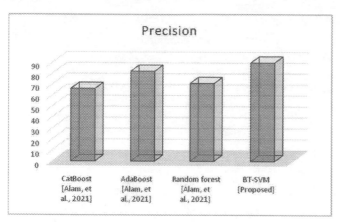

Fig. 47.4 Precision comparison of proposed and existing

sensitivity or real positive rate. Figure 47.5 shows the comparison of the proposed methodology with existing methodologies

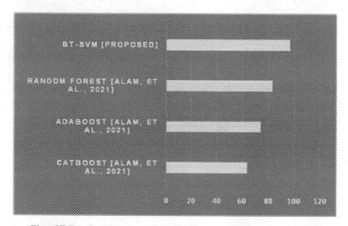

Fig. 47.5 Recall comparison of proposed with existing

5. Conclusion

Even if there are various fraud detection methods, cannot claim that this specific algorithm detects the scam. The final values lead us to the conclusion that the accuracy of the CatBoost, AdaBoost, and Random Forest algorithms is equivalent. The binary tree adaptive SVM method outperforms the other algorithm in terms of accuracy, precision-recall, and F-measures. Thus, to draw the conclusion that binary tree adaptive SVM detects credit card fraud better than the previous methods.

References

1. Zheng, W., Yan, L., Gou, C. and Wang, F.Y., 2021, January. Federated meta-learning for fraudulent credit card detection. In Proceedings of the Twenty-Ninth International Conference on International Joint Conferences on Artificial Intelligence (pp. 4654-4660).
2. Ghauri, F.A., 2021. WHY FINANCIAL SECTORS MUST STRENGTHEN CYBERSECURITY. International Journal of Computer Science and Information Security (IJCSIS), 19(7).
3. Malik, E.F., Khaw, K.W., Belaton, B., Wong, W.P. and Chew, X., 2022. Credit card fraud detection using a new hybrid machine learning architecture. Mathematics, 10(9), p.1480.
4. Trivedi, N.K., Simaiya, S., Lilhore, U.K. and Sharma, S.K., 2020. An efficient credit card fraud detection model based on machine learning methods. International Journal of Advanced Science and Technology, 29(5), pp.3414-3424.
5. Chen, J.I.Z. and Lai, K.L., 2021. Deep convolution neural network model for credit card fraud detection and alert. Journal of Artificial Intelligence, 3(02), pp.101-112.
6. More, R., Awati, C., Shirgave, S., Deshmukh, R. and Patil, S., 2021. Credit card fraud detection using a supervised learning approach. Int. J. Sci. Technol. Res, 9, pp.216-219.
7. Saxena, S., Yagyasen, D., Saranya, C.N., Boddu, R.S.K., Sharma, A.K. and Gupta, S.K., 2021, October. Hybrid Cloud Computing for Data Security System. In 2021 International Conference on Advancements in Electrical, Electronics, Communication, Computing and Automation (ICAECA) (pp. 1-8). IEEE.
8. Refaee, E., Parveen, S., Begum, K.M.J., Parveen, F., Raja, M.C., Gupta, S.K. and Krishnan, S., 2022. Secure and scalable healthcare data transmission in IoT based on optimized routing protocols for mobile computing applications. Wireless Communications and Mobile Computing, 2022.
9. Sailusha, R., Gnaneswar, V., Ramesh, R. and Rao, G.R., 2020, May. Credit card fraud detection using machine learning. In *2020 4th international conference on intelligent computing and control systems (ICICCS)* (pp. 1264-1270). IEEE.
10. Alam, M.N., Podder, P., Bharati, S. and Mondal, M.R.H., 2021. Effective machine learning approaches for credit card fraud detection. In Innovations in Bio-Inspired Computing and Applications: Proceedings of the 11th International Conference on Innovations in Bio-Inspired Computing and Applications (IBICA 2020) held during December 16-18, *2020 11* (pp. 154-163). Springer International Publishing.

Note: All the figures and tables in this chapter were made by the authors.

Advancing Sustainable Science and Technology for a Resilient Future – Sai Kiran Oruganti et al. (eds)
© 2024 Taylor & Francis Group, London, ISBN 978-1-032-79020-6

48

A Data Storage System for Educational Records Using Blockchain Technology

Rishabh Bhardwaj*

Chitkara University Institute of Engineering and Technology, Chitkara University, Punjab, India

Abstract: Education documents that are precise and in-depth are a huge asset for individuals. Education files were digitised recently. Secured data storage is still an issue, though. To provide a trustworthy and secure manner, this article proposes a data storage system for educational records based on blockchain technology. In our study, the blockchain's smart contracts are employed to control the storage process whereas the blockchain itself is employed to guarantee the safety and dependability of data storage. More specifically, the hashed data of the documents are kept on the blockchain, but the unique educational documents remain on off-chain fileservers. To ensure the privacy of data storage, the off-chain documents are regularly linked with the blockchain's hashing information. The results demonstrate how effective this research is at securely storing student records' data in terms of security level, computation time, efficiency, and traceability rate.

Keywords: Education files, Security, Data storage, Blockchain, Student records

1. Introduction

The information is crucial to students, universities, and possible employers because they have the basic feature that allows records to recreate the archival condition. Such documents are crucial to students, education systems, and future employers since records' fundamental character is primitive, which makes it possible for them to recreate historical reality. Education records have been digitized as a result of the growth of information systems (Li and Han, 2019). The statistical outline security technique for changes to databases and extremely intelligent computers is how blockchains are abstracted. Blockchain may seem to be more of a mindset than a technology to address problems with contractual trust. It may be used to identify any useless modifications that occurred in the system and to tell every important component of the functioning of those alterations (Choi, et al., 2022). The educational sector is always evolving. Due to technological advancement, the learning environment transformed from a physical classroom to a virtual one. Each enabling environment is required to keep a student's transcripts and qualifications on file. Certificates serve as documentation of learning for pupils. The open, autonomous ledger created by blockchain technology is unchangeable and facilitates sharing (Priya, et al., 2020). A robust all-encompassing gateway that simplifies the process for all entities in educational institutions to function may be built through the use of several ways that have been investigated. To reduce the amount of paperwork required in the education industry, blockchain technology may be utilized to convert procedures and activities that are performed in institutions' daily operations into reliable software (Arcinas, et al., 2021). Blockchain is often described as a distributed ledger that keeps an ever-growing collection of publicly available data that are blockchain protected against alteration and change. Blockchain technology has improved Internet of Things, intelligent manufacturing, supply chain management, technological platform transfers, and other domains (Hou, et al., 2017).

*Corresponding author: rishabh.bhardwaj.orp@chitkara.edu.in

DOI: 10.1201/9781003490210-48

2. Related Works

Han, et al., (2018) propose a blockchain-based mechanism for empowering individuals to own and share their authorized educational records. their solution integrates the material removal features of blockchain technology to allow teachers to provide verified authorizations that serve as proof of completion or success. The proposed architecture could be able to provide the necessary capability for thread education and a number of other activities. The study introduces a permissioned blockchain-based system to enable institutes to safely and stably transmit and authenticate academic data at the desire of the applicant. For productivity apps, distributed ledgers like Blockchains offer an additional accessible, economical, and secure alternative. The sample containing an appropriate tool for signing up and completing the operation, as well as a repository that stores the hash of the blockchain entries for authentication using Blockchains and Hyperledger Composer (Badr, et al., 2019). The instructional document archive described by Bessa, et al., (2019) uses blockchain technology to maintain and disseminate learning programs for business and management audiences. The BcER2 system enables entities to transmit, exchange, and publish educational records like e-diplomas and e-certificates safely and easily. Alnafrah, et al., (2021) looks at the potential commercialization of its contracts as well as the expenses and advantages of creating this network. The creation of this network would guarantee open and equitable schooling, provide chances for minimal learners to continue studying throughout their lives, and therefore satisfy the four Sustainability Policy objectives. The 'blockchain' is the basic mechanism underpinning the Bitcoin digital payment system. Radanović, et al.,(2018) suggest a blockchain-based permanent distributed record of intellectual work with accompanying reputational rewards that constitutes and democratizes educational status outside of the educational institutions. They are doing exploratory study into a blockchain system for academic data storage, drawing on their earlier brand equity body of research for educational institutions .

System Architecture

The envisaged distributed ledgers system provides efficient record exchange, authentication, and storage. Education institutions, consortium blockchains, storage servers, and the framework service make up the majority of it. The majority of data is stored and shared on the blockchain by the academic system, which is the organization that connects to the cluster members. Blockchain is used by Cooperation blockchain technology is in charge of keeping the summary data of encrypted records. To construct the consortium blockchain, Ethereum is once again used. The use of smart contracts is used to regulate how data is stored and retrieved.

The authentic documents and data are secured on the storage server using encryption. Using the Database management system, the stored server was selected in order to effectively retain and extract information and provide safe file storage. Service Framework includes an array of main components that are readily accessible, don't maintain any data, and offer services to nodes or users through RESTful APIs.

Network of Blockchains for Storage and Sharing

Ethereum will be used to create the consortium blockchain, which has associate nodes for exchanging and maintaining data. The so-called full node really syncs all of the blockchain's data, containing different block bodies, transactional lists, and certain other pertinent data. The networks call for more powerful workstations with more storage and processing power. The light nodes merely maintain the form of a cluster of the active blockchain; they do not take part in downloaded and maintained the blockchain general ledger. They also often do not compete for financial reporting rights. While mobile terminals and embedded devices have less intensive computation and storage equirements than computers, these nodes are also ideally equipped for them.

Authority for Certificates (CA)

A blockchain node must be installed by each educational institution attempting to connect to the Ethereum blockchain. An identity authentication process is added to the system to ensure internet security. Every institution acquires its public and private keys and registers with the CA, to be more exact. The CA will be used to hold authorized organization data. The CA might punish this node and reveal the identity of the offending node. The CA's position as a messaging intermediary between the external node and the blockchain member nodes means that it performs as a key shortest possible time frame and that it has no problem with the block - chain.

3. Proposed Methodology

Design of a smart contract for IISC: An IISC smart contract, which is developed and implemented on the blockchain to control node behavior, is used in the joining process. The organisation's public key, PK(X), which is saved as PublicKey, serves as the institution's identification and is used to index the data about the institution. A distinct blockchain profile address will be given to each member node when it has been properly set up. This address is saved as Block Address. Moreover, msg. the sender is logged for every account address that calls a software application. The suggested system makes use of smart contracts as a means of imposing control over the storing procedure.

Storage of secure data using blockchain: Data stores and blockchain technology are the foundation of the storage process. Although the actual records and files are encrypted before being saved in the repository, the summaries of the records are maintained in the activity item on the blockchain. The blockchain's impartial ecosystem assures that summaries are not improperly modified while being kept. The source records saved on the storage server are periodically updated utilizing data on the blockchain in order to preserve security.

Secure Data Sharing with Blockchain

Historically, educational records have been kept in separate data centers, making it impossible for other institutions to share them due to security regulations or privacy concerns. In rare situations, if there is no suitable data-sharing plan in place, a student who changes to another college may lose access to their academic records from the first institution.

Anti-tampering Inspection Based on Blockchain

There is a circumstance when academic records stored by institutions are altered, and this is only discovered when the records are cross-checked with data from the blockchain To preserve the accuracy of the records in the storage serv0er, an anti-tampering inspection mechanism is used. A security danger will come from the modified records that have been kept on the storage server for a while. In to identify tampering in real time, an edge - preserving approach is added to the scheme, anchoring the off-chain data with corresponding password kept in the chain. The triggering of the mechanism and the design of the particular inspection technique are two components of the mechanism's design. Unless the predetermined circumstances are met, the mechanism is said to be triggered, starting the execution inspection operation. The time-based evaluations are carried out using the booking system, It starts the test accuracy at a certain period of time.

4. Result and Discussion

Evaluation of Performance

Launch procedure analysis: The parallel settings for block creation and data submission allow member node unanimity times to be disregarded. Also, academic documents from various eras are chosen appropriately. This test uses the minimal reaction time (Minimum), the acceptable response time (Average), and the maximal response time (Maximum) as assessment markers. The Apache JMeter 5.2 is the foundation of the load test. 100, 200, 300, and 400 virtual institutions have been simulated by establishing the equivalent quantity of threads. Moreover, a randomly generated integer between 1 and 50 is used to determine

the number of collection operations for each thread. Figure 48.1 depicts the results. The number of virtual institutions is on the x-axis, although the stored application system response is on the y-axis. This chart shows that virtual institutions increase reaction time. The minimum, average, and maximum reaction times rise as the number of virtual institutions increases from 100 to 400. Maximum reaction time is still under 10ms.

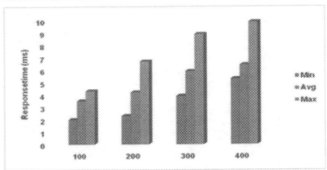

Fig. 48.1 Load analyses of the proposed system

Analysis of Traditional Scheme

An encryption method CP-ABE is often used in cloud storage and sharing scenarios. The properties of the experimentation record are constrained to the same; The interceptor is the number of data packets, and the variate is the running time of encryption and storage time. When more received signal packs arrive, the computational cost of all systems rises progressively until settling. Generally, the computation complexity of all methods is comparable, but our scheme will process data more quickly when the volume of data packages rises to a certain point.

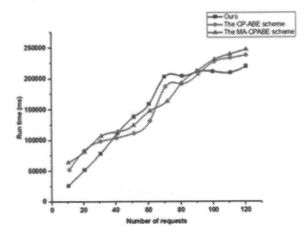

Fig. 48.2 Runtime data packages.

The computing costs of the aforementioned approaches are contrasted under identical both software and hardware setup settings. The findings are shown in Figs 48.2 and 48.3. The abscissa indicates the number of data packs, while the variate indicates the duration of encryption and storage time. These are the only attributes that can be found in the exploratory record. When more received data packets arrive, the computational cost of all systems rises progressively until settling. The computational costs of all approaches are equivalent, but when the amount of data increases above a certain threshold, The methodology will complete the processing more effectively.

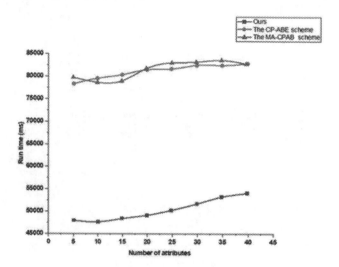

Fig. 48.3 Attribute-based runtime

The coordinate in Fig. 48.3 for encryption and storage time corresponds to Fig. 48.2 ordination for the range of encrypted parameters. The approaches consistently handle a fixed maximum number of packages. Likewise, when the qualities rise, so does the run time for all schemes. The computation time of the suggested system is much lower than that of the other two schemes, as shown in Fig. 48.3. There are proposed strategies that lack attribute level protection.

5. Conclusion

To tackle the need for the conservation and transmission of learnings, a secure storage and sharing system based on the blockchain is being created. Under the proposed concept, the consortium chain between institutions can guarantee the data's confidentiality and safety. It is suggested that a distributed institution authentication system be used to protect blockchain nodes from attack. Blockchain and Storage servers are used to provide secure storage. To maintain the records in the storage server, a system for anti-tampering inspection is also used to improved levels of security, effectiveness, and credibility for the proposed program. A safe infrastructure

to handle these active smart contracts still requires more study. The InterPlanetary File System (IPFS) and Story are examples of decentralized storage systems that will be utilized in the future. The integration of decentralized storage systems for increased functionality and the development of a secure infrastructure for managing active smart contracts might be the focus of future study.

References

1. Li, H. and Han, D., 2019. EduRSS: A blockchain-based educational records secure storage and sharing scheme. IEEE Access, 7, pp.179273-179289.
2. Choi, L.K., Sunarya, P.A. and Fakhrezzy, M., 2022. Blockchain Technology as Authenticated System for Smart Universities. IAIC Transactions on Sustainable Digital Innovation (ITSDI), 4(1), pp.57-61.
3. Priya, N., Ponnavaikko, M. and Aantonny, R., 2020, February. An efficient system framework for managing identity in the educational system based on blockchain technology. In 2020 International Conference on Emerging Trends in Information Technology and Engineering (ic-ETITE) (pp. 1-5). IEEE.
4. Arcinas, M.M., 2021. A Blockchain-Based Framework for Securing Students' Educational Data. Linguistica Antverpiensia, 2021(2), pp.4475-4484.
5. Hou, H., 2017, July. The application of blockchain technology in E-government in China. In 2017 26th International Conference on Computer Communication and Networks (ICCCN) (pp. 1-4). IEEE.
6. Han, M., Li, Z., He, J., Wu, D., Xie, Y. and Baba, A., 2018, September. A novel blockchain-based education records verification solution. In Proceedings of the 19th annual SIG conference on information technology education (pp. 178-183).
7. Badr, A., Rafferty, L., Mahmoud, Q.H., Elgazzar, K. and Hung, P.C., 2019, June. A permission blockchain-based system for verification of academic records. In 2019 10th IFIP International Conference on New Technologies, Mobility and Security (NTMS) (pp. 1-5). IEEE.
8. Bessa, E.E. and Martins, J.S., 2019. A blockchain-based educational record repository. arXiv preprint arXiv:1904.00315.
9. Alnafrah, I. and Mouselli, S., 2021. Revitalizing blockchain technology potentials for smooth academic records management and verification in low-income countries. International Journal of Educational Development, 85, p.102460.
10. Radanović, I. and Likić, R., 2018. Opportunities for use of blockchain technology in medicine. Applied health economics and health policy, 16, pp.583-590.

Note: All the figures in this chapter were made by the authors.

Advancing Sustainable Science and Technology for a Resilient Future – Sai Kiran Oruganti et al. (eds)
© 2024 Taylor & Francis Group, London, ISBN 978-1-032-79020-6

An Efficient Blockchain Based Healthcare Data Management System Using Lightweight Cryptographic Algorithm

49

Pooja Sharma*

Chitkara University Institute of Engineering and Technology, Chitkara University, Punjab, India

Abstract: In addition to implementing blockchain in the medical services, other dispersed approaches were put forth to address the issues with centralised, data island difficulties. Despite the fact that the traditional management of medical records has been considerably altered by blockchain-enabled medical services, certain new issues, like ledger heftiness, data accessible security, and the potential of a quantum assault, are now revealed. Thus, a blockchain-based lightweight cryptographic algorithm (LWCA) is suggested in this research to safeguard medical records. In order to protect user privacy, this research utilized LWCA to transfer data safely. Blockchain technology is also used to store the data safely. The MATLAB tool and a medical dataset are used for the execution. In terms of various performance measures, including average latency, execution time, the simulated outcomes for the suggested LWCA method are contrasted to current methods.

Keywords: Medical service, Blockchain, Security, Data management, Lightweight cryptographic algorithm (LWCA)

1. Introduction

Health files are incredibly essential documents because they serve as digital proof of patients' diagnoses and treatments. The client in a medical malpractice case must demonstrate each component by a majority of the facts. Evidently, the accused is quite motivated to remove or alter the damaging digital information. It is challenging to guarantee the security of clinical information using a centralized medical data management system. With such a method, health information is often kept in a hospital's data set. After getting the necessary database rights, an attacker may remove or edit the data (Tian, et al., 2019).

The security of health data may obviously be protected by distributed blockchain technology. On a blockchain, the medical data are dispersed across several parties' storage facilities. If the data of a few parties are changed or destroyed, it has no impact on the data of other parties. The information on the blockchain is subject to a consensus process.

The privacy and accessibility of data are where the blockchain technology's issues reside. Several times, individuals encrypt medical data to guarantee privacy. A patient's key is used for the encrypting. Whenever encryption information is used, the client provides the key. The executor or administrator of a dead participant's estate is sometimes the plaintiff in a medical malpractice case (Ramani, et al., 2018).

Big data in the field of medicine encompasses a variety of information gathered from IoT-enabled devices, such as patient medical records, hospital records, results of medical lab tests, and many more data. The amount of big data in healthcare is growing rapidly as a result of technological advancements. Thus, managing, extracting, securing, and storing massive data in healthcare requires sophisticated technology. There are currently a number of digital big data sources available on the market, including health information reports, patient reports, clinical data, and virtual medical histories. (Muthulakshmi, et al., 2022, Marbouh, et al., 2020).

Similar system might be even more customized in a manner that allows investigators to monitor specific client clinical information throughout the research procedure with a date (similar to the stated Bitcoin transaction, for example). Cryptocurrency might be used to identify Healthbank users

*pooja.sharma.orp@chitkara.edu.in

DOI: 10.1201/9781003490210-49

who have made a substantial contribution to the success of medical research initiatives and compensate them at a greater rate than is typical. As a result, Healthbank has come to represent patient or end-user empowerment in medicine, coupled with digitization, novel online company structures, and virtual care services. The autonomy of patients and end users in medicine would be further enhanced by cryptocurrencies. (Rajesh, et al., 2019, Khalaf, et al., 2022).

The objective is to address the issues of scalability, data integrity, privacy, and security in health care information handling. The goal of this research is to create a system that protects the privacy of private healthcare information while enabling swift and safe blockchain transactions. The system intends to reduce computational cost and improve performance by employing a lightweight encryption method.

2. Related Works

Christo, et al., 2021 suggested the elliptic curve cryptography (ECC) method, a simple verification strategy to efficiently distribute the information. There are several studies that employ router information sharing, and this one uses Electronic Medical Cards (EMC) to store healthcare information. Information privacy and information identification are two significant challenges in user privacy that are covered in this book. Through adopting the Cryptocurrency record system, only permits morally acceptable individuals can view the information, information security is protected. Lastly, the structures and systems stores the encryption. In order to obtain the data quickly, the test results are stored inside the edge network using edge internet of things IoT. The data may be decrypted and processed quickly by an authorised person.

Satamraju, et al., 2020 suggested a novel approach that combines cryptocurrency technology and IoT connections to solve possible privacy and security issues for data integrity. Agreements play a key role in this implementation phase and are employed to handle information, perform devices identification, permission, and security systems. They also provide a fresh interaction design paradigm for integrating the two technologies, emphasizing its improvements over the previous methods. System design scalability may be improved by including off-chain data storage into the architecture.

Blockchain is a new technology that offers safe administration, assured access control, and authentication for IoT devices. IoT is a Connectivity delivered through the web where customer data is processed and gathered securely. In order to provide modern medical, the institution must also be able to diagnose patients who are located far away. The real time health architecture has serious issues with information security, costs, memory, scalability, trust, and transparency across several platforms. Due to the open internet environment, it is important to control data security and authentication since

the user's validity is questioned. Many tactics are primarily focused on dealing with security issues, such as forgery, timing, denial of service, stolen smartcard attacks, etc. Cryptocurrency upholds the principles of perfect privacy in order to identify the individuals participating in operations. Blockchain, enhanced data sharing, better security, and cheaper overhead costs in virtualization, as well as the lack of a centralized third party are the driving forces behind the adoption of blockchain in health informatics. Together with extra regulatory constraints, healthcare informatics has certain special security and privacy concerns. This research uses a probabilistic model to offer a unique authentication and authorisation system for blockchain-enabled IoT networks (Tahir, et al., 2020).

3. Proposed Methodology

A kind of encrypting known as lightweight cryptography has a tiny model efficiency or impact. Its worldwide standards and guidelines compilation are now under process, and it aims to increase the applicability of blockchain on various measures. This activity of collecting, safeguarding, and evaluating data gathered from various sources is known as medical network security. Medical systems may develop comprehensive images of individuals, customize medicines, collaborate, and enhance medical results by controlling the abundance of healthcare data that is readily accessible.

Encryption

In the encryption process, the plaintext's ASCII value is initially determined. The equivalent numerical value is then determined. Then, choose the predetermined set of numbers 298 for this procedure. The value obtained when the sum of the ASCII values of the key letters was multiplied by 100

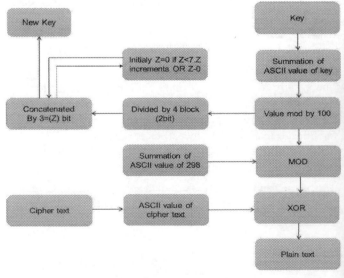

Fig. 49.1 Encryption

is used to modify 298. The outcome is then translated into binary, and the numerical value of the plaintext's ASCII value is then XORed with it. The XOR result is once again translated to arithmetic, and the resulting arithmetic value corresponds to the ciphertext's ASCII value. Fig. 49.1 shows the encryption process used in the research.

Decryption

In the decryption phase, the cipher text ASCII value is first determined, and then its corresponding binary value. Once again, the value discovered when the sum of the ASCII values of the key characters was multiplied by 100 is used to modify the random number 298. The decryption component received this value. This new outcome is binary-encoded and XOR with the cipher text binary representation of the ASCII value. The XOR result is once again translated to decimal form, and the resulting decimal value is the plaintext's ASCII value. The plaintext of the cipher text is obtained by determining the symbol that corresponds to that ASCII value. Figure 49.2 shows the decryption used in the research.

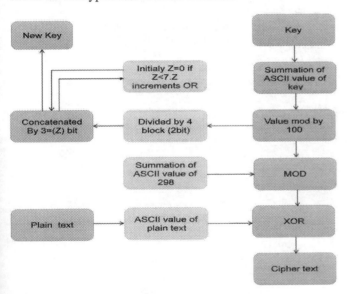

Fig. 49.2 Decryption

4. Result and Discussion

In this section, we discuss in detail about the results of healthcare data management system using lightweight Cryptographic algorithm technique. The algorithm guarantees information exchange, reliability, and confidentiality. Tiny businesses cannot afford to spend a lot of money on pricey antivirus software and other security measures. The proposed algorithm offers them a low-cost solution that works well. The method has been tested on a variety of systems and with a wide range of input data types, making it very trustworthy

and valid. Hence the consumer might get reliable production at a reasonable price.

Average Latency

The average latency of a blockchain network is the interval between the registration of a movement and the program's verification of it, while the latency of an interchange is the interval during which the system processes the operation. In this Fig. 49.3 latency is measured in seconds. It depicts the latency and transaction success rate. B1 average latency have much less latency as settings, while B2 average latency decrease the through put to 3seconds respectively.

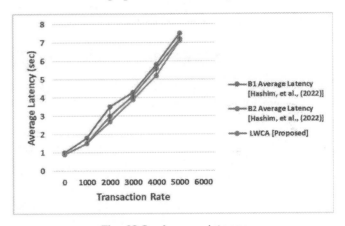

Fig. 49.3 Average latency

Execution Time

Figure 49.4 shows the execution time for existing and proposed methods based on healthcare data management. The execution times for each transaction rate for both existing networks indicate a progressive rise to a larger transaction rate. Blockchain is utilized for health record-keeping, clinical trial, patient monitoring, enhances safety, show information and transparency. It minimizes the time and expense of data translation while maintaining hospital financial statements.

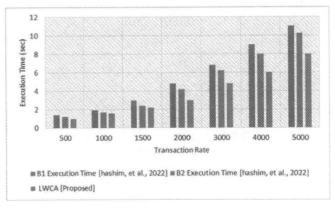

Fig. 49.4 Execution time

5. Conclusion

Blockchain offers decentralised administration, an irreversible independent audit, data provenance, robustness/availability, and security/privacy. We introduced lightweight cryptographic algorithm for blockchain, outlined their advantages over conventional methods for medical and healthcare applications (improving patient records governance, enhancing the insurance claim process, speeding up clinical and biomedical research, and creating an innovative biomedical and healthcare information log, among others). Using blockchain technology may provide a number of recognized possible difficulties such as transparency/confidentiality, speed/scalability, and the risk of attack.

References

1. Tian, H., He, J. and Ding, Y., 2019. Medical data management on blockchain with privacy. Journal of medical systems, 43, pp.1-6.
2. Ramani, V., Kumar, T., Bracken, A., Liyanage, M. and Ylianttila, M., 2018, December. Secure and efficient data accessibility in blockchain based healthcare systems. In 2018 IEEE Global Communications Conference (GLOBECOM) (pp. 206-212). IEEE.
3. Muthulakshmi, S. and Kannammal, A., 2022. Comprehensive Study of Lightweight Block-Chain Model for Data Sharing in Internet of Things for Smart Applications. Mathematical Statistician and Engineering Applications, 71(4), pp.6660-6671.
4. Marbouh, D., Abbasi, T., Maasmi, F., Omar, I.A., Debe, M.S., Salah, K., Jayaraman, R. and Ellahham, S., 2020. Blockchain for COVID-19: review, opportunities, and a trusted tracking system. Arabian journal for science and engineering, 45, pp.9895-9911.
5. Rajesh, N. and Selvakumar, A.A.L., 2019. Association rules and deep learning for cryptographic algorithm in privacy preserving data mining. Cluster Computing, 22, pp.119-131.
6. Khalaf, O.I., Natarajan, R., Mahadev, N., Christodoss, P.R., Nainan, T., Romero, C.A.T. and Abdulsahib, G.M., 2022. Blinder Oaxaca and Wilk Neutrosophic Fuzzy Set-based IoT Sensor Communication for Remote Healthcare Analysis. IEEE Access.
7. Christo, M.S., Jesi, V.E., Priyadarsini, U., Anbarasu, V., Venugopal, H. and Karuppiah, M., 2021. Ensuring improved security in medical data using ecc and blockchain technology with edge devices. Security and Communication Networks, 2021, pp.1-13.
8. Satamraju, K.P., 2020. Proof of concept of scalable integration of internet of things and blockchain in healthcare. Sensors, 20(5), p.1389.
9. Tahir, M., Sardaraz, M., Muhammad, S. and Saud Khan, M., 2020. A lightweight authentication and authorization framework for blockchain-enabled IoT network in health-informatics. Sustainability, 12(17), p.6960.

Note: All the figures in this chapter were made by the authors.

Advancing Sustainable Science and Technology for a Resilient Future – Sai Kiran Oruganti et al. (eds)
© 2024 Taylor & Francis Group, London, ISBN 978-1-032-79020-6

50

A Secured Blockchain-Based Smart Contract for Vehicle Anti-Theft System

Savinder Kaur*

Chitkara University Institute of Engineering and Technology, Chitkara University, Punjab, India

Abstract: The theft of vehicles has significantly increased during the past few years. In our society, it is ostracized. Due to the lack of an effective theft detection method, the effects of car theft have been severely impacting both the social protection and financial strength of the entire world. The few vehicle anti-theft methods that are now in use have serious issues with data protection, personal data leaks, centralized systems, and proper key management. This article suggests a blockchain-based vehicle anti-theft framework (B-VATF) that tackles these issues via smart contracts (SC). The use of Blockchain regarding vehicle safety is also discussed in this article, along with a step-by-step execution of the suggested technique that includes test results and a thorough analysis of crucial comparisons with other current systems. With the assistance of B-VATF, more than one individual might be permitted to operate a car without damaging its data or compromising security.

Keywords: Theft detection, Vehicle anti-theft, Blockchain, Smart contracts (SC), Blockchain-based vehicle anti-theft framework (B-VATF)

1. Introduction

Blockchain was first developed as a framework for backing the well-known cryptocurrency bitcoins. By cryptography, bitcoins were originally put out in 2008 and launched in 2009. The financial market has had tremendous expansion since that time, reaching 10 billion dollars in 2016. A blockchain key used to record the completed activities is what makes up blockchain technology, which is essentially a blockchain (Monrat, et al., 2019). A broad spectrum of socialization is impacted by vehicles. They provide a common and dependable way to get around in the modern culture. Nonetheless, automobile theft is a problem for society in many wealthy nations. Vehicles are protected against theft by anti-theft devices. The anti-theft system is a technique for preventing unauthorized access to any priceless goods (Das, et al., 2021). In current culture, a car may be a reliable form of mobility. Vehicle theft has escalated internationally in recent years. To combat car theft, several vehicle anti-theft technologies were created and put into use. Nonetheless, it remains a difficult challenge to address

the problem of car theft. IoT (Internet of Things), GPS/GSM (Global System for Mobile Communications), dated smartphones, fingerprint verification, and facial recognition technologies have all been used to create several vehicle anti-theft systems. Yet, these technologies impair vehicle security and suffer from data privacy concerns (Jacob, et al., 2023). Several anti-theft control systems have been developed recently. Using wireless connections, an integrated info-security circuit board connects to other cars, wayside equipment, and mobile phones using the CAN bus, LIN bus, Flex Ray, and MOST Bus to exchange data with ECUs and sensors within a vehicle. The system's fundamental flaw is its inability to provide dependable, secured automobile interactions due to connection latency and data consistency. In case a vehicle is taken, there are several wireless router systems developed that may be used to remotely deactivate the vehicle's main automotive devices. Vehicle-vehicle connections must be secured for it too (Vinya, et al., 2022). A new distributed infrastructure and computing paradigm knew as blockchain stores data in an organized chained data model

*Corresponding author: savinder.kaur.orp@chitkara.edu.in

DOI: 10.1201/9781003490210-50

updates that data using a consensus protocol, and ensures data security using cryptographic technologies. Due to the decentralized feature of blockchain, even if a small handful of nodes fail, it won't disrupt the regular functioning of the remaining nodes, which lowers operational costs. According to the blockchain's immutability, incorrect content, vehicle failures, as well as on documentation about traffic accidents would be permanently recorded in the blockchain. This will allow for the crystallization of facts and the resolution of the security of vehicle software problem (Mugariri, et al, 2022). The rest of this manuscript is structured as follows: Section II describes the related work, section III the recommended approach, section IV the results, and section V the manuscript's conclusion.

2. Related Works

The study (Razmjouei, et al., 2020) suggests a robust but very gentle universal authentication method that works with a decentralized database and the radio frequency identification standard to authenticate passive entry passive start. The development of a safe blockchain based on smart contracts that work with the vehicle's multiple access is also underway. The created smart contracts in this study can instantly authorize individuals and their automobiles. The study states for an improved centralized shared mobility system based on the service level agreement (SLA), that specifies company objectives the source is required to satisfy and record the products that providers should supply. The SLA standards are described as the smart contract, which enables inter-cooperation and automates the process without involving a guarantor. Blockchains are utilized to create a specific example of notebook exchange that demonstrates how many technologies may be put to use in a marketplace (Hang, et al., 2019). The Blockchains Composition is used to evaluate the smart contract's functionality. The control center, the grid operator, and the equipment provider all participate in the regulatory process by receiving corrected bits from the blockchain and unencrypted from the

decoding. The suggested method by (Zhang, et al., 2022) provides benefits in processing and transmission costs while still fulfilling security standards for the secrecy and integrity of information, according to the security analysis and performance evaluation. The Internet of Vehicles (IoV) database secured sharing scheme is proposed (Ma, et al., 2020) using the intelligent transportation sensor network as an instance. It uses smart contracts to accomplish automated registrations, quick validation, and trustworthy IoV information sharing. The critical portion of the data is processed by the smart contract using homomorphism encryption and zero-knowledge proofs; this portion is present on the blockchain as cryptography. In the study, (Gautam, et al., 2020) researchers created a reliable automobile anti-theft system that makes use of Blockchain encryption. They proposed containing some detectors to guard against any actual stealing activity. It also features an alarm system that gathers the picture and position of the invaders and sends alerts through text messages and email. To further defend against potential cyber security incidents, three-layered blockchain protection has been offered. For the safety of vehicles, they develop an irreversible, verifiable storage system based on the interplanetary file system and the blockchain.

3. Proposed Methodology

The suggested B-VATF could prevent theft of the vehicle by recognizing improper contact with that. With the B-VATF design, there are six clusters. These sites are the Owners, the Vehicle Seller Agency (VSA), the Vehicle Certification Agency (VCA), the Blockchain System, the Vehicle Transportation Agency (VTA), and the Vehicle. It is possible to take additional clusters if required. Properties are used in the B-VATF design, such as the Authorized Drivers (AD) and the Universal Vehicle Key (UVK). The suggested B-VATF design is displayed in Fig. 50.1, which shows the communication system between the existing nodes in the B-VATF architecture in the following methods.

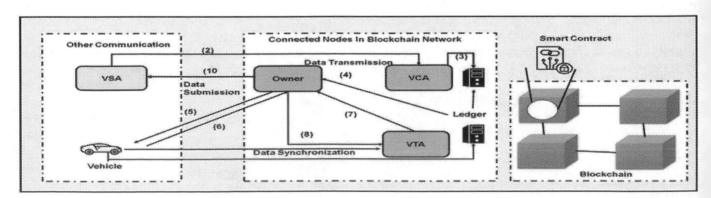

Fig. 50.1   A description of the B-VATF technology's systems

Step 1: The VSA, that's in charge of properly acquiring and validating vehicle owner data, must get identification from every registered user. To register a vehicle, financial compensation to the VSA with the required information upon purchasing a vehicle. To register the newly bought car, the owner must get in touch with the closest or preferred VSA.

Step 2: Together with the dashboard, the VSA sends the seller's identification to the VCA. A UVK is then produced for the registrant by the VCA.

Step 3: On the blockchain database, the VCA maintains the UVK and the seller's accreditations. For the proprietor to examine the blockchain funds, the VCA gives access to an application.

Step 4: A smart contract may only be activated by its proprietor after successful authentication on the bitcoin blockchain. Since his networking authentication, a user may execute a smart contract that VCA has already put on the blockchain

Step 5: To activate the vehicle, a proprietor must provide permission using the biometric system. The proprietor must update and determine whether the project is for the authorized user to the blockchain server when they are permitted to use the vehicle. The seller will destroy any more information after a drive excursion.

Step 6: A notice would be issued to the proprietor and the VTA if an inappropriate individual attempted to enter the vehicle without proper authorization after the vehicle's performance has been checked.

Step 7: The vehicle owner would respond to the VTA's request for the condition of the vehicle after it has been tracked consistently for a time frame.

Step 8: When a vehicle was declared safe or unsafe, the owner provides the VTA an update on its condition.

4. Result

The analyzed smart contract for the B-VATF is presented here along with the research outcomes. To use an internet device called Synthesis integrated development environment, stability programming software for smart contract development was already created. The consequences of the strategy will emphasize and implementation of the smart contract's coding was obtained. To achieve the necessary result, such as the progress of the authentication process, they occasionally must provide data, such as UVK. The input data may theoretically be dynamically extracted from a database using the smart contract.

The Mixture IDE, which provides an interface for executing the smart contract, is used to execute the experimental observations. It is an open-source platform that enables us to create smart contract software on the web using Elegance programming. A smart contract program could be implemented using the popular technology Stature. Software running in the Ethereum virtual machine that facilitates analyzing, building, and deploying the smart contract is used to provide blockchain-specific functions. Table 1 demonstrates the how smart contract's specified functionalities were used in this study. The section gives some experimental contexts.

Table 50.1 Functions of the smart contract utilized

Function Name	User for
• Set UVK	authenticating a vehicle owner
• Removes ADs	removing the AD's data
• Store driver data	adding an AD's data
• Current Driver	setting the current driver

Resources rationalism and components of the environment analytical reasoning define the limits and the challenge differently from computational rationalism. We define the limitations in terms of the computational costs imposed by an imaginary computing framework and considered them to be the optimization of a function that computing cost. The computational cost of the proposed method is compared with the existing method in Fig. 50.2. When compared to an existing method which has an accuracy rate of 71%, the recommended method B-VATF value is 95%. The suggested B-VATF performs well in a vehicle in an anti-theft system and has a low computational cost.

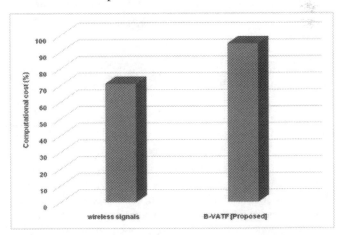

Fig. 50.2 The computational cost of the wireless signals

According to Figure 50.3, user identifying accuracy could reach around 92% and user authorization accuracy must reach over 96%. The accuracy of human authentication may reach around 98% after integrating the trained model, according to their findings after attempting to include the training sets of the verified forgery users.

The outcomes demonstrate that the technology could reach better human authentication accuracy when we continually add the data of legitimate users and illegitimate users into the test dataset. It could be integrated into an internet-of-vehicle framework, and the device could broadcast information about unauthorized users to all vehicles. As per research data, B-VATF successfully identifies humans with a total TPR and FPR of about 96% and 3.7%, and users with an average TPR and FPR of 90% and 2%, accordingly. The distribution of TPR and FPR for user credentials and human attestation is also shown in Figure 50.3. B-VATF effectively issues a 96% alert to the vehicle owner with a TPR of 96% and a 92.8% accurate suggestion to the vehicle users with a TPR of 90%.

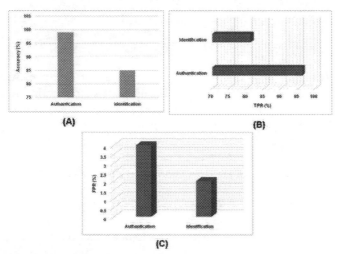

Fig. 50.3 Accuracy, TPR, and FPR for authentication and identification

5. Conclusion

The use of blockchain technology is expanding quickly across several industries due to its efficiency and safety. The blockchain is a technique for rendering the application server more secure. Utilizing blockchain technology and smart contracts, we have defined and constructed a decentralized B-VATF in the article. To address the problems with a central approach, we have chosen a decentralized platform such as blockchain. An automatic user identification procedure may be made more secure and accurate by applying a smart contract. The suggested framework has been put into practice, and it has shown to be capable of delivering a safe, autonomous, accessible, and trusted ecosystem for both privacy protection and vehicle safety. The broad use and maturation of blockchain technologies are still in progress. There might be unknowns, technical difficulties, and shifting standards with any new technology. To overcome technical barriers and guarantee the scalability, safety, and longevity of the proposed B-VATF, further study is required. The blockchain-based smart contract system may undergo comprehensive testing and real-world application as part of future research in cooperation with industry stakeholders. This may assist in identifying practical difficulties, assessing the system's effectiveness, and gathering input in order to develop the system in accordance with actual use cases and needs.

Reference

1. Monrat, A.A., Schelén, O. and Andersson, K., 2019. A survey of blockchain from the perspectives of applications, challenges, and opportunities. IEEE Access, 7, pp.117134-117151.
2. Das, D., Banerjee, S., Ghosh, U., Biswas, U. and Bashir, A.K., 2021. A decentralized vehicle anti-theft system using Blockchain and smart contracts. Peer-to-Peer Networking and Applications, 14, pp.2775-2788.
3. Jacob, R.R., Mahesh, R., Gomez, S., Sahabuddin, S. and Robinson, V., 2023. EverTrack: A System to Track History of Used Cars Using Blockchain. In Inventive Computation and Information Technologies: Proceedings of ICICIT 2022 (pp. 415-423). Singapore: Springer Nature Singapore.
4. Vinya, V.L., Anuradha, Y., Karimi, H.R., Divakarachari, P.B. and Sunkari, V., 2022. A Novel Blockchain Approach for Improving the Security and Reliability of Wireless Sensor Networks Using Jellyfish Search Optimizer. Electronics, 11(21), p.3449.
5. Mugariri, Hanifa Abdullah, Miguel Garcıa-Torres,B. D. Parameshchari,and Khalid Nazim Abdul Sattar, " Promoting Information Privacy Protection Awareness for Internet of Things (IoT)", Hindawi Mobile Information Systems, Volume 2022, Article ID 4247651, 11 pages
6. Razmjouei, P., Kavousi-Fard, A., Dabbaghjamanesh, M., Jin, T., and Su, W., 2020. Ultra-lightweight mutual authentication in the vehicle based on smart contract blockchain: Case of MITM attack. IEEE Sensors Journal, 21(14), pp.15839-15848.
7. Hang, L. and Kim, D.H., 2019. SLA-based sharing economy service with a smart contract for resource integrity in the internet of things. Applied Sciences, 9(17), p.3602.
8. Zhang, J., Wang, Z. and Yan, Q., 2022. Intelligent user identity authentication in vehicle security systems based on wireless signals. Complex & Intelligent Systems, 8(2), pp.1243-1257.
9. Ma, Z., Wang, L. and Zhao, W., 2020. Blockchain-driven trusted data sharing with privacy protection in IoT sensor networks. IEEE Sensors Journal, 21(22), pp.25472-25479.
10. Gautam, R., Shrestha, R., Mishra, S. and Singh, J.K., 2022. Blockchain Enabled Vehicle Anti-theft System. In Security and Privacy in Cyberspace (pp. 209-226).

Note: All the figures and table in this chapter were made by the authors.

Implementation of a Trust-based Blockchain Model for Medical Applications

51

Rahul*

Chitkara University Institute of Engineering and Technology,
Chitkara University, Punjab, India

Abstract: Internet and device accessibility are now crucial factors in societal, institutional, and individual decisions. Internet of Medical Things (IoMT), the result of combining healthcare elements provides a platform for efficient service delivery to the benefit of patients and healthcare professionals. While the privacy, usability, and dependability of In the IoMT, medical data are valued highly semantic gaps and a lack of suitable assets or attributes continue to be barriers to effective information transmission in unified trust management solutions. As a result, we suggest a trust-based blockchain model (TBM) in IoT areas in which a smart contract ensures budget verification and an Indirect Trust Inference System (ITIS) eliminates semantic deficiencies and improves reliable factor estimates through network nodes as well as edges. The TBM-IoMT employs a confidential blockchain network to create reliable communication by verifying function depends upon its interoperable structure, allowing for regulated communication required for fixing fusion and integration difficulties to be enabled by many IoMT architectural zones.

Keywords: Healthcare, Internet of medical things (IoMT), Trust management, Trust-based blockchain model (TBM), Indirect trust inference system (ITIS)

1. Introduction

Medical care uses a collection of networked devices to build IoT networks specifically for medical care assessment in internet of things (IoT) applications. It is commonly acknowledged that individuals with chronic illnesses, such as diabetes, hypertension, or respiratory conditions, need medical, hospital, and emergency services more frequently than those with minor illnesses. Systems called the Internet of medical things (IoMT) use middleware to gather data from various sensing devices (Ratta, et al., 2021). An IoT subgroup devoted to the medical sector is called an Internet of medical things (IoMT). IoMT devices will account for 40% of the IoT market by the end of 2020. Because of the potential for IoMT devices to cut healthcare sector spending, this is expected to increase over the next years. By utilizing IoMT devices more and more, particularly for chronic conditions and telemedicine, this industry may save up to $300 billion. Investors predict that the IoMT market will grow from the current $28 billion in sales in 2017 to $135 billion in sales by 2025 (Ghubaish, et al., 2020). An Internet of Medical Things (IoMT) platform is an intelligent device that primarily consists of Sensors and electronic circuits to collect biomedical signals from patients, a processing unit to maintain the signals, a network device to transfer the biomedical data over a network, a temporary or permanent storage unit, and a data visualization platform with artificial intelligence schemes to analyze the information are all included. Interpret the information make decisions as the doctor chooses (Vishnu, et al., 2020). The blockchain protocol arranges all the data into a series of interconnected blocks, each of which can store transactions related to a single application (Dwivedi, et al., 2019).

Medical data are highly valued in the IoMT, but there are still impediments to effective information transmission in unified trust management solutions due to semantic gaps and a lack of acceptable assets or qualities. In light of this, we propose TBM for IoT spaces, where a smart contract assures budget verification and an ITIS fixes semantic flaws and enhances accurate factor estimates through network nodes and edges.

*rahul.orp@chitkara.edu.in

DOI: 10.1201/9781003490210-51

2. Related Works

CPS detection in the healthcare sector utilizes a classification algorithm and blockchain-based data sharing. To collect data for the model, sensor devices are deployed, and the Deep Belief Network (DBN) model is employed to detect intrusions (Nguyen, et al., 2021). The efficient Lightweight Integrated Blockchain ELIB approach is built as a tiered structure, and the network's whole nodes are controlled by a single public BC that distributes node management (Mohanty, et al., 2020). To use blockchain platforms to increase the efficiency of the decision-making process, using a range of measures, including the Analytical Hierarchy Process AHP, Fuzzy Analytical Hierarchy Process FAHP, and Fuzzy Technique for Order Preference by Similarity to Ideal Solution FTOPSIS, can be analyzed and used to form consistent results (Yang, et al., 2021). EBSMO Exponential Boolean Spider Monkey Optimization is used to further improve the efficiency of the encryption process (Kaushal, et al., 2022). Oaxaca-based Blinder It is designed to use Shapiro Wilk Neutrosophic Fuzzy (BO-SWNF) data analytics for remote medical care. Using the WESAD dataset, data is collected (Khalaf, et al., 2022).

3. Proposed Methodology

Trust-based blockchain model for medical applications

Each block of the chain of transactions that makes up the blockchain is a collection of all recently completed, validated transactions of blocks, and the overall structure of blockchain is depicted in Figure 51.1. Along with the cryptographical connections between each block. Each block contains all of these transactional characteristics as well as a block-by-block computation and storage of a consolidated hash code. As soon as the transaction is validated, this block is added permanently to the blockchain, and the chain continues to extend.

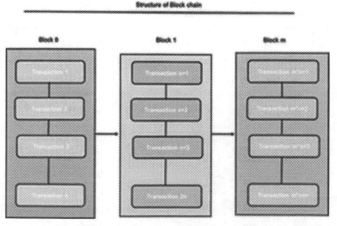

Fig. 51.1 Structure of blockchain

Several essential principles are included in the Blockchain. The Pre Hash Code is one of them. Each block must have the hash code that serves as an identification element for that block. The hashing method used to get this hash is quite complicated. To be a part of that block, the hash information for every transaction that has occurred must be full. The Merkel's Root hex value in the block header contains the transaction information. A blockchain normally has two components: blocks, which record transactions and maintain their unbroken sequence, and transactions, which are the events produced by the system's participants. The diversity of a Blockchain further prevents any authority from unilaterally endorsing transactions or establishing clear requirements for transaction acceptance. This implies that for transactions to be acknowledged, a decision must be reached by all nodes in the chain. Blockchain is a decentralized node network where the data is stored. It is an excellent piece of technology for protecting sensitive data throughout the whole system. Through the application of this technology, sensitive information may be transmitted safely and privately. It is a better decision to safely store all necessary documents in one location. It must also be a flexible, tamper-proof platform where previous records cannot be changed. All network users must confirm a new transaction before it can be added to the chain. They do this through the use of a transaction-verifying algorithm. However, the Blockchain system defines what is considered to be "valid," and these definitions might differ between systems. Thus, a majority of the parties have the authority to decide if the transaction is valid.

Let's analyze an example to get how the Blockchain operates. Let's say A needs to send B money. Every node in the network receives a broadcast of the block that serves as the description of the transaction. Following then, the transaction can be approved if there is a large enough group of miners. The transaction is posted to the Blockchain after receiving permission from miners who finish the proof of concept, and B immediately receives the money. If someone tries to make a slight change to a transaction or block that is a part of a Blockchain, the modified block will not be added or reflected in the Blockchain since the majority of network users still have access to the old Blockchain and will not accept the new block.

When data is shared and saved by all network members and a continuous record of past and present actions is created, it allows reliable communication. This technology can join many networks to provide insights about the value of customized services. Hence, Blockchain's immutability and security may be easily recognized. The most important component of a Blockchain is a block, which is a permanent database that stores all of the most recent transactions. Three technologies combine to form the blockchain. There are several security features and protocols built into the Blockchain. The most popular framework for creating Blockchain programs is Predictability.

Every transaction is verified by the development of a new block in a Blockchain and is recorded together with details about time, date, participants, and amount that is sent along. The entire Blockchain is maintained by each person that is a participant in it. The fields of use for blockchain and IoMT technology were thoroughly investigated. The application of these two revolutionary technologies, namely IoMT and Blockchain, in the healthcare industry was also studied and addressed, along with a number of potential difficulties and problems. Each transaction in the Blockchain is validated by the miner after they have solved a challenging mathematical challenge. Each following block contains a hash, which is a unique fingerprint of the previous block and validates the new block in return. The transaction is confirmed and recorded in the database when the puzzle has been solved.

4. Result and Discussion

In this section, we discuss in detail the implementation of trust based blockchain model for medical applications. Privacy, usability, and dependability are the parameters used to identify the effectiveness of the proposed methodology. Artificial Neural Network (ANN) (Ogundokun, et al., 2021), Backtracking Search Algorithm (BSA) (Moayedi, et al., 2023), Artificial Intelligence Internet of Medical Things (AiIoMt) (Awotunde, et al., 2022). The proposed methodology (TBM-IoMT) achieves the highest efficiency in medical applications than other existing methods.

Privacy

Using Ripple chain based on verified nodes and Health edge, the system's privacy or transparency is achieved which is connected to the transfers between several zones that are controlled by restricted communication. Figure 2 illustrates the proposed method's highest efficiency when compared to the existing method.

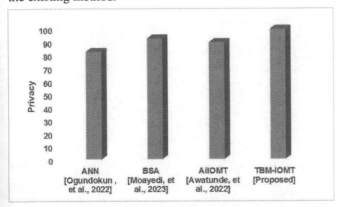

Fig. 51.2 privacy compared to the proposed method

Usability

The simplicity of use of a software program is measured by its usability and appropriateness for the customer base. Cryptography is frequently utilized by cryptocurrencies. Figure 51.3 illustrates the proposed method's highest efficiency when compared to the existing method.

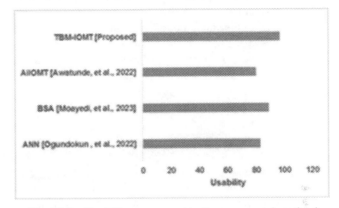

Fig. 51.3 Usability compared to the proposed method

Dependability

Once information is added to the blockchain, it cannot be deleted or changed. Its beneficial functioning may be restricted by this exchange value because frequent modifications or changes are not allowed. This requirement is significantly improved and unwanted reactions are eliminated with the proper usage of blockchain. Figure 51.4 illustrates the proposed method's highest efficiency when compared to the existing method.

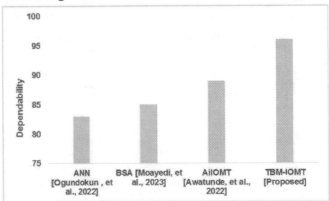

Fig. 51.4 Dependability compared to the proposed method

Compared to previous methods, the proposed method is the most effective for medical applications.

5. Conclusion

Technology based on blockchains has the power to change whole sectors. It can make the current systems very secure and unbreakable. One sector where data is growing quickly is the healthcare sector, where there is a great demand for technology like blockchain to securely store data and enable analysis and effective record-keeping for healthcare improvement. By appreciating blockchain technology, the healthcare sector has a great chance of making technological breakthroughs. The proposed method involved trust based Blockchain model for medical applications in the healthcare industry. The suggested method performs better in terms of privacy, usability, and reliability than other comparable methods. The potential for the future include investigating improved privacy-preserving techniques, interoperability requirements, and incorporation with modern innovations like AI and IoT, and additional study on scalability and efficiency enhancements.

References

1. Ratta, P., Kaur, A., Sharma, S., Shabaz, M. and Dhiman, G., 2021. Application of blockchain and internet of things in healthcare and medical sector: applications, challenges, and future perspectives. *Journal of Food Quality, 2021*, pp.1-20.
2. Ghubaish, A., Salman, T., Zolanvari, M., Unal, D., Al-Ali, A. and Jain, R., 2020. Recent advances in the internet-of-medical-things (IoMT) systems security. IEEE Internet of Things Journal, 8(11), pp.8707-8718.
3. Vishnu, S., Ramson, S.J. and Jegan, R., 2020, March. Internet of medical things (IoMT)-An overview. In 2020 5th international conference on devices, circuits, and systems (ICDCS) (pp. 101-104). IEEE.
4. Dwivedi, A.D., Malina, L., Dzurenda, P. and Srivastava, G., 2019, July. Optimized blockchain model for internet of things-based healthcare applications. In 2019 42nd international conference on telecommunications and signal processing (TSP) (pp. 135-139). IEEE
5. Singh, S., Rathore, S., Alfarraj, O., Tolba, A. and Yoon, B., 2022. A framework for privacy-preservation of IoT healthcare data using Federated Learning and blockchain technology. Future Generation Computer Systems, 129, pp.380-388.
6. Mohanty, S.N., Ramya, K.C., Rani, S.S., Gupta, D., Shankar, K., Lakshmanaprabu, S.K. and Khanna, A., 2020. An efficient Lightweight integrated Blockchain (ELIB) model for IoT security and privacy. Future Generation Computer Systems, 102, pp. 1027-1037.
7. Yang, W., Garg, S., Huang, Z. and Kang, B., 2021. A decision model for blockchain applicability into knowledge-based conversation system. Knowledge-Based Systems, 220, p.106791.
8. Kaushal, R.K., Bhardwaj, R., Kumar, N., Aljohani, A.A., Gupta, S.K., Singh, P. and Purohit, N., 2022. Using Mobile Computing to Provide a Smart and Secure Internet of Things (IoT) Framework for Medical Applications. Wireless Communications and Mobile Computing, 2022.
9. Khalaf, O.I., Natarajan, R., Mahadev, N., Christodoss, P.R., Nainan, T., Romero, C.A.T. and Abdulsahib, G.M., 2022. Blinder Oaxaca and Wilk Neutrosophic Fuzzy Set-based IoT Sensor Communication for Remote Healthcare Analysis. IEEE Access.
10. Ogundokun, R.O., Misra, S., Douglas, M., Damaševičius, R. and Maskeliūnas, R., 2022. Medical internet-of-things-based breast cancer diagnosis using hyperparameter-optimized. Future Internet, 14(5), p. 153.
11. Moayedi, H., Canatalay, P.J., Ahmadi Dehrashid, A., Cifci, M.A., Salari, M. and Le, B.N., 2023. Multilayer Perceptron and Their Comparison with Two Nature-Inspired Hybrid Techniques of Biogeography-Based Optimization (BBO) and Backtracking Search Algorithm (BSA) for Assessment of Landslide Susceptibility. Land, 12(1), p. 242.
12. Awotunde, J.B., Folorunso, S.O., Ajagbe, S.A., Garg, J. and Ajamu, G.J., 2022. AiIoMT: IoMT-based system-enabled artificial intelligence for enhanced smart healthcare systems. Machine Learning for Critical Internet of Medical Things: Applications and Use Cases, pp. 229-254.

Note: All the figures in this chapter were made by the authors.

Advancing Sustainable Science and Technology for a Resilient Future – Sai Kiran Oruganti et al. (eds)
© 2024 Taylor & Francis Group, London, ISBN 978-1-032-79020-6

52

A Conceptual Framework Integrating Blockchain and Supply Chain for Sustainable Development

Ankit Punia*

Chitkara University Institute of Engineering and Technology, Chitkara University, Punjab, India

Abstract: Today, the management of supply chains (SCs) uses block chain technology as an emerging information technology instrument for sustainable development. Despite its benefits in preserving connectedness and dependability among SC associates, its research is rather infrequent in the field on SC cooperation and sustainability management. This research looks at how the integration of block chain on SC operations may affect (raise or decrease) the effectiveness and expansion of SC integrations, which in turn may have an impact on SC performance results. With the help of SC specialists from diverse sectors, this report particularly experimentally verifies a measurement as well as structural equation model. The results demonstrate that the attributes of block chain technology—information integrity, information sharing, and smart contracts—have marginally negative consequences on integration effectiveness and considerable beneficial effects on the integration development. Although integration effectiveness has a negative impact on business performance, integration development has a beneficial impact.

Keywords: Supply Chains (SC), Block chain, Integration, Business, Smart contracts

1. Introduction

An effective and strategic cooperation between customers-among the fundamental requirements throughout supply chain management is that company and its suppliers. Collaboration in the supply chain (SC) involves exchanging crucial data gleaned from market and global network activities, then making quick decisions together based on that data. The shared advantages and hazards of two trade partners can be increased by cooperative efforts to balance supply and demand. Information technology (IT) in particular, including online services, bar codes, and radio-frequency identification, has been vital to the efficient running of SC cooperation(Parmentola, et al., 2022).

Data records is available to all users. BlockChain Technology provides features such as data immutability, information sharing, and support for smart contracts connection and dependability, which are necessary for SC collaborations. Yet, users can share virtually any info special regards block chain tech's decentralization, security, and intelligent application.

BCT offers benefits including knowledge immutability, accessibility, and smart contracts that enhance connection and dependability, all of which are necessary for SC cooperation (Kim and shin, 2019).

In several supply chain businesses, Supply chain is increasingly a necessary need and a key requirement in many industries, including the agree-food sector, pharmaceutical and medical products, and large value consumables. Differentiate. It is simple to lose or change luxury and valuable objects whose provenance would ordinarily depend on paper certifications and invoices (Saberi, et al., 2019, Gupta, et al., 2022). According to some recent studies, block chain may help with sustainable SC management (SCM), which can solve issues with unethical conduct in SCs like the ones mentioned above as well as environmental problems. They contend that by examining block chain implementation to support SC sustainability in underdeveloped regions, we may get fresh perspectives that are not feasible when looking at SCs in developed states (Kshetri, et al., 2021, Saxena, et al., 2021).

*Corresponding author: novakannan@gmail.com

[1]vartikakul@gmail.com, [2]hirechetan@gmail.com, [3]karanjagdale42@gmail.com

DOI: 10.1201/9781003490210-52

2. Related Works

Govindan, 2022 examines the challenges associated with using block chain technology to the re manufacturing industry. A framework has been put out and tested at a Danish firm that re manufactures equipment. The most successful and beneficial obstacles among the common challenges have been identified using a multi-criteria decision-making process. Using a Flexible Delphi technique, confirmed by professionals chosen from the food manufacturing and supply organizations, a further quantitative prioritization of these obstacles is created. By tracking, these solutions let the production or manufacturing company create effective take-back procedures. Nevertheless, the data exchanged in the remanufacturing value chain via these technologies between partners is unreliable. While block chain technology has many benefits, it might be challenging for practitioners to incorporate it into the remanufacturing environment (Okorie, et al., 2021). The Synthetic Furry Algebraic Hierarchical Process (SF-AHP) technique will be used to describe feasible actions, prioritize any risk that may develop in this integration process, and construct the structure to support the incorporation of block chain - based to establish an environmentally friendly tea supply chain. The design that incorporates block chain into all operations is then demonstrated. The whole tea supply chain is covered by the suggested design, which takes into account every aspect of sustainability (Mangla, et al., 2022).

3. Proposed Methodology

Data Collection

Two main English-Korean bilinguals translated the original document into Korean, and then they reversed the translation into English to ensure that it was accurate and equivalent. Seven academics and professionals interested in block chain-enabled SC activities created the initial screening of the Korean instrument. Slight adjustments were made to account for Korean business procedures and terms. A top national market research company managed the final questionnaire in the following stages: listing sample population, selection, and research organization.

Measurement Item Generation

Securing the construct's correctness is the goal of the assessment item production process, which is predicated on research reviews discussions including educational and professional specialists. For a preliminary listing of probable observed variables, a thorough assessment of existing research is advised in order to guarantee that the Measuring elements encompass a construct's scope. According to

research reviews, the operationalization of the SC cooperation and skill level as well as the degree of BC technology applicability for a focal firm (Appendix A). These ranges of better integrity, data, and block chain based ' relevance was used to describe BC technologies use. The supervisors was requested to rank each item on a seven-point Liker scale (1 = strongly disagree, 2 = disagree, 3 = somewhat disagree, 4 = neutral, 5 = somewhat agree, 6 = agree, and 7 = strongly agree) to the extent that they agreed or disagreed with it. The elements gauge how much a focus firm (reply firm) perceives Framework of BCT, SC cooperation, and SC output. Despite its benefits in maintaining connectedness and dependability among SC partners, this study is rather uncommon in the study of environment performance management and SC collaboration.

Hypotheses Information Relationship between Transparency and SC Efficiency and Development

Each member in the block chain network has an imitate of the document continuously and updated similarly with the most recent data. Unclean trying to undermine the validity is easily detectable and recognizable from the whole the sequence any modifications Any that this users can access content.. It is anticipated that information transparency afforded by BCT would increase SC partner collaboration efficiency and encourage the expansion of existing partnerships. Based on the foregoing, this study puts up the following theories: The efficiency of SC partnerships is increased by BCT-enabled information openness, according to hypothesis 1a. The expansion of SC partnerships is positively impacted by BCT-enabled information openness, according to hypothesis 1b.

Immutability of Information and Efficient and SC Relationship Development

Moreover, BCT provides the immutability of any data generated and transferred inside the network. Due to this so-called "data immutability,"without the approval of other network access, data and information on the block chain network cannot be changed or erased. Based on the foregoing, this study puts up the following theories: Information immutability afforded by BCT increases the effectiveness of SC partnerships, according to hypothesis 2a. Information immutability afforded by BCT contributes to the expansion of SC partnerships, according to hypothesis 2b.

Company Providing and SC Cooperation for Growth and Efficiency

The phrase "smart contract" refers to a computerised that transfer approach carries out a contract's provisions. These contracts make it possible to store, share, and enforce contractual terms on a block chain. Despite these challenges, ongoing research, technological advancements, and the emergence of alternative consensus mechanisms like Proof of Stake (PoS) are addressing some of these difficulties and pushing the boundaries of blockchain technology.

The consequence impact of block chain technology on the success and growth of SC collaboration is established according to the following hypothesis. BCT-enabled smart contracts increase the effectiveness of SC partnerships, according to hypothesis 3a. Hypothesis 3b: SC partnership growth is positively impacted by BCT-enabled smart contracts .Convergent and discriminant validity results

Efficiency, Performance, and Development of the SC Partnership

It is believed that effective partnerships that make use of established systems like VMI or CPFR would greatly minimise the time and expenses associated with the trade partner communication process.Hypothesis 4a: SC financial performance will be negatively impacted by BCT-driven SC partnership efficiency. BCT-driven SC partnership efficiency is hypothesised to have a detrimental impact on SC operational performance in hypothesis 4b.The quick expansion of SC collaborations can result in both financial and operational performances through the strategic use of BCT. BCT-driven SC partnership expansion is hypothesised to have a favourable impact in Hypothesis 5a on the business results of SC.Hypothesis 5b. SC operational performance will improve as a result of BCT-driven SC partnership growth.

Measurement Model Assessment

AMOS 24.0 software was utilised to evaluate and validate the survey's validity and reliability prior to conducting the hypothesis testing. Reliability, convergent validity, unit-A correlation analysis is used to evaluate the dimensionality and discriminant validity of the data (CFA). The magnitude of the correlation coefficients for the additional latent variables was evaluated to the sum of squares of an AVE (Table 52.1) in order to evaluate discriminant validity. Except

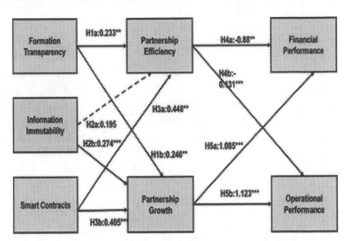

Fig. 52.1 Results of the tests for the hypotheses (*** path significant at 0.001; 0.05** significant at 0.05)

for the relationships between partnership development and financial performance, relationship growth and information atomicity, and success of the companyall correlations met this requirement. These sets of hypotheses were notably more different from one another, according to a chi-square discriminant validity test (p 0.001). Furthermore, for all three sets of constructs, a second CFA with two distinct constructs outperformed a CFA with two integrated constructs in terms of fit index. For instance, a CFA with two distinct constructs for operational success and partnership growth (2/df = 2.039; CFI = 0.99; TLI=0.99; RMSEA=0.05).

Structural Equation Model Result: The suggested model's validity and reliability were first evaluated usinga model of structural equations (SEM), which was then utilised to study the sequential link between BCT, SC partnership, and SC performance. With normed-2 = 2.07, CFI = 0.97, TLI = 0.97, and RMSEA = 0.06, the model demonstrated satisfactory fit. The validity of the model is implied by a good match between the model and the data. Figure 52.1 displays the results of the created SEM analysis. These results generally support the notion that more BCT use promotes improved growth in cooperation ability. Both Hypothesis 1 and Hypothesis

Table 52.1 Convergent and discriminant validity results

	CR	AVE	TRNS	IMM	SMRT	PEFF	PGRW	FPERF	OPERF
TRNS	1.924	1.802	(1.896)						
IMM	1.949	1.823	1.881	(1.907)					
SMRT	1.941	1.830	01.838	1.878	(1.917)				
PEFF	1.943	1.803	1.711	1.719	1.720	(1.897)			
PGRW	1.936	1.781	1.849	1.901	1.853	1.788	(1.886)		
FPERF	1.969	1.886	1.742	1.720	1.746	1.671	1.864	(1.942)	
OPERF	1.951	1.827	1.816	1.812	1.826	1.688	1.922	1.929	(1.900)

3 are significant. Proves that a focal firm is more likely to anticipate greater levels of partnership efficiency and growth the more transparent and applicable the information is (H1a: = 0.233; p 0.05; H1b: = 0.246; p 0.001; H3a: = 0.448; p 0.05; H3b: = 0.405; p 0.001).Yet, Hypothesis 2 shows a contradictory result. Enhanced data integrity is expected to increase the union in economic, legal, and practical terms standpoint (H3b: = 0.274; p 0.001), but it may not always result in more effective allocation of both financial and other resources (H2a: = 0.195; p 0.135). These results generally support the notion that more BCT use promotes improved growth in cooperation ability, transparent and applicable the information is (H1a: = 0.233; p 0.05; H1b: = 0.246; p 0.001; H3a: = 0.448; p 0.05; H3b: = 0.405; p 0.001). Yet, Hypothesis 2 shows contradictory result. Enhanced data integrity is expected to increase the union in economic, legal, and practical terms standpoint (H3b: = 0.274; p 0.001), but it may not always result in more effective allocation of both financial and other resources (H2a: = 0.195; p 0.135).

4. Conclusion

This study looked into how BCT application affected SC performance and partnership. Long-term SC performance improvement in the domain of SC cooperation is likewise anticipated to be significantly aided by BCT.Yet, in the medium run, technological tools like BCT seldom ever completely replace the task carried out by human interaction. When faster decision making is required because of a SC exception, BCT technology qualities might become disruptive if SC collaboration procedures use pricey BCT hardware. Prospect studies should thus examine additional factors that contribute to partnership effectiveness and growth management in order to improve coordination or inter-organizational competencies. Our understanding of the circumstances in which operational and financial performance may advance may be aided by this information. Blockchain technology has the potential to revolutionize supply chain management by improving transparency, efficiency, and security.

References

1. Parmentola, A., Petrillo, A., Tutore, I. and De Felice, F., 2022. Is blockchain able to enhance environmental sustainability? A systematic review and research agenda from the perspective of Sustainable Development Goals (SDGs). Business Strategy and the Environment, 31(1), pp.194-217.

2. Kim, J.S. and Shin, N., 2019. The impact of blockchain technology application on supply chain partnership and performance. Sustainability, 11(21), p.6181.

3. Saberi, S., Kouhizadeh, M., Sarkis, J. and Shen, L., 2019. Blockchain technology and its relationships to sustainable supply chain management. International Journal of Production Research, 57(7), pp.2117-2135.

5. Kshetri, N., 2021. Blockchain and sustainable supply chain management in developing countries. International Journal of Information Management, 60, p.102376.

6. S. K. Gupta, B. Pattnaik, V. Agrawal, R. S. K. Boddu, A. Srivastava and B. Hazela, "Malware Detection Using Genetic Cascaded Support Vector Machine Classifier in Internet of Things," 2022 Second International Conference on Computer Science, Engineering and Applications (ICCSEA), 2022, pp. 1-6, doi: 10.1109/ICCSEA54677.2022.9936404.

7. Saxena, S., Yagyasen, D., Saranya, C.N., Boddu, R.S.K., Sharma, A.K. and Gupta, S.K., 2021, October.Hybrid Cloud Computing for Data Security System. In 2021 International Conference on Advancements in Electrical, Electronics, Communication, Computing and Automation (ICAECA) (pp. 1-8). IEEE.

8. Govindan, K., 2022. Tunneling the barriers of blockchain technology in remanufacturing for achieving sustainable development goals: A circular manufacturing perspective. Business Strategy and the Environment.

9. Okorie, O., Russell, J., Jin, Y., Turner, C., Wang, Y. and Charnley, F., 2022. Removing barriers to Blockchain use in circular food supply chains: Practitioner views on achieving operational effectiveness. *Cleaner Logistics and Supply Chain*, 5, p.100087.

10. Mangla, S.K., Kazançoğlu, Y., Yıldızbaşı, A., Öztürk, C. and Çalık, A., 2022. A conceptual framework for blockchain-based sustainable supply chain and evaluating implementation barriers: A case of thtea supply chain. Business Strategy and the Environment.

Note: All the figure and table in this chapter were made by the authors.

Advancing Sustainable Science and Technology for a Resilient Future – Sai Kiran Oruganti et al. (eds)
© 2024 Taylor & Francis Group, London, ISBN 978-1-032-79020-6

53

Agricultural Data Integrity Using Swarm Optimization Based Blockchain Technology

Komal Parashar*

Chitkara University Institute of Engineering and Technology, Chitkara University, Punjab, India

Abstract: The Internet of Things (IoT) has given smart farming and agriculture a new dimension due to its inherent capacity to delegate user-created activities or to transmit sensor-collected agricultural information to producers for analysis on a variety of terminal devices. The widespread use of precision agricultural technologies has led to a significant increase in the sector's data requirements. Researchers are concerned about the reliability of the data because it is common for agricultural data to be disorganized, especially data from combined yield monitors. The blockchain (BC) assures that data have not been illegally altered or, at the very least, keeps track of what changes have been done by certain individuals, which might be a potential solution to the issue of ambiguous data quality caused by earlier data tampering. As a result, this study suggests a novel BC technology-based swarm optimization (SO) for a secure fish farm platform to ensure agricultural data integrity. Fish farmers will have a secure location to store their large, immutable agricultural data thanks to the established technology. Several fish farm processes are carried out automatically using smart contracts to reduce the potential for mistakes. A proof of concept that combines a conventional fish farm system with the hyper ledger Fabric BC is constructed on top of the suggested architecture. Through a series of tests using different measures, the effectiveness and usefulness of the recommended platform are proven.

Keywords: Agriculture, Internet of things (IoT), Data integrity, Blockchain (BC), Wwarm optimization

1. Introduction

The world's fish sector is growing quickly, and during the previous 50 years, per capita seafood consumption has doubled. Aquaculture, commonly known as fish farming, now provides a livelihood for more than one in ten people worldwide. A complex system that is in charge of distributing agricultural products to the marketplace is known as the agricultural BC. Agricultural commercial resources play a crucial role in ensuring that agricultural goods fulfill consumer demand, retain their quality, and are safe to consume. Commercial agricultural resources might be said to be widely dispersed. In industrialized areas, agricultural business resources are heavily influenced by interests, and there is a clear excess in resource investment. The coverage rate in distant and undeveloped areas, however, is very low, and agricultural businesses virtually ever spontaneously fulfill the demand for agricultural goods in these areas, displaying egregious social inconsistencies (Leng, et al., 2018). As

part of the present agricultural expansion and reform, new techniques and innovations are needed for the agriculture business to become more transparent and accountable. One of the new instruments is BC technology. The BC system offers a secure information structure that makes it a possible number of unreliable parties, in contrast to traditional centralized and monopolistic farm management systems (Lin, et al., 2020). The role of BC in operational traceability is poorly understood in some industries, including agriculture, public services, and others. Determining more about BC applications in operations and how businesses utilize it to develop and acquire commercial value, especially in agriculture (Alobid, et al., 2022). In this study, we presented swarm optimization-based BC technology for the integrity of agricultural data. The remaining segments are as described in the following: Part 2 outlines the relevant works, Part 3 outlines the proposed methods, Part 4 explains the results of the experiment, and Part 5 is the conclusion section.

*ankit.punia.orp@chitkara.edu.in

DOI: 10.1201/9781003490210-53

2. Related Works

Zhao, et al., 2019 reviewed the most current developments in BC technology, as well as its primary applications in the agri-food value chain and difficulties, using comprehensive literature network analysis. The result suggests that BC technology has been utilized, together with cutting-edge information and communications technology (ICT) & IoT, for the enhancement of agri-food value chain management. Si, et al., 2019 suggested minimal BC-based IoT information-sharing security architecture. The findings demonstrate the viability of the framework and the viability of verifying the system's location data for secure storage systems. Rahman, et al., 2020 surveyed the literature and discovered that the main obstacles to using BC for smart agriculture are its performance, scalability, cost, and throughput. The results show that when it comes to average transaction prices, Ethereum is the most costly. Hang, et al., 2020 suggested a BC-based fish farm infrastructure to guarantee the accuracy of agricultural data. The platform's architecture intends to provide fish producer's safe storage for keeping the vast volumes of unalterable agricultural data. Lin, et al., 2017 analyzed ICT-based technical principles related to blockchain. Also, a blockchain-based model ICT e-agriculture system is suggested for application at the local and regional levels. AgriBlockIoT, a distributed BC-based accountability system for controlling the agri-food supply chain, was developed by Caro, et al., in 2018 and allows IoT devices that create and consume digital data to be connected smoothly across the chain. Application of blockchain ethereum for IoT sensor data integrity for irrigation in precision agriculture (Sumarudin, et al., 2022). The result of an experiment demonstrates that an Ethereum-based BC network utilizing consensus proof-of-work can guarantee the reliability of irrigation data sensors.

3. Proposed Overview

The interaction between the blockchain network and the traditional fish farm system is shown in detail in Fig. 53.1.

Inside a single fish tank, farm sensors, and actuators communicate, and when more fish tanks are added, a whole fish farming environment is created. The different IoT data from the fish farming environment. Data from several fish tanks are gathered in time series via the data collecting module. Parameters for controlling the farm's actuators are generated by the control module, which then makes optimal adjustments. The research discusses the issue of the accuracy and integrity of agricultural data, especially data from integrated yield monitors, which often display disarray and the possibility of manipulation. In the context of precision agricultural technology, researchers are worried about the reliability and integrity of agricultural data. **Swarm**

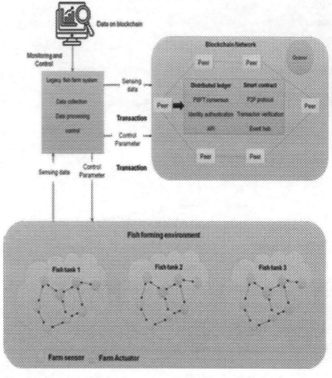

Fig. 53.1 The suggested blockchain-based fish farm platform's interaction diagram

optimization: SO is a kind of swarm intelligence approach that is embedded in evolutionary computation, one of the three pillars of CI. This algorithm's design was inspired by the navigational and foraging behaviors of groups of animals, such as schools of fish and flocks of birds. The SO algorithm is similar in concept to a search algorithm in that it employs a huge population of people, known as particles, to seek the best answer in a multi-dimensional search space. Particles keep exploring hyperspace to find the best solution, and they do so according to these rules: To keep track of the optimum placement for each particle so far using a fitness function that measures how near each particle is to the ideal solution; Observe and document their present movement's directional bias and speed. Based on the above, each particle changes its direction and speed, as illustrated in Equation (1):

$$Y_j(s + 1) = Y_j(s) + Z_j(s + 1) \tag{1}$$

where $Y_j(s + 1)$ and $Y_j(s)$ are the vectors that represent the future and present locations of each particle, respectively, and $Z_j(s + 1)$ is the vector that represents the speed and direction of each particle's future motion. These velocities are adjusted stochastically at each iteration using both the particles' local best locations and the global best position across all particles in the past, as illustrated in Equation (2).

$$Z_j(s + 1) = u.Z_j(s) + \varphi_1.q_1.(O_{j,\text{best}} - Y_j(s)$$
$$+ \varphi_2.q_2.(O_{\text{glob,best}} - Y_j(s)) \tag{2}$$

where q_1 and q_2 are weight factors between $[0, 1]$, φ_1 and φ_2 are two positive values (acceleration constants), $O_{j,best}$ and $O_{glob,best}$ is the local and overall optimal particle locations, respectively. We use the canonical PSO algorithm unless the weight factor w (inertia weight) equals one, in which case we use the original PSO algorithm. The distance traveled by each particle during each iteration is updated by a combination of three variables, as shown in Equation (2). The inertia component $u.Z_j(s)$: seeks to keep each particle moving in the same direction and at the same speed; The cognitive/individual component $\varphi_1.q_1.(O_{j,best} - Y_j(s))$: describes the separation between the present location of each particle and their respective optimal placement; The social component $\varphi_2.q_2.(O_{glob,best} - Y_j(s))$: determines the separation between the particle's present location and the ideal location discovered by the whole swarm.

4. Result and Discussion

System performance analysis: In this part, we give a complete test to validate the proposed platform's performance, making use of some performance indicators and formulas when necessary. Users may evaluate the efficiency of a given blockchain implementation against a benchmark set by running simulations using the open-source benchmark simulation program Hyper ledger Caliper. The throughput of a blockchain system is measured in transactions per second (tps). Table 53.1 displays the set values for the experiment's setup settings.

The block size of Hyper ledger Fabric's prototype network is set to a maximum of 10 transactions by design. The mean amount of transactions that occurred each second during the whole 60-second trial is shown in Fig. 53.2.

The overall average throughput rises as the block size grows, as the graph clearly shows. Because of its high throughput in general, a data block of 512 activities will be chosen for the following tests. Read throughput and transaction throughput are two more buckets into which the throughput may be sorted. In contrast to transaction throughput, which measures the amount of valid transactions processed by the blockchain

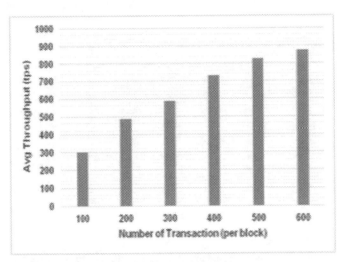

Fig. 53.2 A blocks size has an impact on the average transaction throughput

in a given time frame, read throughput measures the amount of operations that occur within a certain time frame. You may find the formulas for these two measures in Eq. (3) and (4). Importantly, the sum of legitimate transactions must be subtracted from the amount of invalid ones.

$$\text{Read Throughput} = \frac{\text{Number of total read operations}}{\text{total time}} \quad (3)$$

$$\text{Transaction Throughout} = \frac{\text{Number of total valid transaction}}{\text{al time}}$$
$$(4)$$

The experimental findings are shown in Fig. 53.3; the average read throughput was determined by changing the transmission rate from 500 tps to 3000 tps over 60 seconds. The transmit rate of 2500 tps causes the read throughput to peak at 2314 tps. As a result, the best-read throughput transmits rate is determined to be 2500 tps.

Similarly, as the transaction throughput falls constantly with the rising of send rate after 1100 tps (as shown in Fig. 53.4), this value has been determined as the ideal send frequency for transaction throughput.

Table 53.1 Configuration for evaluating BC network performance

Experimental Conditions	Values
Time required to construct a block	2 secs
The maximum number of transactions allowed per block	512
The maximum amount of bytes of a block	99 MB
The percentage of customers that terminate transactions	5
Number of channels	1
Number of peers	4

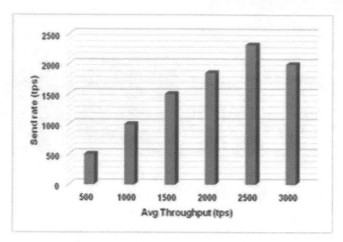

Fig. 53.3 Impact of transmit rate on read throughput on a blockchain network

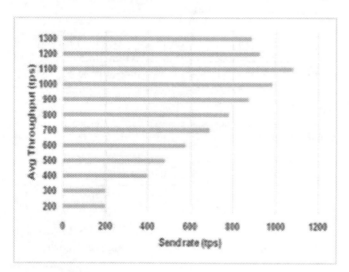

Fig. 53.4 Impact of transaction rate on read throughput on a blockchain network

5. Conclusion

This research aims to provide a workable technique to construct practical BC-based applications, to change agricultural sector advancements, particularly in fish farming. This study suggests a novel BC technology-based swarm optimization (SO) for a secure fish farm platform to ensure agricultural data integrity. The potential for using BC technology in environmental data and agricultural intervention programs may be clarified by further investigation into the use of this technology in real-world agriculture case studies or already-existing agricultural environmental monitoring systems, like our prototype. The intended solution delivers some enhancements, off-chain storage, including scalability, high throughput, and anonymity, when used with a tailored implementation of blockchain. By incorporating the suggested remedy with current agricultural environmental monitoring systems, it may be further investigated. This

connection would make it possible to gather and analyze environmental data alongside data on fish farming, giving a complete picture of how environmental conditions affect fish farm operations. Data transparency and integrity may be guaranteed throughout the whole data collecting and analysis process by connecting blockchain with these platforms.

References

1. Leng, K., Bi, Y., Jing, L., Fu, H.C. and Van Nieuwenhuyse, I., 2018. Research on agricultural supply chain systems with double chain architecture based on blockchain technology. Future Generation Computer Systems, 86, pp.641-649.
2. Lin, W., Huang, X., Fang, H., Wang, V., Hua, Y., Wang, J., Yin, H., Yi, D. and Yau, L., 2020. Blockchain technology in current agricultural systems: from techniques to applications. IEEE Access, 8, pp.143920-143937.Chuang, H. (2020). The impacts of institutional ownership on stock returns. Empir. Econ. 58(2):507 533.
3. Alobid, M., Abujudeh, S. and Szűcs, I., 2022. The role of blockchain in revolutionizing the agricultural sector. Sustainability, 14(7), p.4313. Dasgupta, A., Prat, A. and Verardo, M. (2011). Institutional trade persistence and long-term equity returns. J. Finance. 66(2):635 653.
4. Zhao, G., Liu, S., Lopez, C., Lu, H., Elgueta, S., Chen, H. and Boshkoska, B.M., 2019. Blockchain technology in agri-food value chain management: A synthesis of applications, challenges, and future research directions. Computers in industry, 109, pp.83-99.
5. Si, H., Sun, C., Li, Y., Qiao, H. and Shi, L., 2019. IoT information sharing security mechanism based on blockchain technology. Future generation computer systems, 101, pp.1028-1040.
6. Rahman, M.U., Baiardi, F. and Ricci, L., 2020, December. Blockchain smart contract for scalable data sharing in IoT: a case study of smart agriculture. In 2020 IEEE Global Conference on Artificial Intelligence and Internet of Things (GCAIoT) (pp. 1-7). IEEE.
7. Hang, L., Ullah, I. and Kim, D.H., 2020. A secure fish farm platform based on blockchain for agriculture data integrity. Computers and Electronics in Agriculture, 170, p.105251.
8. Lin, Y.P., Petway, J.R., Anthony, J., Mukhtar, H., Liao, S.W., Chou, C.F. and Ho, Y.F., 2017. Blockchain: The evolutionary next step for ICT e-agriculture. Environments, 4(3), p.50.
9. Caro, M.P., Ali, M.S., Vecchio, M. and Giaffreda, R., 2018, May. Blockchain-based traceability in Agri-Food supply chain management: A practical implementation. In 2018 IoT Vertical and Topical Summit on Agriculture-Tuscany (IOT Tuscany) (pp. 1-4). IEEE.
10. Sumarudin, A., Putra, W.P., Puspaningrum, A., Suheryadi, A., Anam, I.S., Yani, M. and Hanif, I., 2022, December. Implementation of IoT Sensor Data Integrity for Irrigation in Precision Agriculture Using Blockchain Ethereum. In 2022 5th International Seminar on Research of Information Technology and Intelligent Systems (ISRITI) (pp. 29-33). IEEE.

Note: All the figures and table in this chapter were made by the authors.

Advancing Sustainable Science and Technology for a Resilient Future – Sai Kiran Oruganti et al. (eds)
© 2024 Taylor & Francis Group, London, ISBN 978-1-032-79020-6

54

Machine Learning Based big Data Analytics Framework for Healthcare Applications

Aashim Dhawan*

Chitkara University Institute of Engineering and Technology,
Chitkara University, Punjab, India

Abstract: With the expansion of big data (BD) in the health and biomedical sectors, intelligent prediction of medical data promotes early disease identification, patient treatment, and social service. Although when medical information is poor quality or inadequate, analysis accuracy suffers. Also, distinct regional diseases have distinctive traits that vary between regions, which could make it harder to forecast when a disease would spread. In this study, we simplify the machine learning (ML) method termed as a multi-objective firefly optimized improved support vector machine (MFO-ISVM) for accurate chronic illness prediction in places with high disease incidence. We test our suggested strategy using data from actual hospitals. The raw data is first pre-processed using a z-score normalization method. We are aware of no work in the field of healthcare BD analytics that specifically addressed both data types. The suggested MFO-ISVM method has the highest efficiency when compared to other algorithms.

Keywords: Healthcare, Big data (BD), Machine learning (ML), z-score normalization, Multi-objective firefly optimized improved support vector machine (MFO-ISVM)

1. Introduction

Both organized and unstructured data are used by medical institutions. Structured data is comprehensive, freeform, and available in a wide range of formats. BD, or unstructured data, on the other hand, does not fit into the conventional data processing framework. BD is a large collection of data sets that can't be managed, stored, or analyzed using traditional technology. It is still preserved in a file but is not looked at. Due to the challenges associated with accessing and evaluating such data in the lack of a well-defined schema, specialist technologies, and methods are required to add value to it (Batko, et al., 2022). ML applications in the field of healthcare have generated a lot of attention. It is possible to use ML algorithms to enhance healthcare thanks to BD and increased computer capacity (Alanazi, et al., 2022). The healthcare industry is being significantly impacted by ML applications. Countries that are having trouble with

overpopulated health care and a lack of doctors can benefit a lot from ML. ML-based methods help find the very first signs of a disease outbreak. Using ML in the healthcare industry has the potential to revolutionize this industry. It allows healthcare professionals to devote their attention to patients rather than looking for or inputting information (Javaid, et al., 2022). BD technology is now being used and developed in the medical business as a result of the increasing acceleration of medical data collection brought on by the growing integration of information and medical technologies (Dhiman, et al., 2022). We proposed the ML method termed a multi-objective firefly-optimized improved support vector machine (MFO-ISVM). To successfully integrate machine learning in healthcare analytics while respecting patient safety, privacy, and ethical norms, it is necessary for data scientists, healthcare practitioners, policymakers, and regulatory authorities to work together.

*Corresponding author: aashim.dhawan.orp@chitkara.edu.in

DOI: 10.1201/9781003490210-54

2. Related Works

Char, et al., 2020 examined a methodical procedure for locating ML-HCA ethical issues, beginning with a conceptual model of the pipeline of ML-HCA conception, development, and implementation as well as the parallel pipeline of review and supervision activities at each level. The possible benefits, potential hazards, and consequent benefit-to-harm ratio will vary along with the assessed sensitivity, specificity, and determinability. Li, et al., 2021 studied the use of ML methods for BD analysis in the healthcare industry. The strengths and shortcomings of the currently used approaches as well as numerous research difficulties are also emphasized. This research will help government organizations and healthcare professionals stay up to date on the most recent developments in ML-based BD analytics for smart healthcare. Galetsi, et al., 2020 looked at managing high-risk/expensive patients, using BD, Hadoop, and cloud computing in genomics, and creating mobile apps for disease management. A significant study focuses on enhancing illness prediction by studying patients' medical records using sophisticated analysis. Using the task-technology fit paradigm and the technology acceptance model, Shahbaz, et al., 2019 investigated the process of BD analytics adoption in healthcare enterprises. The findings may be used by healthcare organizations to increase psychological empowerment among staff members and to better understand how BD analytics are utilized. Nti, et al., 2022 evaluated a mini-literature evaluation of ML in BDA; a total of 1512 published publications were found. Krishnamoorthi, et al., 2022 investigated the potential applications of machine learning- and BD analytics in the treatment of diabetes. The data analysis shows that the ML-based framework suggested is capable of achieving 86 percent.

3. Proposed Method

Data Set

The collection of medical data and the development of digital health records have produced a strong database for the use of ML in the medical industry. Important information and guidelines for illness diagnosis, therapy, and medical research can only be discovered through the investigation and consideration of enormous volumes of medical data. Some data, however, seem to be unrelated. Data that does not implicate personal privacy when it appears alone may be adequate for analysis after connecting with personal information.

Certain information and patterns that couldn't be found may be disclosed and leaked to unreliable third parties by doing data analysis on medical data. To restrict the mining of sensitive information from big data and to safeguard privacy,

it is required to analyze and process data. The confidentially of patients cannot be inferred from the information, although massive medical data have undergone some cleaning processes. Nevertheless, after mining a significant portion of the data, some sensitive data may have escaped via the processing output.

Data Preprocessing

Clarification of feature statistics is required before applying the pre-processing techniques to choose the one that is best for the dataset. Center, dispersion, lowest, maximum, and proportion of missing data make up feature statistics.

Using Mean Imputation to Replace Missing Data

Based on the distribution of the data, this substitutes the missing value with the mean, average sample, or model. When there are numerous missing values, they are all imputed with the same value, the mean, which changes the distribution's form; the feature with a percentage of more than 50% will be removed from the dataset.

Z-score Normalization

This process is also known as "Standardization." The technique seeks to scale data properties between zero and one. A is standardized to S' using the following equation (1) if B is the mean of the value of attribute A and σ_B is the standard deviation, original value s of B.

$$S' = \frac{s - \overline{B}}{\sigma_B} \tag{1}$$

Provide a mean of zero and a standard deviation of one by applying this standardization to the feature data. With WEKA's "Standardization" filer, the "Age" feature's range is from -3.342 to 7.09. Moreover, the "Whole Insurance Period" option has a range of -1.017 to 4.48.

Multi-objective Firefly Optimized Improved Support Vector Machine

In general, the MOO issue may be described as an optimization problem with n = 1, ..., N concurrently minimized objective functions. As the range of the choice variables is often constrained in most real engineering optimization problems, additional constraints are frequently added to the original set of constraints to guarantee that the solution is physically feasible. MV-SVM can be computationally more demanding than binary SVMs, as it involves training and optimizing multiple hyperplanes. Additionally, selecting appropriate kernel functions and tuning hyperparameters can be challenging in MV-SVM, especially when dealing with high-dimensional data or a large number of classes. These issues are known as constrained multi-objective optimization problems

(CMOPs), and they may be expressed mathematically in equation (2) shown below.

$$\min e_n(Y_j)\ n = 1, ..., N \wedge j = 1, ... M_o$$

$$s.t$$

$$hl(Y_j) \geq 0, l = 1, ..., L$$
$$gk\ (Y_j) = 0, k = 1, ..., K$$
$$Y_{j,i} = [y_i^{Lower}.y_i^{Upper}], i = 1, ..., m) \qquad (1)$$

where N is the total number of optimization-related goal functions, L is the total number of inequality constraints, while $hl(Y_j)$ is the l-th inequality constraint, K is the total number of equality constraints, and $gk\ (Y_j)$ is the k-th equality constraint, y_i^{Lower}, and y_i^{Upper} stands for the search space›s lower and upper boundaries for the decision variables.

Firefly algorithm (FA) contains three guiding principles that characterize firefly behavior: (i) Since they are all gender neutral, fireflies are drawn to one another; (ii) The brighter of any two fireflies will attract the less brilliant one because attraction is inversely proportional to brightness. As there is more space between the two fireflies, the appeal wanes, however (iii) if there isn't another firefly that is brighter than the present one, it will attract randomly. Firefly brightness is related to the fitness function. Based on Cartesian distance, the group of firefly i to another, brighter firefly j may be shown in equation (3).

$$y_j = y_j + \beta_p \times f^{-\gamma q^2_{ji}} \times (y_{i\_} y_j) + \alpha \times (\text{rand} - 0.5) \qquad (3)$$

where β_p denotes the starting attractiveness, and γ is the absorption coefficient, where α is a constant randomization parameter between 0 and 1 that represents the noise of the environment and can be used to create more solutions. rand denotes random number generate from a consistent allocation between 0 and 1 and adjusted to range between [–0.5,0.5]. Lastly, r shows how far apart any two fireflies are (j, i).

$$q_{ji=\|y_j - y_i\|} \qquad (4)$$

Where y_i indicates the position of firefly i.

SVM has been a good method for statistical analysis and categorization because it is based on solid science and can show several important properties that other methods have trouble with. In SVM, the data is split into multiple classes by a hyperplane, which provides the maximum dimensional margin and minimizes the empirical performance of the classifier at the same time. It is also called a maximum margin classifier because of this. The SVM classifier is seen as a way for machines to learn based on statistical learning. This classifier can come up with a way to separate data into different groups. This is done by using the N-dimensional hyperplane, which can be measured with the help of a known preparing dataset. The sample in the training dataset is labeled (y_i, z_i), where I = 1, 2, ... , N, where N is the number of samples, yi is a class of sample, and z_i is the training dataset. The main task of the SVM is to find the maximum distance between a hyperplane and the closest points in a higher-dimensional space. Equation (5) can be used to find the largest margin's boundary function.

$$\textit{Minimize } W(\alpha) = \frac{1}{2}\sum_{j=1}^{M}\sum_{i=1}^{M}z_j z_i \alpha_j \alpha_i l\left(y_j, y_i\right) - \sum_{j=1}^{M}\alpha_i \qquad (5)$$

4. Result

In this study, we proposed the ML method termed a multi-objective firefly-optimized improved support vector machine (MFO-ISVM). For analyzing we use the metrics like Accuracy and F1-score.

F1-score

The proposed model uses the harmonic mean to combine "recall and precision" into a single factor called the "f1-score." The very worst F1-score has a value of 0 or very close to 0 and is measured by the equation (6).

$$F1-score = \frac{(\text{precision}) \times (\text{recall}) \times 2}{\text{precision} + \text{recall}} \qquad (6)$$

Compared to other methods, the suggested method's f1 measure got 96%, while NB got 87% and LR got 83%. As a result, when compared to other methods, our proposed method has a high f1-measure value. Fig. 54.1 shows a comparison of the f1-measure.

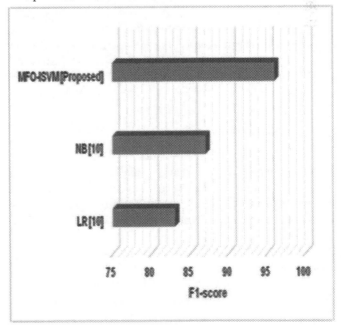

Fig. 54.1   Results of f1-score

Accuracy

The system's accuracy is judged by the number of samples for which the suggested method correctly predicted the results. The accuracy is calculated using equation (7).

$$Accuracy = \frac{TP + TN}{TP + TN + FP + FN} \quad (7)$$

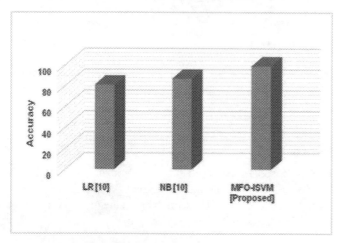

Fig. 54.2 Results of accuracy

Fig. 54.2 displays comparable values for the accuracy metrics and makes it evident that the suggested technique can generate performance results that are better than those generated by the current research approaches. The suggested method's accuracy performs better compared to the results of the current ones, which include LR 81%, and NB 87%. Early detection and prevention: Analytics can help identify patterns and trends in patient data that may indicate the early stages of a disease or potential health risks. By leveraging predictive models and machine learning algorithms, healthcare applications can analyze patient data in real-time and flag anomalies or risk factors.

5. Conclusions and Future Work

This paper gives information about an optimized and secure healthcare platform that changes the industry by giving patients and care teams better information. When this technology is used, the costs of health care go down. In this paper, we simplify the ML method termed as multi-objective firefly optimized improved support vector machine (MFO-ISVM) for accurate chronic illness prediction in places with high disease incidence. The suggested MFO-ISVM method has the highest efficiency when compared to other algorithms. Therefore, future work could focus on trying to combine FA

with the other classification models and comparing it to other approaches based on characteristic variety to figure out how well it works. It's crucial to remember that although these trends have a lot of promise, how quickly they are adopted and integrated into healthcare systems will rely on a number of different things, including infrastructural capacities, legal frameworks, and ethical concerns.

References

1. Batko, K. and Ślęzak, A., 2022. The use of Big Data Analytics in healthcare. Journal of Big Data, 9(1), p.3.
2. Alanazi, A., 2022. Using machine learning for healthcare challenges and opportunities. Informatics in Medicine Unlocked, p.100924.
3. Javaid, M., Haleem, A., Singh, R.P., Suman, R. and Rab, S., 2022. Significance of machine learning in healthcare: Features, pillars, and applications. International Journal of Intelligent Networks, 3, pp.58-73.
4. Dhiman, G., Juneja, S., Mohafez, H., El-Bayoumy, I., Sharma, L.K., Hadizadeh, M., Islam, M.A., Viriyasitavat, W. and Khandaker, M.U., 2022. Federated learning approach to protecting healthcare data over big data scenario. Sustainability, 14(5), p.2500.
5. Char, D.S., Abràmoff, M.D. and Feudtner, C., 2020. Identifying ethical considerations for machine learning healthcare applications. The American Journal of Bioethics, 20(11), pp.7-17.
6. Li, W., Chai, Y., Khan, F., Jan, S.R.U., Verma, S., Menon, V.G., and Li, X., 2021. A comprehensive survey on machine learning-based big data analytics for IoT-enabled smart healthcare system. Mobile networks and applications, 26, pp.234-252.
7. Galetsi, P. and Katsaliaki, K., 2020. A review of the literature on big data analytics in healthcare. Journal of the Operational Research Society, 71(10), pp.1511-1529.
8. Shahbaz, M., Gao, C., Zhai, L., Shahzad, F. and Hu, Y., 2019. Investigating the adoption of big data analytics in healthcare: the moderating role of resistance to change. Journal of Big Data, 6(1), pp.1-20.
9. Nti, I.K., Quarcoo, J.A., Aning, J. and Fosu, G.K., 2022. A mini-review of machine learning in big data analytics: Applications, challenges, and prospects. Big Data Mining and Analytics, 5(2), pp.81-97.
10. Krishnamoorthi, R., Joshi, S., Almarzouki, H.Z., Shukla, P.K., Rizwan, A., Kalpana, C. and Tiwari, B., 2022. A novel diabetes healthcare disease prediction framework using machine learning techniques. Journal of Healthcare Engineering, 2022.

Note: All the figures in this chapter were made by the authors.

Advancing Sustainable Science and Technology for a Resilient Future – Sai Kiran Oruganti et al. (eds)
© 2024 Taylor & Francis Group, London, ISBN 978-1-032-79020-6

55

A Novel Data Mining Approach for Intrusion Detection in Smart Grids

Rahul Thakur*

Chitkara University Institute of Engineering and Technology, Chitkara University, Punjab, India

Abstract: Rebuilding the electrical business has been discussed in light of the rising trend in consumption of electricity, insufficient resources, and the wear and tear on the current grid infrastructure. While there are many advantages to using Internet of Things (IoT) innovation and converting to a Smart Grid (SG), there are also concerns with regard to security. Because an intrusion detection system (IDS) is a potential strategy for fending off cyberattacks. As a result, a unique KMC-GBN (K-means Clustered - Gradient-Adaptive Bayesian Network) technique for intrusion prevention in these kinds of networks is proposed in this research. On the NSL KDD database, tests were performed. Accuracy, f-measure, Kappa, and ROC metrics are used to evaluate the performance of the suggested technique. The outcomes demonstrate that the suggested strategy for IDS for SG outperforms other methods.

Keywords: Smart Grid (SG), Internet of things (IoT), Intrusion detection system (IDS), K-means clustered gradient-adaptive Bayesian network (KMC-GBN)

1. Introduction

Internet of Things (IoT) is a modern technology that emerged from the growth of information and communication technologies (ICT). ICT may be used to the built environment to address issues including medical, efficiency, and transportation issues. IoT is used by individuals in clever surroundings including green infrastructure, clever medical centres, and clever transportation systems. The biggest IoT network and the most important infrastructure in IoTs is the smart grid (SG). Together with many data networks, the SG will include billions of intelligent things. Yet, one of the main problems preventing the widespread development and deployment of the IoT in the power system is safety. IoT and ICT in smart cities have several benefits. Computing systems, their data exchange channels, and the material users handle are the subject of a subset of cybercrime. (Subasi et al., 2018). An intrusion detection system (IDS) is a passive control tool that keeps an eye on any strange network activity in order to spot assaults and information gathering by attackers. It serves as a backup line of protection when cryptographic systems malfunction. AI is used by the smart grid IDS to perform anomaly and attack detection. Digital platforms may identify new assaults more effectively than existing prediction techniques such as grounded in statistics and data extraction, in contrast to requesting fewer effort. Yet, the efficiency of these structures greater precision and latency been the major focus of the expanding research on AI-based smart grid IDS. There has been little, if any, consideration given to the methods' degree of complexity when these IDS algorithms go computational intelligence algorithms to decision trees. We think the moment is opportune for a mobilization since online analytical with a black box is coming under intense investigation each day and because the use of AI-based restrictions is still in its infancy but expanding quickly (Yayla, et al., 2022). The administration and maintenance of the smart grid may benefit greatly from the use of instruments for data mining. Data mining is necessary for transforming information into Information gathering and information extraction are multidisciplinary activities in the setting of the smart grid network. The two key components of a database system are database administration and data mining. The technique of data gathering is discovering and abstracting information for use in judgment, whereas

*Corresponding author: rahul.thakur.orp@chitkara.edu.in

DOI: 10.1201/9781003490210-55

database administration deals with the handling and storing of data. Data analysis and examination using computing is known as data mining. Both clear information extraction and statement processes need data analysis. For the study of many information related to energy production, information transmission, automated delivery, billing, customer engagement, and problem diagnostics in a smart power grid, data mining is required. Aggregation of actual statistics is unavoidable in a smart grid. With the use of various data mining methods, the pattern of enormous data sets may be efficiently retrieved and examined for power generation, dependability, and sole decision processing (Chhaya, et al., 2021, Rajesh, et al., 2019).System activities throughout time are classified by the IDS as unique disruptions, regular control actions, or intrusions. Common routes are collections of important states that serve as a description or signature for each situation. An essential component of the IDS described in this work is a data mining approach that combines audit logs and synchronization measuring information from different system components to identify common pathways. The automated technique can handle extremely large volumes of data and removes the need for human trend analysis and coding (Pan, et al., 2019).

2. Related Works

Salih and Abdulazeez, 2021 demonstrated the outcomes of comparing several classification algorithms in order to create and by respect of the confusion matrix, accuracy, F-measure, kappa,ROC matrix of the IDS system, selectivity, and responsivity. Nonetheless, the confusion matrix and accuracy metric have received the majority of study attention as indicators of identification success. Also, it offers a comprehensive contrast of the dataset, data pre-processing, number of features picked, characteristic method, classification, and assessment of the methods' effectiveness as mentioned in the systems manual.Shukla, et al., 2022 mentions that the two-way, dynamic and multi Internet Technology is the SG network (CPS). Whereas the conventional grid network is a physically one-way system. SG is a component of the most ground-breaking Internet-of-Things application (IoT). It is feasible to transmit and store energy use and supplier information at all times. The SG network is exposed to outside hackers. It is crucial to identify anomalies as soon as possible since failing to do so might result in serious risks, blackouts, and financial losses for the consumer, producer, and community of smart cities.

Mrabet, et al., 2021 mentions that the intrusion detection system (IDS) is crucial for protecting smart grid networks and spotting malicious activity, it has several drawbacks. To address these problems, numerous research papers have been written using a variety of algorithms and methods. Hence,

a thorough comparison of various methods is required. An overview of the data mining methods utilized by IDS in the Smart Grid is provided in this publication.

3. Proposed Methodology

Dataset: The NSL-KDD dataset was used in this study to evaluate and assess several methodologies. One of the most significant intrusion detection datasets accessible in research institutions and utilized by several investigators is the NSL-KDD collection. Each attribute vector in this collection has 43 components, and each class includes 41 different attributes. The sort of assault is determined by element 42, and the difficulty is determined by element 43. This dataset includes contains a variety of DOS, Probe, R26, and U2R assaults. Table 1 displays the number of assaults and the number of typical instances in the dataset. Attack or non-attack detection and attack type determination may be done in two steps. This study discusses detecting attacks or lack thereof for each sample of data.

Table 55.1 Number of attacks and non-attacks

Number of samples	Class	Phase
58630	Attack	Train
67343	Normal	
12833	Attack	Test
9711	Normal	

K-means Clustering: This k-means technique divides a collection of n items across k groups based on the input variable k. Every cluster has a centre, which may be thought of as the centroid or centre of gravity of the cluster. The centre is the average value of the items in the cluster. The benefit of clustering is that we just need to work with the cluster's centres rather than all of the items inside it. This greatly decreases complication.

Gradient-Adaptive Bayesian Network: A Bayesian network (BN) is often built through understanding the variables plus architecture using testing phase. In fact, it is doubtful that the training data could account for all the differences that would be observed, particularly in intrusion detection. As a result, analysing intrusion detection sequences with dynamic settings usually results in BN's efficiency declining. The BN has to be retrained to reflect those modifications in order to handle such differences. Just the conditional probabilities between variables among 2 key BN elements require retraining since the communication network that relates to the reliance between elements is consistent across contexts.

This amount including the re-training information must be kept to a minimum in order to retain an effectiveness. To further refine and tweak the BN, the re-training information should only include the substantial misclassifications. The network will become unpredictable and unstable if conditional probabilities are updated using statistics just from the re-training information. During the re-training process, the conditional probabilities between each concerned node in the network must be updated, unlike when assigning values for input layers. As a result, the suggested technique proliferates variances or mistakes backwards from BN layer to BN layer, much as backward propagation occurs during training of multi-layered perceptron's. A gradient descent approach is used to modify every datatype conditional probabilities after obtaining the propagated errors. An elaborate representation of the suggested technique is provided in Fig. 55.1.

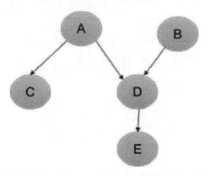

Fig. 55.1 Gradient adaptive Bayesian network

4. Result and Discussion

The use of classification algorithms for IDS in smart grids to classify different types of intrusion detection allows for the efficacy of the suggested strategy to be evaluated. Several measures, including accuracy, f-measure, kappa, and ROC metrics, The categorization method is more accurate and less difficult as a result of the dimension reduction and data preprocessing, which also decreases training and testing time by eliminating the unnecessary characteristics.

Accuracy: In order to demonstrate the quality of the sample categorization via the matrix's shape, the confusion matrix must first be produced. The number of true positives (TP), which refers to the situation when the network's actual and anticipated outcomes are normal, is provided in the first column of the first row of the matrix. The number of false positives (FP), or the actual result that the network has been attacked as opposed to the projected result, is shown in the first column of the second line. The number of false negatives (FN), or instances when the network is normal but the expected outcome is attacked, is shown in the second column of the first line. The number of true negatives (TN), or the number of outcomes that are really the opposite of what was predicted, is shown in the second column of the second line. The effectiveness and accuracy of the model categorization may be determined using the confusion matrix.

After the acquisition of T P, T N, F P, and F N, we may build a variety of the following assessment metrics and associated formulas are used to assess the model's categorization from several perspectives, including accuracy, precision, sensitivity, recall rate, and F1 score.

$$\text{Accuracy} = \frac{TP + TN}{TP + FP + FN + TN} \tag{1}$$

$$F\text{-measure} = \frac{2PR}{P + R} \tag{2}$$

Proposed method KMC-GBN is better than the existing method as shown in Fig. 55.2, which shows the accuracy of existing and proposed method.

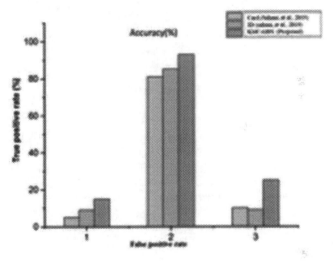

Fig. 55.2 Accuracy

F-measure of the proposed method is better than that existing method. Figure 55.3 shows the F-measure of existing and proposed method.

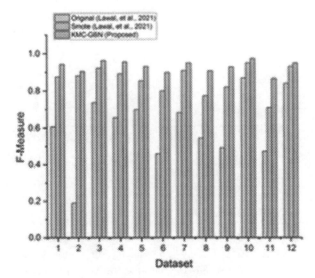

Fig. 55.3 F-measure

Kappa: Most common measure in evaluating structuredvariables is the kappa statistic (k). From Fig. 55.4 the proposed method is better than existing method.

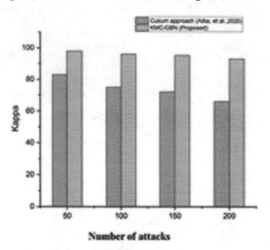

Fig. 55.4 Kappa

ROC Curve

A measurement tool addressing supervised learning issues is the Receiver Operator Characteristic (ROC) curve. In order to examine the capacity to predict a categorical result over a parameter range, receiver operating characteristic (ROC) curves are used. A further indicator of standardized tests is the area under the ROC curve. Figure 55.5 shows the ROC of existing and proposed method.

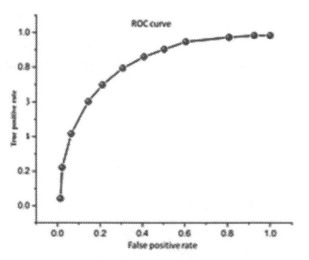

Fig. 55.5 ROC curve

5. Conclusion

In this study, a framework for intrusion detection in smart grids is proposed. The system that is being suggested will collect data and use KMC-GBN for intrusion prevention in

these networks. Accuracy, F-measure, kappa, and (ROC) metrics are used to assess the effectiveness of the proposed approach on the data base. They evaluated IDS for SG intrusion detection techniques and evaluated categorization computational efficiency to put the increased organizational, which will correlate warnings from IDS for SG performs better than other techniques, into practice for next projects. These are just a few suggestions, and there are countless other possibilities for novel data mining approaches. Consider your specific domain or application area, existing gaps in the literature, and emerging trends to identify the most promising directions for your future work.

References

1. Subasi, A., Al-Marwani, K., Alghamdi, R., Kwairanga, A., Qaisar, S.M., Al-Nory, M. and Rambo, K.A., 2018, April. Intrusion detection in smart grid using data mining techniques. In 2018 21st Saudi Computer Society National Computer Conference (NCC) (pp. 1-6). IEEE.
2. Yayla, A., Haghnegahdar, L. and Dincelli, E., 2022. Explainable artificial intelligence for smart grid intrusion detection systems. IT Professional, 24(5), pp.18-24.
3. Chhaya, L., Sharma, P., Kumar, A. and Bhagwatikar, G., 2021. Application of data mining in smart grid technology. In Encyclopedia of Information Science and Technology, Fifth Edition (pp. 815-827). IGI Global.
4. Rajesh, N. and Selvakumar, A.A.L., 2019. Association rules and deep learning for cryptographic algorithm in privacy preserving data mining. Cluster Computing, 22, pp.119-131.
5. Mazhar, T., Irfan, H.M., Haq, I., Ullah, I., Ashraf, M., Shloul, T.A., Ghadi, Y.Y. and Pan, S., Morris, T. and Adhikari, U., 2019. Developing a hybrid intrusion detection system using data mining for power systems. IEEE Transactions on Smart Grid, 6(6), pp.3104-3113.
6. Irudayasamy, A., Ganesh, D., Natesh, M., Rajesh, N. and Salma, U., 2022. Big data analytics on the impact of OMICRON and its influence on unvaccinated community through advanced machine learning concepts. International Journal of System Assurance Engineering.
7. Salih, A.A. and Abdulazeez, A.M., 2021. Evaluation of classification algorithms for intrusion detection system: A review. Journal of Soft Computing and Data Mining, 2(1), pp.31-40.
8. Shukla, S., Thakur, S. and Breslin, J.G., 2022, February. Anomaly detection in smart grid network using FC-based blockchain model and linear SVM. In Machine Learning, Optimization, and Data Science: 7th International Conference, LOD 2021, Grasmere, UK, October 4–8, 2021, Revised Selected Papers, Part I (pp. 157-171). Cham: Springer International Publishing.
9. El Mrabet, Z., El Ghazi, H. and Kaabouch, N., 2019, May. A performance comparison of data mining algorithms based intrusion detection system for smart grid. In 2019 IEEE International Conference on Electro Information Technology (EIT) (pp. 298-303). IEEE.

Note: All the figures and table in this chapter were made by the authors.

Advancing Sustainable Science and Technology for a Resilient Future – Sai Kiran Oruganti et al. (eds)
© 2024 Taylor & Francis Group, London, ISBN 978-1-032-79020-6

Fuzzy Based Big Data Business Analytics for Efficient Decision Making

56

Nimesh Raj*

Chitkara University Institute of Engineering and Technology, Chitkara University, Punjab, India

Abstract: Data analysis, data mining, and machine learning (ML) approaches are used in business analytics to gather knowledge and comprehension of how well business operations are performing. The knowledge and insights acquired assist guide business planning. It would be ideal to have a method to categorise and forecast the wage levels of employees because they have important functions in the business operations. The government or private business might provide employees competitive salaries to attract and keep them by using this prediction and categorization technology. Here, we introduce a new crow search XGBoost (CSX) approach for categorising and forecasting salary levels. To encourage business analysis using big data, fuzzy logic is incorporated into the machine learning technology. Evaluation findings demonstrate our approach suitability for categorising and predicting pay scales in the corporate world, which then in turn improves business intelligence in challenging situations.

Keywords: Business analytics, Machine learning (ML), Big data, fuzzy logic, Crow search XGBoost (CSX)

1. Introduction

Business processes show how various complicated business tasks are carried out inside a certain corporate organisation in a systematic manner. They are representative of enterprise operations. It also makes it possible to clearly see how each party involved in a particular technical or administrative business process interacts with others within the organisation and works together as a team. These days, a lot of data is collected from many areas but also processed by cutting-edge technology like sensors, cell devices, and the online. Getting useful info from all of this data for next innovations or advancements is the difficult aspect of big data. Because the store of the data is often high, the learning in the extraction must be effective and it must be a strategy that can be applied in real time. Machine learning uses a variety of methods to extract interesting patterns and characteristics from trends and facts. By incorporating it into the ML algorithms, spammed and non-spammed emails can be identified in a variety of contexts. Fuzzy logic has assisted in the achievement of significant change. Logical thinking has made work simpler and more cost-,time-, and energy-efficient. Fuzzy logic was first introduced in 1965(Jane, et al 2019). Big data has proven to be extremely beneficial in a

variety of fields, including business, management, economics, science, medical services, and science, to mention a few. The Google Flu Trends (GFT) can anticipate more than twice as many medical visits for influenza like sickness than the Centres for Disease Control, according to a notable example provided by Time (Wang, et al., 2017). These technologies, which also require significant computational power to evaluate enormous datasets gathered from diverse sources, include but are not limited to MySQL, Map Reduce, machine learning and artificial intelligence analytics is the study of how to employ analytical methods like drilling statistics and descriptive analytics to extract usable data from the beginning (Raut, et al., 2021, Irudayasamy, et al., 2022). Decision-making is aided by big data analysis, which permits a methodical compression of data to make it more manageable. Given that choosing a green supplier is a cross challenge and that there are numerous green criteria cited in the literature and studies, big data analysis was used to determine the frequencies densities of these criteria. Starting with the criterion containing the greatest amplitude concentration, the first ten factors were chosen. The providers were then shortlisted after being assessed using hybrid Learning approaches in the distributed setting (Yildizbas, et al., 2022, Khalaf, et al., 2022).

*Corresponding author: nimesh.raj.orp@chitkara.edu.in

DOI: 10.1201/9781003490210-56

2. Related Works

Ahn, et al., 2019 offer a system for categorising and forecasting salary levels. A machine-learning technology that supports business analytics on huge data combines fuzzy logic choosing an appropriate and effective Information System Project(ISP) in today's competitive economy is to minimize financial risk and maximize administration of institution performance by reducing uncertainty. The multiple criteria decision making (MCDM) method has to become an effective strategy to resolve this kind of issue because judgments of this nature typically include a number of requirements, and it is frequently essential to reach a damage between possibly conflicting considerations (Pramanik, et al., 2022).

(Sharma, et al., 2023) examines the conventional as well as cutting-edge approaches used for big data clustering, analytics, and their applications to consumer electronics.

In order to prove business intelligence(BI), big data, and big data analysis(BDA) are not independent techniques but rather a cohesive decision support system, (Jin, et al., 2018) evaluated the literature on each topic. Second, using the processes of sorts and logistical for a typical courier service as a case study, they examine how organizations use big data and BDA practically in conjunction with BI.

Researchers are concentrating on big data analysis, which is utilized to predict future health status and offers a practical means of resolving prediction problems. To enable better decision-making, a lot of research is being done on predictive analytics using machine learning approaches. Big data analysis provides fantastic opportunities to estimate future health issues based on health characteristics and to produce the best outcomes. (Talasila, et al., 2020).

3. Proposed Methodology

Dataset

This study made use of the data set made available by kaggle. The data set has a sample size of 1471, 34 feature variables, mostly grouped into three types of variables: basic information about the individual, employment history, and attendance rate.

XGBoost Algorithm

The XGBoost The standard decision tree's decision criteria are also used by regression trees. In the regression task, each leaf node with scores signifies a decision, and each internal node offers a value for an attribute test. As seen in the example here is the outcome, which is the sum of all scores predicted by trees.

$$\hat{x} = \sum_{l=1}^{l} e_l(y_j), e_l \in E \qquad (1)$$

Where E is the function space containing all the regression trees, e_l is the score for the lth tree, and Y_j is the jth training sample of business. The Gradient Boosting Machine (GBM) and XGBoost both use gradient boosting, while XGBoost somewhat improves on the regularised goal, which penalises model complexity.

$$K = \sum_i l(x_j, x_j) + \sum_l \Omega(e_l) \qquad (2)$$

Thus, is the K overall objective function, and l is a calculable discrete linear loss function that how far apart the forecast x_j, and the actual data x_j are from one another. A regularisation term called the fi is defined as follows.

$$\Omega(e) = \gamma^S + \frac{1}{2}\lambda \|x^2\| \qquad (3)$$

S controls the amount of tree leaves, and w determines each leaf's score, where γ^s and λ are constant factors. In contrast to GBM, XGBoost extends the loss function using second order approximation and replaces the constant element to get the relatively simple objective, which speeds up optimising the objective in general. XGBoost can therefore adapt to a variety of issues.

In addition to the previously described regularised aim, XGBoost uses two further methods to reduce over fitting. In the first, freshly added weights are scaled following each tree-boosting stage by a factor coefficient, which lessens the effect of each particular tree and leads to large for other trees to improve this model. The second approach, the column sub-sampling, inhibits over-fitting more well than the more conventional row sub-sampling does.

Additionally, XGBoost introduces a number of potent techniques to optimise model performance, such as an approximation algorithm for an exact greedy algorithm that stores the information for in-memory blocks for parallel learning and makes use of a production and inventory algorithm with cache awareness to enable out-of-core computation. The approaches mentioned above enable XGBoost to run significantly quicker and handle bigger datasets.

Crow Search

It draws inspiration from the cunning stealing behavior of the crow bird. Crows store excess food in secret locations and collect it as needed. In order to steal it, a crow pursues the one with the better food source. It strives to prevent being a victim in the future by learning from its own experience with theft. The meta heuristic optimization method simulates these crow behavioral traits. The population is made up of the crow swarm (N). The environment is viewed as the search space, the hiding places as specific positions that correspond to a workable solution, and each crow Xi, [i=1, 2,...N] is considered to be a search agent. The fitness function is based on food source quality, with the best food source being the overall best solution.

Fuzzy Logic

The fuzzy logic method differs from "crisp logic," which relies on propositional logic and involves binary judgements and reasoning. In fuzzy logic, variables can vary from 0 to 1, rather than being strictly limited to such binary constraints. An indicator of a variable's level of membership in a fuzzy set is used instead.

4. Result and Discussion

Every experiment is run on a distributed big data analytics platform that has three worker nodes and one master node. A Linux Ubuntu-16.04 system with an Intel coreTM i7-6500 U processor and 8.00 GB of memory runs each individual node. Scala 2.11.8, Spark 2.2.0, and Apache Hadoop distribution 2.7.1 make up the software.

Mean Absolute Error

The mean of the absolute difference that exists between the value that is known and the value that was anticipated is what the MAE attempts to calculate. The MAE, in contrast to the RMSE, is not affected by the presence of outliers in the data. Both the MAE and RMSE have a range of 0 to 1 for their values. A comparison of the MAE can be seen in Fig. 56.1. The currently used approach is inferior to the CGX that has been offered.

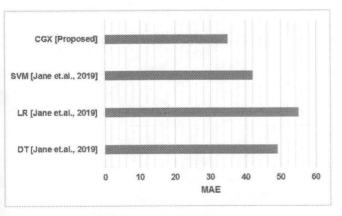

Fig. 56.1 Mean absolute error

5. Root Mean Squared Error

The sample standard deviation, also known as the root mean square error, is the statistic that is used to quantify the degree to which the observed values deviate from the expected values. The relative mean square error is shown in Fig. 56.2. The currently used approach is inferior to the CGX that has been offered.

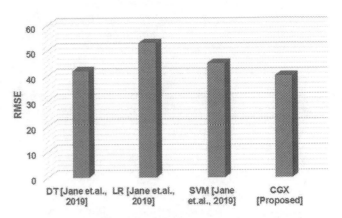

Fig. 56.2 Root mean squared error

6. Conclusion

When dealing with massive data settings, the application of machine learning approaches for decision-making produces favourable results and offers specialists from a number of professions helpful recommendations for upcoming advancements in their respective domains. The proposed approaches have an effective in MAE with 35% and RMSE with 30% when compared to the current methods. Making decisions is aided by fuzzy logic by helping to identify the uncertainties in a situation and adapting to changing environmental conditions. The proposed method in this paper explains that machine learning is a developing topic and its performance in comparison to existing approaches is better. So, we can strengthen the fuzzy extensions on ML algorithms to improve the performance in future.

References

1. Jane, J.B. and Ganesh, E.N., 2019. A Review On Big Data With Machine Learning And Fuzzy Logic For Better Decision Making. International Journal of Scientific & Technology Research, 8(10), pp.1121-1125.
2. Wang, H., Xu, Z. and Pedrycz, W., 2017. An overview on the roles of fuzzy set techniques in big data processing: Trends, challenges and opportunities. Knowledge-Based Systems, 118, pp.15-30.
3. Raut, R.D., Yadav, V.S., Cheikhrouhou, N., Narwane, V.S. and Narkhede, B.E., 2021. Big data analytics: Implementation challenges in Indian manufacturing supply chains. Computers in Industry, 125, p.103368.
4. Irudayasamy, A., Ganesh, D., Natesh, M. et al. Big data analytics on the impact of OMICRON and its influence on unvaccinated community through advanced machine learning concepts.
5. Int J Syst Assur Eng Manag (2022). https://doi.org/10.1007/s13198-022-01735-w

6. Yildizbasi, A. and Arioz, Y., 2022. Green supplier selection in new era for sustainability: a novel method for integrating big data analytics and a hybrid fuzzy multi-criteria decision making. Soft Computing, 26, pp.253-270.

7. O. I. Khalaf et al., "Blinder Oaxaca and Wilk Neutrosophic Fuzzy Set-based IoT Sensor Communication for Remote Healthcare Analysis," in IEEE Access, 2022, doi:10.1109/ACCESS.2022.3207751.

8. Ahn, S., Couture, S.V., Cuzzocrea, A., Dam, K., Grasso, G.M., Leung, C.K., McCormick, K.L. and Wodi, B.H., 2019, June. A fuzzy logic based machine learning tool for supporting big data business analytics in complex artificial intelligence environments. In 2019 IEEE international conference on fuzzy systems (FUZZ-IEEE) (pp. 1-6). IEEE.

9. Pramanik, D., Mondal, S.C. and Haldar, A., 2020. A framework for managing uncertainty in information system project selection: An intelligent fuzzy approach. International Journal of Management Science and Engineering Management, 15(1), pp.70-78.

10. Sharma, A., Singh, S.K., Badwal, E., Kumar, S., Gupta, B.B., Arya, V., Chui, K.T. and Santaniello, D., 2023, January. Fuzzy Based Clustering of Consumers' Big Data in Industrial Applications. In 2023 IEEE International Conference on Consumer Electronics (ICCE) (pp. 01-03). IEEE.

11. Jin, D.H. and Kim, H.J., 2018. Integrated understanding of big data, big data analysis, and business intelligence: A case study of logistics. Sustainability, 10(10), p.3778.

12. Talasila, V., Madhubabu, K., Madhubabu, K., Mahadasyam, M., Atchala, N. and Kande, L., 2020. The prediction of diseases using rough set theory with recurrent neural network in big data analytics. International Journal of Intelligent Engineering and Systems, 13(5), pp.10-18.

Note: All the figures in this chapter were made by the authors.

Advancing Sustainable Science and Technology for a Resilient Future – Sai Kiran Oruganti et al. (eds)
© 2024 Taylor & Francis Group, London, ISBN 978-1-032-79020-6

57

A Novel Data Science Approach for Agricultural Crop Yield Prediction

Nittin Sharma*

Chitkara University Institute of Engineering and Technology,Chitkara University, Punjab, India

Abstract: Weather conditions play a major role in determining agricultural productivity. Rainfall is a key factor in rice farming. To assist the farmers in maximising agricultural production, this situation calls for immediate assistance to forecast future crop yields and an assessment. Crop yield forecasting is a significant issue in agriculture. In this paper, we introduce a new data science technique termed the battle royal optimized Bayesian network (BRO-BN) for forecasting crop production in agriculture. To assess the effectiveness of the suggested BRO-BN approach, this study uses samples of time-series data from crops. The cleaned data with the required features are then obtained using min-max normalisation. Following that, the suggested approach is employed to accurately predict crop yields, and BRO is used to improve BN's performance. The experimental findings show that the suggested strategy outperforms other current methods in terms of crop yield prediction efficiency.

Keywords: Agriculture, Crop yield, Prediction, Data science, Battle royal optimized Bayesian network (BRO-BN)

1. Introduction

One of the most significant goals of livestock science research is to lower the expenses of reproduction and feed by improving and making evaluations in this goal. Crop yield prediction is crucial in food production. The weather environment has a major impact on agricultural produce. Crop yield forecasting is a major agricultural issue. Intentionally, every grower worries about the amount of yield he can anticipate. Crop yield predictions in antiquity were made possible by farmers' prior experience with demanding crops. The amount of data on Indian agriculture is enormous. For many applications, the period when something becomes knowledge is immensely useful. Agriculture provides a relatively large portion of the population with work in addition to food and raw materials. Yield prediction may also benefit for farmers to make decisions. Crop yield prediction is a complicated task due to a variety of complicated elements such as environment, climate, soil, fertilizer application, and seed variation One of the objectives of agricultural yield is to produce the most crops possible at the lowest possible cost (Pant, et al., 2021). Agriculture process parameters differ from region to region and manufacturer to manufacturer. Collecting similar

information in a bigger setting may also be a challenging issue. Nonetheless, the Indian Meteorological Department tallies the data on environmental conditions gathered in the Indian Republic at each 1 sq. m. area in different parts of the district. Huge collections of this kind of data are often used to predict how they will affect the primary crops in a region or location. Crop growth is impacted by climatic conditions, which are crucial to crop production. Low agricultural harvests in India are attributed mostly to a lack of high-producing varieties, inadequate transfer of technology of optimum agricultural methods, and a lack of available water. Both crop production and crop administration must be improved if the agricultural output is to rise. Farmers and other stakeholders may find it helpful to employ technologies and other computer programming methodologies to anticipate agricultural yield under various climatic circumstances when making critical agronomic and crop selection decisions. Data mining allows us to forecast agricultural yields as well. We may advise the farmer to plant a better crop for a higher yield by thoroughly analysing the prior data. Early detection and correction of agricultural yield indicator issues may assist improve yield and consequent profit. Large-scale meteorological events may have a substantial influence on agricultural productivity

*nittinsharma1943@gmail.com

DOI: 10.1201/9781003490210-57

by changing local climate patterns. Farmers have recently been required to produce an increasing number of crops as circumstances and times change rapidly (Ramesh, et al., 2015). Farmers still lack sufficient understanding of new crops and thus are unaware of environmental variables that affect agricultural yield. Crop prediction could help with all these kinds of problems. In India Rainwater has a major effect on agriculture and is unforeseen. Together with potatoes, sugarcane, and oil seeds, wheat and rice are the major crop farmed in India (Sagar, et al., 2018). A novel data science method called battle royal optimized Bayesian network (BRO-BN) has been especially successful in predicting crop yield in agriculture. BRO-BN to predict yield based on a series of remotely detected images.

2. Related Works

Gopal, et al., 2019 proposed MLR-ANN approach. When MLR interception and coefficient were used to establish the input layer weights and bias of the ANN, the presented hybrid model was designed to examine the prediction accuracy. For the prediction of rice crop growth parameters, two kinds of feed-forward back propagation neural network (FFBPNN) were created (Gupta, et al., 2015). FFBPANN I and FFBPANN-II model. FFBPANN-I was created with one input neuron and FFBPANN-II was created with two input neurons. Nevavuori, et al., 2019 Crop detection and weed classification has been successfully implemented using Convolutional Neural Networks (CNNs). With CNNs, there is no need for predetermined features since the convolutional layers of the network handle feature extraction and provide the best possible features. Gupta, et al., 2022 Discussed about methods that make it possible to estimate early outcomes from big data predictive analytics models.

3. Methodology

The nine agricultural fields that were chosen for this research are close to the city of Pori. The fields had a total area of around 90 hectares (222.3 acres). Wheat and malting barley were the main crops produced in the fields, although the model was trained across the fields without distinguishing between the different crop types. These fields' multispectral data were collected over the 2017 growing season (Nevavuori, et al., 2019).

Pre-processing: This min-max normalisation method gives the initial data a linear transformation. Data values are mapped using the Min-Max normalisation to a range between "0" and "1" or "-1" and "1". The benefit of Min-Max normalisation is that it keeps the connection between the original data intact. Nevertheless, since the results might be affected by a few big numbers, this approach is significantly impacted by absolute values or extremes. The data is sized

in the [0.1-0.9] range for D-Min-Max normalisation, which works similarly to Min-Max.

3. Battle Royal Optimized Bayesian Network (BRO-BN)

The Battle Royale online games were the inspiration for the Battle Royal Optimization Algorithm (BRO). In this sort of fighting game, players battle against one another in a controlled area while trying to survive and kill as many opponents as they can. If a player sustains constant injury for a certain period, they will randomly rebirth in a different area of the game map. The initial potential solutions in the BRO algorithm are randomly distributed throughout the whole problem situation, like in Battle Royale video games. Each solution will be compared to those closest to it, and the solution with the higher fitness value will be considered the winner and the inferior solution the loser. The injury level for every solution will be changed throughout this process, with the injury level for the victor becoming reset to 0 and the injury level for the loser becoming raised by 1. A solution will be part of being able in accordance with Eq. 1 based on the issue at hand if it suffers injury repeatedly. In order to proceed towards the best solution so far identified, Eq. 2 will be applied if the injury level doesn't really match. To clarify these equations, r is a randomized integer selected at random from a uniformly distributed between [0, 1]. And represent the injured location and the optimal idea (so far) with respect to dimension d.

$$C_{\text{dam},d} = i(fy_w - py_w) + py_w \tag{1}$$

$$C_{\text{dam},d} = C_{\text{dam},d} + i(c_{\text{best},d} - C_{\text{dam},d}) \tag{2}$$

The technique's fundamental characteristic is that the safe place of the problem situation reduces around the optimal method with each repeat of the search strategy by $\Delta = \Delta + \left[\dfrac{\Delta}{2}\right]$, if $i \geq \Delta$, in other terms, once i approach the value of Δ, the safe zone shrinks. The value of Δ is $\dfrac{\text{MaxCicle}}{\text{round}(\log_1 0(\text{maxCicle}))}$, where MaxCicle is the repetition count to the highest. The space restriction ensures that approaches converge on the most suitable choice. Eq. 3 performs the task of reducing the issue area.

$$\begin{aligned} py_w &= c_{\text{best},d} - SD(c_w) \\ fy_w &= c_{\text{best},d} - SD(c_w) \end{aligned} \tag{3}$$

Bayesian network: Machine learning approaches may infer the framework and conditional probability from training data. The conditional probability may be derived from statistical information and transmitted via the network framework's connections to the target label using Bayesian probability inference. The resulting probability value may be utilised as an indicator for the classifying choice by specifying a level of confidence. The Bayesian formula may be represented mathematically as follows:

$$K\left(S_j V\right) = \frac{K(V \mid S_j) \times K(S_j)}{\sum_{r=1}^{n} K(V \mid S_r) \times K(S_r)}$$

$$q = 1, 2, \ldots, n \quad (4)$$

In accordance with fundamental statistical theory, such as the Chain Rule and the independence relation inferred from the network framework. The joint probability of E may be determined by creating local distributes with its parent nodes,

$$K(V) = \prod_{i=1}^{n} K(V_i | \text{Parent of } (K_1)) \quad (5)$$

4. Results

The performances of the existing methods are MLR-ANN, FFBPNN, and CNN compared with our proposed method. The parameters are "accuracy, precision, and density".

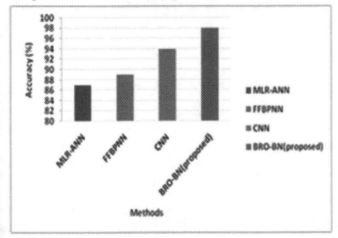

Fig. 57.1 Accuracy

The degree toward which measurements of a quantity are as close as possible to that quantity's exact value is called the accuracy of a measuring system. The accuracy of existing MLR-ANN, FFBPNN, and CNN is 87%, 89%, and 94% while the proposed BRO-BN is 98%. The degree to which repeated measures under the same conditions yield the same findings is the accuracy of a measuring system, which is connected to reproducible and reproducibility. Fig. 57.2 displays the accuracy and precision of existing and proposed methods. The precision of existing MLR-ANN, FFBPNN, and CNN is 86%, 88%, and 93% while the proposed BRO-BN is 97%.

When compared to certain other machine learning experiments, it is stated explicitly that the suggested deep reinforcement learning model can more closely preserve the distribution features of the actual agricultural production data. The density is shown in Fig. 57.3. Conclusion

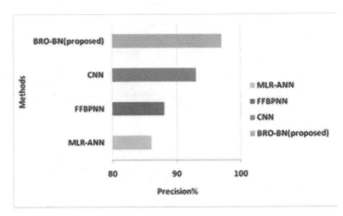

Fig. 57.2 Precision

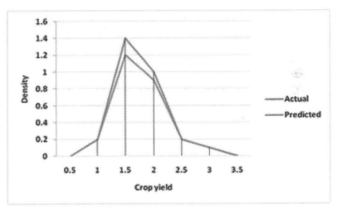

Fig. 57.3 Density

In this study, we proposed the battle royal optimized Bayesian network (BRO-BN), a novel data science method for predicting crop production in agriculture. To determine the proposed BRO-BN method, we used samples of time-series data from crops. The required features are subsequently obtained from the cleaned data using min-max normalisation. BRO was used to improve the BN's performance. In this study, we identified the accuracy; the precision of our proposed method is 98%, 97%. The primary drawback of this study is the small number of measures that were examined. If we can resolve this problem, the research will be improved significantly for further studies.

References

1. Pant, J., Pant, R.P., Singh, M.K., Singh, D.P. and Pant, H., 2021. Analysis of agricultural crop yield prediction using statistical techniques of machine learning. Materials Today: Proceedings, 46, pp.10922-10926.
2. Ramesh, D. and Vardhan, B.V., 2015. Analysis of crop yield prediction using data mining techniques. International Journal of research in engineering and technology, 4(1), pp.47-473.

3. Sagar, B.M. and Cauvery, N.K., 2018. Agriculture data analytics in crop yield estimation: a critical review. Indonesian Journal of Electrical Engineering and Computer Science, 12(3), pp.1087-1093.

4. Gopal, P.M. and Bhargavi, R., 2019. A novel approach for efficient crop yield prediction. Computers and Electronics in Agriculture, 165, p.104968.

5. Gupta, D.K., Kumar, P., Mishra, V.N., Prasad, R., Dikshit, P.K.S., Dwivedi, S.B., Ohri, A., Singh, R.S. and Srivastava, V., 2015. Bistatic measurements for the estimation of rice crop variables using an artificial neural network. Advances in Space Research, 55(6), pp.1613-1623.

6. Nevavuori, P., Narra, N. and Lipping, T., 2019. Crop yield prediction with deep convolutional neural networks. Computers and electronics in agriculture, 163, p.104859.

7. Gupta, S.K., Tiwari, S., Abd Jamil, A. and Singh, P., 2022. Faster as well as Early Measurements from Big Data Predictive Analytics Models. ECS Transactions, 107(1), p.2927.

8. Nevavuori, P., Narra, N. and Lipping, T., 2019. Crop yield prediction with deep convolutional neural networks. Computers and electronics in agriculture, 163, p.104859.

9. Jhajharia, K., Mathur, P., Jain, S. and Nijhawan, S., 2023. Crop Yield Prediction using Machine Learning and Deep Learning Techniques. Procedia Computer Science, 218, pp.406-417.

10. Rajesh, N., Christodoss, P.R. Analysis of origin, risk factors influencing COVID-19 cases in India and its prediction using ensemble learning. Int J Syst Assur EngManag (2021). https://doi.org/10.1007/s13198-021-01356-9

Note: All the figures in this chapter were made by the authors.

Advancing Sustainable Science and Technology for a Resilient Future – Sai Kiran Oruganti et al. (eds)
© 2024 Taylor & Francis Group, London, ISBN 978-1-032-79020-6

Sentiment Classification of Social Media Comments Using Data Mining

58

Takveer Singh*

Chitkara University Institute of Engineering and Technology,
Chitkara University, Punjab, India

Abstract: The accessible nature of the Internet has allowed social media to become deeply embedded in our everyday lives. Among the many social media sites currently in use, Twitter is extremely widespread. Tweets are a popular way for people to share their thoughts and opinions on a wide range of subjects, from politics and sports to the economy and beyond. Data scientists have zeroed in on the vast dataset created by the millions of tweets sent out every day in order to conduct sentiment analysis. The sentiment study concentrated on locating user posts on social media on a particular subject and classifying them as good, negative, or neutral. Artificial neural network-probabilistic neural network (ANN-PNN) is a new hybrid deep learning (DL) technique presented in this paper for the categorization of sentiment in comments on twitter. The suggested technique is more effective than currently used techniques for categorizing the sentiment of tweets, according to the study's findings.

Keywords: Internet, Social media, Twitter, Deep learning (DL), Artificial neural network-probabilistic neural network (ANN-PNN)

1. Introduction

Due to social media's quick development, an increasing number of people are expressing their ideas in writing. This motivates research to leverage social media as a source of data for sentiment analysis, which is a method of automatically deciphering, extracting, and processing text information to acquire sentiment information included in an opinion statement (Fitri, et al., 2019). One can now search and gather opinions from a huge population of people outside of their immediate network thanks to the internet. Also, more people are using the Internet to share their ideas with others. Sentiment analysis, often known as information extraction, is a strategy for examining people's attitudes, sentiments, and feelings concerning an object. Opinions have a significant influence on and give guidance to people, organizations, and online networking during the decision-making process. Opinions are a person's beliefs and the basis of their appraisal, judgment, and analysis of any event (Saxena, et al., 2021).

The basic objective is to develop a representation to predict sentiment by examining word correlation and categorizing them as having a favorable or unfavorable sentiment. Social media are incredibly important to people's daily lives in the modern world. As machine learning grows more and additional significant and popular, much like natural language processing (NLP), we must contract with the representations on these platforms. We must also examine and explore the emotions on these platforms. From "pure" dictionary-based examination to "more serious" deep learning, neural networks, there are many different approaches to approach a subject. We work to assign the right emotional polarity to the relevant tweets by developing learning algorithms and classifiers (Nemes, et al., 2021).

One of the text mining research areas that are currently active is sentiment analysis (SA) (Zad, et al., 2021). It uses a computer technique and systematic approach to identify textual excerpts, quantify them, and handle emotive states and subjectivity (Rajagopal, N.K, et al., 2022). Among the various applications of SA, gaining insight into public view

*Corresponding author: takveer.singh.orp@chitkara.edu.in

DOI: 10.1201/9781003490210-58

on a variety of socio-political subjects through the analysis of media content from social media, computer controlled examination of the historical number of content, and research of customer reviews for the reason of obtaining genuine feedback from customers and buyer sale prediction are crucial.

Customer feedback research, managing a brand, marketing assessment, forecasting the stock market, and statistical analysis of trends are just few of the many fields that might benefit from sentiment categorization of Twitter reviews. Companies can use this technology to boost customer satisfaction and gain an edge over competitors by making decisions based on empirical data. Constraints on tweet length, unstructured language, background noise, and sarcasm all work against the development of an effective sentiment classification algorithm for Twitter reviews. To facilitate trustworthy assessment of the public's views and sentiment patterns on Twitter, this study seeks to develop a hybrid DL framework to reliably determine sentiment.

2. Related Works

Drus,et al., 2019 contributes the following three ideas. Initially, they demonstrate how sentiment analysis in social media is done. Second, they determine the most popular category of social media sites from which data may be extracted for sentiment analysis. Third, they show how sentiment analysis is used in social media.

To determine the polarity of the tweet in this study, a variety of methodologies were employed. Without the need for the user to read through individual reviews, smart systems can be developed that can give users means a better of movies, products, services, etc. The user can then make decisions according to the outcomes that the intelligent systems provide without having to read individual reviews by Baid, et al., 2017.

Xu, et al., 2019 proposed a BiLSTM-based approach for sentiment analysis of comments to the job of sentiment analysis of comments. The sentiment information commitment degree is incorporated into the TF-IDF algorithm of the word weight calculation in response to the shortcomings of the word developed analytical in the existing research, and a group led technique of word vector depending on the improved term weight calculation is suggested.

Dhaoui, et al., 2017, put two well-known sentiment analysis methods-lexicon-based and machine learning to the test. Second, we show that integrating the various techniques enhances positive sentiment categorization efficiency in terms of recall and precision.

Based on user postings on social media, study's of Ahmad, et al.,2019proposed a phrase extremist classification system.

The proposed work is divided into three modules: (I) collecting content from subscribers; (ii) processing of content data; and (iii) classifying data into extreme and non-extreme modules using the LSTM + CNN architecture as well as other ML and DL classifiers.

3. Proposed Overview

This investigation gathered 4,500 tweets about health from the Twitter API (Application Programming Interface).

4. Artificial Neural Network (ANN)

Inspired by the human brain, artificial neural networks are parallel and dispersed information processing structures. They are made up of processing components, each of which has its own memory, coupled to one another by weighted connections. In other words, ANN is computer programmers that resemble biological brain networks. Three elements make up the construction of ANN: the neuron (artificial nerve cell), the links, and the learning algorithm. The fundamental processing unit of an ANN is the neuron. The network's neurons take in one or more inputs depending on the variables influencing the problem and generate the number of outcomes anticipated by the issue. An ANN is created by connecting neurons. Neurons congregate to create layers in a broad ANN system by moving in the same direction.

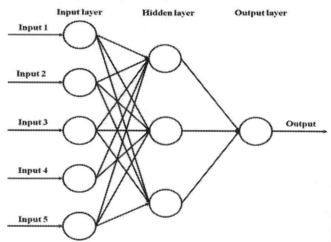

Fig. 58.1 Artificial neural network architecture

Probabilistic Neural Network (PNN)

The model developed by the appraiser mechanism of SOLID opportunities is called a Probabilistic Neural Network (PNN). Because there is just one stage to the classification process in this model, it is good and quick. In this study, PNN was employed, and it has three layers: the input layer, the pattern layer, and the summation layer. Each text has several terms that are frequently utilized as input layers and in pattern layers. The amount of neurons in the pattern layer depends on how many words are in each document. In this study, the PNN approach only had one training session. When compared to other approaches like the neural network feed-forward neural network, the training procedure is quite quick. A document that needs to be categorized is input on the input layer. The layer pattern is produced by weighing the document's words, which is utilized as input in the calculation. For Equation(1). The summation layer is for the following stage. Using the equation's formula (2), sum up all the values from the preceding stage for each category in this layer.

$$U_{ki}(w_q = \frac{1}{2\pi\sigma^2}\exp\left(\frac{w_j^n d_k, r^{-1}}{\sigma^2}\right) \tag{1}$$

$$K_i\left(w_J\right) = \sum_{k=1}^{M_l} U_{ki}\left(w_J\right), r$$
$$= 1, 2, 3, \ldots x \tag{2}$$

4. Result and Discussion

MATLAB Simulink tool is used in this study for sentiment classification. This section evaluates the ANN-performance PNNs regarding sensitivity, f-measure, accuracy, and precision. The suggested ANN-PNN is also compared to several existing FL, MNB, and KMC techniques. The precision and accuracy of the results for the suggested and existing methodologies The percentage of information for which the proposed (ANN-PNN) correctly predicted the outcome is used to measure the model's efficacy. Precision is the extent to which the proposed method only selects the most important information for further examination. Precision is determined by dividing the number of positive instances by the combined true and false positives.

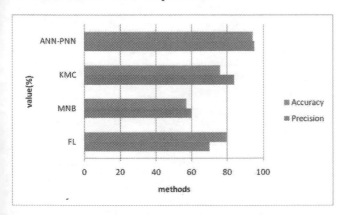

Fig. 58.2 Accuracy and precision

According to this graph, the suggested technique outperforms FL (accuracy=80%; precision=70%), MNB (accuracy=59%;

precision=60%), and KMC (accuracy=79%; precision=82%) in terms of accuracy (95.87%) and precision (96.76%)

Fig. 58.3 displays the results of the sensitivity and f1-score tests for both proposed and current techniques. A model's sensitivity measures how well it makes accurate guesses for all crucial classes. By calculating the mean of both, the f1-measure and sensitivities of the proposed model are combined into a single component.

We deduce from this figure that the suggested ANN-PNN technique outperforms existing approaches FL (sensitivity=62%; f1-measure=59%), MNB (sensitivity=79%; f1-measure=82%], and KMC (sensitivity=82%; f1-measure=75%)] in terms of sensitivity (99 %) and f1-measure (95%).

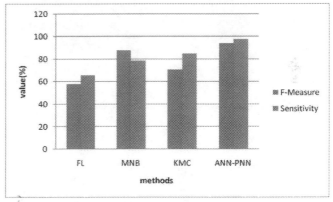

Fig. 58.3 Sensitivity and F-measure

5. Conclusion

This paper employed ANN-PNN classifier methods. Results were attained in this way. On the Twitter dataset, classification tests were carried out. The experiment performance results were then computed to assess the performance, the metrics accuracy, precision, sensitivity, and f-measure were analyzed. Analysis of all the experimental findings reveals that the proposed method achieves better performance than existing methods in classifying twitter comments. For every performance criterion, ANN-PNN performed the best. It is intended to deploy more advanced models for categorization in the next studies.

References

1. Fitri, V.A., Andreswari, R. and Hasibuan, M.A., 2019. Sentiment analysis of social media Twitter with case of Anti-LGBT campaign in Indonesia using Naïve Bayes, decision tree, and random forest algorithm. Procedia Computer Science, 161, pp.765-772.
2. Saxena, S., Yagyasen, D., Saranya, C.N., Boddu, R.S.K., Sharma, A.K. and Gupta, S.K., 2021, October. Hybrid Cloud

Computing for Data Security System. In 2021 International Conference on Advancements in Electrical, Electronics, Communication, Computing and Automation (ICAECA) (pp. 1-8). IEEE.

3. Nemes, L. and Kiss, A., 2021. Social media sentiment analysis based on COVID-19. Journal of Information and Telecommunication, 5(1), pp.1-15.

4. Zad, S., Heidari, M., Jones, J.H. and Uzuner, O., 2021, May. A survey on concept-level sentiment analysis techniques of textual data. In 2021 IEEE World AI IoT Congress (AIIoT) (pp. 0285-0291). IEEE.

5. Rajagopal, N.K., Qureshi, N.I., Durga, S., Ramirez Asis, E.H., Huerta Soto, R.M., Gupta, S.K. and Deepak, S., 2022. Future of business culture: an artificial intelligence-driven digital framework for organization decision-making process. Complexity, 2022.

6. Drus, Z. and Khalid, H., 2019. Sentiment analysis in social media and its application: Systematic literature review. Procedia Computer Science, 161, pp.707-714.

7. Baid, P., Gupta, A. and Chaplot, N., 2017. Sentiment analysis of movie reviews using machine learning techniques. International Journal of Computer Applications, 179(7), pp.45-49.

8. Xu, G., Meng, Y., Qiu, X., Yu, Z. and Wu, X., 2019. Sentiment analysis of comment texts based on BiLSTM. Ieee Access, 7, pp.51522-51532.

9. Dhaoui, C., Webster, C.M. and Tan, L.P., 2017. Social media sentiment analysis: lexicon versus machine learning. Journal of Consumer Marketing, 34(6), pp.480-488.

10. Ahmad, S., Asghar, M.Z., Alotaibi, F.M. and Awan, I., 2019. Detection and classification of social media-based extremist affiliations using sentiment analysis techniques. Human-centric Computing and Information Sciences, 9, pp.1-23.

Note: All the figures in this chapter were made by the authors.

Advancing Sustainable Science and Technology for a Resilient Future – Sai Kiran Oruganti et al. (eds)
© *2024 Taylor & Francis Group, London, ISBN 978-1-032-79020-6*

59

Fake News Detection on Social Media under the Background of Data Mining

Saksham Sood*

Chitkara University Institute of Engineering and Technology,
Chitkara University, Punjab, India

Abstract: The introduction of the World Wide Web and rapid proliferation about social media platforms (like Facebook and Twitter) laid foundations for a level information transmission that has never been seen before in the annals of human history. Consumers are producing and disseminating more content than ever because proliferation of social media platforms; some need to be more accurate and accurate. This dissemination of false information harms how people see a significant activity; thus, it must be addressed in a contemporary manner. The automated identification of misinformation in a text article is a considerable task. In this study, we gather information on true and false news items from various individuals through Twitter and media sites like PolitiFact. Finally, we provide a novel cheetah chase optimized fuzzy k-nearest neighbor algorithm (CCO-FKNNA) to recognize false news. Compared to other techniques, we conclude that the suggested network utilized in this study had the highest accuracy in identifying false statements.

Keywords: Social media platform, Fake news, Fake news identification, Cheetah chase optimized fuzzy k-nearest neighbor algorithm (CCO-FKNNA)

1. Introduction

The approach journalism is generated, shared, and received has been significantly affected by social media systems, leading to unexpected possibilities and challenging obstacles. The fact that social media has grown into a platform for misinformation campaigns that threaten its precision of every news environment is a significant issue nowadays (Lilhore, et al., 2022). Although sharing, liking, and commenting news with social media is more practical and affordable than through traditional news organizations, this kind of social media news still needs to catch up to other mainstream media outlets. The deliberate transmission of false information may affect people's capacity to tell what is true and what is wrong, among other things (Kesarwani, et al., 2022). Fake news may be identified using various techniques, such as statistics, data mining, or other systematic IT-focused methods. Instead of depending on technology fixes, it investigates variables that might affect how customers perceive trustworthiness (Gimpel, et al., 2020). People increasingly prefer using social media

for news consumption over conventional news outlets as it gains popularity. When discussing a news piece with their online systems, news people resort to news spreaders (Liu, et al., 2020). The engagement with websites like Facebook and Twitter is growing, and these media significantly impact the views and opinions. This false information may influence people's choices, cause substantial losses in money, and disgrace itself. Much work has gone into combating the creation and spread of incorrect information (Mansouri, et al., 2020).

2. Related Work

A conceptual description for thoroughly understanding and detecting fake news is needed to attract and unify scholars in adjacent fields to perform phony news investigations. Demonstrating fake news identification from different viewpoints, (Zhou, et al., 2019) use data mining and machine learning approaches, typically incorporating news articles and other data on Facebook and Twitter, natural language

*Corresponding author: sakshamsood8234@gmail.com

DOI: 10.1201/9781003490210-59

processing, information retrieval, and social networking. Finding false news is an important task that ensures that internet users get the correct information and contributes to maintaining a reliable news environment. Most existing statistical approaches rely on finding clues in news articles, which is often ineffective since fake news is sometimes crafted purposely to deceive readers by replicating basic information (Shu et al.,2019). spread of false information is a severe problem in the rapidly evolving network. Due to their inability to effectively categorise material, current classification methods for phony news detection have only partially prevented the spread, which leads to an elevated false alert rate. A stacked ensemble of three algorithms for identifying fake news was developed. Akinyemi, et al.,(2020) suggest that the suggested classifier has a superior detection performance, decreases frequency of false alarms in news, which helps to more precisely identify bogus news. Most incorrect news detection techniques now in use are unsupervised, which necessitates a significant investment in the effort and time needed to create a consistently serviced are using user behavior on social media to determine how consumers feel about the reliability of news (Yang, et al., 2019).

Data Set

Twitter posts and their affiliated labels make up the fake news detection dataset.

Tweets are categorized as either fake label. The dataset accept prespecified train, experiment, also confirmation separated. There are 6420 samples in the data, 2140 samples in the test data, and 2140 samples in validation data, for a combined 10,700 media stories and articles that were collected design various platforms. The fake tweets were gathered from reality online sites like Politifact, News Checker. Tools like Google fact-check-explorer. For gathering authentic tweets verified Twitter usernames were utilized. (Wani, et al., 2021).

Fuzzy K-Nearest Neighbor Algorithms

The K-Nearest Neighbor approach is a well-known supervised method that may be used for various regression-classification issues. Unusual occurrence detection methods that are successful include FKNN classifying. These social media fake news detection algorithm FKNN is a straightforward method for creating technique of categorization that identifies a category occurrences of an issue. Every news report is put to the test at various depth levels. We move a window measuring 30×15×7 down the depth axis with a 50% overlap between 2 successive signatures. To determine a probability estimate, use a fuzzy K-NN-based rule. To create a single probability estimate, an order statistics operator is used to merge the ten key streams. To determine the assurance level for a specific test result.

$$Conf(S_r) = \frac{\sum_{K=1}^{K} u^M\left(S_T^k\right) \times \dfrac{1}{dist\left(S_r, S_T^k\right) - D}}{\sum_{k=1}^{K} 1/dist\left(S_r, S_T^k\right)} \quad (1)$$

According to this variation, the quantitatively analyzed is influenced by the K closest neighbors' calculated distance. The total confidence rating will increase with examples that are near one another. In the latest revision of the K-NN, we calculate the probability estimate using.

$$Conf(S_r) = \sum_{K=1}^{K} u^M\left(S_T^k\right) \times \frac{1}{1 + \max\left(0, \dfrac{dist\left(S_r, S_T^k\right) - D}{N}\right)} \quad (2)$$

A probabilistic form of the K-NN, represented by Equation 2, is one in which the distance measure between the prototypes and the closest neighbours determines the total confident value. Test signatures with low confidence levels will be remote from all prototypes.

The CCA-FKNNA Algorithm

The goal of hybrid in the specific instance of optimization technique is to combine the most vital aspects of two or more algorithms to create a new algorithm predicted to perform better than its predecessors on a broad range of optimization problems. The computational techniques, which combine local search benefits with search strategy, provide a comprehensive framework for domestication. The benefits of regional and global search are incorporated in evolutionary computation algorithms, providing an overall framework for hybridization. Moreover, there are many instances when integrating multiple or more global optimisation techniques has enhanced efficiency while increasing accuracy and resilience. The ability of hybridised environment algorithms to handle various real-world challenges, including complexity, ambiguity, and misinterpretation, is gaining popularity. Based on CCA and FKNNE, a brand-new hybrid algorithm is suggested. While the FKNNE method retains a feature subset and a fast local convergence rate, it may experience an overfitting problem when solving heterogeneous challenges. Two of CCA's most essential characteristics are the capacity to discover regional and personality solutions. CCA can occasionally become stuck in optimization algorithms and has a slow convergence rate for classifier functions. Something may benefit from the differing data being passed along. The suggested CCA-FKNNA can improve the unsatisfactory solutions by combining the advantages of both techniques, which quickens integration. The proposed hybrid method appropriately uses the FKNNA modification and the CCA investigation capabilities. As a result, CCA overcomes the inadequacy of extraction. The suggested process requires relatively little computing work compared to the initial CCA to identify three solutions and determine the recombination amplification factor. The CCA-FKNNA proposed method successfully regulates the genetic variability to produce the desired mutation. This makes it possible for the process to get out of the global optimization trap issue and accelerate completion.

3. Result and Discussion

Accuracy

A difference between the result and the correct number is caused by inadequate precision. The percentage of actual outcomes reveals how balanced the data is overall. Accuracy is assessed using an equation (3).

$$Accuracy = \frac{TP + TN}{TP + TN + FP + FN} \tag{3}$$

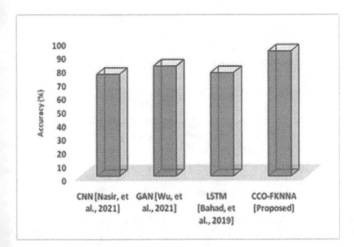

Fig. 59.1 Comparison of accuracy

Figure 59.1 shows the comparable values for the accuracy measures. Compared to existing methods like CNN, which has an accuracy rate of 75%, GAN, which has an accuracy rate of 81%, and LSTM, which has an accuracy rate of 76%, the recommended methods CCO-FKNNA value is 92%. The suggested CCO-FKNNA performs well in detecting a piece of fake news and has higher accuracy than other methods.

Precision: The most crucial standard for accuracy is precision; It is clearly defined as the percentage of adequately categorised cases in all instances of predictively positive data. Equation (4) is used to compute the precision.

$$Precision = \frac{TP}{TP + FP} \tag{4}$$

Comparable values for the precision measures are shown in Fig. 59.2. This proves the suggested strategy may provide performance results superior to those obtained by

the current study methods. The precision of the proposed approach is 95%, which performs better than existing outcomes. CNN, GAN, and LSTM precision rates are 79%, 73%, and 80%.

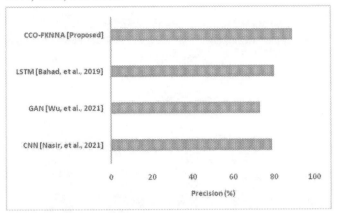

Fig. 59.2 Comparison of precision

Recall: A model's potential to identify each critical sample within a data collection is known as recall. The recall is calculated using equation (5).

$$Recall = \frac{TP}{TP + FN} \tag{5}$$

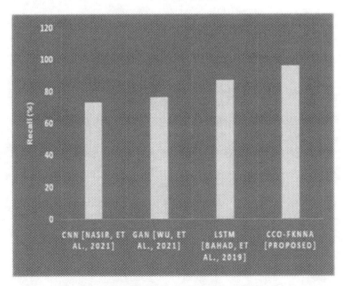

Fig. 59.3 Comparision of recall

Figure 59.3 shows the comparative data for the recall metrics. Recall rates for CNN 73%, GAN 76%, LSTM 87%, and CCO-FKNNA 96%. The proposed method performed better than the current results, with a recall of 96%.

F1-score

The harmonic mean of the proposed model is computed to merge "recall and precision" into a single component called the f1-score. Equation (6) is used to determine the f1-score.

$$F1\text{-score} = \frac{2 \times \text{precision} \times \text{recall}}{\text{precision} + \text{recall}} \qquad (6)$$

Figure 59.4 shows the comparative data for the recall metrics. Recall rates for CNN 71%, GAN 77%, LSTM 70%, and CCO-FKNNA 85%. The suggested method outperforms current results with an F1-score of 85%.

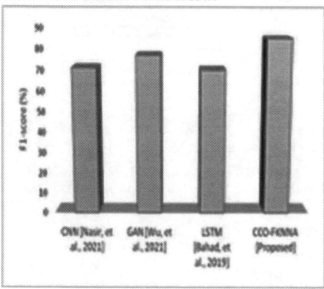

Fig. 59.4 Comparison of F1-score

4. Conclusion

The study is about accurate and inaccurate news stories from many people using Twitter and other outlets like Fake news websites. Throughout this research, we suggest an innovative fuzzy k-nearest neighbor approach (CCO-FKNNA) for cheetah chase optimization to detect fake news. Participant observation findings indicate that the suggested strategy offers more useful requirements than the traditional approaches.

References

1. Lilhore U.K., Imoize A.L., Lee C.-C., Simaiya S., Pani S.K., Goyal N., Kumar A., Li C.-T. "Enhanced Convolutional Neural Network Model for Cassava Leaf Disease Identification and Classification."Mathematics. 2022; 10(4). Article No.: 580. DOI: 10.3390/math10040580.
2. Kesarwani, A., Chauhan, S.S. and Nair, A.R., 2020, June. Fake news detection on social media using a k-nearest neighbor classifier. In 2020 international conference on advances in computing and communication engineering (ICACCE) (pp. 1-4). IEEE.
3. Gimpel, H., Heger, S., Kasper, J. and Schäfer, R., 2020, January. The Power of Related Articles-Improving Fake News Detection on Social Media Platforms. In HICSS (pp. 1-10).
4. Liu, Y. and Wu, Y.F.B., 2020. Fned: a deep network for fake news early detection on social media. ACM Transactions on Information Systems (TOIS), 38(3), pp.1-33.
5. Mansouri, R., Naderan-Tahan, M. and Rashti, M.J., 2020, August. A semi-supervised learning method for fake news detection in social media. In 2020 28th Iranian Conference on Electrical Engineering (ICEE) (pp. 1-5). IEEE.
6. Zhou, X., Zafarani, R., Shu, K. and Liu, H., 2019, January. Fake news: Fundamental theories, detection strategies, and challenges. In Proceedings of the twelfth ACM international conference on web search and data mining (pp. 836-837).
7. Shu, K., Wang, S. and Liu, H., 2019, January. Beyond news contents: The role of social context for fake news detection. In Proceedings of the twelfth ACM international conference on web search and data mining (pp. 312-320).
8. Akinyemi, B., Adewusi, O. and Oyebade, A., 2020. An improved classification model for fake news detection in social media. Int. J. Inf. Technol. Comput. Sci, 12(1), pp.34-43.
9. Yang, S., Shu, K., Wang, S., Gu, R., Wu, F. and Liu, H., 2019, July. Unsupervised fake news detection on social media: A generative approach. In Proceedings of the AAAI conference on artificial intelligence (Vol. 33, No. 01, pp. 5644-5651).
10. Wani, A., Joshi, I., Khandve, S., Wagh, V. and Joshi, R., 2021. Evaluating deep learning approaches for covid19 fake news detection. In Combating Online Hostile Posts in Regional Languages during Emergency Situation: First International Workshop, CONSTRAINT 2021, Collocated with AAAI 2021, Virtual Event, February 8, 2021, Revised Selected Papers 1 (pp. 153-163). Springer International Publishing.

Note: All the figures in this chapter were made by the authors.

Advancing Sustainable Science and Technology for a Resilient Future – Sai Kiran Oruganti et al. (eds)
© 2024 Taylor & Francis Group, London, ISBN 978-1-032-79020-6

60

Network Anomaly Detection Using a Hybrid Machine Learning Framework

Prerak Sudan*

Chitkara University Institute of Engineering and Technology,
Chitkara University, Punjab, India

Abstract: Network anomaly detection system (NADS) is widely used in a variety of sectors and allows for the monitoring of computer networks that react differently from the network protocol. Yet, the issue comes when several application areas possess various defining environmental abnormalities. These factors make it tough and complex to select the optimal algorithms that fit and satisfy the criteria of particular domains. Furthermore, the problem of centralization might result in the catastrophic collapse of a network when strong malicious software is introduced. We therefore offer a novel hybrid machine learning (ML) approach for NADS in this study termed long/short-term memory and support vector machine (LSTM-SVM). The results of the experiments demonstrate that, in terms of identifying network abnormalities, the suggested LSTM-SVM method outperforms all other methods currently in use.

Keywords: Network anomaly detection system (NADS), Computer networks, Malicious software, Hybrid machine learning (ML), Long/short-term memory and support vector machine(LSTM-SVM)

1. Introduction

Machine learning (ML) has proved enticing potential for supporting improved resource optimization, quality-of-service (QOS) guarantees, and scalability in optical networks. In particular, ML allows network administrators to accomplish knowledge-based autonomy services providing by modeling complex network patterns and developing effective internet-providing strategies using complex network activities that are unsolved using traditional theoretical techniques (Chen, et al., 2019). The anomaly-based techniques exhibit more adaptability and better generalization, and can succeed also when confronted with the categorization of unexplained network traffics. A good network anomaly detection system should be able to identify discovered anomalous in the modern day of novel network attacks (Lin, et al., 2019). Several elements that influence the application of machine learning in network intrusion detection are highlighted in the research. Machine learning has recently gained popularity and had significant success. Machine learning has so far been widely used in computer science for audio, picture, and

video identification (Haider, et al., 2020). Network anomaly detection use machine learning methods to create a system which can determine among anomalous occurrences out from remaining information. Finding patterns in the data that differ from other observations is the objective of anomaly detection. Networks anomalies are construed in the framework of cyber attacks by IT administrators, that categorize them as any occurrences that differ from the usual, i.e., the anomaly is a sign of hostile activity in the network activity (Said Elsayed, et al., 2020). Using single-processor machine learning methods, the identification of anomalies has been effectively completed. Due to their inability to constantly keep up with practical processing demands, scattered machine learning approaches are also required for the detection of network anomalies. This is a result of the increase in bitrate. Also, in machine learning-based detection approaches, dataset preparation has been crucial in collecting data since it not only increases the accuracy of the findings but also lowers computing costs by eliminating duplicate and unnecessary data samples Kaur (2020).

*prerak.sudan.orp@chitkara.edu.in

DOI: 10.1201/9781003490210-60

The remainder of this document is organized as follows: Section II summarizes the related study, Section III offers a proposal, Section IV presents the findings, and Section V offers a conclusion.

2. Related Works

The research presents a hybrid machine learning technique that combines convolutional neural network (CNN) architecture with a memory storage device in order to boost effectiveness in identifying and recognizing anomaly connections. The NSL-KDD database was used for the tests, and the outcomes of conventional machine learning are examined. (Suganthi, et al., 2022). A control plane-based orchestration is suggested by (Malik, et al., 2020) for a number of very complicated issues and attacks. The recommended approach makes use of a hybrid Cuda-enabled DL-driven architecture that combines CNN's predictive capabilities with LSTM's (Long Short-Term Memory) for quick and efficient cross-strikes and attack detection. Utilizing the most current state-of-the-art database, CICIDS2017, and recognized effectiveness evaluation techniques, the proposed strategy has been examined and validated. As part of the network intrusion detection system (NIDS) in the Software-defined Networking (SDN) management, the study shows how to analyze network information using machine learning algorithms to uncover suspected attacks. Many conventional and cutting-edge cluster machine learning methods, such as regression with support vectors, Outcome Networks, are used to demonstrate intrusion detection. As it serves as the reference database for various cutting-edge NIDS methods, the NSL-KDD dataset is utilized for developing and testing the suggested methodologies. The information is subjected to several comprehensive preparation methods to retrieve the data in its optimal configuration, producing results that are better compared to other methods (Alzahrani, et al., 2021). Kwon, et al., 2019, they provide a summary of deep learning approaches, comprising limited Boltzmann machine-based neural networks, profound neural networks, recurrent neural networks, and also machine learning methods pertinent to network anomaly detection. The detailed analysis in the paper also highlights the most recent work that used machine learning methods with a focus on network anomaly detection (). The inapplicability of the current anomaly detection methods to systems, their processing complexity, and their large inaccurate optimistic frequencies make them ineffective for the circumstance. The Grey Wolf Optimization (GWO) and CNN are therefore combined to create a hybrid computational framework for network anomaly detection in the study (Garg, et al, 2019).

3. Proposed Methodology

The suggested approach combines two separate machine learning models, such as LSTM-SVM, to develop a hybrid machine learning strategy for identifying abnormalities in networking information. The presented features are categorized using the hybrid LSTM-SVM network.

4. Combined LSTM-SVM

The based scheme LSTM-SVM classification has four layers: inputs layer, hidden layer, FC layer, and output layer. Before transferring the input to the hidden layer, the LSTM input layer gets 80% of the parameters that were selected through the prediction analysis. A gated cell or gated unit is the name for the LSTM's hidden layers. It contains four layers, each of which influences all the previous ones to determine the state and outputs of the unit. Those two items were subsequently transferred to the following hidden layers. In contrast to RNN, which only has each output vector layer for each layer of the neural network, LSTM has one hyperbolic tangent layer and three logistic sigmoid levels. These valves decide which information may be discarded and what needs to be preserved for the subsequent unit. When the hidden layer process has been completed, the FC layer receives the output and uses it to determine the precise complete architectures of the features gathered by either of the network's higher layers. LSTM and SVM are two approaches that are combined in the developed framework. The hybrid LSTM-SVM architecture is shown in Fig. 60.1 for explanation objectives.

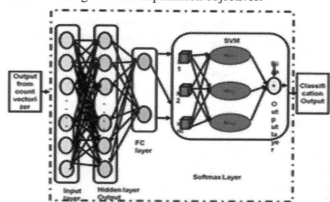

Fig. 60.1 Hybrid LSTM-architectural SVM's design

They often reside around the base of the networking structure, when the input has been reduced to a compact depiction of characteristics. Because of this, each network in the FC layer is given a unique set of weights for each unit in the layer. The softmax layer is then given the output from the FC layer. Usually, the softmax layer is the bottom layer of an LSTM receiver. Using an expression that is extremely similar to the exponential increase used in suppressor, the softmax layer is a convolution layers function that is often employed for multi-class categorization. The softmax layer may be used as a prediction if the categories are completely restricting. The recommended model enhances the LSTM network's softmax layer and employs an SVM classifier for precise analysis. The SVM carries out the classification by finding the hyper planes to discriminate the categories. Throughout the diagnosing and conditioning phases described below, all techniques were applied. The substance for research is supplied into the LSTM network; the statistical model of the LSTM neural networking is created, also acquired features is formed utilizing the input parameters from the pre-existing components for each component in the validation data. It has been finished by the SVM classifier. Every vector that must be categorized is described using the same processes, and the training data LSTM network generates an integrative values. The trained model is tested using 20% of the features extraction database when the developing process is complete. The SVM classifier subsequently combines the embedding variables with additional classifying parameters to get the projected categorization.

5. Results and Discussions

The study was completed using a sample collected, from the NSL KDD spreadsheet, to assess the performance of the suggested technique. The KDD Cup 1999 dataset was upgraded into this one. The NSL KDD database generated a reasonable amount of preparation and testing material while also successfully resolving the KDD Cup 1999 dataset's inherent recurrent flaw. It contains four typical anomalies, including DOS, probe, R2L, and U2L, as indicated in Table 60.1. A network's data traffic is compiled from various data packages and is gathered at regular intervals. Every datagram is made up of a series of data characters with 41 characteristics and one class label. The first to ten features in the list of 41 data features cover the most fundamental information, followed by features 11 to 22 that deal with content and features 23 to 41 that deal with traffic. Denial of Service (DoS), Root Local (R2L), User to Root (U2R), and probing attacks (probe) are the categories into which the affected classifiers are categorized after examining the characteristics of the attacks.

6. Training and Testing

According to the study, typical parameters including accuracy, sensitivity, and specificity were utilized to evaluate the effectiveness of the suggested technique. The true positive, true negative, false positive and false Negative variables from the dataset are utilized to create those parameters. The results of this comparison are provided in Fig. 60.2 after the effectiveness of the suggested method has already been evaluated against several machine learning algorithms. The experiment takes place in separate contexts, including testing and training as stated in Fig. 60.2, examination of the data as

Table 60.1 The associated explanation of the NSL KDD dataset contains

Characteristics	DOS	R2L	Normal	U2L	Probe	Total
Testing	7458	2754	9711	200	2421	22544
Training	45927	995	67343	52	11656	125973

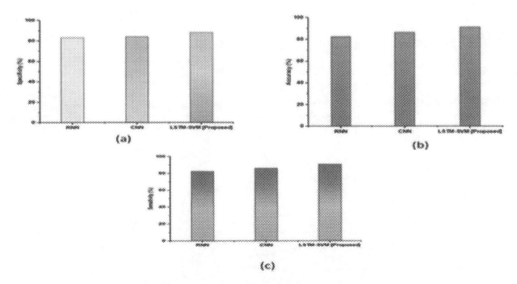

Fig. 60.2 Performance analysis of LSTM-SVM

displayed in table 1. 125973 and 22544 records of training and testing data were used in the research.

Figure 60.2 presents the results of experiments on several conventional machine learning models and the suggested technique.

The recommended strategy, LSTM-SVM, outperforms all other machine learning models with in terms of outcomes; the proposed method produces the highest performance.

7. Conclusion

The study being described demonstrates that the hybrid machine learning model could detect abnormalities in networking data. To increase the effectiveness of the classification results in anomaly detection, this research applies the sequential dependence of the Long Short Term Memory model and the deep shallow extraction of features of the multilayer neural network framework together with SVM. Using the NSL KDD dataset, experiments were conducted in two separate situations to gauge the effectiveness of the suggested technique and compare it to more traditional machine learning methods. The upcoming program calls for hybrid machine-learning models to analyze abnormalities in streaming data. The material given does not specifically address the study's shortcomings. The unique dataset or network environment utilized for the experiment, possible biases in the data, and the generalizability of the conclusions to other network settings are common restrictions in this context. To fully comprehend the limitations, it is crucial to refer to the original research. The efficacy and viability of the hybrid architecture will be confirmed by conducting comprehensive experiments and assessments in real-world network situations. The framework may be improved and its efficacy evaluated in a variety of network environments via collaborations with industry partners and network security experts.

References

1. Chen, X., Li, B., Proietti, R., Zhu, Z. and Yoo, S.B., 2019. Self-taught anomaly detection with hybrid unsupervised/supervised machine learning in optical networks. Journal of Lightwave Technology, 37(7), pp.1742-1749.

2. Lin, Peng, Kejiang Ye, and Cheng-Zhong Xu. "Dynamic network anomaly detection system by using deep learning techniques." In Cloud Computing–CLOUD 2019: 12th International Conference, Held as Part of the Services Conference Federation, SCF 2019, San Diego, CA, USA, June 25–30, 2019, Proceedings 12, pp. 161-176. Springer International Publishing, 2019.

3. Haider, S., Akhunzada, A., Mustafa, I., Patel, T.B., Fernandez, A., Choo, K.K.R. and Iqbal, J., 2020. A deep CNN ensemble framework for efficient DDoS attack detection in software-defined networks. Ieee Access, 8, pp.53972-53983.nt upgrading.

4. Said Elsayed, M., Le-Khac, N.A., Dev, S. and Jurcut, A.D., 2020, November. Network anomaly detection using LSTM based autoencoder. In Proceedings of the 16th ACM Symposium on QoS and Security for Wireless and Mobile Networks (pp. 37-45).

5. Kaur G. A comparison of two hybrid ensemble techniques for network anomaly detection in spark distributed environment. Journal of Information Security and Applications. 2020 Dec 1;55:102601.

6. Suganthi, J., Nagarajan, B. and Muhtumari, S., 2022. Network Anomaly Detection Using Hybrid Deep Learning Technique. In Advances in Parallel Computing Algorithms, Tools and Paradigms (pp. 103-109). IOS Press.

7. Malik, J., Akhunzada, A., Bibi, I., Imran, M., Musaddiq, A. and Kim, S.W., 2020. Hybrid deep learning: An efficient reconnaissance and surveillance detection mechanism in SDN. IEEE Access, 8, pp.134695-134706.

8. Alzahrani, A.O. and Alenazi, M.J., 2021. Designing a network intrusion detection system based on machine learning for software-defined networks. Future Internet, 13(5), p.111.

9. Kwon, D., Kim, H., Kim, J., Suh, S.C., Kim, I. and Kim, K.J., 2019. A survey of deep learning-based network anomaly detection. Cluster Computing, 22, pp.949-961.

10. Garg, S., Kaur, K., Kumar, N., Kaddoum, G., Zomaya, A.Y. and Ranjan, R., 2019. A hybrid deep learning-based model for anomaly detection in cloud data center networks. IEEE Transactions on Network and Service Management, 16(3), pp.924-935.

Note: All the figures and table in this chapter were made by the authors.

Advancing Sustainable Science and Technology for a Resilient Future – Sai Kiran Oruganti et al. (eds)
© 2024 Taylor & Francis Group, London, ISBN 978-1-032-79020-6

61

Cancer Prediction Model using Meta-Heuristic Optimization based Machine Learning Algorithm

Manisha Chandna*

Chitkara University Institute of Engineering and Technology,
Chitkara University, Punjab, India

Abstract: Skin cancer (SC) is one of the top three deadly varieties of cancer produced by damaged DNA that might cause mortality. There exist several investigations for the computerized analysis of malignancy in skin lesion images. Light reflections from the skin's surface, differences in color lighting, and the varying forms and sizes of the lesions make the interpretation of these images difficult. So, in this study, a novel sea lion-optimized scalable linear loss support vector machine (SLO-SLLSVM) method is introduced for precise SC classification. Before normalizing the input images we initially highlight the important locations with a bilateral filter (BF). Using the gray level co-occurrence matrix (GLCM), significant features are identified from the preprocessed images. Finally, accurate classification is accomplished using our suggested approach. Our empirical findings demonstrate that the proposed model is more reliable and durable than earlier models.

Keywords: Skin cancer, Cancer prediction, Contrast enhancement algorithm (CEA), Grey level co-occurrence matrix (GLCM), Sea lion optimized scalable linear loss support vector machine (SLO-SLLSVM)

1. Introduction

The creation of DNA-microarray nanotechnology has had a big impact on biological research. The biological behavior and importance of hundreds of genes can be examined by scientists through parallel screening. It aids in the initial detection of several severe disorders, including cancer (Sharma, et al., 2019). The patient survival rate can be significantly increased by early cancer identification. The best strategy to lower the risk of mortality from cancer is through screening. The most important sources of relevant information are medical photographs. For the diagnosis of cancer, several imaging modalities, including mammography, and ultrasound (US) imaging are frequently used (Bourouis, et al., 2022). The rate of this detection and treatment is steadily rising. The third most frequent kind of skin cancer and one of the malignant malignancies is melanoma. Melanoma, a type of cancer that alters the color of the skin as a result of aberrant pigment-producing cells, is also known as skin cancer. Melanin granules build up and spread to the skin's outermost layer, resulting in

the illness (Bi, et al., 2021).For many years, one of the most important applications of bioinformatics research has been the classification of cancer. Researchers are working hard to find new genetic indicators that could assist differentiate between cancerous and non-cancerous tissues. The inherent difficulty of cancer treatment at the biological level necessitates the classification of malignant patterns depending on their genomic characteristics (Christodoss, et al., 2022). Most conventional methods of classifyingcancers are focused on the anatomic origin of the tumor, however research has shown that genes are crucial in the microenvironment of the tumors. One goal in the classification of cancer is to identify the minimal gene set that can shed light on the underlying mechanism of tumor growth. The research has suggested several methods for categorizing malignant profiles using microarray data (Khalaf, et al., 2022).The greatest cause of death worldwide is cancer, which is a collection of complex disorders. In cancer research, cancer diagnosis and prognosis are crucial because they can raise survival rates and lower patient mortality rates. Although the suggested method yields

*Corresponding author: manishachandna968@gmail.com

DOI: 10.1201/9781003490210-61

encouraging findings, its applicability to actual environments and various datasets needs to be confirmed. The model's performance could differ when used with unobserved data from other sources, so a careful evaluation of the technique's usefulness in clinical practice is required.

2. Related Works

In order to diagnose breast cancer, (Bouzoukis, et al., 2022) proposed a CAD model that combined a wavelet neural network with the grey wolf optimization technique. In this system, the images' contrast was enhanced with a sigmoid filter, and the speckle noise was eliminated with IDAD. The preprocessed image was used to produce a ROI, and then the morphological and surface features were retrieved and merged.

Xiao, et al.,2018 introduced a multi-model ensemble strategy based on deep learning for cancer forecast .They specifically examined gene expression data collected from the lung, stomach, and breast tissues. They used the DESeq method to find differential expression genetic information between normal and tumor phenotypes to prevent over-fitting in categorization.Enhancing prediction accuracy and substantially saving computing time.

When training a deep network to predict the outcome of colorectal cancer based on images of tumor tissue samples, it is best to combine the convolution and recurrent architectures (Bychkov, et al., 2018).

Xiao, et al., 2018 suggested layered sparse auto-encoder (SAE). To be more specific, they preprocessed the gene expression information from the breast, stomach, and lung tissues using a differentially expressed analytic technique. The suggested SSAE-based deep learning model is then used.

Sakellaropoulos, et al., 2019provides and evaluation is usually of DNN models that have been trained on cell-line data to predict medication response based on gene expression. It would appear that DNN is better able to capture complex biological interactions than the other machine learning frameworks that are currently considered to be state-of-the-art.

3. Proposed Overview

Bilateral filter (BF)

The bilateral filter (BF) performs a summation that is weighted according to the distance in both space and intensity between the pixel resolutions in a local neighborhood. The weights for this summation are determined by the BF. The edges are kept in good shape while the noise is averaging nearly out using this method. The result of the BF is determined mathematically as follows at a pixel location x, where (x) is a point in the spatial vicinity of x, (x) is a parameter determining the weights' drop

in the spatial and intensity domains, respectively, and C is the normalization constant.

$$\tilde{R}(C) = \frac{1}{X} \sum_{b \in M(c)} v^{\frac{-\|b-c\|^2}{2\sigma_w^2}} v^{\frac{-|R(b)-R(c)|^2}{2\sigma_i^2}} R(b) \quad (1)$$

$$X = \sum_{b \in M(c)} v^{\frac{-\|b-c\|^2}{2\sigma_2^2}} v^{-\frac{|R(b)-R(c)|^2}{2\sigma_i^2}} \quad (2)$$

Gray Level Co-occurrence Matrix (GLCM)

GLCM known as the "grey level co-occurrence matrix," is among the most popular techniques for texture analysis. It predicts that second-order statistics and picture attributes are connected. The number of occurrences of the pair of grey levels spread apart by a given distance in the original image is represented by each element in the GLCM. Only a subset of these features was chosen to reduce the amount of computational complexity involved in the process of estimating the degree of similarity between various grey level co-occurrence matrices. A measurement of an image's textural uniformity is its energy, which is also referred to as its angular second moment. When the grey level distribution takes on whether it's a constant or periodic form, the energy level reaches its maximum potential value .The matrix for a homogeneous image will contain few entries of bigger magnitude because a homogeneous image has a small number of dominant grey tone changes, which leads to a high result for the picture's energy feature.

Sea Lion-optimized Scalable Linear Loss Support Vector Machine (SLO-SLLSVM)

Sea lion optimization

The phase of finding and following prey is referred to as the "detecting and tracking" phase. As was mentioned earlier, sea lions use their whiskers to determine their prey. Sea lions are better able to sense their prey and determine their location when the direction of their whiskers is perpendicular to the direction in which the water waves are moving. However, compared to when they were oriented in the same manner as they are now, the whiskers vibrated less. Sea lions can locate their prey and summon other members of their subgroup to help them pursue and capture it. This sea lion is regarded to be the leader of this hunting mechanism, and the other members of the group are responsible for updating their positions to the prey they are pursuing. Sea lions are classified as amphibians due to their vocalization phase. To put it another way, sea lions can be found both in the ocean and on dry land. Their vibrations travel through the water at a speed that is four times faster than through the air. Sea lions can communicate with one another through a variety of vocalizations, and this is especially true when they are pursuing prey or hunting in groups.

Scalable Linear Loss Support Vector Machine (SLLSVM)

Suggest the SLLSVM to increase the classification algorithm's training set efficiency. SVM, LS-SVM, and LSVM empirical risks should be compared. They differ while having the same aim moving the result plane away from each class due to the various loss functions. In actuality, In contrast to LS-SVM, which uses the soft margin loss function, SVM tries to create classes at least one distance from the hyperplane by using the least - square loss function. A linear loss function is also used by LSVM to separate the subclasses from the hyperplane.

Taking into account the practice data using the proper penalty factors, which is normally selected during validation, create vectors. Decide on the slack values. Compute the appropriate penalty parameters, the weight vectors, and then group the new point.

4. Result and Discussion

In our analysis of skin cancer, we made use of the MNIST HAM-10000 dataset, which can be found on kaggle. It has 10015 different photos of skin hues that are organized into seven different categories. This section assesses the sensitivity, accuracy, f-measure, and specificity of the SLLSVM. Also, the proposed SLLSVM is contrasted with several current SSAE, LSSVM, and SASLO approaches. The suggested and existing approaches' sensitivities and accuracy of the results of the effectiveness of the proposed SLLSVM are expressed as the proportion of data for which the outcome was accurately predicted.

According to this graph Fig. 61.1, the suggested technique outperforms SSAE (accuracy=65%; sensitivity=66.58%), LSSVM (accuracy=69%; sensitivity =72%), and SASLO (accuracy=79.99%; sensitivity =82.55%) in terms of accuracy (92.57%) and sensitivity (95.56%).

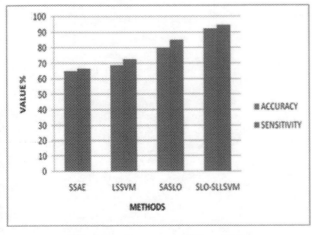

Fig. 61.1 Accuracy and sensitivity

Display the outcomes of the f1-measure and specificity testing for both the suggested and used strategies. How well a model makes reliable assumptions for all significant classes is measured by its specificity. The f1-measure and specificity of the suggested model are integrated into a single component by computing the mean of both.

Deduce from this Fig. 61.2 that the suggested SLLSVM technique outperforms existing approaches SSAE (specificity =70%; f1-measure =71.22%), LSSVM (specificity =74.55%; f1-measure=76.25%), and SASLO (specificity =82.22%; f1-measure=86.65%) in terms of specificity (93.55 %) and f1-measure (96.56%).

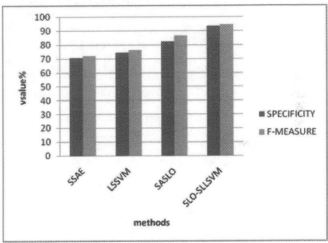

Fig. 61.2 Specificity andF-measure

5. Conclusion

In this paper, we proposed a novel sea lion-optimized scalable linear loss support vector machine (SLO-SLLSVM) for providing the best skin cancer diagnosis system. We use a bilateral filter to emphasize the significant places (BF). Significant characteristics are extracted from the preprocessed images by using the gray level co-occurrence matrix (GLCM). Eventually, adopting the method that we suggested allowed for correct classification to be completed. The results of our empirical research indicate that the model that we offer is more dependable and long-lasting than models developed in the past for the classification of the cancerous areas to improve the system's accuracy. To display the effectiveness of the recommended technique, the outcomes are contrasted with other recent research. SLO-SLLSVM has the best performance across the board for every metric of evaluation.In the future, additional research can be conducted on the identified genes in order to improve the detection of disease-specific biomarkers and the understanding of how these genes interact with one another. Convolutional neural networks (CNNs), for example, could be used in future research to improve the performance of skin cancer categorization. CNNs have demonstrated promising performance in image processing tasks and may enhance the resilience and accuracy of the suggested approach.

References

1. Sharma, A. and Rani, R., 2019. C-HMOSHSSA: Gene selection for cancer classification using multi-objective meta-heuristic and machine learning methods. Computer methods and programs in biomedicine, 178, pp.219-235.

2. Bourouis, S., Band, S.S., Mosavi, A., Agrawal, S. and Hamdi, M., 2022. Meta-heuristic algorithm-tuned neural network for breast cancer diagnosis using ultrasound images. Frontiers in Oncology, 12, p.834028.

3. Bi, D., Zhu, D., Sheykhahmad, F.R. and Qiao, M., 2021. Computer-aided skin cancer diagnosis based on a new meta-heuristic algorithm combined with support vector method. Biomedical Signal Processing and Control, 68, p.102631.

4. Christodoss, P.R. and Natarajan, R., 2022. Deep-CNN Model for Acute Lymphocytic Leukemia (ALL) Classification Using Microscopic Blood Images: Global Research. In Handbook of Research on Technologies and Systems for E-Collaboration During Global Crises (pp. 1-14). IGI Global.

5. Khalaf, O.I., Natarajan, R., Mahadev, N., Christodoss, P.R., Nainan, T., Romero, C.A.T. and Abdulsahib, G.M., 2022. Blinder Oaxaca and Wilk Neutrosophic Fuzzy Set-based IoT Sensor Communication for Remote Healthcare Analysis. IEEE Access.

6. Bourouis, S., Band, S.S., Mosavi, A., Agrawal, S. and Hamdi, M., 2022. Meta-heuristic algorithm-tuned neural network for breast cancer diagnosis using ultrasound images. *Frontiers in Oncology, 12*, p.834028.

7. Xiao, Y., Wu, J., Lin, Z., and Zhao, X., 2018. A deep learning-based multi-model ensemble method for cancer prediction. Computer methods and programs in biomedicine, 153, pp.1-9.

8. Bychkov, D., Linder, N., Turkki, R., Nordling, S., Kovanen, P.E., Verrill, C., Walliander, M., Lundin, M., Haglund, C. and Lundin, J., 2018. Deep learning based tissue analysis predicts outcome in colorectal cancer. Scientific reports, 8(1), p.3395.

9. Xiao, Y., Wu, J., Lin, Z., and Zhao, X., 2018. A semi-supervised deep learning method based on stacked sparse auto-encoder for cancer prediction using RNA-seq data. Computer methods and programs in biomedicine, 166, pp.99-105.

19. Sakellaropoulos, T., Vougas, K., Narang, S., Koinis, F., Kotsinas, A., Polyzos, A., Moss, T.J., Piha-Paul, S., Zhou, H., Kardala, E. and Damianidou, E., 2019. A deep learning framework for predicting response to therapy in cancer. Cell reports, 29(11), pp.3367-3373

Note: All the figures in this chapter were made by the authors.

Advancing Sustainable Science and Technology for a Resilient Future – Sai Kiran Oruganti et al. (eds)
© *2024 Taylor & Francis Group, London, ISBN 978-1-032-79020-6*

Machine Learning-Based Identification of Water Bodies Using Remote Sensing Images

62

Ramesh Saini*

Chitkara University Institute of Engineering and Technology,
Chitkara University, Punjab, India

Abstract: There are numerous current breakthroughs in both the machine learning (ML) and image processing domains. Several activities depend on being able to identify water bodies and having a basic understanding of the geology of the area. This understanding also aids in emergency situations, such as rescue operations, as well as in future planning for the region's growth. The entire purpose of this study is to locate water bodies using the information provided by photographs, and then to determine the space or area across where they're dispersed. Using remotely sensed images, the water bodies in this area were identified using the randomly search whale optimised XGBoost (RWO-X) approach. Accuracy, average intersection over union (AIOU), and frequency scaled IOU (FSIOU) metrics are used to evaluate how well the suggested strategy performs. The experimental results show that, when used to identify water bodies via remote sensing images, the suggested method outperforms all other methods by a wide margin.

Keywords: Water bodies, Remotely sensed images, Machine learning (ML), Randomly search whale optimised XGBoost (RSWO-X)

1. Introduction

One of the most essential and necessary natural resource for urban areas is water, which is important for human existence, growth in society, and addressing climate change. For a healthy ecosystem to survive on the earth, water is an essential resource. It makes a considerable contribution to the maintenance of global warming, the carbon cycle, and a healthy environment. In many scientific areas, accurate data on the geographical distribution of open surface water is essential. For instance, the evaluation of current and future water supplies, climate predictions, and the appropriateness of agriculture depend greatly on water information. Surface water era, growth, contraction, and disappearance are significant environmental and regional climate change factors (Alvim, et al., 2020). Water also plays an essential ingredient in economy growth since it has long-term issues on numerous in several agricultural, environmental, and biological challenges. As a result, speedy and reliable extraction of water resource information may offer critical data for water resource

study, flood surveillance, wetlands preservation, and disaster avoidance and reduction. The fate of antibiotic resistance genes (ARGs) released by urban wastewater treatment plants (WWTP) has garnered more interest in recent years. There is global agreement that raw municipal wastewater, processed wastewater, and wastewater mud, in which environmental microorganisms with genetically distinct "compositions," interact with one another, and hotspots for resistance to antibiotics development and propagation. The occurrence of black, foul-smelling water is a serious issue in aquatic ecosystems that are impacted by urbanization. Water bodies that turn black and have an unpleasant stench are said to be "black-odorous," a sensory term for polluted water that has an impact on ecosystems, human health, and urban environments. Large-scale water-body monitoring is now possible because of the fast growth in remote sensing (RS) technology, expanding picture availability, and the emergence of the era of big data RS (Dang,et al., 2021). The need for automated water-body recognition from RS imagery has grown significantly in order to pursue timely massive water-

*Corresponding author: ramesh.saini.orp@chitkara.edu.in

DOI: 10.1201/9781003490210-62

body monitoring. Since water-body regions include water bodies of many sorts, such as small and wide regions, and since there are significant variances among these water-body categories, the water-body regions in optical RS images often exhibit multiscale features. Malware detection framework based on machine learning in IoT using a Genetic Cascaded Support Vector Machine (GC-SVM) classifier (Gupta, A et al., 2022). Saleem Raja, et al., 2022 proposed several ways to detect the malicious URLs but, new attack vectors that are introduced by cyber criminals can easily bypass the security system. Therefore, it is still challenging to identify water bodies accurately from RS photos. Despite the fact that synthetic aperture radar (SAR) images have been shown to have the capacity to monitor water bodies, it is still crucial to properly monitor water bodies using optical RS images since they are easier to gather than SAR images (Sekertekin, et al., 2018).

2. Related Works

Sivanpillai, et al., 2018 used advanced Spaceborne Thermal Emission and Reflection Radiometer (ASTER) images to identify small water bodies in the Powder River Basin in contrast to more popular moderate-resolution Land sat images. Dang, et al., 2021 provided a DSSN method of segmenting water bodies using both their spectral and spatial data, to reduce water border pixel classification mistakes and inaccuracies. Yu, et al., 2017 provided a new CNN-based spectral-spatial deep learning system for extracting water bodies from Landsat images. This is the very first time CNN has been used to extract a water body from Landsat data. From the input data, CNN hierarchically selects meaningful high-level features and then classifies them using the logistic regression (LR) classifier. They used distributed processing to speed up their classification of pixel labels using a bag of visual phrases to gather nearby data. They also used this data to find abnormal changes. Although many researchers are working on identifying water bodies from RS imagery, it is still challenging to simultaneously collect the various forms, sizes, some of the water-body scales Oceans, lakes, small ponds, and narrow streams. Additionally, some weakly supervised learning approaches have been proposed (Wu, et al., 2021). Ajay Kumar, et al., 2020 used RS, GIS, and AHP techniques in a part of Deccan Volcanic Province (DVP) to identify groundwater potential zones

Randomly Search Whale Optimized XGboost

In recent years, XGboost has become known as an effective machine learning algorithm with high accuracy and running period and it is frequently utilized in issues with classifications. When employing XGboost classified, the trainer's settings must be adjusted to increase performance.

The XGboost's efficiency is determined on the parameters used. The grid search technique is the most often used parameter modification method; however, its search range is too limited, which makes it hard to find the best values. Based on Randomly search whale optimisation parameters RSWOP, we propose an XGboost classification approach. Python tool XGboost is chosen to optimise three key parameters in XGboost classification: learning rate, maximum tree depth, and sample sampling rate. The learning rate is while upgrading leaf node the weight will increase by ETA. The promotion calculation procedure becomes more cautious by lowering the weight of the trait. The generally used value range is [0, 1], and 0.3 is the default. Maximum tree depth regulates the selection tree's complexity. The model becomes more sophisticated as the value increases, but overfitting occurs. Sample sampling rate in order to successfully avoid overfitting, XGboost chooses the sample ratio for the initial spanning tree based on the ratio of the training dataset. The default is 1.

Stage 1: The gathered data were preprocessed using a normalised approach.

Stage 2: To simplify the following training of the XGboost model, PCA dimension reduction is applied to the preprocessed data. This results in data with low- dimension.

Stage 3: In RWO-X the methods' starting parameters are specified, with each whale in a three-dimensional environment.

Stage 4: Select the lower and higher of the XGboost method parameters that must be tuned in order to construct each whale's location within a sufficient range.

Stage 5: Compute the fitness of every whale location to the collected whale population and the XGboost.

Stage 6: Organize the collected fitness values to find the whale position among the present whale population and maintain it as the present best position globally.

Stage 7: Altering the position of the whale.

Stage 8: Stage 3-5 should be repeated, when the amount of iteration count reaches a maximum, the loop is terminated and the best parameters are obtained.

Stage 9: After training, enter the best parameters acquired into the XGboost to produce the best.

3. Results

The existing methods are MNDWI, AWEI, and ASTER. The parameters are accuracy, average intersection over union (AIOU), and Frequency scaled intersection over union (FSIOU).

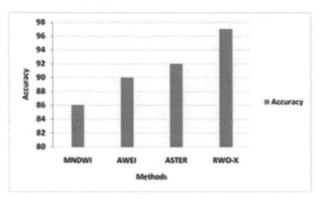

Fig. 62.1 Accuracy

In Fig. 62.1 we calculate the accuracy between existing and our proposed method. The accuracy of existing MNDWI, AWEI, and ASTER is 86%, 90%, and 92% while the proposed RWO-X is 97%.

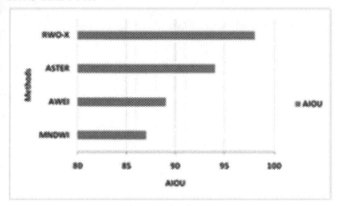

Fig. 62.2 AIOU

Fig. 62.2 displays the performance for AIOU of existing methods MNDWI, AWEI, and ASTER is 87%, 89%, and 94% compared with our proposed RWO-X is 98%.

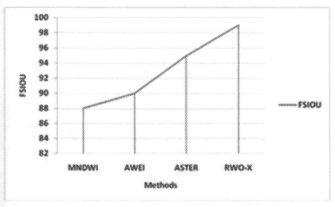

Fig. 62.3 FSIOU

Frequency scaled intersection over union (FSIOU) to measure water-body detection system works shown in Fig. 62.3.

Existing method's MNDWI, AWEI, and ASTER (FSIOU) performance is 88%, 90%, and 95% while our proposed method RWO-X is 99%.

4. Conclusion

In this study, we proposed the randomly search whale optimized XGboost (RWO-X) a new machine learning method to detect water bodies using the data provided by images, and then to establish the space or region across where they're scattered. In this study, we identified the accuracy, average intersection over union (AIOU), and Frequency scaled intersection over union (FSIOU) of our proposed method is 97%, 98%, and 99%.And the limitation of this research is Gradient boosting is particularly sensitive to outliers since every classifier should correct the mistakes, which is a common fact that is sometimes overlooked. Furthermore, in the future, a more in-depth investigation should be carried out with a specific focus on the causes of waterlogging in the study region and the development of a methodical strategy for its prevention and mitigation.

References

1. Alvim, C.B., Mendoza-Roca, J.A. and Bes-Piá, A., 2020. Wastewater treatment plant as microplastics release source–quantification and identification techniques. Journal of Environmental Management, 255, p.109739.
2. Dang, B. and Li, Y., 2021. MSResNet: Multiscale residual network via self-supervised learning for water-body detection in remote sensing imagery. Remote Sensing, 13(16), p.3122.
3. S. K. Gupta, B. Pattnaik, V. Agrawal, R. S. K. Boddu, A. Srivastava and B. Hazela, "Malware Detection Using Genetic Cascaded Support Vector Machine Classifier in Internet of Things," 2022 Second International Conference on Computer Science, Engineering and Applications (ICCSEA), 2022, pp. 1-6.
4. S. R. A, M. R, R. N, S. L and A. N, "Survey on Malicious URL Detection Techniques," 2022 6th International Conference on Trends in Electronics and Informatics (ICOEI), 2022, pp. 778-781.
5. Sekertekin, A., Cicekli, S.Y. and Arslan, N., 2018, October. Index-based identification of surface water resources using Sentinel-2 satellite imagery. In 2018 2nd International Symposium on Multidisciplinary Studies and Innovative Technologies (ISMSIT) (pp. 1-5). IEEE.
6. Sivanpillai, R. and Miller, S.N., 2018. Improvements in mapping water bodies using ASTER data. Ecological Informatics, 5(1), pp.73-78.
7. Dang, B. and Li, Y., 2021. MSResNet: Multiscale residual network via self-supervised learning for water-body detection in remote sensing imagery. Remote Sensing, 13(16), p.3122.
8. Yu, L., Wang, Z., Tian, S., Ye, F., Ding, J. and Kong, J., 2017. Convolutional neural networks for water body extraction from

Landsat imagery. International Journal of Computational Intelligence and Applications, 16(01), p.1750001.

9. Wu, Y., Han, P. and Zheng, Z., 2021. Instant water body variation detection via analysis on remote sensing imagery. Journal of Real-Time Image Processing, pp.1-14.

10. Ajay Kumar, V., Mondal, N.C. and Ahmed, S., 2020. Identification of groundwater potential zones using RS, GIS and AHP techniques: a case study in a part of Deccan volcanic province (DVP), Maharashtra, India. Journal of the Indian Society of Remote Sensing, 48, pp.497-511.

Note: All the figures in this chapter were made by the authors.

Advancing Sustainable Science and Technology for a Resilient Future – Sai Kiran Oruganti et al. (eds)
© 2024 Taylor & Francis Group, London, ISBN 978-1-032-79020-6

63

An Automated Healthcare Data Analysis Framework Usingmachine Learning

Girish Kalele*

Chitkara University Institute of Engineering and Technology, Chitkara University, Punjab, India

Abstract: Automated Machine Learning's (AML) primary goal is to offer smooth ML incorporation across sectors, enabling excellent performance in routine jobs. AML is currently being utilized in healthcare in more specific contexts with structured information, including tabulated lab tests. AML must still be used to interpret the vast amounts of daily medical text. AML in clinical note assessment, a disregarded study area that fills a hole in ML study, is a feasible method to accomplish this. This article addresses this gap and offers AML for clinical documentation based on a unique random tree-adaptive k-nearest neighbor (RT-KNN) technique. The suggested RT-KNN method outperforms current methods in terms of efficiency when analyzing clinical notes based on the study's findings.

Keywords: Automated machine learning (AML), Healthcare, Clinical note, Random tree-adaptive k-nearest neighbor (RT-KNN)

1. Introduction

The healthcare industry creates a large amount of data. Using dynamic, virtual big data platforms equipped with the most cutting-edge techniques and technology may be beneficial to enhance patient safety (Ali et al., 2018). The phrase "mobile health" (or "m-health") refers to health monitoring via cell devices, tools for monitoring patients, etc. It's been referred to as the most significant technological advance of our time. M-health has utilized big data analytics and artificial intelligence (AI) to offer a successful healthcare system. Advanced medical research has used electronic health records (EHRs), medical images, and complicated terminology is also varied, challenging, and frequently unorganized. (Khan, et al., 2020). Lightweight health data security and privacy are essential for the Internet of Medical Things (IoMTs), a foundational component of contemporary healthcare with stringent inventory levels. The first and most particular approach is "machine learning," which uses algorithms to automate the application of rules in modeling. This cutting-edge method has several practical applications in the healthcare industry (Khalaf et al., 2017). The need for Automated Machine Learning (AML) to analyze the massive amounts of everyday medical text, particularly in the setting of clinical notes, is the issue stated in the introduction. While AML has been used in healthcare with structured data, such as tabulated lab tests, it has not yet been applied to unstructured medical language. Due to this, there is a gap in the literature on machine learning that must be filled.

2. Related Works

Wang et al., 2018 used the healthcare industry as a case study to analyze the links between considerable data analytics skills, IT-enabled transformation techniques, and rewards. A reliable literature review maps the scientific area. Our theoretical framework uses resource-based theory to discover generated organizational values and focus on data, research methodology, and technological advancements. Based on query protocol, (Pirbhulal, 2019) created Alarm-net for intelligent home automation. This method was not only vulnerable to adversarial privacy assaults, which may identify an inhabitant's whereabouts. Conversely, implementing security takes longer because it uses so many resources. Chen et al., 2018 suggested developing an algorithm as follows that is built on an ontology that texts patients based on monitoring data. Yet, it might be challenging to monitor continuously, depending on the

*Corresponding author: girish.kalele.orp@chitkara.edu.in

DOI: 10.1201/9781003490210-63

information from the internet about their patients. Accurately anticipating the patient's mood requires efficient, simple recovery, analytical methods, classification techniques, and ML algorithms. For depression disorder, need plenty of text analytics was provided. Globally, technology has an impact on financial reporting systems. The study by Jayesh et al. 2022 focuses on investigating and examining the financial reporting practices that have sparked upheaval.

Four categories are used to describe the literature review that is connected to this article: work that is related to Using MLPs, LSTM, CL, Spark, Machine, and DL. The all-purpose clustering processing framework Spark from Apache was created to be quick, fault-tolerant, high-level, and consistent with Hadoop. A tool for distributed computing and massive data sets processing is provided by Spark (Barquero, J.B 2018). By utilizing Natural Language Processing (NLP) techniques, this methodology collects information from the internet and analyses user sentiment. To do this, the text's subjective material is eliminated, opinions are analyzed, and Jithendra P.2022 is evaluated favorably and unfavorably.

3. Methodology

In May of 2016, on the online platform KAGGLE, where data scientists compete to solve various prediction problems, a challenge was launched to anticipate the number of patients who would not attend appointments. The dataset consists of 110,527 selections tagged as scheduled (Noever, D. 2021).

This research focuses on developing and validating a combined RT-KNN methodology to improve seasonal projections. In the section titled "Introduction," it was mentioned that the RT and KNN methods are capable of providing a decent representation of the data. Despite this, the performance of these methods is not at its best when given tiny datasets. In order to circumvent this restriction, we have combined the aforementioned versions in a composite fashion. On the entire training set, the two base residuals, KNN and RT, are fitted, and then using the test set, the models produce predictions independently. After that, the findings are averaged to get the ultimate conclusion.

$$\mu(x) = \frac{1}{N}\sum_{n=1}^{N}\omega_n\rho_n(x) \qquad (1)$$

Where (x) represents the final weighted average outcome of the proposed method, represents the weight assigned to the nth classifier, which is determined by the accuracy performance, ρ_n represents the prediction from the nth model, and represents the sample data points. The prediction accuracy of the ensemble, which is based on the RT-KNN approach, is improved in the following factors:

- It encourages using a smaller sample size to depict the data dispersion accurately.
- It helps to reduce the risk of making generalizations.
- Controls variance in a limited dataset.
- Reduces the amount of work required for analyzing prediction accuracy.

Within the framework of the RT-KNN model, all estimation methods receive the exact weighting. The fundamental RT model is configured to carry out bootstrapping on the training subset, lessening the degree to which trees are similar. So, the model's performance is improved because of this, provided that only a limited number of training samples are available. The KNN base model is configured to use the distance between data points as the criteria for determining closeness. The leaf size, metric, random state, n jobs, n neighbors, , and weights factors are the tuning dimensions for the RT-KNN algorithm.

4. Result and Discussion

This section of the article will examine the machine learning-based automated methodology for healthcare data analysis. Using parameters such as "F-measure, Accuracy, Precision, and Recall," the proposed technique, RT-KNN, is verified. Figure 63.1 shows the result of the F-Measure. The F-Measure is often used to evaluate algorithms for machine learning, especially those utilized by information retrieval systems like search engines and systems that process natural languages. It is possible to change the F-Measure such that accuracy is valued more highly than recall or the other way around. The algorithms CNN, MLP, RT-KNN, and F-Measures provide 87, 94, and 99.

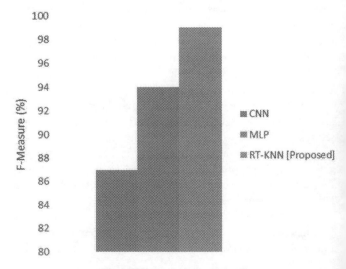

Fig. 63.1 F-measure result.

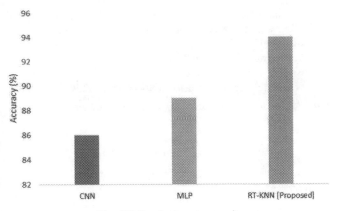

Fig. 63.2 Accuracy result.

The accuracy response is shown in both Fig. 63.2. Accuracy relates to how closely a measurement matches the agreed-upon value. Accuracy is "the degree to which the outcome of a survey corresponds to the proper estimate or a standard." CNN, MLP, and RT-KNN are the approaches, and their accuracy achieves 86, 89, and 94, respectively.

The degree of agreement between two or more measurements is called the precision of composites. If you weigh the same thing five times and get 3.2 kg each time, your height is precise but not necessarily practical. Precision does not need accuracy. Fig. 63.4 shows the result of precision. Precision achieves of 87, 90, and 97 are obtained using the methods CNN, MLP, and RT-KNN.

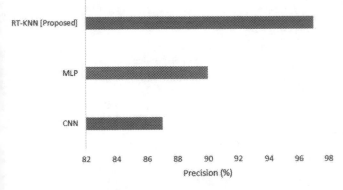

Fig. 63.3 Precision result.

A recall is the proportion of data samples an algorithm for learning correctly identifies as relating to a class of interest, the suitable type from the entire selection for that category. Figure 63.4 shows the result of the recall. The algorithms include CNN, MLP, RT-KNN, and 86, 95, and 98 recall achieves.

Conclusion

The utilization of AML in healthcare is now taking place in situations that are simpler and have more uniform

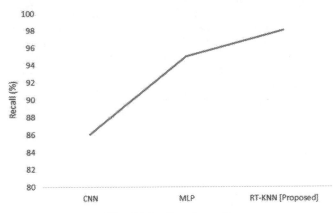

Fig. 63.4 Recall result.

information, such as tabulated laboratory findings. The use of AML is still necessary to comprehend the enormous amounts of medical text that are written each day. AML in clinical note evaluation, a neglected study topic that fills a need in ML study, is a strategy that may be used effectively to accomplish this goal. This research addresses this gap and suggests AML for clinical documentation based on a one-of-a-kind random tree-adaptive k-nearest neighbor (RT-KNN) algorithm. Specifically, this study will focus on the clinical documentation side of things. The clinical documentation will serve as the primary focal point of the article. The outcomes of the study indicate that the RT-KNN method is superior to other approaches presently being used in terms of the amount of time and effort saved during the examination of clinical notes.

References

1. Ali, O., Shrestha, A., Soar, J. and Wamba, S.F., 2018. Cloud computing-enabled healthcare opportunities, issues, and applications: A systematic review. International Journal of Information Management, 43, pp.146-158.
2. Khalaf, M., Hussain, A.J., Keight, R., Al-Jumeily, D., Fergus, P., Keenan, R. and Tso, P., 2017. Machine learning approaches to the application of disease modifying therapy for sickle cell using classification models. Neurocomputing, 228, pp.154-164.
3. Khan, Z.F. and Alotaibi, S.R., 2020. Applications of artificial intelligence and big data analytics in m-health: a healthcare system perspective. Journal of healthcare engineering, 2020, pp.1-15.
4. Wang, Y., Kung, L., Wang, W.Y.C. and Cegielski, C.G., 2018. An integrated big data analytics-enabled transformation model: Application to health care. Information & Management, 55(1), pp.64-79.
5. Barquero, J.B. Getting Started with Spark.
6. Pirbhulal, S., Samuel, O.W., Wu, W., Sangaiah, A.K. and Li, G., 2019. A joint resource-aware and medical data security framework for wearable healthcare systems. Future Generation Computer Systems, 95, pp.382-391.

7. Chen, Y., Zhou, B., Zhang, W., Gong, W. and Sun, G., 2018, June. Sentiment analysis based on deep learning and its application in screening for perinatal depression. In 2018 IEEE Third International Conference on Data Science in Cyberspace (DSC) (pp. 451-456). IEEE. Data Sci. Cyberspace, DSC 2018. (2018) 451– 456.

8. Nayak, J.P. and Parameshachari, B.D., 2022. Defect Detection in Printed Circuit Boards Using Leaky-LeNet 5. International Journal of Software Innovation (IJSI), 10(1), pp.1-13.

9. Jayesh, G.S., Novaliendry, D., Gupta, S.K., Sharma, A.K. and Hazela, B., 2022. A Comprehensive Analysis of Technologies for Accounting and Finance in Manufacturing Firms. ECS Transactions, 107(1), p.2715.

10. Noever, D. and Noever, S.E.M., 2021. Virus-MNIST: A benchmark malware dataset. arXiv preprint arXiv:2103.00602.

Note: All the figures in this chapter were made by the authors.

Advancing Sustainable Science and Technology for a Resilient Future – Sai Kiran Oruganti et al. (eds)
© 2024 Taylor & Francis Group, London, ISBN 978-1-032-79020-6

64

Effective Intrusion Detection in Wireless Sensor Networks Using a Machine Learning Approach

Sanjay Bhatnagar*

Chitkara University Institute of Engineering and Technology,
Chitkara University, Punjab, India

Abstract: The Internet's popularity and scope have been expanding steadily during the last decade. Several industries and sectors have found uses for wireless sensor networks (WSN), including firefighting, the military, the petroleum industry, the security sector, environmental monitoring, and more. Due to its need for intermediate or other nodes, low battery power supply, limited bandwidth support, distributed architecture, and self-organization, the WSN node is vulnerable to many security-related attacks. Due to its inherent complexity, a WSN might be vulnerable to a wide range of threats, including those that compromise its core functionality. Hence, in this research, we present a novel intrusion detection method based on generative polynomial logistic regression (GPLR). To improve intrusion detection, data acquired from wireless sensor nodes are first preprocessed using a normalization technique. According to experimental findings, the suggested technique outperforms traditional intrusion detection systems in terms of accuracy.

Keywords: Wireless sensor network (WSN), Attack, Intrusion detection, Generative polynomial logistic regression (GPLR)

1. Introduction

A strategy to defend against network systems threats is intrusion detection. Systems, software, or anyone who attempts to gain access to an information system or carry out prohibited conduct is referred to as an intrusion. An intrusion is defined as a series of acts that aim to affect the security, accessibility, or validity of computer systems. A technology or computer program described as an intrusion detection system produces data while actively scanning a computer or network system for suspicious activities (Subbiah, et al., 2022). The provision of security to wireless sensor networks (WSN) is of crucial relevance given the large and continuing growth of internet technology and the fact that these networks are often installed in the inaccessible environment and suffer many security threats. These issues may be resolved and network data security could be ensured by intrusion detection. Effective systems for detecting and alerting about suspected intrusions into networks or other systems include intrusion detection. But still, if online commerce grows in volume and complexity, intrusion detection is faced with

additional difficulties. This is because as online commerce develops, so do the network features (Zhiqiang, et al., 2022). Smart farming, condition monitoring, the health of structures, manufacturing tracking, disaster response, rescue efforts, wild animal tracking, resounding victory tracking, smoke detectors, surveillance, and tracking in border states, along with other tasks are carried out using WSNs in both civilian and protection system environments. A crucial use of WSNs is the detection of incursion in borders and improper disclosure in confined spaces and installations (Singh, et al., 2022). The development of high-tech gadgets and network technologies creates a lot of information, which steadily reduces intrusion detection systems' detection capability. Because the networks changing nature and resource needs to analyze huge volumes of information from remote environments, detecting an invasion with high detection accuracy becomes challenging. Moreover, the authentication process, authority, and handling of suspicious activity depend on intrusion detection systems (Gowdhaman, et al., 2021). Machine learning is a technique that operates without computer vision and automatically improves or learns from knowledge or practice. The value of

*Corresponding author: sanjay.bhatnagar.orp@chitkara.edu.in

DOI: 10.1201/9781003490210-64

machine learning is its capacity to supply generic solutions through architecture that can adapt and perform better. Its diverse nature makes it essential in a variety of fields, including technology, medicine, and computers. Many problems with WSNs have been resolved using recent ML advancements. Using ML restricts human intervention and reprogramming while simultaneously enhancing WSN performance (Kumar, et al., 2019).

2. Related Works

To efficiently identify and detect attackers in wireless technologies, a device intrusion detection system has been suggested by (Riyaz, et al., 2020) to ensure security for data sharing. The study introduces a manufacturer technique for selecting features called unsupervised spectrum and sequential comparison transfer function feature selection process. The above default values the most significant features and specifies which ones should be used with the modern convolution neural network. Data sampling and feature selection are the two components of the intrusion detection systems, useful IDS that they suggest (Ren, et al., 2019) employing hybrid data optimization. In data sampling, the isolation forest is used to remove anomalies, the bayesian network is used to improve the sample selection ratios, and the randomized vegetation classifiers are used as a form of assessment to get the ideal training dataset. In the study, experimental results on securing communication are performed to analyze the security of IoT networks based on quality-of-service metrics for the implementation of intrusion detection systems. The network output is then measured by contrasting the findings with the most security-related statistical data. For enhancing communication security, (Santhosh Kumar, et al., 2023) suggest fresh and efficient intrusion detection systems (IDS) that make use of a fuzzy CNN-based machine learning-based categorization technique. Due to its short storage capacity, bandwidth consumption support, data transfer via several networks, dependence on the gateway and other base stations, dispersed nature, and alone within, the WSN node is susceptible to many access control attacks. Across every layer of the operating system interconnection model, WSN attacks are shown. Due to this, wireless sensor nodes suffer a wide range of issues. Several of these problems are functional, while others are caused by assaults. To detect attacks and prevent them, a defense and network monitoring system should be developed. To find internal threads and provide alerts about attacks, IDS is crucial (Abhale, et al., 2020). In the research, (Dwivedi, et al., 2018) examine approaches for detecting outliers that are focused on machine learning. Between these techniques, the network model appears to have an improvement over the other techniques. A Predictive classification method is a

useful tool for determining the conditional dependence of the networks that are accessible in WSN.

3. Proposed Methodology

In determining the kinds of intrusions in WSNs, we use a machine learning technique, using a subset of the possible determinants that are currently accessible. Wireless Sensor Network is referred to as WSN. It is a network of multiple autonomous sensors that are placed across a space to keep track of environmental or physical parameters. These sensors have wireless communication capabilities that enable them to speak with one another as well as a centralized base station or gateway.

Dataset: The NSL-KDD and UNSWNB15 databases, which are extensively used in WSN intrusion detection, have been used in the study to verify the performance of the intrusion detection programs. This NSL-KDD dataset consists of 34 values of the variables, 7 symbol characteristics, and one-dimensional labeling for each specimen. There are five different specimens kind, comprising four different forms of attacking information and regularize. Two training data sets (KDDTrain+ and KDDTrain+ 20%) and one collection of testing datasets (KDDTest+) are included in the NSL-KDD system. Within the collected data used for training, there are 21 distinct types of attacks, and the experimental data also contains 17 more. It is a stronger measure of real network connections since UNSW-NB15 offers a more modern database than NSL-KDD. There are 2540044 database samples in all, along with 100 GB of the actual data transfer. This dataset's properties vary from those of NSL-KDD while remaining in line with the present network paradigm. A typical category and nine attacking categories make up its four classes.

4. Z-Score Normalization

The procedure known as "Z-score normalization," often referred to as "zero-mean normalization," normalizes each input descriptor by analyzing the normalized mean and standard deviation for every parameter across a selection of test sets. Each characteristic is given a standard deviation and mean. In the generalized equation, the substitution is stated:

$$v' = \frac{v - \mu_A}{\sigma_A} \tag{1}$$

Where A represents the attribute's μ means and A represents its standard deviation. This results in information where each parameter has a zero variation and zero means. The Z-Score normalization process is initially applied to each training sample in the data set before generating a retraining set and starting the training process. The average, variation, and statistical significance for each statistic throughout a training data collection must be computed, recorded, and used as weights in the final system design. In the structure of neural networks, it is a preprocessing level. The outputs of the neural network can be quite different from the normalized information as it was trained on a different kind of dataset. This statistical normalization provides the advantage of decreasing the consequences of data anomalies by minimizing their volume.

5. Generative Polynomial Logistic Regression

A methodological approach described as logistic regression predicts the probability of a binary parameter. It is expected that there is a significant negative correlation between the main variables and the derivatives' scrolling probability. In generative polynomial logistic regression with k classes, every division is chosen as the "pivot," and k -1 individual binaries logistic regression models are constructed. The category k's structure will be applied if it is selected as the center.

$$In \frac{\Pr(Y=k)}{\Pr(Y=k)} = b_k.X \tag{2}$$

Where Xa vector of observed weather information is, γ is a randomized procedure that chooses the results, and b_k is a collection of regresses connected to category k. In light of this, the probability that x belongs to class k is expressed as follows:

$$Pr(Y=k) = Pr(Y=k)e^{b_k.X} \tag{3}$$

Since the probabilities of x belonging to class add up to 1, class k is more probable.

$$= 1 - \sum_{K=1}^{K-1} Pr(Y=k)$$

$$Pr(Y=k) = 1 - \sum_{K=1}^{K-1} Pr(Y=k)e^{b_k.X} \tag{4}$$

$$= 1 - Pr(Y=k)\sum_{K=1}^{K-1}e^{b_k.X}$$

We may modify the equations above as follows to remove the probability on the side:

$$Pr(Y=k) = \frac{1}{1 + \sum_{k=1}^{K-1}e^{b_k.X}} \tag{5}$$

The following Eq. (3)

$$Pr(Y=k) = \frac{e^{b_k.X}}{1 + \sum_{k=1}^{K-1}e^{b_k.X}} \tag{6}$$

Polynomial logistic regression produces a predicted classy that is such that given experimental data x

$$y = \frac{\arg\max Pr(Y=k)}{k} \tag{7}$$

Generally uses the greatest probability approach to estimate the regression analysis b_k. The performance of the classifier and instabilities are minimized by using the ridges estimate. The transformation of a discrete variable with p categorical values yields p propensity scores (0 or 1), and each of them indicates whether the variable's statistical value drops into a certain category. Nevertheless, several modern data have shown that component weighing may be included in polynomial logistic regression, which generally utilizes the available components.

6. Results and Discussion

In this section, we show the outcomes of the proposed method's ability to identify anomalies in wireless sensor networks using the simulated settings. We suppose that the wireless channel remains constant during the propagation of a complete packet, but that it varies randomly and autonomously from packet to packet. To determine intrusion for various packet arrivals, we use the log-normal shadowed route degradation models. Figures 63.1 and 64.2 show demonstrations of the detecting method's effectiveness across variations in the user's real transmitted data and for

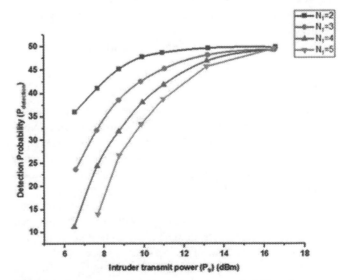

Fig 64.1 Intrusion strength against the probability of detecting

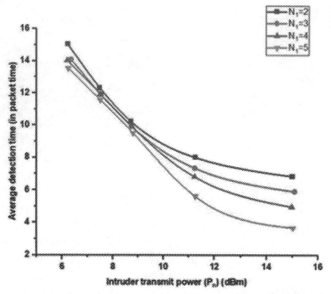

Fig 64.2 Detection frequency against intrusion ability

various intrusion buffer lengths (N_1). Another training stage is necessary for such research to inform every detection network of the typical intrusion data of its neighbor. After that, the transmitter's input power is raised, and the detecting percentages and detection times are recorded. The criterion at which we declare that the detection has failed (N_3) is set at 25, representing the quantity of detected suspicious communications.

As predicted, the chance and time required to discover an event improve as it grows in magnitude. Moreover, reduced intrusion buffer lengths (N_1) result in improved detecting probabilities and detection delays, but they also result in a higher proportion of false alarms. The software security requirements must be taken into consideration when evaluating this exchange in the wired network.

7. Conclusion

The challenge of intrusion detection in wireless sensor networks is very complex. Most of the existing WSN intrusion detection models include machine learning techniques; however they only use one technique throughout the whole network. This study presents an intrusion detection system that utilizes generative polynomial logistic regression. In recent, machine learning approaches have been employed extensively in intrusion detection systems, but there is still potential for enhancing detection accuracy and other performance data. The intrusion detection technique used at each node was taken into account independently in our implementations. Supportive reduced techniques may enhance the detection and containment process. One of the

primary difficulties that have to be rapidly resolved in real-world WSN systems is intrusion detection. By leveraging advanced machine learning techniques, big data analytics, behavior-based detection, real-time response mechanisms, threat intelligence integration, and addressing emerging challenges like IoT and cloud security, future work can significantly enhance the effectiveness and efficiency of intrusion detection systems, ultimately strengthening the security posture of computer networks and systems.

References

1. Subbiah, S., Anbananthen, K.S.M., Thangaraj, S., Kannan, S. and Chelliah, D., 2022. Intrusion detection technique in wireless sensor network using grid search random forest with Boruta feature selection algorithm. Journal of Communications and Networks, 24(2), pp.264-273.
2. Zhiqiang, L., Mohiuddin, G., Jiangbin, Z., Asim, M. and Sifei, W., 2022. Intrusion detection in wireless sensor network using enhanced empirical-based component analysis. Future Generation Computer Systems, 135, pp.181-193.
3. Singh, A., Amutha, J., Nagar, J., Sharma, S. and Lee, C.C., 2022. Lt-fs-id: Log-transformed feature learning and feature-scaling-based machine learning algorithms to predict the k-barriers for intrusion detection using a wireless sensor network. Sensors, 22(3), p.1070.
4. Gowdhaman, V. and Dhanapal, R., 2021. An intrusion detection system for wireless sensor networks using deep neural network. Soft Computing, pp.1-9.
5. Kumar, D.P., Amgoth, T. and Annavarapu, C.S.R., 2019. Machine learning algorithms for wireless sensor networks: Information Fusion, 49, pp.1-25.
6. Riyaz, B. and Ganapathy, S., 2020. A deep learning approach for effective intrusion detection in wireless networks using CNN. Soft Computing, 24, pp.17265-17278.
7. Ren, J., Guo, J., Qian, W., Yuan, H., Hao, X. and Jingjing, H., 2019. Building an effective intrusion detection system by using hybrid data optimization based on machine learning algorithms. Security and communication networks, 2019.
8. Santhosh Kumar, S.V.N., Selvi, M. and Kannan, A., 2023. A Comprehensive Survey on Machine Learning-Based Intrusion Detection Systems for Secure Communication in Internet of Things. Computational Intelligence and Neuroscience, 2023.
9. Abhale, A.B. and Manivannan, S.S., 2020. Supervised machine learning classification algorithmic approach for finding anomaly type of intrusion detection in a wireless sensor network. Optical Memory and Neural Networks, 29, pp.244-256.
10. Dwivedi, R.K., Pandey, S. and Kumar, R., 2018, January. A study on machine learning approaches for outlier detection in a wireless sensor network. In 2018 8th International Conference on Cloud Computing, Data Science & Engineering (Confluence) (pp. 189-192). IEEE.

Note: All the figures in this chapter were made by the authors.

Advancing Sustainable Science and Technology for a Resilient Future – Sai Kiran Oruganti et al. (eds)
© 2024 Taylor & Francis Group, London, ISBN 978-1-032-79020-6

65

Predictive Maintenance of Industrial Machines Using Novel Machine Learning Algorithm

Raman Verma*

Chitkara University Institute of Engineering and Technology,
Chitkara University, Punjab, India

Abstract: Electric drive process control and preventative analysis help the industry avert significant financial losses brought on by unforeseen motor breakdowns and significantly increase the reliability of the system. This research presents a revolutionary Jaya-optimised sequential long-short-term memory (JO-SLSTM) approach-based machine learning (ML) framework in preventive analytics. By creating the data collecting and data assessment, using the suggested JO-SLSTM method, and contrasting it with the simulation platform analysis, the system was evaluated using a real-world industry scenario. Data analysis tools have access to data that has been gathered by several different sensors, machine PLCs, and protocols. The suggested JO-SLSTM methodology predicts various machine conditions with a high degree of accuracy in comparison to current techniques, according to preliminary findings.

Keywords: Industrial machines, Predictive maintenance, Machine learning (ML), Jaya optimized sequential long-short term memory (JO-SLSTM)

1. Introduction

The original industrial automation strategy has undergone several changes thanks to Company 4.0. With the introduction of cognitive automation and the subsequent implementation of the idea of intelligent production, which results in smart products and services, the Internet of Things and Cyber Physical System technologies play important roles in this context. The challenges of an increasingly dynamic environment are brought on by this innovative approach for businesses. Many of these businesses are not prepared to handle this new situation, where the presence of a lot does not always work together to boost productivity (Dalzochio et al., (2020)). Germany's energy policy mandates that the country's electrical infrastructure improve efficiency, protect the environment, and provide everyone with access to affordable energy. At the same time, the use of the grid is significantly impacted by the coming mobility revolution (Sundas et al., (2020)). Our distribution grid system's characteristics will change significantly as a result. Just on the generation side, fossil fuels will eventually be largely phased out of the production of conventional energy. As a result,

energy production will become less centralized. Petroleum-based power plants that were centrally located offered reliable and manageable energy production. Many renewable sources, such as wind and solar, on the other hand, have an unpredictable mode of production that is heavily reliant on the weather, which can destabilize production. Manufacturing workloads and gaps will both happen (Hoffmann et al., (2020)).The term "preventative maintenance" describes the application of machine learning and data analytics algorithms to track the condition of mechanical devices and foretell when repair work will be required. With this strategy, it is possible to schedule maintenance on time, which minimizes equipment downtime and maintenance expenses while maximizing machine uptime and equipment efficiency(Cinar et al., (2020)).Due to the increasing amount of data generated by today's sensor-monitored industrial machinery, technicians, product managers, and data analysts have great expectations for potentially insightful information. The widely used strategy of predictive maintenance resulted from the ongoing industrial digital transformation. Its central tenet is to constantly monitor and evaluate a system to foresee maintenance requirements and prevent total failures (Zenisek

*Corresponding author: raman.verma.orp@chitkara.edu.in

DOI: 10.1201/9781003490210-65

et al., (2019)). The rest of this manuscript is structured as follows: Section II describes the related work, section III the recommended approach, section IV the results, and section V the manuscript's conclusion

2. Related Works

Arena et al., (2022) explained the manufacturing environment is marked by increased competition, quick responses, cost-cutting, and production reliability to satisfy customer demands. As a result, interest in Industry 4.0's new industrial paradigm has grown globally, inspiring many manufacturers to undergo substantial digitalization. Data-driven decision-making processes now have a novel approach thanks to digital technologies, where information from factories equipped with sensor technology is examined for understanding. Ferreira et al., (2022) describe Predictive maintenance as a key area that will benefit from the opening of Sector 4.0. With the majority of the research presuming an expert-based ML modeling, there have been several recent attempts to apply machine learning. In contrast to those studies, this one looks at a model that only uses machine learning and employs two main techniques. Beginning with supervised training as our primary focus, they modify and contrast ten fully open machine learning algorithms. Ouadah et al., (2022) examine that Utilizing information from process performance measurements and monitoring equipment data, proactive maintenance refers to the process of predicting malfunctions. The data from device detection systems are frequently analyzed using methods of machine learning and algorithms. A device can work extra precisely through the process of machine learning, which involves gathering and analyzing data. Methods used in machine learning commonly use acquire the knowledge that is fed labeled data. Silvestrin et al., (2019) and Sharma et al., (2021) represent maximizing the accessibility of engineering as the goal of predictive maintenance. Machine learning has begun to play a significant role in the field of machine failure prediction over the past ten years, which helps with predictive maintenance. Numerous methods have been put forth to take advantage of machine learning using sensory data acquired from engineering systems. Traditionally, these have been created using a traditional machine learning algorithm and extraction of features from the data. Recently, instantaneous feature extraction methods for deep learning have also been applied.

3. Recommended Approach

Data acquisition

To increase productivity and prevent quality failures, the main objective of this job is to provide accidental or deliberate for industrial machines. We have information from an account of

the particular that plays back and cutbacks packaging films for various customers.

4. Machine Specifics

The information was generated by a 14-arm characterization of the particular that creates packaging rolls of various sizes out of a large wound roll. Unwinding the roll from the twisty machine, straightening it, and feeding it to the slitter for cutting into the preferred roll length and width are the first steps in the procedure. Depending on the needs, knives or blades cut the packaging film. After researching the procedure, we discovered that the deterioration of the labeling roll was more influenced by tension and pressure. The operator supplies the machine with a predetermined tension and pressure. Depending on the roll radius, roll width, and roll length, the machine maintains this predetermined tension and pressure. To maintain the predetermined rate as the roll radius grows, tension must be decreased. In a similar vein, as the cycle progresses, pressure values rise. This is accomplished by the onboard programmable logic controller, which sends signals to the necessary actuators.

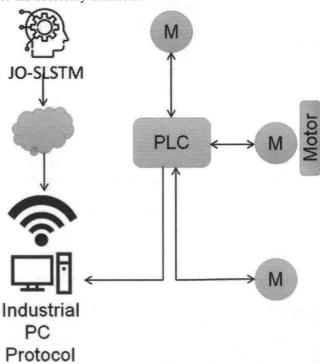

Fig. 65.1 Layout of the system

5. Summary of the Data

Sensors were used to gather and push data produced by the slitting machine to the cloud. This information was gathered for a month, sampling once every second. The system keeps the information as a CSV file with 5 columns: Time Postage, Width, Pressure, Diameter, and Tension.

6. System Configuration

Data for different parameters which must be monitored are stored in the machine's PLC. The Daisy Chaining idea is used to minimize the wiring between machines and PLCs. Data is sent over an RS485 port to the adaptor, which transforms it into TCP format before feeding it to the industrial personal computer. The IPC connects to the internet and sends data packets in the form of the MQTT protocol to the cloud.

LSTM

A variation of the common RNN is the long - short-term memory (LSTM) network. Using LSTM units in place of the basic neurons in the hidden layer in RNN, the issue of potential to eliminate and the explosion of long-term dependencies can be handled more effectively.

$$h_{(t)} = \sigma\left(Ux_{(t)} + Wh_{(t-1)} + b\right) \quad (1)$$

$$y_{(t)} = \sigma\left(Vh_{(t)} + c\right) \quad (2)$$

$$f_{(t)} = \sigma\left(W_{fx}x_{(t)} + W_{fh}h_{(t-1)} + b_f\right) \quad (3)$$

$$i_{(t)} = \sigma\left(W_{ix}x_{(t)} + W_{ih}h_{(t-1)} + b_i\right) \quad (4)$$

$$O_{(t)} = \sigma(W_{ox}x_{(t)} + W_{oh}h_{(t-1)} + b_0) \quad (5)$$

Where σ represents a nonlinear activation function. The gates can often be activated using the sigmoid function. An intermediary state C (t) is produced inside the LSTM as

$$C_{(t)} = \tanh\left(W_{cx}x_{(t)} + W_{ch}h_{(t-1)} + b_c\right) \quad (6)$$

Following that, this LSTM's memory cell and concealed state are modified as

$$C_{(t)} = f_{(t)} \odot C_{(t)} + i_{(t)} \odot C_{(t)} \quad (7)$$

$$h_{(t)} = O_{(t)} \odot \tanh(C_{(t)}) \quad (8)$$

Jaya Algorithm

The Jaya algorithm randomly creates a random number (P) while adhering to the top and lower constraints of the processing parameters. Then, Equation (1) is used to probabilistically update each variable of each potential solution in P.

$$V_{i,j,k}^{new} = V_{i,j,k}^{old} + r1_{i,j,k}(V_{j,best,k}$$
$$- abs\left(V_{i,j,k}^{old}\right) - r2_{i,j,k}(V_{i,worst,k} - abs\left(V_{i,j,k}^{old}\right)) \quad (1)$$

Where j is a potential solution from the current population and i is the repetition number. The updated value of the kth variable in the jth candidate solution is $V_{i,j,k}^{new}$ new, whereas the jth potential solution's old value for the kth variable is $V_{i,j,k}^{old}$ old. Similar to this, g is the maximum number of years, P is the population size, and d is the number of variables. r_1 and r2 are randomly initialized with a range of [0, 1], and g is the largest number of generations. $V_{j,best,k}$ is this same kth component of the strongest candidate solutions obtained in the ith iteration, and $V_{i,worst,k}$ is the ith iteration's worst solution space's kth factor. The random values r1 and r2 serve as scale factors and guarantee thorough search space exploration. The algorithm's capacity for exploration is further improved by the consideration of the variable's absolute value ($abs(V_{i,j,k}^{old})$ in Equation (1). A candidate solution in the Algorithms goes away from the most undesirable answer while also getting closer to the best answer with each generation. This results in a successful exploration and utilization of the search space.

7. Result and Discussion

The profound neural network model performed the data modeling more effectively. The frequency of low-quality cycles, in contrast to high-quality cycles, is however relatively low. The model therefore continuously updates the weights while actively learning from the newly arriving data. Predictive modeling is used to reduce small production cycles and plan maintenance activities. The design is used to project values through the conclusion of the manufacturing cycle to foretell the quality of the work that is produced.

Table 65.I Comparison of various prediction models

Supervised Model	Prediction Accuracy (%)
Support Vector Machine	95.23
Naive Bayes	96.63
CART	94.45
Deep Neural Network	99.68

The following accuracy was demonstrated by the supervised models that received training on the dataset, which was split into three sections: training, cross-validation, and testing (Table 65.1).

8. Conclusion

An innovative machine learning (ML) framework for preventive analytics based on the Jaya-optimized serial long-short-term memory (JO-SLSTM) technique is presented in this study. The system was assessed using an actual industry scenario by developing the data gathering and data evaluation, comparing it with the simulated platform analysis, and applying the suggested JO-SLSTM approach. Data collected by several sensors, machines PLCs, and protocols are accessible to data analysis tools. According to preliminary results, the proposed JO-SLSTM methodology outperforms existing methods in predicting a variety of machine situations. Applications of machine learning and artificial intelligence for preventive service and maintenance management are still being developed in this field. Studying the effects of additional features, the effects of applying other methodologies, utilizing the remaining useful lifespan as the predictor variables, and providing examples using different data sets are possible future tasks. This effort adds to the amount of published research on the use of ML to enhance decision-making because ML is essentially an area that is still developing. To prove that the proposed system works and can be used in many different situations, future work could involve applying the JO-SLSTM method to different data sets from different businesses or situations. This would help measure how well the system works in different areas and give a better idea of how it can be used.

References

1. Dalzochio, J., Kunst, R., Pignaton, E., Binotto, A., Sanyal, S., Favilla, J. and Barbosa, J., 2020. Machine learning and reasoning for predictive maintenance in Industry 4.0: Current status and challenges. *Computers in Industry*, *123*, p.103298.

2. Hoffmann, M.W., Wildermuth, S., Gitzel, R., Boyaci, A., Gebhardt, J., Kaul, H., Amihai, I., Forg, B., Suriyah, M., Leibfried, T. and Stich, V., 2020. Integration of novel sensors and machine learning for predictive maintenance in medium voltage switchgear to enable the energy and mobility revolutions. Sensors, 20(7), p.2099.

3. Çınar, Z.M., AbdussalamNuhu, A., Zeeshan, Q., Korhan, O., Asmael, M. and Safaei, B., 2020. Machine learning in predictive maintenance towards sustainable smart manufacturing in industry 4.0. Sustainability, 12(19), p.8211.

4. Zenisek, J., Holzinger, F. and Affenzeller, M., 2019. Machine learning based concept drift detection for predictive maintenance. Computers & Industrial Engineering, 137, p.106031.

5. Sundas A., Panda S.N. "IoT Based Integrated Technologies for Garbage Monitoring System." ICRITO 2020 - IEEE 8th International Conference on Reliability, Infocom Technologies and Optimization (Trends and Future Directions). 2020; 57-62. Article No.: 9197846.

6. Arena, S., Florian, E., Zennaro, I., Orrù, P.F. and Sgarbossa, F., 2022. A novel decision support system for managing predictive maintenance strategies based on machine learning approaches. Safety science, 146, p.105529.

7. Ferreira, L., Pilastri, A., Romano, F. and Cortez, P., 2022. Using supervised and one-class automated machine learning for predictive maintenance. Applied Soft Computing, 131, p.109820.

8. Ouadah, A., Zemmouchi-Ghomari, L. and Salhi, N., 2022. Selecting an appropriate supervised machine learning algorithm for predictive maintenance. The International Journal of Advanced Manufacturing Technology, 119(7-8), pp.4277-4301.

9. Silvestrin, L.P., Hoogendoorn, M. and Koole, G., 2019, December. A Comparative Study of State-of-the-Art Machine Learning Algorithms for Predictive Maintenance. In SSCI (pp. 760-767).

10. Sharma A., Kumar V., Babbar A., Dhawan V., Kotecha K., Prakash C. "Experimental investigation and optimization of electric discharge machining process parameters using grey-fuzzy-based hybrid techniques." Materials. 2021; 14(19). Article No.: 5820.

Note: All the figure and table in this chapter were made by the authors.

Advancing Sustainable Science and Technology for a Resilient Future – Sai Kiran Oruganti et al. (eds)
© 2024 Taylor & Francis Group, London, ISBN 978-1-032-79020-6

66

A Conceptual Framework for Remote Patient Monitoring Using the Internet of Things

Tannmay Gupta*

Chitkara University Institute of Engineering and Technology,
Chitkara University, Punjab, India

Abstract: Any procedure that must be closely watched to guarantee that the accepted norms and practices are carefully followed must be monitored, and clinical trial management is no different. Given that there are people and subjects involved, this is among the procedures that need to be watched over the most. Technological approaches can be applied to clinical trial monitoring to accelerate the procedure and, as a result, increase accuracy. This study develops a unique mayfly-optimized kernel-adaptive artificial neural network (MFO-KANN) algorithm and a conceptual model for clinical testing surveillance leveraging physiological records from wearable sensors. For pre-processing and feature extraction, accordingly, min-max normalization and principal component analysis (PCA) are used. The proposed approach is then utilized to decide whether or not to enable a person to complete the experiment. This paper provides guidelines for remotely monitoring clinical trials, which the research group can utilize to improve its decision-making.

Keywords: Clinical trial, Remote monitoring, Principal component analysis (PCA), Mayfly optimized kernel-adaptive artificial neural network (MFO-KANN)

1. Introduction

Technology and services in the healthcare sector are developing constantly. The development of remote patient monitoring in this area offers several advantages as the world's population grows and health problems become more serious. Technology has improved to the point where the patient may go on with daily activities at home while being maintained under observation utilizing modern communications and monitoring systems. This is made possible by comparatively easy software to maintain surveillance on people surrounding hospital wards (Iranpak, et al., 2021). Remote healthcare is a growing field of research as the world moves towards remote monitoring and accurate, fast diseases detection. The terminology "remote healthcare" refers to a variety of practices that use technological devices to track patients outside of the hospital. Patients who have been diagnosed with long-term diseases, patients who have limited mobility or another kind of handicap, patients who have just had surgery,

newborns, and older patients are just a few of the patient groups that are targeted by remote patient monitoring. There are diseases associated with every one of these patient groups that benefit from ongoing monitoring. Aiming to make every patient's daily life as pleasant as possible, smart medicine aims to help patients in their everyday routines (Ratta, et al., 2021). Based on the Internet of Things (IoT), one of the most popular mobile health (mHealth) applications offers in-home monitoring system in addition to preventive, proactive healthcare technology solutions. The application of modern computer sciences in "big data," genetics, and artificial intelligence are just a few examples of the growing fields that fall under the wide umbrella term of "digital health," which also encompasses mHealth. Mobile technology, medical device, and communications technologies for health care are together referred to as "mHealth" or "mobile health" (Philip, et al., 2021). Many modern technologies, including mobile apps, smart gadgets, biosensing, wearable technology, home virtual agents, and blockchain-based electronic medical

*Corresponding author: tannmay.gupta.orp@chitkara.edu.in

DOI: 10.1201/9781003490210-66

record networks, are heralding the dawn of a paradigm in healthcare. By 2022, it's predicted that the IoT would be used in the health industry to the tune of $409.9 million. The industry experts at Digital trends anticipate a CAGR of almost 37% for the worldwide IoT market in the healthcare sector by 2020 (Akkas, et al., 2020). One way to get over the obstacles in the health sector is via a remote patient monitoring system. Using IoT offers remote people competent healthcare providers. The patient's therapy may benefit from a potential medical and well-being assessment. With remote patient monitoring technology, the doctor may check on the patient's health from anywhere at any time. When the condition has been identified, the treatment may begin (Yew, et al., 2020). Develop a comprehensive incident response plan that outlines the steps to be taken in the event of a security breach or privacy incident. This plan should include procedures for notifying affected parties, investigating the incident, and implementing corrective measures.

2. Related Works

In the study (Malasinghe, et al., 2019), sensitive information in the IoT is prioritized using an optimization method, and long short-term memory (LSTM) artificial neural network are used in cloud technology to categorize and remotely monitor patients' conditions, which may be regarded as an essential novel component of the research. Sensor information is transmitted by the IoT platform to the cloud using the fifth generation of the Network. The basis of cloud technology is the LSTM neural network training technique. To create a system that enables real-time and remote health monitoring based on IoT infrastructure and connected to cloud technology, (Kaur, et al., 2019) used a variety of machine learning techniques and took into consideration standard datasets of healthcare stored on the internet. The technology will be permitted to make suggestions using historical and experimental data that is stored in the network. In the research, a smart senior care system based on the IoT is presented to monitor health indicators and identify biological and behavioral modifications. By using a sensing system via IoT devices, it offers a health monitoring system that allows the associated medical teams to continually monitor and evaluate a handicapped or old person's behavioral activities as well as the biological parameters (Hosseinzadeh, et al., 2020). The research model for remote health monitoring uses a gentle blocks cryptosystem to provide security for health and medical data in an IoT context. This model utilizes data collection techniques to analyze physiological data collected by remote medical IoT devices to forecast crucial circumstances and assess the health status of the patients.

Valuable patient information is secured using a lightweight secure block encryption approach (Akhbarifar, et al., 2020). The study provides many methods for remote monitoring, but none of the research suggests a unique, comprehensive, and integrated architecture that employs a COVID-19 beginning system that's also reliable and adaptable enough to be safely employed in patient rooms and at homes. The objective of (Paganelli, et al., 2022) is to present a comprehensive IoT-based theoretical model that explains the fundamental requirements of incorporation, perspective of remote patient monitoring in hospitals.

3. Proposed Methodology

A wireless device to record physiological information obtained from medical study participants including blood pressure, blood oxygen saturation level, electrocardiogram, respiratory rate, heartbeat, and respiratory rate transparency. In the proposed monitoring program, in which the prediction assessment is carried out, these physiological datasets pass through a wireless connection. Datasets are preprocessed and turned into extracted features that are utilized to teach the classifiers to operate. The database receives the data from the monitoring system over a network of wireless communications for remote monitoring. Figure 1 depicts the flow of the overall methodology.

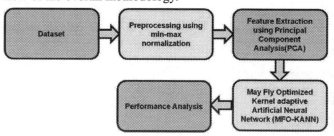

Fig. 66.1   The flow of the proposed methodology

Min-max normalization: Normalization is a preprocessing technique. Furthermore, preprocessing is a method used to clean up the initial information and improve the effectiveness of learning algorithms. In other terms, anytime data is acquired from various sources, it is done so in an unprocessed way that renders processing and machine learning impractical. Many well-known normalizing techniques are available, including Simple Feature Scaling, Min-Max, and others. Initially, using the continuity formula (1), the Simple Feature Scaling approach is generally used:

$$Y_{norm} = \frac{Y_{old}}{Y_{max}} \tag{1}$$

As a consequence of the data being reduced to a specified range in the Min-Max method typically 0 to 1, the standard deviations will decrease as a function of the constrained span.

Equation (2) is the Min-Max scale equation.

$$Y_{norm} = \frac{Y_{old} - Y_{min}}{Y_{max} - Y_{min}} \quad (2)$$

Principal component analysis (PCA): The PCA is a method of data feature reduction. The PCA has been extensively used in a variety of sectors since it is uncomplicated, simple to explain, and does not have any variable restrictions. To map n-dimensional features to k-dimensional features (k ≤ n), which is the basic goal of the PCA. The initial n-dimensional characteristics are transformed into new perpendicular main perspectives as principal components, which are the k-dimensional data. The main goal of the PCA is to minimize data duplication with the assumption that minimal details would be lost as feasible to meet the goal of feature extraction.

Kernel adaptive artificial neural networks: Artificial neural networks (ANNs) are parallel information processing techniques that may represent complex interactions and complicated utilize multiple input-output network parameters from the observational result. Accurate data preprocessing, the right choice in structure, and a strong network training option are all essential to the success of establishing resourceful and adaptable networks. The most accurate learning model for ANNs is the kernel neural network that was created. The network parameters are changed during back propagation (BP) training to lower the output values. At first, the system utilizes network parameters that are determined at random. Any data that are sent between the input and output layers in a feeds forward-BP algorithms. The networks subsequently calculate a desirable result, and the difference between the intended and observed outcomes is instead calculated. The difference between the actual and desired values is used to identify the network error. Back propagating the calculated error updates each weight individually. To minimize the mistake, this process is repeated. These are the stages that represent the other patterns situation for a feeds forward-BP algorithm. For the other pattern in the neural network, the j_{th} neuronal vendor's values of x_{oi}. The j_{th} molecule's net input for patterns o in the buried layers is;

$$net_{oi} = \sum_{j}^{M} U_{ji} P_{oj} \quad (3)$$

Where U_{ji} indicates the weight between neurons j and i. Every unit's i's outputs, (e_j), represent the level of decision making. According to this nonlinear function, e_j:

$$e(net) = \frac{1}{(1 + f^{-Lnet}}; (0 < e(net) < 1) \quad (4)$$

The ith neuron in the hidden layer produces the following outcome:

$$P_{oi} = e_i (net_{oi}) \quad (5)$$

The following are the outputs layer's l^{th} neuron's network input:

$$net_l = \sum_{i}^{M} U_{li} Y_{oi} \quad (6)$$

This U_{li} is the weighting factor between the jth hidden layer and the lth output layer.

$$P_{ol} = e_l (net_l) \quad (7)$$

The error function for a pattern p is called E_p.

$$F_o = \frac{1}{2} \sum_{l}^{M} (s_{ol} - P_{ol})^2 \quad (8)$$

When patterns o on layer l have intended and actual outputs, s_{ol} and P_{ol}, accordingly.

4. Mayfly Optimization

While using the MO algorithm, male and female mayflies in swarming would've been segregated. And male mayflies would always be powerful, making them do more in optimization. The people in the MO algorithm would change their placements under their actual positions o_l (s) and velocity u_i(s); similar to the particle swarm optimization (PSO) methods:

$$o_i(s+1) = o_i(s) + u_i(s+1) \quad (9)$$

Equation (9) could be employed to modify the placements of all male and female mayflies. They could, however, increase this velocity in various ways.

5. Results

We have conducted a comparative analysis of several classification techniques, including Radar SVM, Feature Fusion SVM, and KANN classifications. The analysis of principal components is used to eliminate noise and outliner from the wireless product's gathered data during processing so that the classification technique could make utilize the resulting valuable data. They are utilized to quickly spot defects and probable problem sites, allowing for a timely and accurate outcome. The efficiency of the recommended approach was assessed using common measures including accuracy, and precision. The comparison's findings are shown in Fig. 66.2 as the accuracy and precision rate for various classifiers. After employing around 30 features for the radar SVM separately and 10 features for the fusion SVM, the accuracy profiles as a function of feature number stabilize. The average accuracy and precision using those

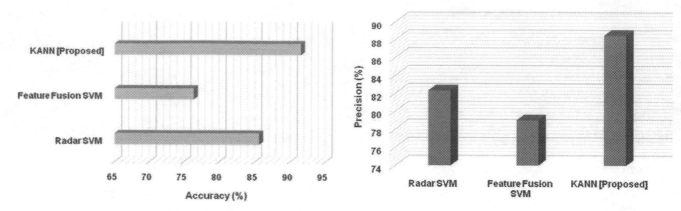

Fig. 66.2 Comparison of radar SVM, feature fusion SVM, and KANN

characteristics are in the range of 85.6-82.3% for the radar SVM and 76.4-79.0% for the feature fusion SVM. The categorization will not significantly benefit from the addition of more characteristics, and if all the features are employed, it may even conclude in decreased effectiveness. The accuracy and precision that result exceed the instances of using both sensors separately. Findings for radar SVM and feature fusion SVM seem to be fairly comparable; the suggested technique has the greatest result.

6. Conclusion

In the study, we suggest an IoT-based patient monitoring system. The IoT system is developed utilizing a microprocessor associated with a multitude of smart technologies and the internet. Devices are used to gather input data, which are then sent to the cloud for storage, in which the warning message is generated. With wearable devices, health professionals, caregivers, academics, specialists, etc. may reliably track participants' and patients' responses to a recently found drug from a distance, which can improve participant quality of life. Our proposed methodology serves as the foundation for remote patient monitoring. The main goal of our future study will be to suggest a motion sensor-based technique to monitor patients' activities.

References

1. Iranpak, S., Shahbahrami, A. and Shakeri, H., 2021. Remote patient monitoring and classifying using the internet of things platform combined with cloud computing. Journal of Big Data, 8, pp.1-22.
2. Ratta, P., Kaur, A., Sharma, S., Shabaz, M. and Dhiman, G., 2021. Application of blockchain and internet of things in healthcare and medical sector: applications, challenges, and future perspectives. Journal of Food Quality, 2021, pp.1-20.
3. Philip, N.Y., Rodrigues, J.J., Wang, H., Fong, S.J. and Chen, J., 2021. Internet of Things for in-home health monitoring systems: Current advances, challenges, and future directions. IEEE Journal on Selected Areas in Communications, 39(2), pp.300-310.
4. Akkaş, M.A., Sokullu, R. and Çetin, H.E., 2020. Healthcare and patient monitoring using IoT. Internet of Things, 11, p.100173.
5. Yew, H.T., Ng, M.F., Ping, S.Z., Chung, S.K., Chekima, A. and Dargham, J.A., 2020, February. IoT-based real-time remote patient monitoring system. In 2020 16th IEEE international colloquium on signal processing & its applications (CSPA) (pp. 176-179). IEEE.
6. Malasinghe, L.P., Ramzan, N. and Dahal, K., 2019. Remote patient monitoring: a comprehensive study. Journal of Ambient Intelligence and Humanized Computing, 10, pp.57-76.
7. Kaur, P., Kumar, R. and Kumar, M., 2019. A healthcare monitoring system using random forests and the internet of things (IoT). Multimedia Tools and Applications, 78, pp.19905-19916.
8. Hosseinzadeh, M., Koohpayehzadeh, J., Ghafour, M.Y., Ahmed, A.M., Asghari, P., Souri, A., Pourasghari, H. and Rezapour, A., 2020. An elderly health monitoring system based on biological and behavioral indicators in the internet of things. Journal of Ambient Intelligence and Humanized Computing, pp.1-11.
9. Akhbarifar, S., Javadi, H.H.S., Rahmani, A.M. and Hosseinzadeh, M., 2020. A secure remote health monitoring model for early disease diagnosis in a cloud-based IoT environment. Personal and Ubiquitous Computing, pp.1-17.
10. Paganelli, A.I., Velmovitsky, P.E., Miranda, P., Branco, A., Alencar, P., Cowan, D., Endler, M. and Morita, P.P., 2022. A conceptual IoT-based early-warning architecture for remote monitoring of COVID-19 patients in wards and at home. Internet of Things, 18, p.100399.

Note: All the figures in this chapter were made by the authors.

Advancing Sustainable Science and Technology for a Resilient Future – Sai Kiran Oruganti et al. (eds)
© 2024 Taylor & Francis Group, London, ISBN 978-1-032-79020-6

67

Implementation of a Smart Education System Based on IoT Enabled Decision Making Process

Sachin Mittal*

Chitkara University Institute of Engineering and Technology,
Chitkara University, Punjab, India

Abstract: The idea of smart education entails thoroughly modernizing all educational procedures as well as the techniques and tools applied to them. Users of smart education as well as the variety of programs available are both growing quickly. There are questions about how to raise the caliber of materials and distribution strategies. In pursuit of enhancing the smart education programs at higher learning institutions, this research will investigate the possible utilization of the Internet of Things (IoT) to gather data and analyze it. We suggest a unique greedy-adaptive random forest (GRF) strategy for decision-making in smart education in this research. Based on the samples of acquired data, the proposed approach is examined. The experiment results presented in this research demonstrate the effectiveness of the suggested GRF technique in the decision-making process for smart education.

Keywords: Smart education, Internet of things (IoT), Decision making, Greedy-adaptive random forest (GRF)

1. Introduction

The Internet of Things (IoT) has grown in prominence recently since it provides worldwide networks for connecting objects and gadgets to the data centers. As a consequence, the IoT makes it possible for people and things to interact at every time and from any location, enabling the worldwide identification and integration of information and understanding in addition to the generation of new information. Educational institutions should promote the inclusion of operations such as continuous instruction, investigation, and training, as well as their connection with the institutions of educational policy and the tools that link them to the labour market. In parallel, the IoT will bring about a number of changes in the field of academic achievement, including those related to technology (Web Coding, planning and testing, smartphone applications), educational reform, modifications in the way students are educated, useful and exploratory adjustments, changes in the institution, improvements to privacy protection, modifications to standards and morality, adjustments of a commercial intermediary, and some kinds of modifications (Mircea, et al., 2021).

The term "smart education" refers to a modern learning methodology for the digital age. It has captured the interest of many professionals from many scientific domains in latest generations. Digital education has enormous growth potential since it can be conducted anytime, anyplace, utilizing gadgets. The method for delivering and subscribing to the internet educational software's offerings may be a key element the achievement of intelligent instruction. Using the Internet of things (IoT) may be the best option in this case. Combining these ideas may enhance learning capacity by offering genuine possibilities for growth (Khujamatov,et al., 2020).

Academic achievement may best be described as the transfer and acquisition of information via a comprehensive connection among instructors and students, particularly at the primary level of education that starts in the classroom. During the decades, there have been many modifications to the teaching and learning processes. A new age of online teaching and learning has begun as a COVID-19 pandemic's effect. Education is performed in instructional settings utilizing basic instruments like a chalkboard, a cleaner for the chalk,

*Corresponding author: sachinm84824@gmail.com

DOI: 10.1201/9781003490210-67

manuals, and homework. In conventional education, pupils must enroll themselves and the teacher's attendance in the classroom is required. Another crucial element for keeping the courses going was regularly taking physical attendance of the pupils. There was no other resource accessible to learn more. Learners would have been denied a foundational knowledge if they had been unable to make it to the session (Pandey, et al., 2022, Irudayasamy, et al., 2022).

2. Related Works

The purpose of the analysis, according to (Zeeshan et al., 2022), is to demonstrate the capabilities of IoT applications in educational settings. This research article examines current work on IoT applications in education and provides a thorough analysis of the topic from three separate perspectives, namely those of classroom management, instructors, and learners. The present research shows that IoT has been applied to help educators, children, and educational authorities as evidenced in recent studies. The research also analyzes challenges for Iot systems and points out such impediments to IoT deployment in educational contexts as safety, secrecy, sustainability, dependability, and systematic oppression.

A synergistic framework was presented by (Iqbal et al., 2020) by fusing the essential components of smart technology with a flow theory approach, which can increase overall personalized learning experience while also preserving a smart educational and workplace environments. The provided framework incorporates a wide range of technical and communication tools to energize the learning experience and modify or sustain the needs of various pupils below one roof.

The purpose of this study is to understand the contributions that the growth of IoT makes to education, especially in aiding the decision-making processes and tackling some of the issues that the educational system finds, according to (Silva et al., 2020). This investigation employs qualitative methodology and bibliographic overview techniques to support "internet of things" notions in education and the decision-making process in education. It also conducts a situation theory-based analysis of the phenomenon. As a result, the latter research evaluated various situations for the use of this technology for instruction and a more effective management of educational organisational structures, as well as pointing to potential areas for the construction of alternatives when making decisions that have previously been hampered by the lack of available information devices.

Internet of things is a theme both (Rukmana and Mulyanti, 2020 and Bali, et al., 2021) analyze and collect information regarding. When web-based learning is the main objective, IoT can be used as a learning administration system. A study of the research from IoT-related conferences and web-based learning for smart school systems is the methodology adopted. The results show that operational services, service management, data processing, and training services can all be integrated with the Internet of Things.

Khujamatov et al., 2019 explore issues with the educational system and solutions based on Internet of Things (IoT) and Network equipment. An methodology for carrying out a contemporary learning experience known as "IoT based Centralized Double Stage Education," that combines contemporary technologies and conventional instructional techniques, has been devised.

3. Proposed Methodology

Several investigations regarding the application of the Internet of Things (IoT) in smart education demonstrate the value of the suggested GRF approach to growth in the decision-making process for smart education in the suggested methodology.

Dataset

Data is gathered by a variety of sensors and smartphones, and after going through a gateways, it is connected to the internet and kept in one location. Data will be kept in the cloud and subjected to analysis to draw out important information before being sent to the viewer.

Adaptive random forest

When employing the bagging approach to combine classification, the adaptive random forest method will generate numerous training sample sets that are distinct from one another. In this study, the random forest algorithm for decision-making in smart education was characterized by strong aptitude and excellent reputation in the capacity to classify. A set of decision-making in smart education is described as the adaptive random forest technique.

Greedy optimization

Here, we include a dataflow explanation of the suggested greedy weighting method. There are two inputs required: a vector z with predetermined values to be weighted for the columns in y, and a matrix z with each column representing a single feature for weighting.

Algorithm 1

Input: matrix Y, vector, convergence thresholds;

Output: vector u of optimized weights for each column

$m, c \leftarrow$

Number of rows and columns of Y;

$u \leftarrow v \leftarrow Nulivector\ of\ length\ c;$

O Nullvector of length n;

$t \leftarrow 0$;

While not converged do

    $t \leftarrow t + 1$

    For $i = 1$ to $c\ d\ o$

    $Y|i| \leftarrow metricy(z.(O + Y|:,i|)/t)/t)$;

    End

$$i_{max} \leftarrow \arg \overset{max}{\underset{i}{}} (v|i|);$$

    $O \leftarrow O+U[:,imax]$;

    $U[imax] \leftarrow u[imax]+1$;

End

Return u/t;

The unnormalized weighted sum is solved greedily in each step of the procedure, and the object O holding those results is created with zero entries. The overall number of weights as well as each column's weights are both started at zero. The total number of weights is increased in each iteration, first. After being normalized by the total number of weights, each sum of o with a column of y is then assessed independently using the evaluation measure. One weight factor is applied to and added to O for the highest value column. Iterations of this process are performed. In most cases, an iteration number of 100 is sufficient; larger values result in more accuracy but longer calculation times. A vector of length c that has the number of columns of y and weights for each column that add up to 1 is what the method outputs.

4. Result and Discussion

Applying classification algorithm in the procedure of decision making allows for the evaluation of the effectiveness of the suggested technique. Based on the number of metrics, such as accuracy, error, and ROC curve, our suggestion made by the proposed design would solve the issues with requirements like smart education systems based on decision-making procedures.

Accuracy

Accuracy is the degree of closeness between a measurement and its real value. Smart education is defined as the "efficient and consistent use of information and communication technology to achieve a teaching goal while utilizing an appropriate educational approach". Proposed method GRF is better than the existing method as shown in Fig. 67.1, which shows the accuracy of existing and proposed method.

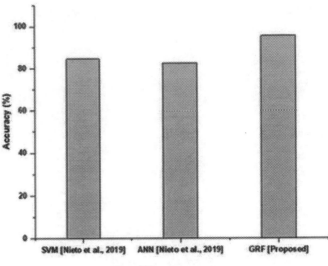

Fig. 67.1 Accuracy

Error

An error graph is a graphic representation of how changing parameters or variables affect a model's error or tested features. Figure 67.2 shows the proposed method is better than existing method.

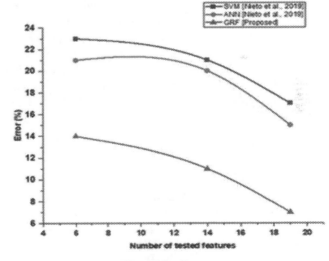

Fig. 67.2 Error

ROC curve

A graph demonstrating a categorization effectiveness of the algorithm at all modeling tools is known as a ROC curve, or receiver's operator characteristic curve. This graph shows the two parameters. Figure 67.3 shows the ROC curve where the proposed method is better than that existing method.

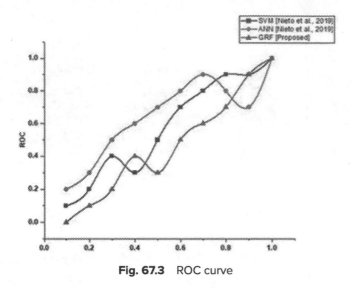

Fig. 67.3 ROC curve

5. Conclusion

The IoT is being used to provide a framework for collecting, storing, and sending information to a single network. The management will be able to learn from the data analysis about areas that need to be addressed in order to raise the standard of students' complete assessment. By putting forth fresh models, we will implement IOT in a number of significant industries in the future, including education and agribusiness. Additionally, we will expand this study concept to include all academic institutions. Also, as a future possible improvement with the additional features of educational system, this application could be fitted with internet of things (IoT) capabilities.

References

1. Mircea, M., Stoica, M. and Ghilic-Micu, B., 2021. Investigating the impact of the internet of things in higher education environment. IEEE Access, 9, pp.33396-33409.
2. Khujamatov, K., Reypnazarov, E., Akhmedov, N. and Khasanov, D., 2020, November. IoT based centralized double stage education. In 2020 International Conference on Information Science and Communications Technologies (ICISCT) (pp. 1-5). IEEE.
3. Pandey, D., Singh, N., Singh, V. and Khan, M.W., 2022. Paradigms of smart education with IoT approach. Internet of things and its applications, pp.223-233.
4. Irudayasamy, A., Christotodoss, P.R. and Natarajan, R., 2022. Multilingual Novel Summarizer for Visually Challenged Peoples. In Handbook of Research on Technologies and Systems for E-Collaboration during Global Crises (pp. 27-46). IGI Global.
5. Eeshwaroju, S., Jakkula, P. and Ganesan, S., 2020, September. Iot based empowerment by smart health monitoring, smart education and smart jobs. In 2020 International Conference on Computing and Information Technology (ICCIT-1441) (pp. 1-5). IEEE.
6. Khalaf, "Blinder Oaxaca and Wilk Neutrosophic Fuzzy Set-based IoT Sensor Communication for Remote Healthcare Analysis," in IEEE Access, 2022,
7. Zeeshan, K., Hämäläinen, T. and Neittaanmäki, P., 2022. Internet of Things for sustainable smart education: An overview. Sustainability, 14(7), p.4293.
8. Iqbal, H.M., Parra-Saldivar, R., Zavala-Yoe, R. and Ramirez-Mendoza, R.A., 2020. Smart educational tools and learning management systems: supportive framework. International journal on interactive design and manufacturing (IJIDeM), 14, pp.1179-1193.
9. Bali, M.S., Gupta, K., Koundal, D., Zaguia, A., Mahajan, S. and Pandit, A.K., 2021. Smart architectural framework for symmetrical data offloading in IoT. Symmetry, 13(10), p.1889.
10. Silva, R., Bernardo, C.D.P., Watanabe, C.Y.V., Silva, R.M.P.D. and Neto, J.M.D.S., 2020. Contributions of the internet of things in education as support tool in the educational management decision-making process. International Journal of Innovation and Learning, 27(2), pp.175-196.
11. Rukmana, A.A. and Mulyanti, B., 2020, April. Internet of Things (IoT): Web learning for smart school system. In IOP Conference Series: Materials Science and Engineering (Vol. 830, No. 3, p. 032042). IOP Publishing.
12. Sundas, A. and Panda, S.N., 2020, June. IoT based integrated technologies for garbage monitoring system. In 2020 8th International Conference on Reliability, Infocom Technologies and Optimization (Trends and Future Directions) (ICRITO) (pp. 57-62). IEEE.
13. Khujamatov, K., Reypnazarov, E., Akhmedov, N. and Khasanov, D., 2020, November. IoT based centralized double stage education. In 2020 International Conference on Information Science and Communications Technologies (ICISCT) (pp. 1-5). IEEE.

Note: All the figures in this chapter were made by the authors.

Advancing Sustainable Science and Technology for a Resilient Future – Sai Kiran Oruganti et al. (eds)
© 2024 Taylor & Francis Group, London, ISBN 978-1-032-79020-6

68

Enhancing Quality of Service in IoT Based Intelligent Transportation System

Abhiraj Malhotra*

Chitkara University Institute of Engineering and Technology,
Chitkara University, Punjab, India

Abstract: Applications have evolved due to the Internet of Things (IoT), and they can now be used in numerous areas of a modern city due to intelligent technologies. When the volume of the information collected increases, machine learning (ML) techniques are utilized to further enhance a user's knowledge and capabilities. The field of intelligent transportation, which has been studied using both ML and IoT techniques, has attracted a number of academics. Yet, the quality of service (QoS) in smart transport networks is still lacking. Hence, in this research, we introduce a neural-encoded long/short-term memory (N-LSTM). We have acquired a traffic flow dataset to analyze the effectiveness of the suggested N-LSTM approach. The experiment is conducted with the aid of the MATLAB tool, and the outcomes of the proposed strategy are examined. The findings demonstrate that the suggested approach maximizes QoS efficiency in the intelligent transportation network.

Keywords: Internet of things (IoT), Intelligent transportation, Quality of service (QoS), Neural-encoded long/short-term memory (N-LSTM)

1. Introduction

IoT connects the physical world's objects to the digital world and makes it possible to stay connected at any time and from any location using any device that has a physical on/off switch. It gives rise to a world in which real-world devices and living beings, in addition to data and environments created entirely in virtual space, interact with one another. The IoT is based on the idea of transforming a wide variety of everyday items into high-tech ones. In the development of successful system, this enormous amount of data needs to be manage and transformed into information that can be put to good use. Big Data Analytics is a crucial component in the process of deriving actionable insights from the data that is produced (Muthuramalingam et al.,2019). Technology has improved the lives of the average person. Still, due to high mobility and the source of energy nature of compact devices at the edge of IoT-based information technology, (QoS) for multimedia on smart phones and portable IoT devices has been compromised. To facilitate multimedia transmission in ITS, it is necessary

to develop a Data processing Green, Durable, Believable, and Available algorithm. (Sodhro, ,et al.,2019) According to popular belief, intelligent surroundings will improve QoS available to society and will optimize these services via unbiased data collecting. Several sensors and webcams are installed in the intelligent transport environment, producing a wide range of data known as Big Data. This information aids in the processing, analysis, observation, and prediction of the previously unknown patterns used to deliver services to society (Babar,et al.,2019) Future networks specifically aim to create a platform that is green, sustainable, dependable, and accessible while also meeting demands from end users and the network infrastructure. By sending huge data rates across wide throughput at any time and everywhere, computing may also be leveraged to enable the high - speed with a flexible and adaptive multimedia platform (Gupta, et al.,2022). Information and communication technology (ICT) advancements in the fields of hardware, software, and telecommunications have opened up new potential for the creation of a smart, environmentally friendly transportation

*Corresponding author: abhiraj.malhotra.orp@chitkara.edu.in

DOI: 10.1201/9781003490210-68

system. The adoption of Intelligent Transportation Systems (ITS), which place a strong emphasis, will be made possible by the integration of communication technologies with the transportation infrastructure. ITS will improve and make travelling safer. These guiding principles will be crucial to attaining the access and mobility, environmental protection, and economic growth goals of intelligent transportation systems (Kaushal, et al.,2022).

By fusing cutting-edge technologies, data analytics, and communication systems, intelligent transportation systems seek to improve the effectiveness, safety, and sustainability of transportation networks. They make it possible to monitor traffic in real-time, make intelligent decisions, and allocate resources efficiently. As a result, traffic management is improved, congestion is decreased, safety is increased, and the whole transportation experience is improved.

2. Related Works

In the most important Smart Cities scenarios, the most considerable Internet of Things (IoT) systems are examined for ITS (Brincat et al., 2019). Javed et al., 2018 emphasized the connection between safety and service level for ITS applications. By discussing the fundamental communications architecture of ITS, they look at the main design challenge related to safety and QoS. Gaber, et al.,2018 offered a bioinspired and trust-based method for selecting cluster heads in WSN used for ITS. Gillam, et al.,2021 offered a classification of use cases and a thorough analysis of the research pertaining to specific use cases. Ning, et al.,2017 built a CQS system focusing on service quality promotion and reliability assurance in SIoVs. As a first step in addressing the impact of the active network change, they investigate a dynamically access service valuation system.

3. Proposed Overview

NLSTM

LSTM has been successfully applied to many different areas, including music production, image captioning, speech recognition, and machine translation. On the foundation of RNN's hidden-layer cell, LSTM enhances it. Cell enhancements can help solve RNN's disappearing gradient issue the forget gate, input gate, and output gate are all memory units that NLSTM employs. The information can be selectively retained or discarded based on the preferences of the memory units. They need to consider both the downstream and upstream traffic patterns when making predictions about traffic at the current site. The data dimensionality will rise, and the computation will become overly complex if all of the downstream and upstream traffic data are added to the prediction model. The autoencoder is used to extract features of the preceding and following traffic flows, thereby resolving the issue. That is to say; there is a lessening in the number of dimensions in the traffic data. The learned properties are incorporated into the prediction network's training set. This not only takes into account the effects of both downstream and upstream traffic flows but it also keeps the data dimension from growing too large. NLSTM takes in two types of information: the current position's traffic history (mt) and the characteristics of the upstream and downstream flows (nt). Current flow mt, upstream and downstream characteristics nt, and the previous unit state make up the forgotten gate's input data. The forgetting gate decides what data can safely be forgotten. The input gate's data is analogous to that of a forgetting gate. Like the forgetting gate, the input gate takes in data from the outside world. The purpose of the input gate is to determine what data should be fed into the system. The cell's input, denoted by the symbol S_t, is summed with the existing state of the cell. Any value of S_t after a cell's status has been modified. This requires a net loss of cellular state from the prior time period and a net gain of cellular state in the present. Output gate input data is analogous to forget gate input data.

$$g_t = \sigma\left(\omega_{b1}.m_t + \omega_{b2}.n_t + \omega_{b3}.p_{t-1} + c_f\right) \tag{1}$$

$$l_t = \sigma\left(\omega_{l1}.m_t + \omega_{l2}.n_t + \omega_{l3}.p_{t-1} + c_l\right) \tag{2}$$

$$\overline{s_t} = \tan h\left(\omega_{s1}.m_t + \omega_{s2}.n_t + \omega_{s3}.p_{t-1} + c_s\right) \tag{3}$$

$$S_t = g_t.s_{t-1} + l_t.\overline{s_t} \tag{4}$$

$$u_t = \sigma\left(\omega_{u1}.m_t + \omega_{u2}.n_t + \omega_{u3}.p_{t-1} + c_u\right) \tag{5}$$

$$p_t = u_t.\tan h\left(s_t\right) \tag{6}$$

The information to be transmitted is chosen at the output gate. Lastly, tanh is applied to the cell state and multiplied by the product of the output of the output gate, where $\sigma(m) = \dfrac{1}{1+e^{-m}}$ and $\tanh(m) = \dfrac{e^m - e^{-m}}{e^m + e^{-m}}$. We present an NLSTM prediction model to take into account the impact

of both upstream and downstream traffic flows without significantly increasing the number of calculations required. We begin by using the AutoEncoder to learn from what happened upstream and what happened downstream. After that, we make transportation forecasts using an NLSTM model. To extract features from upstream and downstream traffic flow data, the AutoEncoder's encoder is employed as a feature extractor. The prediction network receives the extracted features. The quality of forecasting can be enhanced by taking into account the effect of locations upstream and downstream on the flow of traffic at the present position.

NLSTM's input combines upstream and downstream traffic flow characteristics and the present position's traffic flow data. The NLSTM model anticipates the following moment's traffic flow data. A form of unsubstantiated learning, the AutoEncoder can be implemented as a data feature extractor. Before training, we determined the starting amount of the weight matrix. The weight matrix's significance to the network cannot be overstated. After learning the weight matrix, we'd like the data to still exhibit its original properties. When the extracted feature faithfully recreates the original data, the weight matrix successfully preserved the original data's characteristics. After the AutoEncoder has been trained, it is separated into an encoder and a decoder. The encoder is an element of the NLSTM prediction model that serves to extract data features. After training is complete, the decoder is thrown away after having been utilized to ensure the accuracy of the retrieved features. To create the NLSTM prediction model, we appended an NLSTM model to the end of the encoder.

The network's final output layer has been activated using the Rectified Linear Unit (ReLU) function. Other network layers have their activation functions set to tanh. The loss function in this paper is

$$M\left(a, \overline{a}\right) = \frac{\sum_{n=1}^{N}\left(\overline{a_n} - a_n\right)}{2N} \tag{7}$$

N is the sum number of predictions, *an* is the observed value, and is the forecast value.

4. Result and Discussion

In this section, an analysis of the NLSTMs through put, delay, coverage ratio, and quality of service optimization is carried out. In addition, the MFIS, LSTM, and FDL-IDF methods that are currently in use are contrasted with the NLSTM that has been proposed. The quality of service optimization and coverage ratio of the outcomes of the efficacy of the proposed NLSTM is expressed as the proportion of data for which the outcome was successfully anticipated. These ratios are derived from both the recommended and current techniques. According to Fig. 68.1, the suggested technique outperforms LSTM (QoS optimization = 75.66%; coverage ratio = 83.69%), MFIS (QoS optimization = 78.66%; coverage ratio = 79.35%), and FDL-IDF (QoS optimization =85.63%; coverage ratio = 88.12%) in terms of QoS optimization (91.56%) and coverage ratio (90.54%).

Show the results of the throughput testing suggested and the actual tactics and throughput (in kilobytes/s kbps). During our analysis of the efficiency results in relation to throughput, we made the discovery that the throughput of the system increases proportionately with the magnitude of the operation. Deduce from this Fig. 68.2 that the suggested

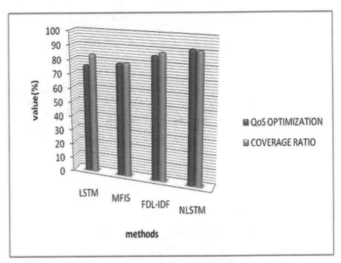

Fig. 68.1 QoS optimization and coverage ratio

NLSTM technique outperforms existing approaches LSTM (throughput =1200 kbps), MFIS (throughput =1856 kbps), and FDL-IDF (throughput =2356 kbps) in terms of throughput (2945 kbps).

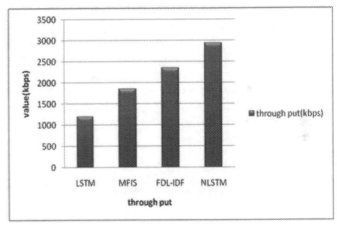

Fig. 68.2 Throughput

When there is a higher chance that the message will be successfully transmitted, the propagation delay will be shorter. When there is a one hundred percent chance that the demonstration will be successful, the time allotted is at its shortest. Hence, the propagation delay will be reduced if there is a more significant possibility that the signal will be successfully transmitted. Deduce from this Fig. 68.3 that the suggested NLSTM technique outperforms existing approaches LSTM (delay=0.45 s), MFIS (delay =0.65 s), and FDL-IDF (delay=0.48 s) in terms of delay (0.11 s).

5. Conclusion

QoS optimization methodology for ITS in the IoT was proposed in the article. It is absolutely necessary to manage

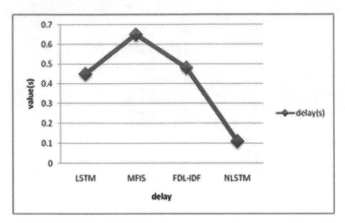

Fig. 68.3 Delay

and monitor the QoS from both the network and the user's point of view due to the more dynamic and resource-sharing characteristics of transportation. Images of users and networks that are crystal clear, well-lit, and of a high resolution are required for the use case that is being examined for QoS optimization when intelligent transmission computing platforms are in operation. In this regard; we presented the NLSTM algorithm as a solution that is ideally suited for QoS in IoT-based ITS. The suggested method has been shown to work with real-time data sets. Its performance has been evaluated and compared to that of more traditional ways of competition in terms of several performance indicators. The results that were collected confirmed that the NLSTM scheme is a promising option for optimizing the QoS during the operation of an ITS. In the future, we intend to create intelligent transportation networks that take the QoS into consideration.

References

1. Muthuramalingam, S., Bharathi, A., Rakesh Kumar, S., Gayathri, N., Sathiyaraj, R. and Balamurugan, B., 2019. IoT based intelligent transportation system (IoT-ITS) for global perspective: A case study. Internet of Things and Big Data Analytics for Smart Generation, pp.279-300.

2. Sodhro, A.H., Obaidat, M.S., Abbasi, Q.H., Pace, P., Pirbhulal, S., Fortino, G., Imran, M.A. and Qaraqe, M., 2019. Quality of service optimization in an IoT-driven intelligent transportation system. IEEE Wireless Communications, 26(6), pp.10-17.

3. Babar, M. and Arif, F., 2019. Real-time data processing scheme using big data analytics in internet of things based smart transportation environment. Journal of Ambient Intelligence and Humanized Computing, 10, pp.4167-4177.

4. Gupta, S.K., Pattnaik, B., Agrawal, V., Boddu, R.S.K., Srivastava, A. and Hazela, B., 2022, September. Malware Detection Using Genetic Cascaded Support Vector Machine Classifier in Internet of Things. In 2022 Second International Conference on Computer Science, Engineering and Applications (ICCSEA) (pp. 1-6). IEEE.

5. Kaushal, R.K., Bhardwaj, R., Kumar, N., Aljohani, A.A., Gupta, S.K., Singh, P. and Purohit, N., 2022. Using Mobile Computing to Provide a Smart and Secure Internet of Things (IoT) Framework for Medical Applications. Wireless Communications and Mobile Computing, 2022.

6. Brincat, A.A., Pacifici, F., Martinaglia, S. and Mazzola, F., 2019, April. The internet of things for intelligent transportation systems in real smart cities scenarios. In 2019 IEEE 5th World Forum on Internet of Things (WF-IoT) (pp. 128-132). IEEE.

7. Javed, M.A., Hamida, E.B., Al-Fuqaha, A. and Bhargava, B., 2018. Adaptive security for intelligent transport system applications. IEEE Intelligent Transportation Systems Magazine, 10(2), pp.110-120.

8. Gaber, T., Abdelwahab, S., Elhoseny, M. and Hassanien, A.E., 2018. Trust-based secure clustering in WSN-based intelligent transportation systems. Computer Networks, 146, pp.151-158.

9. Arthurs, P., Gillam, L., Krause, P., Wang, N., Halder, K. and Mouzakitis, A., 2021. A taxonomy and survey of edge cloud computing for intelligent transportation systems and connected vehicles. IEEE Transactions on Intelligent Transportation Systems.

10. Ning, Z., Hu, X., Chen, Z., Zhou, M., Hu, B., Cheng, J. and Obaidat, M.S., 2017. A cooperative quality-aware service access system for social Internet of vehicles. IEEE Internet of Things Journal, 5(4), pp.2506-2517.

Note: All the figures in this chapter were made by the authors.

Advancing Sustainable Science and Technology for a Resilient Future – Sai Kiran Oruganti et al. (eds)
© 2024 Taylor & Francis Group, London, ISBN 978-1-032-79020-6

69

An Automated Smart Agricultural System Using Internet of Things

Amanveer Singh*

Chitkara University Institute of Engineering and Technology,
Chitkara University, Punjab, India

Abstract: The agricultural sector is crucial to the national economy. The internet of things (IoT) is a recent technological advancement in the digital age, whereby devices are able to connect with one another and processes are automated and controlled with the aid of the internet. This research suggests a smart irrigation system that forecasts a crop's water needs using machine learning algorithms in order to reduce crop loss during harvest or post-harvest. The three most important variables to consider when estimating the amount of water needed in each agricultural crop are moisture, temperature, and humidity. In this setup, sensors for measuring temperature, humidity, and moisture are placed in a farm's fields and data is sent through a microprocessor to a cloud-based IoT device. The data collected in the field is fed into a novel machine learning method called a binary seagull optimized dynamic decision tree (BSODDT), which is then used to make accurate predictions.

Keywords: Agricultural sector, Internet of things (IoT), Smart irrigation system, Binary seagull optimized dynamic decision tree (BSODDT)

1. Introduction

The agriculture sector's sustainability is essential for providing food security and the reduction of poverty for the world's continually growing population. The appearance of various food safety controversies and mishaps in the food business, such as dioxin in poultry, has also made a well-documented tracking system a requirement for QC in the food supply chain. It will be particularly difficult to manage water sustainably because of water shortages and weather and climate change circumstances in the following years. Due to these elements, it is imperative to create a strategy shift that moves away from the current paradigm of increased agricultural output and toward agriculture. Cloud computing is a paradigm for collaborative and social computing that is gaining popularity because it provides users with on-demand access to various services (Mahalakshmi, et al., 2020).

Krishna, et al., 2017 despite its computing structure, this computing paradigm is scalable and capable of operating in diverse applications. It offers a wide range of services, with software, architecture, and databases standing out. The novel application model put forward in this study is created as a symmetric key encryption service to store the data collected from the field-installed built-in sensors. Computer resources are available as a service and wherever you need them, thanks to the cloud, a fast-evolving computing paradigm in the age of innovation. The scalability and ease of access that cloud computing offers, together with the shared resource pool it provides, distinguish it as a successful computing paradigm (Bali, et al., 2021). The integrity of the data is protected based on the security guidelines provided by cloud service providers. Although being outsourced, receptive data should maintain a high level of security, including a patient medical background in the form of scanners and reports, insurance plans, and data. In relation to this, one of the most intriguing application services provided by the cloud computing paradigm that advises customers to encrypt the data obtained from the built-in sensors using application models is encryption-as-a-service (Sundas, et al., 2020). Tensiometric and volumetric approaches are used for irrigation management dependent

*Corresponding author: amanveer.singh.orp@chitkara.edu.in

DOI: 10.1201/9781003490210-69

on soil moisture. Although these techniques are relatively straightforward, the quantities they measure are associated with soil water curves that are unique to each kind of soil. Also, for optimal functionality, the sensors that are employed must undergo regular maintenance. Using a moisture sensor, an intelligent autonomous plant irrigation system provides consistent watering to plants without the need for human supervision. By incorporating IoT technology to monitor and manage numerous agricultural operations, the system's primary goal is to increase productivity and efficacy in agriculture.

2. Related Works

Nuvvula, et al., 2017 proposed Controlled Environment Agriculture (CEA) that may manually change the temperature to automate the operation. The farmer might adjust air conditioning based on device humidity measurements. Goap, et al., 2018 used a smart system based on an open-source algorithm that uses the information from sensors and weather forecasts to figure out how wet the soil will be in the next few days. Sharma, et al., 2022 applied an image-based normalization approach. Thus the network cannot be directly employed to solve prediction issues in intelligent agricultural systems. The simulation of irrigation parameters, using user-friendly interfaces, is used to complete the decision whenever there is a change in the climate of the agricultural environment (Rowshon, et al., 2017). Pavithra, et al., 2017 utilized a Global System for Mobile Communication (GSM) module for monitoring the water level tank and suggesting the exact amount of water that was required for the crops. Also, the system monitors temperature and humidity in order to maintain an appropriate level of nutrients in the soil, which is essential for the growth of plants.

3. Methodology

As in Table 69.1, the master node table receives real-time sensed data that are facts and stores them in the cloud for eventual use in the decision (Munir, et al., 2019). The microcontroller communicated the sensed information from the light, Moisture, sensors, Temperature, and Humidity, which was received at the server side.

A Binary Seagull Optimized Dynamic Decision Tree (BSODDT)

We determine the new search agent (SA) position using an additional variable, A, to prevent collisions between nearby SA.

$$D_t = B \times O_t \tag{1}$$

Where represents the current position of the SA and indicates the position of the SA that avoids interacting with other SA, x denotes the current iteration, while denotes the SA's movement patterns within a specific search space

$$B = e_d - \left(w \times \left(e_d \: / \: Max_{iteration} \right) \right) \tag{2}$$

e_d is used to regulate the frequency of using variable , which is gradually decreased from e_d to 0_{at}. After avoiding neighbourly collisions, the SAs turn toward the best neighbor.

$$N_t = A \times \left(O_{at}(w) - O_t(w) \right) \tag{3}$$

Where stands for the locations of SA, in relation to the most appropriate SA . The random behaviour of is in charge of maintaining a good equilibrium between exploration and extraction. is determined as:

$$A = 2 \times B^2 \times qc \tag{4}$$

Table 69.1 Data collected from sensors

Light	Moisture	Temperature	Humidity	created_at
34	343	31.5	77.7	2017–10–12T15:27:19.0570000
48	1011	25.7	73.2	2017–10–13T08:21:55.8730000
3	3	1	3	2018–07–16T19:07:05.0900000
12	12	12	12	2018–07–16T18:19:11.0630000
9	6	6	9	2018–07–14T20:12:54.9570000
49	982	25.7	71.7	2017–10–13T08:23:17.3530000
1023	42	29.2	97.7	2018–07–13T20:23:41.6200000
12	19	10	12	2018–07–12T20:47:27.9600000
12	13	10	12	2018–07–12T10:26:49.7000000
12	13	10	12	2018–07–11T14:27:38.9900000

Where [0, 1] is the range for the random number rd. Finally, the SA might revise its ranking as the best SA by

$$C_t = |\ D_t + N_t\ | \qquad (5)$$

The distance between the SA and the best-fit SA is represented by C_t.

A supervised learning algorithm called the decision tree classifier is employed in this experiment. It is CART-based and works to build classification and regression trees. Several of the concepts described in the CART model are implemented by the R programming package r-part. When separating the tree's nodes using r-part function, many splitting criteria can be used.

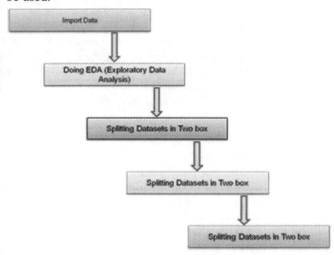

Fig. 69.1 Flow diagram of dynamic decision tree

4. Results

The existing techniques are RASN, SWS, and WSN compared with our proposed approach binary seagull optimized dynamic decision tree (BSODDT) and accuracy, precision, and recall utilized.

The degree to measurement corresponds to its actual value is referred to as its accuracy as in Equation 6. Figure 69.4 depicts the accuracy comparison. RASN, SWS, and WSN these are the existing technique which have values of accuracy 91%, 92.7%, and 97.3%, and our proposed approach BSODDT is 99.4% accuracy. It demonstrates that our proposed method is more accurate when compared to existing approach. Where (P) denotes true positives, (Q) denotes false positives, (R) represents true negatives, and (S) represents false negatives.

$$Accuracy = \frac{Q+P}{Q+P+S+R} \qquad (6)$$

Precision is the ratio of the number of positive samples that can be expected to the number of positive samples that can be reliably predicted, as in Equation 7. Figure 69.3 depicts

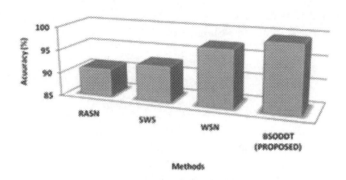

Fig. 69.2 Accuracy results

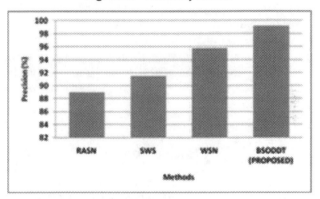

Fig. 69.3 Precision

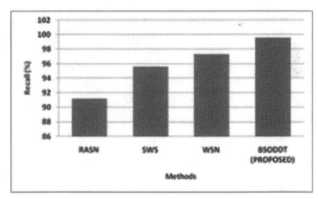

Fig. 69.4 Recall results

precision comparison for existing and proposed approach. The existing approaches, RASN, SWS, and WSN, which have 89%, 91.5%, and 95.8%, and our proposed BSODDT has 99.2%.It illustrates that our proposed technique is higher in precision

$$Precision = \frac{P}{P+R} \qquad (7)$$

The recall is calculated from the proportion of A using the following Equation 8. The comparison of recall between the proposed approach and the existing techniques is shown in Fig. 69.4. The existing techniques RASN, SWS, and WSN, which have values of 91.2%, 95.6%, and 97.3% and the proposed approach BSODDT of 99.6%.

$$Recall = \frac{P}{P+S} \qquad (8)$$

5. Conclusion

Agriculture is essential to the national economy. In this study, we proposed a smart irrigation system which predicts a crop's water requirements using a machine learning method called a binary seagull optimized dynamic decision tree (BSODDT). The Values of results metrics for our proposed method were obtained in terms of accuracy (99.4%), precision (99.2%), and recall (99.6%). Our proposed method have some drawbacks, such as limited population diversity, a tendency to enter local optimum, and low convergence accuracy, particularly for high dimensional. Furthermore, if a water pipeline were to leak or become disorganized remotely, water leaking may be monitored.

References

1. Mahalakshmi, J., Kuppusamy, K., Kaleeswari, C. and Maheswari, P., 2020. IoT sensor-based smart agricultural system. In *Emerging Technologies for Agriculture and Environment: Select Proceedings of ITsFEW 2018* (pp. 39-52). Springer Singapore.

2. Krishna, K.L., Silver, O., Malende, W.F. and Anuradha, K., 2017, February. Internet of Things application for implementation of smart agriculture system. In *2017 International Conference on I-SMAC (IoT in Social, Mobile, Analytics and Cloud)(I-SMAC)* (pp. 54-59). IEEE.

3. Naresh, M. and Munaswamy, P., 2019. Smart agriculture system using IoT technology. *International journal of recent technology and engineering*, 7(5), pp.98-102.

4. Rajeswari, S.K.R.K.A., Suthendran, K. and Rajakumar, K., 2017, June. A smart agricultural model by integrating IoT, mobile and cloud-based big data analytics. In *2017 international conference on intelligent computing and control (I2C2)* (pp. 1-5). IEEE.

5. Kitouni, I., Benmerzoug, D. and Lezzar, F., 2018. Smart agricultural enterprise system based on integration of internet of things and agent technology. *Journal of Organizational and End User Computing (JOEUC)*, 30(4), pp.64-82.

6. Nuvvula, J., Adiraju, S., Mubin, S., Shahana, B. and Valisetty, V., 2017. Environmental smart agriculture monitoring system using internet of things. *International Journal of Pure and Applied Mathematics*, 115(6), pp.313-320.

7. Goap, A., Sharma, D., Shukla, A.K. and Krishna, C.R., 2018. An IoT based smart irrigation management system using Machine learning and open source technologies. *Computers and electronics in agriculture*, 155, pp.41-49.

8. Bali, M.S., Gupta, K., Koundal, D., Zaguia, A., Mahajan, S. and Pandit, A.K., 2021. Smart architectural framework for symmetrical data offloading in IoT. *Symmetry*, 13(10), p.1889.

9. Sundas, A. and Panda, S.N., 2020, June. IoT based integrated technologies for garbage monitoring system. In *2020 8th International Conference on Reliability, Infocom Technologies and Optimization (Trends and Future Directions) (ICRITO)* (pp. 57-62). IEEE.

10. Munir, M.S., Bajwa, I.S. and Cheema, S.M., 2019. An intelligent and secure smart watering system using fuzzy logic and blockchain. Computers & Electrical Engineering, 77, pp.109-119.

Note: All the figures and table in this chapter were made by the authors.

Advancing Sustainable Science and Technology for a Resilient Future – Sai Kiran Oruganti et al. (eds)
© 2024 Taylor & Francis Group, London, ISBN 978-1-032-79020-6

70

Design of a Novel Secured Information Transmission System in Industrial Internet of Things

Aseem Aneja*

Chitkara University Institute of Engineering and Technology, Chitkara University, Punjab, India

Abstract: To do various types of analytics for the Industrial Internet of Things (IIoT), vast amounts of data are processed at the server and edge. For critical examination and decision-making, the process of learning in such an analytical solution must adhere to a dependable and trusted life span. Similarly to this, the process of learning must be dependable and trustworthy taking into account the security flaws in differing stages of an IIoT system. For safe data transmission in the IIoT, we suggest a unique manta ray foraging optimized preventive deep neural network (MFO-PDNN) based on deep learning (DL). The cluster head (CH) screening process in this case uses the MFO technique. The PDNN technique is then used to send data to the base station via the chosen CH. Out of these methodologies, the suggested PDNN-based routing algorithm has got excellent network life.

Keywords: Industrial internet of things (IIoT), Data analysis, Deep learning (DL), Manta ray foraging optimized preventive deep neural network (MFO-PDNN)

1. Inroduction

The connecting of real items outfitted with sensors, software, and other technologies for information exchange and communication with other devices and networks through the Internet is referred to informally as the Internet of Things (IoT). The encryptions Transport Layer Security (TLS) and Secure Sockets Layer (SSL), which it replaced, provide secure Internet connections for actions including web surfing, e-mail, texting, faxing online, and other data transfers (Ramlowat, et al., 2019). The industrial Internet of Things (IIoT) was created in recent decades by the fusion of traditional production engineering, automation, and intelligent computing systems IoT. Industrial control systems, production systems, and factories are incorporating an increasing number of computing components. The broad IoT development includes the IIoT. However, since it must integrate programmable logic controllers (PLC) with systems for acquiring data and exercising control, it has difficulties that set it apart from other IoT devices and services (Kaushal, et al., 2022). One of the newest technologies is the IoT, which uses networked sensors and actuators to allow autonomous machines, instruments, and gadgets. This technology is also a component of the infrastructure for sustainable and smart cities, particularly in the IIoT. Industrial devices are fulfilling their functions in the IIoT to improve industrial operations' productivity, performance, and maintenance (Qureshi, et al., 2020). The main focus of smart cities is the industrial sector, which integrates smart technology to boost the efficiency and dependability of conventional sectors. The IIoT can be most effectively integrated into commercial techniques with strict security requirements by using various industry norms and guidelines for manufacturing devices, interaction protocols, and cybersecurity products and services. Regarding company structure needs and recent connection technologies, the goal of this research is to summarise the level of liberty and safety in IIoT communications systems (Gebremichael, et al., 2020). Using deep learning in the data center to facilitate IIoT systems would therefore lead to congestion control and have an impact on IIoT systems that are frequently delayed because of the huge amount of data transferred between servers and intelligent systems. Data sets are typically necessary to obtain trustworthy outcomes (Bali, et al., 2021). This process of learning must be dependable and trustworthy taking into account the security flaws in differing stages of an IIoT system. The rest of this manuscript is structured as

*Corresponding author: aseem.aneja.orp@chitkara.edu.in

DOI: 10.1201/9781003490210-70

follows: Section II describes the related work, section III the recommended approach, section IV the results, and section V the manuscript's conclusion

2. Related Work

Serror, et al., 2020 defined the specific security objectives and difficulties that the IIoT faces, which, in contrast to consumer installations, mostly stem from efficiency and hazard demands. Although having many parallels to the retail IoT, the IIoT presents a unique set of difficulties and possibilities for security, mostly because its components have a longer lifespan and bigger networks. Refaee, et al., 2022 method examined the utilization of the dynamic secret-sharing mechanism based on the blockchain. The power blockchain sharing approach, which may also share power trade books, allows for the realization of a trustworthy trading hub. Power data transmission security matching is made possible by the power data consensus technique and dynamically linked storage. Tests demonstrate the excellent security and dependability of the optimized Fabric power data transfer and storage. Tang, et al., 2020 discussed the problem of secure transmissions for the wireless energy harvesting Internet of Medical Things (IoMT) based on cognitive radio. In such devices, a main transmitter transmits sensitive medical data to a primary receiver through a secondary transmitter with several antennas, where we assume that a possible listener may hear the PT's sensitive data. Sundas, et al., 2020 presents a dependable Next Generation Cyber Security Architecture for the IIoT atmosphere that identifies and counters cybercrime threats and vulnerabilities. It aids in automating the procedures of sharing crucial real-time data across technologies that don't involve humans. It provides an analytical approach that might be used to secure wireless IIoT connections between businesses and internet activities. Pal, et al., 2021 examined and contrast the security problems that may exist in an IIoT system. It presents some security concerns from logical, technical, and architectural angles while taking into account the various IIoT security standards. To investigate such security issues methodically, they also to investigate such security issues methodically, they also they demonstrate whether various security problems influence the operation of various levels of an IIoT architecture and provide a list of suggested prevention methods. The investigation additionally offers a collection of possible upcoming fields of study for the creation of a substantial, reliable, and secure IIoT system.

3. Proposed Methodology

Mantaray foraging optimized preventive deep neural network (MFO-PDNN)

The following is a discussion of the suppositions that were taken into account during the creation of the MFO-PDNN model:

Network Model

- The source and SNs are both static.
- Data gathering from the CH is done using one sink.
- The SNs are divided into three categories: advanced, intermediate, and normal nodes based on their diverse character.
- A sink should be a super node that is updated with information about all the SNs.
- CH gathers information for SNs and delivers the combined information to the base station.
- In this strategy, the inter-data communication mechanism is used to carry out the data transfer through CH.
- A node is deemed dead when its battery level falls to zero.

Mantaray Foraging Optimization Algorithm

This suggested design evaluates the multi-objective performance for CH selection using the MFO algorithm. Here is a discussion of the MFO algorithm's mathematical modeling.

This research focuses on the information transmission system in IIoT using the theory of deep neural networks for machine learning. It is crucial to group the detectors into a set of endpoints before beginning the transportation procedure. Since energy-efficient transmitting, which increases the fatality rate of systems and uses the least amount of energy, depends heavily on the clustering approach. The CH is chosen to utilize the optimization method following the distribution of SNs across cluster members. To increase network lifetime, it is crucial to choose the CH properly. Therefore, the MultiObjective MFO method is presented to choose the CH from a group to do this. To solve actual engineering issues, a bioinspired optimization method was recently designed.

Cluster head selection using the MFO algorithm

In addition to the probabilistic analysis, CH selection is used to determine whether a node in the cluster will act as the CH.

For CH selection, it evaluates a variety of factors, including raffic density, power, latency, and length. Vertices use extra energy while collecting, transmitting, and receiving nformation. The CH node is going to consume a lot more energy than the other networks because it is in charge of transmitting and gathering information gathered. Consequently, choosing the networks that can maintain greater power when carrying out all of these duties is crucial. These networks must be chosen as CH.

Multi-objectives for CH selection

The node nearest to the user, with the highest energy, coverage, and lowest cost, will be selected as the CH.

Preventive deep neural network

The portion deals with the preventive of understanding an idea that a deep neural network has received. A PDNN is a group of neurons arranged in a series of layers, where the neurons use the previous layer's neuron action potentials as input and do a simple calculation. The network's neurons work together to accomplish a complicated translation from input to output. That used a method known as the error training algorithm, the weights of each neuron are modified after learning this transfer from the information. A neuron in the upper layer often represents the notion that has to be processed. While the source domain of the PDNN is often interpreted, upper neurons are abstract. In the following sections, we'll go through the process of creating an interpretable model of the learned general concept in the domain of the input. Under the scope of activity maximizing, the construction of the prototype may be specified.

The suggested MFO-based PDNN is built with an effective intrusion detection mechanism to identify intrusions more accurately. The MFO method is used in this instance to train the deep neural network.

4. Result and Discussion

In order to assess the efficiency of the proposed MFO-PDNN method, many assessment requirements are utilized. We do analysis using metrics such as Accuracy, Precision, and Recall.

Accuracy

The evaluation method for machine learning models is the one that is more frequently used. Accuracy is precisely described as the proportion of accurately predicted positive to predicted negative outcomes to complete the machine learning model's outcomes.

$$\text{Accuracy} = \frac{TP + TN}{TP + TN + FP + FN} \quad (1)$$

Figure 70.1 shows the comparable values for the accuracy measures. When compared to existing methods like SVM, which has an accuracy rate of 98.31%, DT, which has an

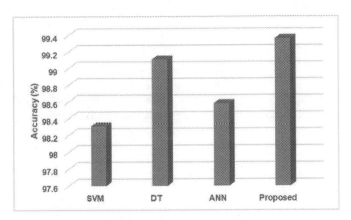

Fig. 70.1 Comparison of accuracy

accuracy rate of 99.11%, and ANN, which has an accuracy rate of 98.58%, the recommended method's MFO-PDNN value is 99.36%. The suggested method performs well in classifying IIoT and has higher accuracy than other methods.

Precision

According to Eq. 2, it is a ratio between results that were projected to be positive and results that were true and falsely positive.

$$\text{Precision} = \frac{TP}{TP + FP} \quad (2)$$

Figure 70.2 shows the comparable values for the accuracy measures. When compared to existing methods like SVM, which has an accuracy rate of 98.21%, DT, which has an accuracy rate of 99.12%, and ANN, which has an accuracy rate of 98.07%, the recommended method's MFO-PDNN value is 99.14%. The suggested method performs well in classifying IIoT and has higher Precision than other methods.

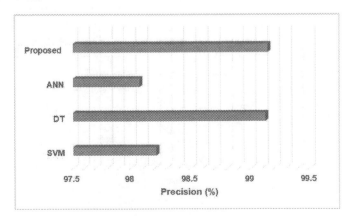

Fig. 70.2 Comparison of precision

Recall

According to Eq. 3, it represents the relationship between true positive predictions and true positive and false negative predictions.

$$Recall = \frac{TP}{TP + FN} \qquad (3)$$

Figure 70.3 shows the comparative data for the recall metrics. Recall rates for SVM are 98.10%, DT 99.14%, ANN 98.49%, and MFO-PDNN 94%. The proposed method performed better than the current results with a recall of 99.39%.

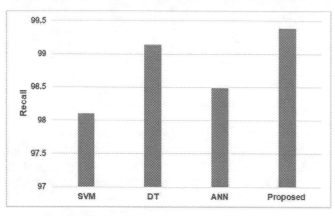

Fig. 70.3 Comparison of recall

5. Conclusion

The approach aims to simplify the modeling and simulation layer development of secure IIoT systems. It provides a revolutionary classification approach named manta ray foraging optimized preventive deep neural network (MFO-PDNN) technology to determine the styles of different creative works. Thus, as compared to other current methods, the performance of the information transmission system in IIoT. The testing results (accuracy, precision, recall) demonstrate the suggested optimization's greater efficacy in comparison to other well-liked optimization techniques. The results suggest that the recommended strategy may greatly and more accurately increase the categorization impact of IIoT. The upcoming IIoT technology versions are being developed to enable having to cut technologies that are presently in development, including artificial intelligence, the cloud, machine learning, portability, virtual reality, as well as immersive.

References

1. Ramlowat, D.D. and Pattanayak, B.K., 2019. Exploring the internet of things (IoT) in education: a review. In Information Systems Design and Intelligent Applications: Proceedings of Fifth International Conference INDIA 2018 Volume 2 (pp. 245-255). Springer Singapore.
2. Kaushal, R.K., Bhardwaj, R., Kumar, N., Aljohani, A.A., Gupta, S.K., Singh, P. and Purohit, N., 2022. Using Mobile Computing to Provide a Smart and Secure Internet of Things (IoT) Framework for Medical Applications. Wireless Communications and Mobile Computing, 2022.
3. Qureshi, K.N., Rana, S.S., Ahmed, A. and Jeon, G., 2020. A novel and secure attacks detection framework for smart cities industrial internet of things. Sustainable Cities and Society, 61, p.102343.
4. Gebremichael, T., Ledwaba, L.P., Eldefrawy, M.H., Hancke, G.P., Pereira, N., Gidlund, M. and Akerberg, J., 2020. Security and privacy in the industrial internet of things: Current standards and future challenges. IEEE Access, 8, pp.152351-152366.
5. Bali, M.S., Gupta, K., Koundal, D., Zaguia, A., Mahajan, S. and Pandit, A.K., 2021. A smart architectural framework for symmetrical data offloading in IoT. Symmetry, 13(10), p.1889.
6. Serror, M., Hack, S., Henze, M., Schuba, M. and Wehrle, K., 2020. Challenges and opportunities in securing the industrial internet of things. IEEE Transactions on Industrial Informatics, 17(5), pp.2985-2996.
7. Refaee, E., Parveen, S., Begum, K.M.J., Parveen, F., Raja, M.C., Gupta, S.K. and Krishnan, S., 2022. Secure and scalable healthcare data transmission in IoT based on optimized routing protocols for mobile computing applications. Wireless Communications and Mobile Computing, 2022.
8. Tang, K., Tang, W., Luo, E., Tan, Z., Meng, W. and Qi, L., 2020. Secure information transmissions in wireless-powered cognitive radio networks for internet of medical things. Security and Communication Networks, 2020, pp.1-10.
9. Sundas, A. and Panda, S.N., 2020, June. IoT based integrated technologies for garbage monitoring system. In 2020 8th International Conference on Reliability, Infocom Technologies and Optimization (Trends and Future Directions) (ICRITO) (pp. 57-62). IEEE.
10. Pal, S. and Jadidi, Z., 2021. Analysis of security issues and countermeasures for the industrial internet of things. Applied Sciences, 11(20), p.9393.

Note: All the figures in this chapter were made by the authors.

Advancing Sustainable Science and Technology for a Resilient Future – Sai Kiran Oruganti et al. (eds)
© 2024 Taylor & Francis Group, London, ISBN 978-1-032-79020-6

71

A Novel Trust aware Energy Efficient Routing Protocol for IoT Networks

Danish Kundra*

Chitkara University Institute of Engineering and Technology,
Chitkara University, Punjab, India

Abstract: Wireless Sensor Network (WSN) bridges the gap between the Internet of Things (IoTs) physical network and its information network. The two most important components in ensuring steady network connectivity are energy and trust. The main problem in IoT is that during multicast routing, the base station (BS) is responsible for securely delivering the data to several destinations via intermediary nodes. Hence, this paper proposes an innovative trust-aware energy-efficient routing protocol based on multilateral chaotic crow swarm search optimization (MCCSSO), which uses the trust and energy factors of the nodes to develop the objective function. MCCSSO initially evaluates node trust and energy to determine optimal paths. This ideal data transmission channel updates the energy and trust of individual nodes at the completion of each transmission, allowing secure nodes to be chosen and improving network security. According to experimental findings, the suggested technique performs better than traditional approaches.

Keywords: Wireless sensor network (WSN), Internet of things (IoTs), Energy, Trust, Multilateral chaotic crow swarm search optimization (MCCSSO)

1. Introduction

Now next version of the Internet, or the extension of the Internet and the World Wide Web, is known as the Internet of Things (IoT). where a significant number of linked objects will enable direct Machine-To-Machine (M2M) communication. Nearly all IoT components, including hardware and software, are significant, but sensors the IoT's ears and eyes and the building blocks of wireless sensor networks are by far the most essential component (WSNs) (Sethi, et al., 2021).

WSNs are self-organizing network domains that can accommodate many inexpensive nodes with capability for wireless data processing and communication. WSN is the name of the adaptive, low-cost, emerging technology which is employed for environmental monitoring. This is composed of a number of tiny devices called group of sensors that are scattered around the area to identify plan for the cellular connection gathering and communication. These sensor nodes have limited battery life, little storage, and low computational complexity (SureshKumar and Vimala, 2021).

The difficult problem of choosing the best path to transmit data, is the focus of the routing protocol. The numerous network features, including techniques, network type, and transmission range, have a significant impact on the choice of an ideal path. Because the Base Station (BS) and small sensing devices (nodes) SNs are closer together in smaller IoT networks, direct communication may occur in a single hop. While direct connection with the BS may not be possible, large-scale IoT networks instead employ multi-hops for communication. Many factors, like radio strength, bandwidth, energy, or memory, may be at responsibility for it though (Kakkar, et al., 2019).

One of the newest technologies in use today, IoT lays the path for several services that drive technical advancements. It is built on combining features of the Internet of Things (IoT), such as IoT terms for semantics, identification, sensing, connection, computation, and services applications come in a variety of categories, such as healthcare, smart cities, agriculture, smart metres, etc. Today's network infrastructure is based on ubiquitous computing, also known as pervasive

*danish.kundra.orp@chitkara.edu.in

DOI: 10.1201/9781003490210-71

networking. This enables the dispersion of intelligent things equipped with wireless connection and a distributive infrastructure (Gopika and Panjanathan, 2020). The rest of this manuscript is structured as follows: Section II describes the related work, section III the recommended approach, section IV the results, and section V the manuscript's conclusion

2. Related Works

Jain, et al., 2019 present the safe and energy-conscious architectural design for the IoT-WSN background. Their entire approach involves employing IOT-based systems to implement Energy Efficient Secure Route Adjustment (ESRA), safety at the user and reports given. Despite efforts to solve these problems, wireless IoT networks still are unable to provide adequate network lifetimes and extensive sensor coverage. Furthermore, the theories put forward in the literature are complex and difficult to use in actual circumstances. That calls for the creation of a straightforward though effective routing technique for wireless Internet of Things sensor networks. (Vashishth, et al., 2018).

They must create an effective protocol for route that not only boosts not only connections, but also customer satisfaction in order to overcome these obstacles. Jaiswal and Anand, 2021 build an energy-efficient routing adapter for IoT applications based on wireless sensor networks that experience unfairness in networks with large traffic loads.

In spite of that trust-based technologies are now able to handle a variety There are still a number of attacks, energy-guzzling nodes, and nodes with communication issues among the unwanted node behaviours. delay. Hu, et al., 2021 propose a novel trust-based secure and energy-efficient routing protocol to address these problems (TBSEER).

3. Proposed Methodology

The proposed protocol's primary goal is to protect the network from escalating hotspot security problems and lengthen the network's life cycle in the context of 5G. This entails addressing concerns with IoT communication device energy consumption, network durability, and effective data transfer.

Trust aware energy efficient routing protocol

The degree of trust needed and the amount of energy needed by nearby nodes that act as packet relays between the sink and source. Yet, Attackers keep taking parts in data respond routing and rerouting packets away from the sink. For multihop wireless sensor networks, the trust aware routing framework (TARF) is developed to address this problem. It detects rogue nodes and blocks routing information based

on the trust level. replay. TARF is an expanded version of the previously described TARF that takes into account the neighbour node's need for energy and confidence level in order to forward packets to the sink. With a higher packet delivery rate and a lower latency, this method achieves improved energy efficiency and latency. These equations (1), (2), and (3), respectively, define the performance meter, the trust factor, and the data flow report. Together, these factors improve network throughput and efficiency while enabling more secure connection.

$$D_f = D_f - \frac{No\ of\ packet\ received\ by\ the\ sink}{No\ of\ packets\ sent\ from\ the\ source} \quad (1)$$

$$A_k(j) = \frac{Packet_{size} \times D_a}{distance\,(j, sink)} \quad (2)$$

$$B_g = M_b \quad (3)$$

Users take into account four key characteristics within a particular wireless sensor network with randomly placed nodes: Trust level (D_f), energy requirements (A_k), dynamic secret key (DSK), and packet flow status (B_g) are all taken into account when calculating a secure link from a data packet to a sink. Each node in the network is first given a common trust value, which the trust monitor(TM) monitors. As the node discards packets intended for transmission from the source to the sink, 's value lowers. Each node is given a DSK that is dynamically produced and assigned over a specified time period, or least update time interval (LUI). Energy monitor (EM) tracks, which stands for the least amount of energy needed by to carry the packets through each node in the direction of the sink., which shows the number of packets transferred between each node and the sink during regular time periods, is used to represent the packet flow status of each node. In order to calculate the secure path, valid neighbour nodes must be identified using the combined values of the highest level of trust, the combining the highest level of trust's ideals EM, the highest packet flow, and the DSK matching. After iterating through all potential the sender and sink's legitimate nodes, the message transfer finally takes place over the secure way. The proposed algorithm is shown below

---
Algorithm 1: Trust aware energy efficient routing protocol

Step 1: Generate nodes randomly

Step 2: Dynamic Secret Key (DSK)

for j = 0 to M do

  if time > LUI then

    if node-Id then

      DSK node-Jt ← int(rand( ) × 250)

end if

end if

if node-Jt then

$\qquad D_f(node\text{-}Id) = 0.5$

end if

end for

Step 3: Routing Table Maintenance

Routing_Table_Maintenance( )

Step 4: Event Generation

Step 5: Identify Neighbor nodes

Step 6: Secure Path Calculation

Secure_Path_Computation( )

Step 7: Repeat step 6 and 7 until it reaches sink

Step 8: Forward the packet through the secured path

for each neighbour node do

$\qquad if\ O_f < O_{threshold}\ \&\&\ max\ K_e\ \&\&\ min\ t\ \&\&\ min\ T$

then

$\qquad sendMessage( )$

$\qquad else\ if\ O_f < O_{threshold}\ then$

$\qquad\qquad select\ next\ node$

$\qquad\qquad end\ if$

$\qquad\qquad\qquad end\ for$

Step 9: Repeat above step until the reaches sink.

Multilateral Chaotic Particle Swarm Optimization Algorithm

Each particle's position indicates a solution, and each particle remembers its current location, speed, and individual best position inside the area being searched. The particles travel to new positions by introducing multiple a velocity vector based on its attraction to the particle's best position as well as the best position found by any other element in the search space, i.e., the particle's position. j The location vector at the $^{d\,th}$ iteration determines c_j^{d+1} and the velocity vector y_j^{d+1} Equations displays the updating formulas for particle velocity and position. (1) and (2):

$$c_j^{d+1} = uc_j^d + v_1 k_1\left(bp_j^d - y_j^d\right) + v_2 k_2 (sp^d$$

$$y_j^d), j = 1,2,\ldots.,mc_{min} \leq c_j^{d+1} \leq c_{max}, \qquad (4)$$

$$y_j^{d+1} = y_j^{d+1} y_{min} \leq y_j^{d+1} \leq y_{max}, \qquad (5)$$

Where m is the majority's amount of particles u, is the inertia weight coefficient, v_1 and v_2 are referred to as acceleration coefficients, and they represent the particle's level of understanding in terms of both the individual and the society, respectively, k_1 and k_2 are random numbers in bp_j^d represents

the current optimal position of i particle, sp^d is the perfect placement for the current population, c_{min} and c_{max} are, alternately, the bottom and higher limits of atom updated velocity; in the paper $c_{min} = c_{max}$. y_{min} and y_{max} are minimum and maximum positions of particles, respectively.

$$\begin{cases} Y_j^{d+1} = y_j^d + k_3.lf.\left(bp_i^d - y_j^d\right), k_i \geq EB, \\ \qquad a\ random\ position\ else, \end{cases} \qquad (6)$$

Where k_i are independent, uniformly distributed different numbers. 0 and 1, *lf* is the length of a crow's flight i, *and EB* is the estimated risk of a crow. *bp* is the location where the current crow i saves food, which is equivalent to the present crow's previous global optimal. i has fount.

$$c_j^{d+1} = vc_j^d + v_3 k_3\left(bp_i^d - Y_j^d\right) + v_2 k_2\left(sp^d - Y_j^d\right), k_i \geq EB,$$

$$c_j^{d+1} = vc_j^d + v_1 k_1\left(bp_i^d - Y_j^d\right) + v_2 k_2\left(sp^d - Y_j^d\right), else \qquad (7)$$

where v_2 reflects the extent of a person's effect j on individual and is a selected at random. The equation's update rate (7) is also limited by c_{min} and c_{max}.

Multilateral crow search algorithm

The characteristics of crow search algorithm (CSA) include simple implementation, minimal parameter configuration, and a somewhat robust capability for development during the search process. In contrast, CSA also has limited search precision, a high likelihood of encountering notably for situations involving multidimensional optimization, the local optimum and early convergence.

4. Result and Discussion

Network coverage area: The cluster radius is fixed at 50 meters in both BEE and HEED. The 15 meter sensing radius has been configured. We divide the network region into 25m*25m grids rather than circles to simplify our problem. In this way, the sensing field is divided into 100 cells. We assume that this region can be tracked by the system as indicated in Fig. 71.1 if a sensor is still functioning in each cell. To guarantee network coverage, a clustering technique should be able to encompass as many cells as possible. The initial coverage with 400 sensors is only 94 cells due to the random distribution of the sensors throughout the network.

Figure 71.1 depicts the Network coverage area comparison. The proposed (MCCSSO) is better than existing method.

Network delay: In an effort to reduce message delivery delays, the trust aware energy efficient routing protocol presents the allocation of resources issue of the routing problem. A data is forwarded in the network less times in order to make best use of the available resources. Using inter-contact latency and variance, it evaluates the delivery usefulness of encountered nodes, which is subsequently used to forecast

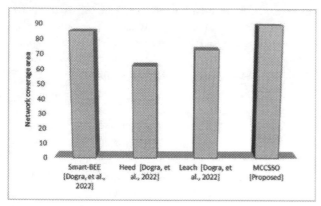

Fig. 71.1 Network coverage area

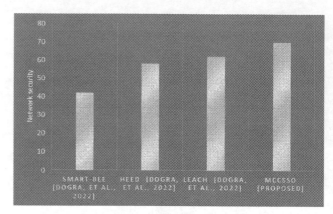

Fig. 71.3 Network security

relay nodes in the network. Figure 71.2 depicts the Network delay comparison. The proposed (MCCSSO) is better than existing method.

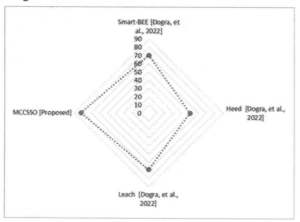

Fig. 71.2 Network delay

Network security: Technologies that safeguard the networking and the programs that use it make internet security architecture. Many automated, scalable lines of defence are used in efficient routing security systems. Each defence layer imposes a set of security policies that the user has chosen. Several algorithms' security performance is assessed and compared to the effectiveness of other methods. The proposed MCCSSO algorithms outperformed existing approaches in terms of security performance. (Fig. 71.3).

Conclusion

To safeguard the network from growing hotspot concerns and increase that network's lifespan. in the context of 5G, this research aims to propose a centralized solution called (MCCSSO) for grouping IoT communication devices into an uneven structure of composite clusters. In particular, the proposed protocol has distinguishing characteristics like creating multi-hop topologies, creating communication clusters with various topologies, balancing energy usage across cluster heads, and minimizing energy usage of cluster nodes, and selecting the most suitable IoT application system transmission interfaces.

References

1. Sethi M., Ahuja S., Rani S., Bawa P., Zaguia A. "Classification of Alzheimer's Disease Using Gaussian-Based Bayesian Parameter Optimization for Deep Convolutional LSTM Network." Computational and Mathematical Methods in Medicine. 2021; 2021. Article No.: 4186666.
2. SureshKumar, K. and Vimala, P., 2021. Energy efficient routing protocol using exponentially-ant lion whale optimization algorithm in wireless sensor networks. Computer Networks, 197, p.108250.
3. Kakkar L., Gupta D., Saxena S., Tanwar S. "An analysis of integration of internet of things and cloud computing." Journal of Computational and Theoretical Nanoscience. 2019; 16(10):4345-4349. DOI: 10.1166/jctn.2019.8523.
4. Gopika, D. and Panjanathan, R., 2020. Energy efficient routing protocols for WSN based IoT applications: A review. Materials Today: Proceedings.
5. Jain, J.K., 2019. Secure and energy-efficient route adjustment model for internet of things. Wireless Personal Communications, 108, pp.633-657.
6. Vashishth, V., Chhabra, A., Khanna, A., Sharma, D.K. and Singh, J., 2018. An energy efficient routing protocol for wireless Internet-of-Things sensor networks. arXiv preprint arXiv:1808.01039.
7. Jaiswal, K. and Anand, V., 2020. EOMR: An energy-efficient optimal multi-path routing protocol to improve QoS in wireless sensor network for IoT applications. Wireless Personal Communications, 111, pp.2493-2515.
8. Hu, H., Han, Y., Yao, M. and Song, X., 2021. Trust based secure and energy efficient routing protocol for wireless sensor networks. IEEE Access, 10, pp.10585-10596.
9. Umesh Kumar Lilhore et al., "A depth-controlled and energy-efficient routing protocol for underwater wireless sensor networks ," in International Journal of Distributed Sensor Networks,Volume: 18 issue: 9, 2022, doi: https://doi.org/10. 1177%2F15501329221117118.
10. O. I. Khalaf et al., "Blinder Oaxaca and Wilk Neutrosophic Fuzzy Set-based IoT Sensor Communication for Remote Healthcare Analysis," in IEEE Access, 2022, doi: 10.1109/ ACCESS.2022.3207751.

Note: All the figures in this chapter were made by the authors.

Advancing Sustainable Science and Technology for a Resilient Future – Sai Kiran Oruganti et al. (eds)
© 2024 Taylor & Francis Group, London, ISBN 978-1-032-79020-6

72

A Novel Biometric Authentication System for Healthcare Data Management in Cloud

Abhinav Rathour*

Chitkara University Institute of Engineering and Technology, Chitkara University, Punjab, India.

Abstract: The security risk to the ever-increasing e-medical data is a direct result of healthcare technology advancements. Patient information is recorded in several forms across the healthcare data management system, resulting in a large volume of unstructured data. Hospitals may also have a number of branches in various regions. There are occasions when it becomes necessary to combine patient health data that is kept in several places for research reasons. In light of this, a healthcare management system that is hosted on the cloud may be a viable option for an effective solution to the efficient management of health care data. But the major concern of a cloud-based healthcare system is the security aspect. In this case, biometric authentication is adequate and accurate for safe data access and retrieval. Hence, an innovative tri-stage biometric signature-based authentication system (T-BSA) for cloud-based healthcare data management is offered in this research. The recommended health care data management approach outperforms current state-of-the-art algorithms.

Keywords: Healthcare data, cloud, biometric authentication, tri-stage biometric signature- based authentication system (T-BSA)

1. Introduction

Due to a variety of factors, fraudsters now find healthcare data is an extremely alluring victim. One of the causes is that healthcare data is easier to get than any other kind of financial data since the data storage is not secured. The health sector has been the target of the most information leaks during the last two years. Privacy concerns for patients should always come first when implementing a new healthcare system. Thus, a system is needed for safe access to and recovery of medical data. Moreover, storage of these enormous data is a problem. Also, well-known healthcare facilities have several locations, and access to medical data on a worldwide scale is essential. So, cloud computing presents itself as an effective remedy in this situation, with the extra benefit of offering these services in a dynamic, adaptable, and reliable way, along with access from any place (Shakil, et al., 2020).

Healthcare data management has experienced disruptive changes the emergence of various healthcare data management concepts as a consequence of the evolution of information supports this transition technology. The objectives of biomedical treatment and research were often out of step with those healthcare data management systems. In order to close this gap, a new framework for the health data management system must be created. Big data analytics must be integrated into healthcare data management systems due to the growing awareness of the use of medical data for governance and investigation. Data aggregation and pre-processing from many sources to produce conclusions, information security, and privacy to deal with a rising number of information leaks and hacker events are additional issues brought on by this, nevertheless (Ismail, et al., 2020).

Using biometrics may thus present hopeful answers to streamline healthcare delivery and radically alter the way that healthcare data is stored. With the development of splitting technologies like Cloud Computing and the Internet of Medical Things, clients can now gather their own individual healthcare data at residence using gadgets (like handsets and fitness trackers) and communicate it on cloud environments, which healthcare professionals can immediately connect to

*Corresponding author: abhinav.rathour.orp@chitkara.edu.in

DOI: 10.1201/9781003490210-72

evaluate clinical notes and provide prompt medical support. With the use of this innovative healthcare service, doctors can remotely monitor patients and give ambulatory treatment in the patients' homes, which not only streamlines healthcare delivery but also helps patients financially (Nguyen, et al., 2019 and Hazela, et al., 2022).

Healthcare professionals may be able to monitor and follow up on patient healthcare data if efficient computer applications are used in the healthcare settings. The delivery of successful, efficient, and equitable healthcare is another objective of healthcare services. Evaluation of IT technologies recently, such as health information systems, has helped to avoid common mistakes, safeguard client security, and offer enough storage for patient data. The delivery of healthcare information has been affected in a number of ways by the accelerating advancements concerning cloud services. The implementation of electronic health systems still has to overcome a number of challenges in terms of client help, cost, internet connection, and emergency recovery. Yet, using cloud computing to process healthcare data may provide great benefits for improving healthcare services. A system based on cloud processing, which includes processing and maintaining medical records in a dispersed healthcare setting, might be thought of as cloud computing (Meri, et al., 2019 and Kaushal, et al., 2022).

2. Related Works

Yan, et al., 2020 the retrieval and storage-based indexing framework (RSIF) described in this article is created to enhance concurrent user and network operator access to medical information kept in the cloud. Concurrent access to saved data is made possible by ongoing, replication-free indexing and time-constrained retrieval. Classifiers is used for all storage systems to categorize the restrictions for data augmentation and updating. The learning process establishes the approximate indexing and ordering for storage and retrieval, accordingly, by dependent evaluation. As long as the processes are independent, this helps to shorten the time for access and retrieval occurring at the same time.

Altowaijri, 2020 introduced cloud technology and the cloud in healthcare. This introduction of cloud computing security challenges, particularly in the context of the healthcare cloud, follows. Lastly, various strategies for enhancing healthcare cloud protection are explored together with their suggested design.

Singla, et al., 2021 introduced a number of crucial fields, including artificial intelligence, machine learning, big data, and cloud computing. With the constant technological advancement, the idea of smart healthcare has gained traction. By evaluating data from IoT devices using AI and machine learning, it is possible to forecast the future

course of illness and create safer environments in which they can be fought. They examine the major concepts while utilizing the most recent innovations in smart healthcare data management technology. It typically targets the local community as well as kids and their families and features a variety of performers in various genres.

3. Proposed Methodology

The primary goal of this research is to create a model that is capable of correctly assessing the biometric authentication in healthcare organizations in the future using healthcare data from selected data management in cloud.

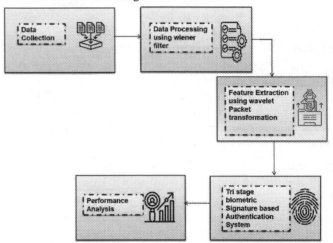

Fig. 72.1 Flow chart for healthcare data management

Dataset

Information was gathered using a variety of portable devices, including iPhone, iPad, touchscreens, and personal digital assistants (PDAs). The samples of fingerprints came from 9000 users. Workers and patients were included in the group of users. The information was gathered throughout five discrete temporal events, each lasting one week and including 1500 participants. 7000 patients and 2000 staff members made up the data set. Each user provided a total of 40 samples, of which 25 were real and 15 were false.

Data processing using wiener filter

The intended process and the estimated random method. The Wiener filter seeks to compute a data analysis approximate of an unspecified pulse by use an information in order as an input and filtering it to produce the approximate as an outcome.

Assume the distorted input data are as follows:

$$v(m) = c(m) + x(m) \qquad (1)$$

where c(m) is the real input data

$$F[v(m), x(m)] = 0 \qquad (2)$$

Where F [.] the assumption function

Feature extraction using wavelet packet transform (WPT)

WPT has been offered as a solution to the issue that Discrete Wavelet Transform (DWT) is ineffective for evaluating input data. Every message is separated into high-frequency and low-frequency elements in this research using WPT. It's having multiple i and the location of the WPT coefficients in the decomposition tree at that level j are used to properly designate the WPT coefficients as $C_{i,j}(l)$. Calculations for the corresponding signal energy are as follows:

$$F_{i,j} = \sum_l c^2_{i,j}(l) \qquad (1)$$

Expressions (2) and (3), which represent the total energy in level I and the relative energy distribution in each level, correspondingly

$$F_{Total}(i) = \sum_j F_{i,j} \qquad (2)$$

$$O_{i,j} = \frac{F_{i,j}}{F_{Total}(i)} \qquad (3)$$

Level j total energy and every level's relative energy distribution are represented in the energy distributions at specified scales are consistently altered for data with various damage modes. $F_{i,j}$ is thus regarded as a crucial factor towards segmentation.

Tri-stage biometric signature based authentication system

Tri-stage biometric signature method, several physical traits of a human body are measured and examined for reliability. Individuals can identify someone from a variety of qualities regardless of their ages. An expanding scientific field with many potential applications is the tri-stage biometric signature technology-based verification system. Three primary stages can be identified according to the typical authentication phase for biometric authentication systems. The following processes: biometric enrollment,tri satge biometric authentication, and biometric deenrollment. An (active) sensor system's sensing process is utilized, which provides a suitable human-sensor-system interface for recording or analyzing a person's biological features. Depending on the sensor system used for a particular biometric method, the recording process produces biometric raw data and calibrating information, which are referred to as biometric characteristics. The algorithms for biometric enrollment, recognition, or derollment receives the data after it has been captured. For the purposes of authentication, it is assumed that authorized users have already been correctly enrolled, meaning that the calculated biometric templates have been saved in a secure biometric database.

4. Result and Discussion

Data about a given healthcare-related behaviour is gathered, assessed, and evaluated to generate healthcare outcome measures. A number of metrics, such as precision, average trust value, are assessed by comparing the proposed with the existing method.

Precision

Precision indicates how much data is included in a number appears characters help analyze closely two or more measures are related to one another. Figure 72.2 shows precision of proposed method is better than existing method.

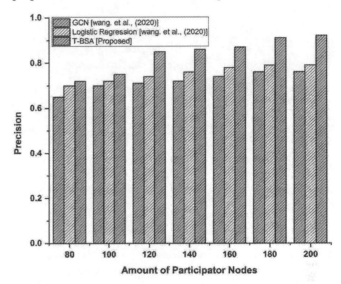

Fig. 72.2 precision

Average trust value

Researchers observe their median confidence measured values of such aberrant sites across transaction epochs (TE). We also keep track of other ordinary regions' average values. Over TE, we can see that aberrant nodes' bad intentions are made clear. Over the first few TEs, the average trust value of normal nodes drops and subsequently acquires continuous oscillation. And it goes without saying that abnormal nodes have a lower value than normal nodes. The test findings match what we anticipated from the concept. By doing this, designers can guarantee that the information transfers in the following phase are founded on a trustworthy standard. Figure 3 shows that average trust value of proposed method is better than existing method.

5. Conclusion

For healthcare data management in Cloud, a cloud-based biometric authentication system called T-BSA is introduced

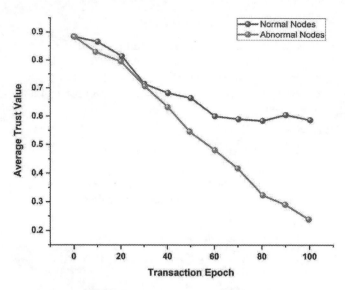

Fig. 72.3 Average trust time

to organize the continuously growing amounts of information related to the healthcare industry and to provide security to varied users. Its approach controls the enormous volumes of data that are generated daily and takes care of the security issue. The results show that T-BSA shows better performance than current approaches and speeds up processes. T-BSA Healthcare has put in place a preference approach that handles crucial documents in a way that makes the network more secure. T-BSA Healthcare is widely applicable to a variety of other fields, including the military, business, nonprofits, and education. The difficulties of applying the T-BSA system in a real-world cloud-based healthcare environment are not discussed in the paper. Future research should take into account elements like compatibility with current infrastructure, integration with healthcare systems, and scalability to manage big user populations and data volumes. The proposed tri-stage biometric signature-based authentication system (T-BSA) might be thoroughly tested in a real-world cloud-based healthcare environment as part of further study. This assessment might involve putting the system to the test on a sizable dataset, taking into account different scenarios and user profiles, and comparing the

system's performance to that of competing authentication techniques.

References

1. Shakil, K.A., Zareen, F.J., Alam, M. and Jabin, S., 2020. BAMHealthCloud: A biometric authentication and data management system for healthcare data in cloud. Journal of King Saud University-Computer and Information Sciences, 32(1), pp.57-64.
2. Ismail, L., Materwala, H., Karduck, A.P. and Adem, A., 2020. Requirements of health data management systems for biomedical care and research: scoping review. Journal of medical Internet research, 22(7), p.e17508.
3. Nguyen, D.C., Pathirana, P.N., Ding, M. and Seneviratne, A., 2019. Blockchain for secure ehrs sharing of mobile cloud based e-health systems. IEEE access, 7, pp.66792-66806.
4. Hazela, B., Gupta, S.K., Soni, N. and Saranya, C.N., 2022. Securing the Confidentiality and Integrity of Cloud Computing Data. ECS Transactions, 107(1), p.2651.
5. Kaushal, R.K., Bhardwaj, R., Kumar, N., Aljohani, A.A., Gupta, S.K., Singh, P. and Purohit, N., 2022. Using Mobile Computing to Provide a Smart and Secure Internet of Things (IoT) Framework for Medical Applications. Wireless Communications and Mobile Computing, 2022.
6. Meri, A., Hasan, M.K., Danaee, M., Jaber, M., Safei, N., Dauwed, M., Abd, S.K. and Al-bsheish, M., 2019. Modelling the utilization of cloud health information systems in the Iraqi public healthcare sector. Telematics and Informatics, 36, pp.132-146.
7. Yan, S., He, L., Seo, J. and Lin, M., 2020. Concurrent healthcare data processing and storage framework using deep-learning in distributed cloud computing environment. IEEE Transactions on Industrial Informatics, 17(4), pp.2794-2801.
8. Altowaijri, S.M., 2020. An architecture to improve the security of cloud computing in the healthcare sector. Smart Infrastructure and Applications: Foundations for Smarter Cities and Societies, pp.249-266.
9. Singla, D., Singh, S.K., Dubey, H. and Kumar, T., 2021, December. Evolving requirements of smart healthcare in cloud computing and MIoT. In International Conference on Smart Systems and Advanced Computing (Syscom-2021) (pp. 102-109).2

Note: All the figures in this chapter were made by the authors.

Advancing Sustainable Science and Technology for a Resilient Future – Sai Kiran Oruganti et al. (eds)
© 2024 Taylor & Francis Group, London, ISBN 978-1-032-79020-6

73

An Innovative Strategy for Implementing an Online Education Platform Based on Cloud Computing

Anubhav Bhalla*

Chitkara University Institute of Engineering and Technology, Chitkara University, Punjab, India

Abstract: The most recent and quickly developing technology, cloud computing, has given forth new developments and possibilities in the IT and education industry. As a result, Online education places a greater emphasis on technology to modify and offer education and training to learners. The use of a cloud computing platform in an online education system introduces an efficient and effective learning technique. In this essay, we provide a brief discussion of the merits, drawbacks, and efficiency of cloud-based online education. The scalability, costs, usability, and other factors of numerous cloud-based online education applications have been studied. An experimental result shows that our cloud computing in online education platform is more efficient than conventional methods.

Keywords: Online education, Cloud computing, Education industry

1. Introduction

Online education refers to a type of learning where students use the internet to access educational materials, communicate with instructors and classmates, and complete coursework remotely, typically from anywhere with an internet connection (Rădulescu, et al., 2016). Online education can range from fully online programs to hybrid models that combine online and in-person instruction. It has become increasingly popular in recent years due to its flexibility, convenience, and accessibility. Bhaskaran, et al., 2019 explain about the online education can be found at all levels of education, from primary school to higher education and professional development. Before implementing the online education platform, it's important to identify the specific needs and requirements of the target audience, such as students, teachers, or administrators. This information will help to design the platform with relevant features and functionality. Choose a cloud computing platform. The next step is to select a cloud computing platform that can support the requirements of the online education platform, such as amazon web services and microsoft azure, etc. Design and develop the platform should be designed with user experience and accessibility in mind. Cloud computing should be easy to navigate, provide relevant content, and offer engaging learning opportunities (El-Attar, et al., 2019). Cloud computing provides a scalable infrastructure that allows you to add or remove resources based on demand. Ensure that your platform is designed to scale up or down seamlessly to handle fluctuations in traffic. Implementing data analytics can help you track user behavior, identify trends, and improve the effectiveness of your online education platform. Continuously monitor the performance of the online education platform and optimize it for better user experience and cost efficiency. Kakkar, et al., 2019 the online education platform can be integrated with other tools and systems, such as learning management systems (LMS), video conferencing software, and assessment tools, to enhance its functionality and provide a comprehensive learning experience. Once the platform is developed and integrated with other systems, it should be thoroughly tested to ensure its functioning as expected. After testing, the platform can be deployed to the cloud and made available to users. To ensure the platform's sustainability, ongoing support, and maintenance should be provided. This includes troubleshooting, updating features, and fixing any bugs or issues that arise. Create a detailed plan for the implementation of the platform. This should include timelines, milestones, and a budget. Identify the resources and personnel needed for each stage of the project. Set up the cloud infrastructure required for the platform, such as virtual machines, storage, networking, and security. Configure the

*Corresponding author: anubhav.bhalla.orp@chitkara.edu.in

DOI: 10.1201/9781003490210-73

settings for performance and scalability (El Ala, et al., 2017). The online education platform should be designed to handle large volumes of traffic and users. The cloud platform should be able to scale automatically based on the demand. Use a combination of open-source and commercial software to develop the platform. Create a platform that is customized to your particular requirements. Use automated testing tools to identify and fix any bugs or issues. Once the platform is ready, deploy it on the cloud infrastructure. Continuously monitor the platform's performance and usage. Use analytics tools to track student engagement, retention, and satisfaction. Optimize the platform's settings and features to improve its performance and effectiveness. Provide technical support and training to students and instructors using the platform. Create a knowledge base or help center to answer common questions and issues (Pocatilu, et al., 2016).

Cloud computing, the latest and fastest-growing technology, has expanded IT and education opportunities. Technology is becoming more important in online education as it evolves. Cloud computing systems in online education are efficient and productive, but they have downsides that must be addressed. This essay examines the pros, cons, and efficiency of cloud-based online education, including scalability, costs, usability, data security, internet connectivity, and digital literacy requirements. Recognizing and assessing these disadvantages helps us comprehend cloud-based online education's consequences and effectiveness in offering high-quality and accessible learning experiences.

2. Related Works

Kumar, et al., 2020 was proposed a technique for dynamically balancing the load, which reduces the makespan time of tasks while balancing the load among VMs. It uses the task migration technique rather than the virtual machine migration technique. Mikroyannidis, et al., 2019 focused on the various scenarios in which the adoption of Blockchain technology can make online education more accessible and decentralized, while also putting learners in charge of their learning experience and associated data. Jasim, et al., 2021 proposed an IoT to build an electronic platform in which students' test results and daily attendance are monitored by their teacher. Where it can clarify the potentials of creating an Internet link between the students and their parents. Dima, et al., 2022 proposed bibliometric analysis which is used in cloud-based online education platforms.

3. Cloud Computing

Cloud computing refers to the delivery of computing resources, such as software, storage, processing power, and networking infrastructure, over the internet as a service.

In cloud computing, users can access these resources on-demand, without the need for owning or maintaining the physical infrastructure themselves. Cloud computing is made possible by large-scale data centers that host vast amounts of computing hardware and software, which can be accessed by users through the internet. To use computing resources such as application software, you must accept the service of a third-party entity. Cloud-based learning systems, on the other hand, require rapid and dependable Internet access. Fig. 73.1 depicts how cloud-based online education operates. However, there are also challenges associated with cloud computing, such as concerns around data privacy and security, potential vendor lock-in, and the need for reliable and fast internet connectivity. These challenges need to be carefully considered and addressed to fully realize the merits of cloud computing.

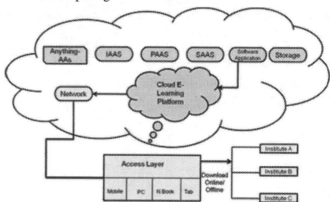

Fig. 73.1 Online education based on cloud

4. Methods

Online education is based on cognitive science concepts and is a good alternative for learners who want to study through multimedia using online educational technologies. Online education is a quick and efficient method of learning that is also cost-effective in terms of user convenience and organization profitability.

5. Online Education Patterns

The efficiency of online education has gradually increased as a result of the dramatic rise in smartphone users in recent years. Users increasing nearly everyday, and want to be more convenient. Smartphone users do not wish to use windows PCs to access multiple applications. Mobile Online education is gaining an increasing number of users and can help learners improve their knowledge at any time and from any location. As a result, corporations are becoming more attentive to this learning tool.

6. Forum-Based Learning

Forum-based learning which learners share ideas, solutions, and issues on a common platform. Expert gives several thoughts or concepts in reaction to questions, which helps to build the learner's talents and boost their confidence.

7. Socialized Online Education

Social Online education is a current technology for effective online education and the latest craze for learners in 2017-2018. The student has the ability to participate and express his opinions and experiences, as well as explore and pursue appropriate learning. Learners find this tool to be quite rich and reliable for improving the understanding and productivity.

8. Cloud Computing Merits

Cloud computing-based online education platforms offer several merits, including low cost, instant software updates, enhanced security, increased employee productivity and performance, seamless collaboration in distributed workforces, improved document format compatibility, and reduced reliance on internal IT support.

10. Result and Discussion

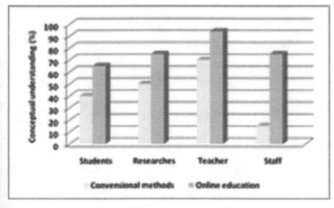

Fig. 73.2   Conceptual understanding

Figure 73.2 shows online education strategies develop 65% of conceptual knowledge in students, while traditional teaching techniques develop just 25%. Similarly, other groups awarded more marks for conceptual understanding using online education tools than traditional techniques. Similar to this, several organizations gave online education strategies more credit for conceptual understanding than traditional methods.

Actual learning is shown in Fig. 73.3. Online education strategies develop 70% of genuine learning, whereas traditional teaching techniques develop just 15%. While other groups expressed greater appreciation for the effectiveness of online education techniques than for traditional ones.

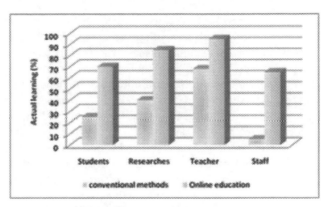

Fig. 73.3   Actual learning ability

According to the Fig. 73.4, only 25% of students preferred traditional teaching methods, whereas 75% of students gave online education strategies positive feedback. Similarly, only 25% of researchers gave good feedback for traditional approaches, while 80% preferred online education.

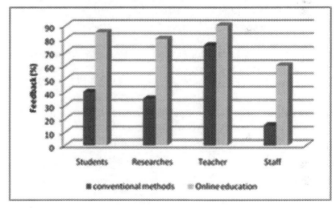

Fig. 73.4   Feedback

11. Conclusion

To improve the learning abilities of employees and students, institutions and organizations are investigating and placing increasing emphasis on online education systems. For a genuine and trustworthy learning experience, various businesses and institutes are supporting cloud-based online education. Everyone is benefiting and taking advantage of the developing online education solutions in a variety of ways, including researching knowledge outside of learning. All the crucial and priceless information uploaded by different institutions and organizations will be simply shared by the users of that network or cloud with the aid of cloud for online education. Users of cloud-based online education do not require high specification hardware; they also do not need to install any software on the system; all they require is an Internet connection to browse or share information and hardware. Cloud computing platform in an online education system is more efficient than traditional techniques. All of the cloud-based tools are not perfect. One of the main disadvantages of cloud computing for online education is the potential for connectivity issues. The study looks at a variety of aspects of cloud-based online education programs, including scalability, pricing, usability, and efficiency. This thorough analysis offers a broad grasp of the subject. The disadvantages or difficulties of cloud-based online education are not specifically discussed in the essay. A more complete understanding of the subject and insights into potential areas for improvement or caution would result from addressing any potential limits. Furthermore, with the rise of mobile devices, online education platforms will need to be optimized for mobile use. Cloud computing provides the necessary infrastructure to deliver online education content to any device, anywhere, and at any time.

Reference

1. Rădulescu, Ş.A., 2016. A perspective on E-Learning and Cloud Computing. *Procedia-Social and Behavioral Sciences*, *141*, pp.1084-1088.
2. Bhaskaran, S. and Santhi, B., 2019. An efficient personalized trust based hybrid recommendation (tbhr) strategy for e-learning system in cloud computing. *Cluster Computing*, *22*, pp.1137-1149.
3. El-Attar, N.E., El-Ela, N.A. and Awad, W.A., 2019. Integrated learning approaches based on cloud computing for personalizing e-learning environment. *International Journal of Web-Based Learning and Teaching Technologies (IJWLTT)*, *14*(2), pp.67-87.
4. Kakkar, L., Gupta, D., Saxena, S. and Tanwar, S., 2019. An analysis of integration of internet of things and cloud computing. Journal of Computational and Theoretical Nanoscience, 16(10), pp.4345-4349.
5. El Ala, N.S.A., Awad, W.A. and El-Bakry, H.M., 2017. Cloud computing for solving E-learning problems. International Journal of *Advanced Computer Science and Applications*, *3*(12).
6. Pocatilu, P., Alecu, F. and Vetrici, M., 2016, November. Using cloud computing for E-learning systems. In *Proceedings of the 8th WSEAS international conference on Data networks, communications, computers* (Vol. 9, No. 1, pp. 54-59). Stevens Point, WI: World Scientific and Engineering Academy and Society.
7. Kumar, M. and Sharma, S.C., 2020. Dynamic load balancing algorithm to minimize the makespan time and utilize the resources effectively in cloud environment. *International Journal of Computers and Applications*, *42*(1), pp.108-117.
8. Mikroyannidis, A., Third, A. and Domingue, J., 2019. Decentralising online education using blockchain technology.
9. Jasim, N.A., AlRikabi, H.T.S. and Farhan, M.S., 2021, September. Internet of Things (IoT) application in the assessment of learning process. In IOP Conference Series: Materials Science and Engineering (Vol. 1184, No. 1, p. 012002). IOP Publishing.
10. Dima, A., Bugheanu, A.M., Boghian, R. and Madsen, D.Ø., 2022. Mapping Knowledge Area Analysis in E-Learning Systems Based on Cloud Computing. Electronics, 12(1), p.62.

Note: All the figures in this chapter were made by the authors.

Advancing Sustainable Science and Technology for a Resilient Future – Sai Kiran Oruganti et al. (eds)
© 2024 Taylor & Francis Group, London, ISBN 978-1-032-79020-6

74

Distributed Denial of Service Attack Detection in Cloud Using Machine Learning Algorithm

Ravi Kumar*

Chitkara University Institute of Engineering and Technology,
Chitkara University, Punjab, India

Abstract: Individuals are provided with on-demand solutions through the Internet due to cloud technology. The solutions are available at all times and from any location. Even while the paradigm offers beneficial services, security problems can still arise. The viability of cloud computing is impacted by a Distributed Denial of Service (DDoS) assault, which also poses security risks to cloud computing. In order to ensure that services are available to authorized users, DDoS attacks must be detected. A machine learning (ML) technique for identifying DDoS attacks in cloud services is presented in this article. In this study, a novel linear stochastic support vector machine (LS-SVM) is presented. Datasets are used in the study to analyze the effectiveness of the suggested LS-SVM approach in terms of several performance indicators. The findings of this study demonstrate that the suggested LS-SVM method outperformed existing methods in detecting DDoS attacks in the cloud.

Keywords: Cloud service, Distributed denial of service (DDoS) attack, Machine learning (ML), Linear stochastic support vector machine (LS-SVM)

1. Introduction

Distributed denial of service (DDoS) refers to a planned attack on a system's capacity to provide its services to one or more victims by way of several secondary targets that have been hacked. On February 7, 2000, a website called yahoo.com was the target of one such significant attack. Similar attacks were reportedly made against other profit - making businesses such CNN.com, e-Bay, Amazon, etc. A number of important systems, including an Ohio safety surveillance system, hundreds of Bank of America-operated automated teller machines, and the South Korean stock market's online trading, were shuttered as a consequence of yet another attack, the slammer worm (Parekh, et al., 2023). As the Internet has developed, cloud computing was becoming storage, administration, processing, and networking. The expanding technological trends and the fast change in company computer network are giving cloud computing more popularity. In a short period of time, DDoS attacks will possibly resulting a cloud application's whole resources available, including

broadband and processing power. Due to its still on solutions, pay-as-you-go pricing model, and cost-effective capabilities, cloud computing is increasingly replacing conventional IT installations. Increasingly, businesses and government agencies are using cloud computing for their IT infrastructure. Compared to fixed network IT deployments, it has a number of benefits, including as on-demand available resources, an absence of requirement for expensive investment projects, and a reduction of maintenance costs (Kushwah, et al., 2019). Cloud computing has become a universally recognized technique. Effectively, it provides a range of services. The accessibility of the solutions, which means that they must always be available to all users, is one of the aspects from these technologies. Users of cloud solutions could access them because the DDoS attacks. Several computers are utilized to initiate the attack in this case Ranga, (2021). As cloud computing offers so many user-friendly features, it is commonly regarded as an excellent system for conveying a great deal of data. With cloud computing, each user is given a separate set of devices that are only used for data analysis.

*Corresponding author: ravi.kumar.orp@chitkara.edu.in

DOI: 10.1201/9781003490210-74

Table 74.1 Datasets information

Dataset name	Number of features	Number of samples					
		Training			Testing		
		Normal	Attack	Total	Normal	Attack	Total
NSL KDD	41	10000	15000	25000	3000	30000	6000

The safety of transferred information from attacks is among cloud computing greatest significant issues since the great majority of data already available is thought to be accessible.

2. Related Works

The study includes experimenting with various machines learning models and working on methods to incorporate them into the DDoS detection system. Many categorization methods that are often used in machine learning-based DDoS attack detection in software defined networks were examined using the CIC-DDoS 2019 dataset (SDN) (Jha, et al., 2023). The study enhance and strengthen the safety and reliability of SDN processes against attack or interference, that used a machine learning model to detect attack traffic and classify traffic of SDN as (attack or normal), and an optimization technique (genetic algorithm) to optimize the precision of the categorization (Kamel, et al., 2022). The study analyzing the various DDoS attack types included using the CIC-DDoS2019 database. The CICFlowMeter-V3 system was used in the dataset's development. To assess whether machine learning techniques have evolved, a traditional internet traffic dataset that includes NetBIOS, Portmap, Syn, UDPLag, UDP, and normal harmless communications was used. (Aldhyani, et al., 2023). As it is so simple to use, the device's ID3 (Iterative Dichotomiser 3) Maximum Multifactor Dimensionality Posteriori Method (ID3-MMDP) is used to recognize DDoS attacks and address their issues. The suggested (Dhiyanesh, et al., 2023) ID3-MMDP technique firstly utilizes cloud platform technologies and then applies attack detection technology based on information entropy to identify DDoS attacks. The findings, using CIC-IDS2017 benchmark database is used for training and testing purposes. The preparation phase includes K-Fold cross validation. The top model for early DDoS attack detection is then selected after training and testing each of the models using K-Fold classification technique (Amrish, et al., 2022).

3. Proposed Methodology

The effectiveness of the suggested approach on the linear stochastic SVM has been assessed by a range of research. NSL KDD is one of the standard datasets that have been utilized in studies. There is 41 features total in the databases. This database entirety is utilized for research. All specimens

of DDoS attacks are included in the datasets. Table 74.1 provides further information on these databases. The suggested methodology's flow is shown in Fig. 74.1.

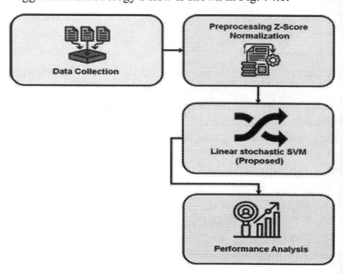

Fig. 74.1 Datasets information

Z-Score Normalization

Z-score normalization, also commonly referred to as zero-mean normalization, normalizes any output descriptors by examining the normalized mean and standard deviation for every parameter over a variety of test datasets. The mean and standard deviation for each attribute are provided. The following substitute is given in the generalized formula 1:

$$v' = \frac{v - \mu_A}{\sigma_A} \tag{1}$$

During which A stands for the attribute's μ means σ_A = and A differences in values" stands for the attribute's standard deviation. As a consequence, every characteristic possesses zero variation and zero meaning in the dataset. Each training sample in the data set is first subjected to the Z-Score normalization procedure before creating a trainee collection and beginning the training procedure. The average, variation, and statistical significance for each statistic throughout a training data collection must be computed, recorded, and used as weights in the final system design. In the structure of neural networks, it is a preprocessing level. The outputs of the neural network can be quite different from the normalized information as it was trained on a different kind of dataset. This statistical normalization provides the advantage of decreasing the consequences of data anomalies by minimizing their volume.

4. Linear Stochastic SVM

Assume training instance-label pairs are $(X_i, y_i), i=1,...,l,$ where $X_i \in R^n$ and $y_i \in \{1,-1\}$

We consider the following SVM problem with a penalty parameter $C > 0$:

$$\underset{w}{min} \frac{1}{2}w^T w + C\sum_{i=1}^{l}\max(1-y_i w^T \phi(X_i),0) \qquad (2)$$

In order to handle data that cannot be separated linearly, the function $\phi(x)$ translates an instance to a higher dimensional space. We call such a situation a nonlinear SVM setup. For some applications, $\phi(x)$ can already properly separate data; we call such cases linear SVM. Many SVM studies consider $w^T X_i + b$ instead of $w^T X_i$ in (2). In general this bias term does not affect the performance much, so here we omit it for the simplicity.

SVM is often resolved using the double solution with the kernels technique because to the large multiplicity of $\phi(x)$ and the potential challenge of obtaining the explicit form of $\phi(x)$:

$$\underset{\infty}{min} \qquad \frac{1}{2}\infty^T Q \infty - e^T \infty$$

$$subject\ to\ \ 0 \le \infty_i \le C, i = 1,....,l, \qquad (3)$$

Where $Q_{ij} = y_i y_j K(x_i, y_j) = y_i y_j \phi(x_i)^T \phi(x_j)$ and $e = [1,...,1]^T$. $K(x_i, y_j)$ is called the kernel function and α is the dual variable.

The stochastic gradient descent (SGD) algorithm is among the most straightforward and well-liked stochastic optimization techniques. Every curved functional could be optimized using SGD over a concave area if only impartial estimations of the value are available. SGD is well suited for complex learning issues since it is both very basic and scalability. Several studies have supported the SGD's completion. The SDG approach has seen extensive theoretical and algorithmic advancements. For instance, looked at how well SGD performed while making nontrivial smoothness assumptions, and used a moving average approach to turn the iterations of SGD into a solution with the best possible accuracy. In addition to an extensive investigation and field support, offered the first parallel stochastic gradient descent method.

5. Performance Evaluation

Accuracy, sensitivity, and specificity measures are used to assess the effectiveness of the suggested method. The existing

methods include Software-Defined Networking (SDN) Alsaeedi, et al., 2019, Decision tree (DT) Mienye, et al., 2019. Equations 4, 5 and 6 define and describe them, accordingly. The greatest accuracy obtained by Linear Stochastic SVM, SDN and decision tree in the mentioned range of 97%, 91% and 87% respectively.

$$Accuracy = \frac{TP+TN}{TP+FP+TN+FN}\times100 \qquad (4)$$

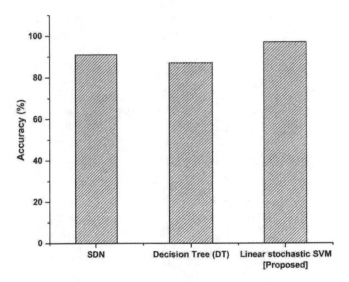

Fig. 74.2 Comparison of accuracy.

The number of samples that the classification properly detected as being part of the attacks is known as the "true positives," or TP, in this context. The quantity of samples that the classifier properly recognized as belonging to the normal distribution is known as the TN, or quantity of true negatives. The number of samples that the classifiers inaccurately recognized as an attack even though they belonged to the regular population is known as the "number of false positives," or FP. FN stands for "number of false negatives," which is the quantity of attack-related data that the classification misconstrued for dataset. The databases' sequential sensitivity results were 93%, 87%, and 85%.

$$Sensitivity = \frac{TP}{TP+FN}\times100 \qquad (5)$$

$$Specificity = \frac{TN}{FP+TN}\times100 \qquad (6)$$

The databases, 95%, 90%, and 81%, of specificity were attained. Figure 74.2, 74.3 and 74.4, illustrates a comparison of the outputs of linear stochastic SVM with SND and decision tree. In all databases, Linear Stochastic SVM has the maximum detection performance, as can be seen in the figure. Moreover, it demonstrates the greatest sensitivity and specificity.

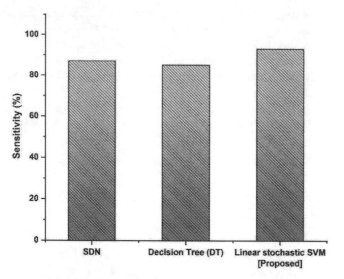

Fig. 74.3 Comparison of sensitivity

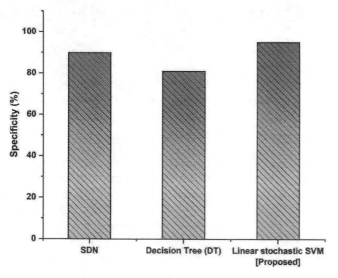

Fig. 74.4 Comparison of specificity

6. Conclusion

The suggested approach is utilized to create a system for detecting DDoS assaults in cloud computing in this research as a solution to the problem of DDoS attack detection. Datasets have been used in a number of studies to assess the effectiveness of the suggested method. The NLS KDD dataset was utilized in this study to train and evaluate the suggested model. To identify the DDoS assault, we used the linear stochastic SVM method. The linear stochastic SVM environment is where the classification module is installed.

To differentiate between legitimate and malicious traffic data, the proposed approaches are utilized. Our test results demonstrate that the suggested approach performs better than the existing methods in the simulated environment. Future research pertaining to the suggested strategy is mentioned in the article. The technology may need to be improved further, evaluated using more datasets, tested in real-world scenarios, and even integrated with current security measures in cloud computing settings.

References

1. Parekh, V. and Saravanan, M., 2023, March. An Empirical Overview on DDoS: Taxonomy, Attacks, Tools and Attack Detection Mechanism. In Proceedings of International Conference on Recent Trends in Computing: ICRTC 2022 (pp. 151-161). Singapore: Springer Nature Singapore.
2. Kushwah, G.S. and Ali, S.T., 2019. Distributed denial of service attacks detection in cloud computing using extreme learning machine. International Journal of Communication Networks and Distributed Systems, 23(3), pp.328-351.
3. Ranga, V., 2021. Distributed denial of service attack detection in cloud computing using hybridextreme learning machine. Turkish Journal of Electrical Engineering and Computer Sciences, 29(4), pp.1852-1870.
4. Jha, A., Das, B., Reddy, D., Jitendra, D.V. and Rezni, S., 2023. Distributed Denial of Service Attack Detection in SDN using Machine Learning. Journal of Advances in Computational Intelligence Theory, 5(1), pp.1-11.
5. Kamel, H. and Abdullah, M.Z., 2022. Distributed denial of service attacks detection for software defined networks based on evolutionary decision tree model. Bulletin of Electrical Engineering and Informatics, 11(4), pp.2322-2330.
6. Aldhyani, T.H. and Alkahtani, H., 2023. Cyber Security for Detecting Distributed Denial of Service Attacks in Agriculture 4.0: Deep Learning Model. Mathematics, 11(1), p.233.
7. Dhiyanesh, B., Karthick, K., Radha, R. and Venaik, A., 2023. Iterative Dichotomiser Posteriori Method Based Service Attack Detection in Cloud Computing. Computer systems science and engineering, 44(2), pp.1099-1107.
8. Amrish, R., Bavapriyan, K., Gopinaath, V., Jawahar, A. and Kumar, C.V., 2022. DDoS detection using machine learning techniques. Journal of IoT in Social, Mobile, Analytics, and Cloud, 4(1), pp.24-32.
9. Alsaeedi, M., Mohamad, M.M. and Al-Roubaiey, A.A., 2019. Toward adaptive and scalable OpenFlow-SDN flow control. IEEE Access, 7, pp.107346-107379.
10. Mienye, I.D., Sun, Y. and Wang, Z., 2019. Prediction performance of improved decision tree-based algorithms: a review. Procedia Manufacturing, 35, pp.698-703.

Note: All the figures and table in this chapter were made by the authors.

Advancing Sustainable Science and Technology for a Resilient Future – Sai Kiran Oruganti et al. (eds)
© 2024 Taylor & Francis Group, London, ISBN 978-1-032-79020-6

75

Optimal Task Scheduling Algorithm Based on Swarm Optimization in Cloud Computing Environment

Rishabh Bhardwaj*

Chitkara University Institute of Engineering and Technology,
Chitkara University, Punjab, India

Abstract: An essential component of cloud computing is task scheduling. Different jobs are allocated to the adequate resource nodes in accordance with quality of service (QoS) requirements and with the use of appropriate techniques, resulting in an NP-hard issue. Due to the fact that this challenging optimization problem is an NP-complete problem, numerous task scheduling approaches have been proposed by various academics to address it. Heuristic techniques have been found to be useful for obtaining optimal or nearly optimal solutions. A brand-new multi-task swarm-adaptive whale optimization (MSAWO) technique is developed in this research. The outcomes also demonstrated that the suggested model performed better than the conventional approaches, drastically shortened the execution time, and required fewer computing resources for analysis.

Keywords: Cloud computing, Task scheduling, Optimization, multi-task swarm-adaptive whale optimization (MSAWO)

1. Introduction

Distributed computing models, such as peer-to-peer (P2P), grids, and cloud computing, among others, needed a number of computing resources that could be made available via the connection to satisfy the need of carrying out large-scale and increased operations. One of the aforementioned Cloud technology, dispersed processing has emerged as the most popular information computing paradigm in modern times, finding use in a variety of fields, including science, commerce, and everyday calculation tasks. The pay-per-usage structure was the only one used to provide products under the cloud-based computing paradigm, which is solely service-oriented. Via the introduction of computing as a web-based service, this model significantly changed how services are offered in the computer business (Agarwal and Srivastava, 2021).

Cloud computing is a recent, well-liked technical innovation that affects computers, the use of infrastructures, storage, networking, freely accessible resources, and software in many different fields. A large financial pay-per-use computing infrastructure that is internet-based, elastic, adaptable, monitored, and extensible is provided by the cloud. Service-level agreements (SLAs), which are negotiated between cloud service providers and service users, are frequently what motivate people to use cloud services. The physical resources can be enclosed as amorphous organizations that offer consumers outside the cloud varying degrees of service. Cloud computing is a term used to describe both the infrastructure and software in the data centers that host the programs and is supplied as services via the Internet. Software-as-a-Service (SaaS) is a common term for the products that are provided, and Infrastructure-as-a-Service (IaaS) and Platform-as-a-Service (PaaS), correspondingly, are used to describe the hardware and operating system that are presented. When these services are made accessible to many groups of individuals, it is referred to as a cloud. Finally term "public cloud" refers to cloud services that are offered to customers under a pay-as-you-go pricing structure. Community clouds cater to the unique demands of a region, whereas private clouds are made available solely to an institution. Hybrid clouds, on the other hand, combine the advantages of both public and private clouds (Mishra, and Majhi, 2021, Kakkar, et al., 2019).

---

*Corresponding author: rishabh.bhardwaj.orp@chitkara.edu.in

DOI: 10.1201/9781003490210-75

The fundamental issue with cloud computing (CC) is task scheduling. The benefit of CC is that all resources are used effectively, and this can only happen when task scheduling is done correctly. Thus, it has been assumed that all activities must execute both task scheduling and resource assignment. At the moment, anyone with a web connection can obtain intelligence from anywhere at any time without knowing anything about the host infrastructure. These computing frameworks have various capabilities and technologies that are within Cloud Service Provider (CSP's) control. The potential of host infrastructure that can employ internet services is enhanced by CC. Using cloud services, CSP increases efficiency by helping its customers (CS) (Devaraj, et al.,2020 and Hazela, et al., 2022).

In order to improve the virtualization produced in a dynamic environment for individual consumers and to deliver dependable and trustworthy service, cloud computing is a novel computer technology. A metering product of varying characteristics delivered through virtualization networking to numerous end customers is called cloud computing. It is a quickly developing piece of computer technology that improves the virtualization of all IT equipment and systems. In the past ten years, the field of cloud computing has experienced rapid growth. It is a model an estimate that makes resources available to subscribers so they can satisfy a variety of demand through the Internet. The most widely used cloud computational resources pools include Hadoop structure, Google File System (GFS), Google Mail, and Aws Storage. One of the most well-known sequential optimisation in cloud computing is the channel assignment issue. Such as a method of providing services, cloud companies typically demand an efficient automatic task scheduling solution based on splitting up computing tasks (Jana,et al., 2019 and Saxena, et al., 2021).

2. Related Works

The improved particle swarm optimization (IPSO) algorithmic suggestion by (Wang, et al., 2019) in this study relies on the adjustable 12 inertia weight and random factor correlations. According to numerical simulations, IPSO method consumes less resources than sequence routing protocol, wasteful method assuming similar conditions, the correlation particle swarm optimization (CPSO) method and the novel adaptive inertia weight based particle swarm optimization (NewPSO) algorithm (which involves energy and virtualized costs).

Adil, et al. 2022 claimed that the Content-Aware Load Balancing using the Particle Swarm Optimization algorithm (PSO-CALBA) proposed in this research is a ground-breaking technique. PSO-CALBA is a scheduling strategy for categorizing information from a source utilizing meta-heuristics and machine learning. Text, audio, video, and image-based jobs performed by humans are categorized using SVM technology into several content categories. For transfering client requests to the cloud, a meta-heuristic approach based on Particle Swarm Optimization is applied (PSO).

Potu et al. 2021 established the enhanced particle swarm optimization (EPSO) strategy using a second gradients technique to address the task planning issues in virtualized environments. Their key objectives are to speed up the use of resources and decrease work completion times. They thoroughly tested for actual price and computation time using the iFogSim simulation.

3. Proposed Methodology

The primary goal of this research is to create a model that is capable of correctly assessing the optimal task scheduling in cloud computing in the future using MSAWO from swarm optimization in cloud environment.

Multi task swarm adaptive whale optimization

1) Multi task swarm optimization

In MSO, a swarm of particles searches a C-dimensional region of interest in quest of the best answer. Every particles j has two current vectors: a current position vector $W_j = [W_{i1}, W_{j2}, \ldots W_{jC}]$ and a current velocity vector $U_j = [U_{j1}, U_{j2}, \ldots U_{jC}]$ where C is the number of dimensions. U_j and W_j are dynamically updated when the MSO process begins. Furthermore, at every round, particles j is guided to modify its velocity and location by (1) and (2) using the best positions that particle J $Obest_j = [Obest_{j1}, Obest_{j2}, \ldots Obest_{jC}]$ and the entire swarm $Hbest = [Hbest_1, Hbest_2, \ldots Hbest_C]$ have identified.

$$u_{jc}(s+1) = u_{jc}(s) + d_1 q_1 (Obest_{jc}(s) - y_{jc}(s))$$
$$+ d_2 q_2 (Hbest_c(s) - y_{jc}(s)) \quad (1)$$
$$y_{jc}(s+1) = y_{jc}(s) + u_{jc}(S+1) \quad (2)$$

From which q_1 and q_2 are two different distribution variables created inside the [0; 1] region, and d_1 and d_2 consist of the neural and social acceleration coefficients. Shows the algorithm 1

Algorithm 1 Multi task swarm optimization
1: Initialization
2: Define the swarm size S and the number of dimensions D
3: for each particle $j \in [1..S]$
4: Randomly generate Y_j and U_j and evaluate the fitness of denoting it as (Y_j)
5: Set $Hbest_j = Y_j$ and$e(Obest_j) = e(Y_j)$
6: end for

7: Set Hbest= $Obest_i$ and e(Hbest)=e($Obest_1$)

8: for each particle j ∈ [1.., T]

9: if e($Obest_j$)<e(Hbest)then <

10: e(Hbest)=e($Obest_j$)

11: end if

12: end for

13: While s maximum number of iterations

14: for each particle j ∈ [1.., T]

15: Evaluate its velocity u_{jc} (s+1)using Equation (1)

16: Update the position y_{jc} (s+1) using Equation (1) Equation (2)

17: if e(y_j (s+1)) < e($Obest_j$) then

18: $Obest_j$=y_j (s+1)

19: f (O_{bestj})) = e (y_j (s + 1)

20: end if

21: f (O_{bestj})) < f (H_{best}) then

22: H_{best} = O_{bestj})

23: e (H_{best}) = e (O_{bestj})

24: end if

25: end for

26: s = s + 1

27: end while

28: return H_{best}

2) *Adaptive whale optimization*

The adaptive whale optimization (AWO) technique is a newest meta-heuristic method updated. The AWO is modelled after the particle targeting technique used by humpbacks. It favour hunting tiny species or plankton schools that are near to the level. In order to make characteristic waves across a circular or "9" shaped path, humpbacks must move surrounding their victim in a shrinking circle and along a curve path at the same time. The whales then swim up inside the bubbles to gulp the captured species. AWO has a either the spiral model or the shrinking encircling mechanism has a 50% chance of being chosen. Stimulate this phenomenon while updating the location of the whales throughout enhancement. Algorithm 2 shows the AWO procedure.

Algorithm 2 Adaptive whale optimization

Set up all whale species as Y_j (j = 1, 2 ... *m*)

$Y_j = Y_j^{min} + rand (0,1) * (Y_j^{max} - Y_j^{min})$

Compute every searching owner's efficiency. Pick the most effective search agent

While (s <maximum number of iterations)

Every iteration of the algorithm (s maximum number of iterations), b,B,D,Hand update O q

If 1 (O_q< 0.5) where the search agent is at the moment has been updated

else (p≥0.5)

Equationally update the position of the current iteration of the algorithm

end if 1

end for

Examine to see if any search agents are going outside of the search space and make any necessary corrections.

Calculate the leap rate K_ (q)

if rand (0, 1) <K_q

Equationally calculate the semi individuals

end if

When a greater option exists, determine every searching agent's fitness and update Y*.

s=s+1

End while

Return Y^*

4. Result and Discussion

To produce optimal task scheduling outcome measures, information regarding a specific behavior relevant to swarm optimization is obtained, analyzed, and evaluated. The challenges with criteria like optimal task scheduling processes would be resolved by the suggested architecture using a number of metrics, including resource load and execution time.

Resource load

Every resource load rate in Fig. 75.1 increases as the number of tasks rises. Nonetheless, this algorithm's variation in resource load rate is minimal and stays at a lesser level compared to GA and ACO. This algorithm's resource load rate is lower than that of GA and ACO by 28% and 24.1%, respectively. It is also lower than that of ACO and MQoS by 24.1% and 20.3%. As an outcome, the allocation of available resources tends to be more sensible and effective, and the institution as a whole is becoming more load balanced.

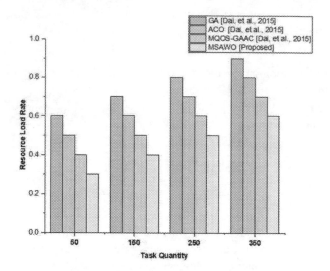

Fig. 75.1 Resources load rate

Execution time

The number of tasks and execution time have a positive association in Fig. 75.2. When there are more than 500 tasks, the efficiency of execution gets much higher as the number of tasks rises. As a result, the efficiency of the execution time is influenced by the number of tasks. Proposed method's execution time is substantially faster than that of GA, ACO, and MQoS-GAAC, and it is kept at a reasonable level. Consequently, MSAWO is superior to a conventional GA, ACO, or MQoS-GAAC in terms of both time efficiency and the effectiveness of identifying the most suitable strategy.

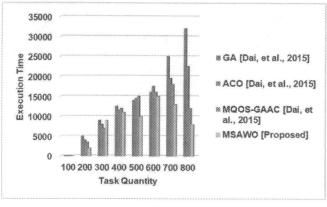

Fig. 75.2 Execution time

5. Conclusion

Heuristic approaches have been successfully proved to facilitate the ability to look for solutions that are optimal or nearly. During this study, a novel multi-task swarm-adaptive whale optimization (MSAWO) method is devised. The results also showed that the proposed method significantly

outperforms more traditional methods, substantially reduced execution time, and required less computational capabilities during evaluation. This study's primary flaw is that it does not compute measures like energy usage and scalability (the capacity to handle larger job sets). The effectiveness of the research might be improved if these metrics could be calculated.

References

1. Agarwal, M. and Srivastava, G.M.S., 2021. Opposition-based learning inspired particle swarm optimization (OPSO) scheme for task scheduling problem in cloud computing. Journal of Ambient Intelligence and Humanized Computing, 12(10), pp.9855-9875.
2. Mishra, K. and Majhi, S.K., 2021. A binary Bird Swarm Optimization based load balancing algorithm for cloud computing environment. Open Computer Science, 11(1), pp.146-160.
3. Devaraj, A.F.S., Elhoseny, M., Dhanasekaran, S., Lydia, E.L. and Shankar, K., 2020. Hybridization of firefly and improved multi-objective particle swarm optimization algorithm for energy efficient load balancing in cloud computing environments. Journal of Parallel and Distributed Computing, 142, pp.36-45.
4. Hazela, B., Gupta, S.K., Soni, N. and Saranya, C.N., 2022. Securing the Confidentiality and Integrity of Cloud Computing Data. ECS Transactions, 107(1), p.2651.
5. Jana, B., Chakraborty, M. and Mandal, T., 2019. A task scheduling technique based on particle swarm optimization algorithm in cloud environment. In Soft Computing: Theories and Applications: Proceedings of SoCTA 2017 (pp. 525-536). Springer Singapore.
6. Saxena, S., Yagyasen, D., Saranya, C.N., Boddu, R.S.K., Sharma, A.K. and Gupta, S.K., 2021, October. Hybrid Cloud Computing for Data Security System. In 2021 International Conference on Advancements in Electrical, Electronics, Communication, Computing and Automation (ICAECA) (pp. 1-8). IEEE.
7. Wang, Q., Fu, X.L., Dong, G.F. and Li, T., 2019. Research on cloud computing task scheduling algorithm based on particle swarm optimization. Journal of Computational Methods in Sciences and Engineering, 19(2), pp.327-335.
8. Adil, M., Nabi, S. and Raza, S., 2022. PSO-CALBA: Particle swarm optimization based content-aware load balancing algorithm in cloud computing environment. Computing and Informatics, 41(5), pp.1157-1185.
9. Potu, N., Jatoth, C. and Parvataneni, P., 2021. Optimizing resource scheduling based on extended particle swarm optimization in fog computing environments. Concurrency and Computation: Practice and Experience, 33(23), p.e6163.
10. Kakkar, L., Gupta, D., Saxena, S. and Tanwar, S., 2019. An analysis of integration of internet of things and cloud computing. Journal of Computational and Theoretical Nanoscience, 16(10), pp.4345-4349.

Note: All the figures in this chapter were made by the authors.

Advancing Sustainable Science and Technology for a Resilient Future – Sai Kiran Oruganti et al. (eds)
© 2024 Taylor & Francis Group, London, ISBN 978-1-032-79020-6

76

Design of an Energy-Efficient Resource Allocation Algorithm for Cloud Computing

Pooja Sharma*

Chitkara University Institute of Engineering and Technology,
Chitkara University, Punjab, India

Abstract: Cloud computing (CC) is a type of computing where resources are given as a service through the Internet, dynamically scaled, and other virtualized. Considering the makespan and energy usage of the resources employed is essential. This research suggests an energy-aware multitask swarm-intelligent whale optimization (EMSWO) technique regarding the time cost and energy usage assumptions in a CC context. Using the Cloud Sim toolbox, we examined how well the EMSWO strategy performed. The testing findings reveal the enormous potential of our technique, offering notable improvements in makespan, demonstrating high potential for improving data center energy management, and successfully meeting the service-level agreements the clients need.

Keywords: Cloud computing (CC), Internet, resource allocation, Energy efficiency, Energy-aware multitask swarm-intelligent whale optimization (EMSWO)

1. Introduction

Cloud computing is a contentious topic in the computer industry since it is a recently emerging computing model. This style of computing makes resources that are virtualized and dynamically scalable accessible as services over the Internet. It integrates well-known computer and network technologies with additional products, such as virtualization, load balancing, parallel computing, cloud services, grid computing, and network storage technologies(Mittal, et al., 2022). Actual hardware and software structures must be set up in a data center to give universal access to a shared pool of flexible computational power through the network on demand (Kakkar, etal., 2019). This is known as cloud computing. The central technology behind cloud computing is resource allocation, which uses the network's processing power to make it easier to carry out challenging activities that call for many calculations. The distribution of resources must consider various variables, including load balancing, makespan, and energy use. Due to the rapid growth of

cloud computing and network communication technologies, several computer service providers, including Google, Microsoft, Yahoo, and IBM, are quickly creating data centers globally to offer cloud computing services.(Park, et al., 2022).Yet, there are considerable operating costs and environmental carbon footprints due to data centers' massive electrical energy consumption that serve cloud applications. In the process of developing a cloud computing business application, the amount of energy used and the time required to use the resources supplied should be taken into account. Resource distribution needs to be carefully thought out and mutually optimized in order to establish an energy-efficient schedule (Shan, et al., 2022).To offer a cloud computing data center resource allocation algorithm that is energy-efficient to promote the sustainability of cloud computing. A cloud computing platform reduces energy consumption while maximizing processing effectiveness and infrastructure utilization (Chen, et al., 2022).Cloud computing seeks to provide rising, low-power computing infrastructure to meet an energy-efficient and secure service mode.

*Corresponding author: pooja.sharma.orp@chitkara.edu.in

DOI: 10.1201/9781003490210-76

2. Related Works

Le, et al., 2020 provided an optimization framework for blockchain-enabled IIoT systems. The proposed problem is formulated as a Markov decision process (MDP). A power optimization approach utilizing Lagrangian duality is suggested to enhance energy efficiency (Zhang, et al., 2020). In the multi-user, multi-BS scenario, Fang, et al., 2021 provided a low-complexity approach for user association. Simulations show that the suggested method can produce a performance that is significantly better than the traditional Orthogonal Multiple Access (OMA) techniques. Proactive Hybrid Pod Autoscaling (ProHPA) responds immediately to irregular workloads and lessens resource allocation(Jeong, et al., 2023). Ali, et al.,2020 introduce the Power Migration Expand resource allocation algorithm (PowMigExpand) .Also, they propose the Energy Efficient Smart Allocator (EESA), a low-cost system that makes use of deep learning to allocate requests to the best servers in terms of energy efficiency.

3. Methodology

The analysis is predicated on the idea that there is a set of jobs, each of which includes several subtasks with strict priority requirements. Each subtask is permitted to be completed using any given resource that is available. A cloud resource (such as a CPU, memory, network, or storage) has a specific capacity. The provided resources are continually accessible, and each subtask is executed on a single resource at a time.

Inputs for each job may be carried out on any subset of the available resources since all of the resources are unconnected and parallel.

Output: The result is a resource allocation plan that is both effective and efficient, with activities being scheduled according to resources' availability and timelines.

Constraints: Each task's execution time on a resource relies on the current circumstances, and the duration cannot be determined in advance. Once a job is started, Services can only work on one subproblem at a time, and it must be completed uninterrupted.

Aims: Their main objective is to decrease make-up time and boost the data center's energy efficiency to develop a power schedule.

As many actual design or decision-making difficulties include the simultaneous optimization of many objectives, we created a resource allocation optimization method that fully incorporates the two features of energy-efficient optimization and makespan optimization.

Energy-Aware Multitask Swarm-Intelligent Whale Optimization (EMSWO) technique

The encircling, bubble net, and hunt for prey concepts of WOA are somewhat reflected in the mathematical model of EMSWO.

Encircling Prey

The position of the search space is not known in advance for this method; instead, the WOA assumes that the present work is the best option, and further position updates are controlled by the following equations.

$$\vec{A} = |\vec{K}.\vec{J^*}(t) - \vec{J}(t)| \tag{1}$$

$$\vec{j}(t+1) = |\vec{J}*(t) - \vec{B}.\vec{A}| \tag{2}$$

Where $\vec{j}$ is the current location and t, denotes the current iteration. The best position thus far is represented by the $\vec{j}*$. The coefficient vectors for $\vec{B}$ and $\vec{K}$ are provided by

$$\vec{B} = 2\vec{a}.\vec{r} - \vec{a} \tag{3}$$

$$\vec{k} = 2.\vec{r} \tag{4}$$

$\vec{a}$ is a random vector in the range [0, 1], andis linearly decreasing in the range [2, 0].

Bubble-net Attacking

Using a global solution in the search space is what this step entails. It consists of two subphases: a spiral updating location and a shrinking encircling mechanism. In both subphases, humpback whales circle their prey. Hence, choosing one of the two subphases has a 50% chance of happening. The single equation provided expresses both subphases mathematically.

$$J(t+1) = \begin{cases} \vec{A'}.e^{bl}.\cos 2\pi l + \vec{j}*(t) & if \ p \geq 0.5 \\ \vec{J*}(t) - \vec{B}.\vec{A} & if \ p < 0.5 \end{cases} \tag{5}$$

An is the reduced form of [2, 0], and is a random number in the range [-a, a]. The spiral's state is defined by the constant b, whereas the value of l is chosen randomly between [-1, 1]. The source of the is

$$(\vec{A'} = |\vec{J}.(t) - \vec{J}(t)| \tag{6}$$

Search for Prey

To find the answer, this phase must explore the search space. The Humpback whales randomly hunt for prey while recognizing where other whales are. Instead of using the best outcomes, the fitness function is updated using a randomly picked value.

$$\vec{A} = |\vec{K}.\vec{J}_{rand} - \vec{J}| \tag{7}$$

$$\vec{x}(t+1) = \vec{J}_{rand} - \vec{B}.\vec{A}n \tag{8}$$

Where the location vector $\vec{J}_{rand}$ was created at random.

4. Results

The performances of the existing methods are Genetic Algorithm(GA) and Dynamic Virtual Machine Relocation(DVMR)compared with our proposed method. The parameters are makespan, accuracy, and energy consumption.

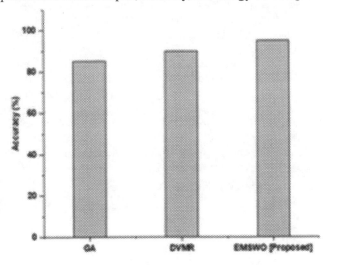

Fig. 76.1 Accuracy

The accuracy of a measuring system refers to how nearly measurements of a quantity and an equalamount are precise values. Fig. 76.1 debates the accuracy results.The accuracy of existing GA and DVMR methods is 85% and 90%, while the proposed EMSWO is 95%.

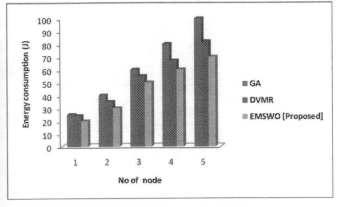

Fig. 76.2 Energy consumption

The degree of energy consumption is the total energy consumed by the end user. Fig. 76.2 displays the energy consumption of existing and proposed methods. The energy consumption of existing GA and DVMR methods is 100J and 82J, while the proposed EMSWO is 70J. The energy consumption of the proposed approach is low when compared to the existing method.

The central performance of a scheduling algorithm is the total execution time of the exit task called makespan. Makespan result is depicts in Fig. 76.3. Themakespan of existing GA and DVMR methods are 75s and 65s, while the proposed EMSWO is the 60s.

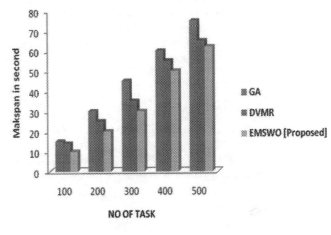

Fig. 76.3 Makespan

5. Conclusion

An energy-aware multitask swarm-intelligent whale optimization (EMSWO) method was suggested in this paper. We investigated the performance of the EMSWO method using the Cloud Sim tool. The testing results demonstrate our method's colossal potential, significantly improving makespan, showing great promise for enhancing data center energy management, and effectively achieving the service-level agreements the clients require. In this study, we determined the energy consumption, makespan, and accuracy of our suggested approach, which are 90J, the 60s, and 95%, respectively. The results and recommendations of the study can be restricted to the particular environment in which they were carried out. The EMSWO approach may function differently in other cloud computing environments or under other workload conditions. The EMSWO technique for energy-aware multitask optimization in cloud computing is introduced in this research. Future research could look into more improvements, optimizing the algorithm, and adding more swarm intelligence or optimization techniques. This would seek to increase effectiveness.

Reference

1. Mittal, S., Bansal, A., Gupta, D., Juneja, S., Turabieh, H., Elarabawy, M.M., Sharma, A. and Bitsue, Z.K., 2022. Using identity-based cryptography as a foundation for an effective and secure cloud model for e-health. Computational Intelligence and Neuroscience, 2022.

2. Kakkar, L., Gupta, D., Saxena, S. and Tanwar, S., 2019. An analysis of the integration of the Internet of things and cloud computing. Journal of Computational and Theoretical Nanoscience, 16(10), pp.4345-4349.

3. Park, S.W., Cho, K.S., Hoefter, G. and Son, S.Y., 2022. Electric vehicle charging management using location-based incentives for reducing renewable energy curtailment considering the distribution system. Applied Energy, 305, p.117680.

4. Zhang, L., He, D., He, Y., Liu, B., Chen, Y. and Shan, S., 2022. Real-time energy saving optimization method for urban rail transit train timetable under delay condition. Energy, 258, p.124853.

5. Chen, J., Xie, Y., Mu, X., Jia, J., Liu, Y. and Wang, X., 2022. Energy efficient resource allocation for IRS assisted CoMP systems. IEEE Transactions on Wireless Communications, 21(7), pp.5688-5702.

6. Yang, Le., Li, M., Si, P., Yang, R., Sun, E. and Zhang, Y., 2020. Energy-efficient resource allocation for blockchain-enabled Industrial Internet of Things with deep reinforcement learning. IEEE Internet of Things Journal, 8(4), pp.2318-2329.

7. Zhang, H., Duan, Y., Long, K. and Leung, V.C., 2020. Energy-efficient resource allocation in terahertz downlink NOMA systems. IEEE Transactions on Communications, 69(2), pp.1375-1384.

8. Fang, F., Wang, K., Ding, Z. and Leung, V.C., 2021. Energy-efficient resource allocation for NOMA-MEC networks with imperfect CSI. IEEE Transactions on Communications, 69(5), pp.3436-3449.

9. Jeong, B., Baek, S., Park, S., Jeon, J. and Jeong, Y.S., 2023. Stable and efficient resource management using deep neural network on cloud computing. Neurocomputing, 521, pp.99-112.

10. Ali, Z., Khaf, S., Abbas, Z.H., Abbas, G., Muhammad, F. and Kim, S., 2020. A deep learning approach for mobility-aware and energy-efficient resource allocation in MEC. IEEE Access, 8, pp.179530-179546.

Note: All the figures in this chapter were made by the authors.

Advancing Sustainable Science and Technology for a Resilient Future – Sai Kiran Oruganti et al. (eds)
© 2024 Taylor & Francis Group, London, ISBN 978-1-032-79020-6

77

Novel Encryption Technique for Cloud Computing Access Control

Savinder Kaur*

Chitkara University Institute of Engineering and Technology, Chitkara University, Punjab, India

Abstract: Cloud computing (CC) has become a critical computing paradigm in the IT sector. With this emergent concept, pay-per-use utilization of computer assets like applications, equipment, connectivity, and memory is possible anywhere in the world. Under this paradigm, it is challenging to store confidential material on unreliable servers in any case. Conventional encryption techniques have ensured the secrecy and adequate security systems of outsourced personal data. Nevertheless, because they need more adaptability, stability, and fine-grained access control, these control mechanisms are not viable in cloud computing. Thus, we introduce a novel multilevel attribute-based key-search encryption with the spider monkey optimization (MAKE+SMO) method. We show through the securing data that our approach is protected from user agreement attacks. Compared to other systems, the performance analysis shows our effectiveness.

Keywords: Cloud computing (CC), confidentiality, Encryption, Access control, Multilevel attribute-based key-search encryption with spider monkey optimization (MAKE+SMO)

1. Introduction

Cloud computing has gained colossal appeal since it allows cloud service providers and consumers to make more money at reduced expenses. It is now widely accepted and offers consumers and suppliers cloud computing at reduced prices. Cloud computing makes the minimal cost of supplying and maintaining massive data centers possible. Protection has become the greatest obstacle to a quicker and broader deployment. Today's cloud computing technology makes many mission-critical calculations susceptible to problems such as service availability, data confidentiality, reputational destiny spreading, etc. There is currently no network security capability available from cloud services (Mittal, et al., 2022). According to the most recent world data corporation analysis, between 2019 and 2023 there will be a large increase in worldwide expenditure on free cloud services (CS) and its architecture. Cloud computing environments help to reduce an organization's consumption of quantity change by reducing the complexity of operating software or hardware by cloud users. The fast growth of cloud computing has resulted in a number of issues for both consumers and providers. The Information Technology (IT) industry is drawn to the advancements of cloud computing technology and is moving its IT activities there. In essence, cloud computing refers to the instant provision of computer resources or services, namely computational capacity and storage systems that do not require external perspective or technology. Here, many dispersed systems are brought together to offer various services, including a website, memory, and a lot more. Among other advantages, pay-per-use, on-demand services, limitless storage space, and adaptability make cloud computing particularly desirable. However, it has several drawbacks, including encryption, security systems, restricted regulation, and outages. The extent and intensity of cloud computing advancements and research, as well as worries about cloud security, have proven to be a key obstacle to the rise of cloud computing. One of cloud computing more complex elements is access control. The cloud security alliance (CSA) listed 14 privacy emphasis areas in a document titled "Security Guidelines for Crucial Areas of Attention in Cloud Computing," released in July 2017. The study field is employing access control to prevent unauthorized individuals from accessing or taking

*Corresponding author: savinder.kaur.orp@chitkara.edu.in

DOI: 10.1201/9781003490210-77

resources stored in the cloud (Yang, et al., 2020). The use of cloud computing technology may help customers address the problem of the exponential growth in data sharing by offering both computation and storage services. However, computers are often run by a third person independent of users. The company's data could be spread out over many systems and accessed by other users. The users should encrypt the saved information before sending it to the external servers to guarantee that it is not accessed by other users or malicious networks (Li, et al., 2019).

2. Related Works

In the study, (Morales, et al., 2020) discuss the security problems with cloud storage on the assumption that users would encrypt data before outsourcing it, would share it with other users, and would be able to retrieve the encrypted data by contacting the supplier. They provide a security strategy for attribute-based encryption (ABE), which enables methods for controlling access over the encrypted data and the information retrieval process using searching network access for storing, distributing, and retrieving encrypted data in the cloud. The research (Shukla, et al., 2021) proposed encryption-based cloud computing approach. The recommended method has been compared to a number of other popular encryption methods, most notably DES, AES, and Two fish. The performance of the innovative technique has been examined using some factors, including encryption time for frames of different sizes and the impact of an explosion on text files. Access control and data secrecy are provided by (Agrawal, et al., 2019), the suggested solution employing variable characteristics encrypting. In order to address the network connectivity issues, distributed application combinations are used. The anonymously encrypt technique is used to distribute the private key while maintaining the user's confidentiality. The method is put into practice in a real-world mobile cloud environment, and a number of metrics are used to gauge how well it works. Develop a hybrid decryption approach in which they encrypt health information using the improved key generation scheme of RSA (IKGSR) algorithm and secrets using the Blowfish method. They use steganography-based access control for key sharing using a partial scanning and information retrieval approach to recover the encrypted material effectively (Chinnasamy, et al., 2022). The objective of this paper is to provide a state-of-the-art approach for image encryption that will enable the advanced encryption standard (AES) to be extended to the Galois field of any features. The innovation modifies each of the four stages in the fundamental method with a digital attribute. They increased the number of options in (Khan, et al., 2019) suggested substitute boxes, which implies that they increased the ability to confuse and broadened the preexisting notions. On top of that, they have used the predicted approach for digital picture encryption.

3. Proposed Methodology

The attribute-based encryption (ABE) approach is often used in cloud computing to guarantee encrypted access control. We suggested the multilevel attribute-based key-search encryption with spider monkey optimization (MAKE+SMO) technique.

4. Multilevel Attribute-Based Key Search Encryption (MAKE)

MAKE is an ABE system in which the cryptosystem is tagged with characteristics, and the higher employment is contained in the users' secret information. A person may decode a cryptosystem when the latter properties match the key's authentication scheme. Figure 77.1 depicts the setup, encryption, key generation, and decryption techniques make up the MAKE-ABE system.

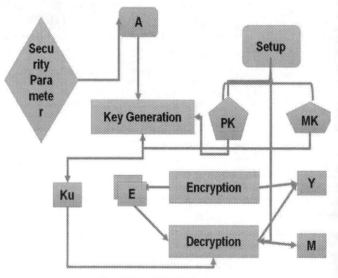

Fig. 77.1 KP-ABE scheme

The optional encryption parameter is an input for the Setup procedure, which uses randomization. It produces a decryption MK together with the public parameters PK. A collection of characteristics, a signal (Y), the public parameters PK, and a random number generator are the encryption scheme's inputs. It produces the encrypted text E. Management challenges A, a decryption MK, and public parameters PK are the Key Generation randomization method inputs. A decryption key Ku is produced by it. The decryption method accepts the encrypted secret code E, the decryption key Ku corresponding to the access control structure A, and the public parameters PK as inputs. If meets A, (y) produces the text M.

5. Spider Monkey Optimization (SMO)

An optimization strategy has been proposed by the authors to replicate the fission-fusion social structure (FFSS)-based foraging behavior of spider monkeys. The FFSS's main attributes are shown below.

- Monkeys with a fission-fusion social structure are sociable and exist in packs of 40–50 people. When breaking the cluster into smaller groups to search for food, the FFSS of swarms may reduce forage rivalry among the individuals.

- Typically, a female (global leader) takes charge of the unit and is in charge of searching for sources of food. If she cannot provide the gang with adequate food, she splits it up into smaller divisions (which may have anywhere between 3 and 8 members) that go out foraging on their own.

- A female (local leader) tasked with leading a comment thread is also expected to decide on the best foraging path to take daily.

- The members of these subgroups then communicate both inside and outside of their grouping in order to maintain topographical stability and exchange data on food availability.

SMO is a collaborative, incremental approach that depends on experimentation and failure, comparable to other inhabitants' techniques. Six steps comprise the SMO procedure: Regional leaders' part, global leader phase, local leader learning phase, global leader learning phase, local leader decision phase, and global leader decision phase. The Gbest-guided Artificial Bee Colony (ABC) and a modified version of ABC inspired the positioning updating procedure in the global leader stage.

6. Result and Discussions

The effectiveness of the suggested multilevel attribute-based key-search encryption with the spider monkey optimization (MAKE+SMO) has been assessed using a variety of factors, and it is defined. The technique is applied in Microsoft Azure, a cloud computing platform that microsoft has provided. In this part, the outcomes are provided. DMABE (Yahya, et al., 2020) and IBPM (Gobi, et al., 2022) are currently used methods that are compared with proposed method.

Figure 77.2 displays the findings from testing the efficiency of access restrictions for different methods at random numbers.

Figure 77.3 illustrates the results of measurements made on the effectiveness of different encryption techniques at various user densities.

According to tests conducted on a range of users, the effectiveness of several decryption techniques is shown in Fig. 77.4. The suggested MAKE+SMO algorithm outperformed existing approaches in terms of production.

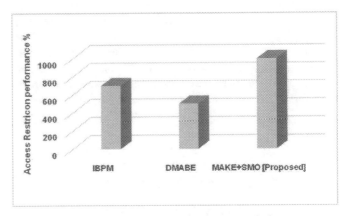

Fig. 77.2 Application of access restrictions

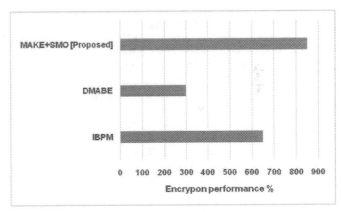

Fig. 77.3 Application of encryption

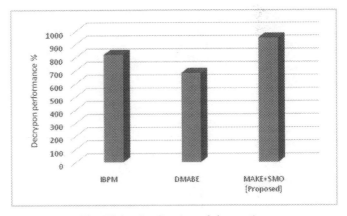

Fig. 77.4 Application of decryption

7. Conclusion

Cloud computing is a technological invention that is still comparatively new but has tremendous potential for worldwide impact. It provides a broad variety of benefits to businesses and customers. By concentrating more on the business itself and spending less on maintenance and software upgrades, it reduces running costs for companies, for example. Before being made accessible in the computer, the data has been encrypted using MAKE+SMO. The cloud data has been decrypted using MAKE+SMO associated with the dataset's characteristics. In order to enhance data access management, this paper explains a search model on protected data for cloud computing. The combination of searchable encryption and access control technologies assures data secrecy across the cloud and user levels while limiting data access to authenticated persons.

References

1. Mittal S., Bansal A., Gupta D., Juneja S., Turabieh H., Elarabawy M.M., Sharma A., Bitsue Z.K. "Using Identity-Based Cryptography as a Foundation for an Effective and Secure Cloud Model for E-Health". Computational Intelligence and Neuroscience. 2022; 2022. Article No.: 7016554.

2. Yang, C., Tan, L., Shi, N., Xu, B., Cao, Y. and Yu, K., 2020. AuthPrivacyChain: A blockchain-based access control framework with privacy protection in cloud. IEEE Access, 8, pp.70604-70615.

3. Li, J., Chen, N. and Zhang, Y., 2019. Extended file hierarchy access control scheme with attribute-based encryption in cloud computing. IEEE Transactions on Emerging Topics in Computing, 9(2), pp.983-993.

4. Morales-Sandoval, M., Cabello, M.H., Marin-Castro, H.M. and Compean, J.L.G., 2020. Attribute-based encryption approach for storage, sharing and retrieval of encrypted data in the cloud. IEEE Access, 8, pp.170101-170116.

5. Shukla, D.K., Dwivedi, VK and Trivedi, M.C., 2021. Encryption algorithm in cloud computing. Materials Today: Proceedings, 37, pp.1869-1875.

6. Agrawal, N. and Tapaswi, S., 2019. A trustworthy agent-based encrypted access control method for mobile cloud computing environment. Pervasive and Mobile computing, 52, pp.13-28.

7. Chinnasamy, P. and Deepalakshmi, P., 2022. HCAC-EHR: hybrid cryptographic access control for secure EHR retrieval in healthcare cloud. Journal of Ambient Intelligence and Humanized Computing, pp.1-19.

8. Khan, M. and Munir, N., 2019. A novel image encryption technique based on generalized advanced encryption standard based on the field of any characteristic. Wireless personal communications, 109, pp.849-867.

9. Luo, J., Qu, S., Chen, Y. and Xiong, Z., 2019. Synchronization of memristor-based chaotic systems by a simplified control and its application to image en-/decryption using DNA encoding. Chinese Journal of Physics, 62, pp.374-387.

10. Gobi, M. and Arunapriya, B., 2022. A Survey on Public-Key and Identity-Based Encryption Scheme with Equality Testing over Encrypted Data in Cloud Computing. JOURNAL OF ALGEBRAIC STATISTICS, 13(2), pp.2129-2134.

Note: All the figures in this chapter were made by the authors.

Advancing Sustainable Science and Technology for a Resilient Future – Sai Kiran Oruganti et al. (eds)
© 2024 Taylor & Francis Group, London, ISBN 978-1-032-79020-6

78

Optimal Path Planning Algorithm with Optimization Technique for Mobile Robots

Rahul*

Chitkara University Institute of Engineering and Technology, Chitkara University, Punjab, India

Abstract: Throughout the last three decades, research into mobile robots has emerged as a promising new field. Mobile robots, on the other hand, are not very good at avoiding obstacles in real-time; thus, their local path planning is less efficient, and their global path planning results in longer routes. A mobile robot exploring an unfamiliar area faces several potential dangers, including the existence of both static and moving obstacles, as well as the possibility of encountering other mobile robots. Without prior knowledge of the environment, the goal of path planning is to steer the mobile robot toward an objective while navigating through it. In this study, a brand-new bilateral frequency bat swarm search optimization (BFBSSO) for planning robot paths is introduced. In comparison to conventional approaches, experimental findings show that the suggested method achieves the best neighborhood path optimization obstacle avoidance forecasting accuracy in real-time.

Keywords: Mobile robot, Path planning, Optimal safe path, Bilateral frequency bat swarm search optimization (BFBSSO)

1. Introduction

More automated environments are starting to use mobile robots. There are several possible applications for mobile robots, including robots that can assist the elderly, motorized guided vehicles for moving goods in factories, autonomous robots that can defuse bombs and robots that can explore distant planets (Campbell et al., (2020)). A mobile robot's navigation includes environment observation, map construction, cognition, path planning, and motion control. Finding its pose or configuration in the environment is a localization and map construction, whereas perception relates to processing its sensory data. Before achieving the desired trajectory by regulating the motion, planning the path following the task utilizing cognitive decision-making is crucial (Sun et al.,(2022) and Grover et al., (2022). The robot, however, is very knowledgeable about its environment. Due to its extensive path planning, it may take a predetermined path to its destination. Applications for arranging/planning the path of mobile robots include medical and surgical procedures, personal assistance, security, stockroom, and circulation,

as well as applications for automated guided vehicles used to transport goods inside a plant (Zafar and Mohanta et al.,(2018)). Mobile robots are used extensively and frequently in various applications, including agriculture, humanitarian demining, security, environmental monitoring, and military use. Mobile robots are distinguished by their capacity to move about in various settings. Thus, a fundamental task for mobile robots is path planning. The main objective of designing a path through an environment is to avoid collisions a challenging situation, moving from a specific starting point to a particular goal location while fulfilling specific requirements. Generally speaking, there are two fundamental sorts of path-planning algorithms. These algorithms rely on information gathered by sensors and data from the robot as it moves through the surroundings (Mavrogiannis et al.,(2021)and Sharma et al .,(2021)). In complicated and dynamic surroundings, navigation must be effective and safe. Mobile robots are able to carry out duties successfully and autonomously in a variety of fields, including logistics, surveillance, and search and rescue, by minimizing travel time, energy consumption, and avoiding obstacles.

*rahul.orp@chitkara.edu.in

DOI: 10.1201/9781003490210-78

2. Related Work

Ajeil et al. (2020) deal with swarm intelligence-based smart route optimization methodologies for robotic systems functioning in both stationary and moving environments are being developed. Two improvements to the basic bat application's search process are suggested using two new algorithms. Patel et al., (2019) presents a thorough analysis of existing mobile robot navigation methods. Here, the development of path-planning strategies in different environments is examined step-by-step, along with research gaps, in both classical and reactive approaches. Li et al. (2019)and Mittal et al.,(2022) gives the clear and detailed description of a development configuration optimizer that will generate the best routes for a spot automaton moving through a tough world full of obstacles. The obvious similarity between such a robot's movement and a thermal path provided the inspiration. The feasible regions that a robot can move through are described as the design world on which conduction of heat occurs; obstacles are modeled as insulating material; the beginning and main objective points are viewed as a source of heat and a heat sink, respectively; and so on (Yang et al., (2019)).

3. Proposed Methodology

The goal of path planning is to steer the mobile robot toward a target while navigating through the environment with no prior knowledge of the surroundings. In this study, a brand-new bilateral frequency bat swarm search optimization (BFBSSO) for planning robot paths is introduced. Fig. 78.1 depicts the proposed methodology's flow.

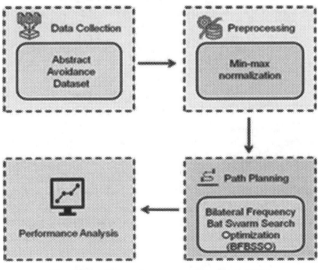

Fig. 78.1 Flow of proposed methodology

4. Data collection
Obstacle Avoidance Dataset

The University of California Irvine's open dataset, Wall-Following Robot Navigation Data, provides information on avoiding obstacles. The "simplified distance" is the minimal level measurement of the distance from the target that the 60-degree arc of the ultrasonic sensor's detection ranges. Usually, front, left, right, and back area are the characteristic values when "simple variety" serves as the parameter value systems and "robot movable direction" as the label value. The Move-Forward, Slight-Right, Sharp-Right, and Slight-Left label values are the most common (Song et al., (2021)).

Data preprocessing using Min-max normalization

To more efficiently manage the data for each audio file, normalization is performed to lessen the negative effects of irregular sample data and to confine the resultant data to a specified range. The following formula is used for determining the level of normality:

$$A' = \frac{a - \min(a)}{\max(a) - \min(a)},$$ (1)

When the beat spectrum computation yields an amplitude value A, normalizing it to a lower value A yields the normalized value A'.

Path planning using Bilateral Frequency Bat Swarm Search Optimization (BFBSSO)

BFBSSO is recently, one of the most well-liked optimization algorithms created a powerful meta-heuristic-based soft computing method. It demonstrated a swarm intelligence optimization meta-heuristic technique developed for the most efficient numerical optimization. This algorithm is inherently influenced by bat social behavior. These bats' echolocation abilities are an excellent method for detecting prey, dodging obstacles, and they use sense to find their roosting niches in the dark distances. The treatment for bats is automatically improved by locating a better solution. Bats move quickly and randomly, with constant frequency positioned y_j, and varying wavelength, as well as loudness from B_p to B_{min} to search for its prey. By altering the pulse emission rate, the frequency can change on its own. $q \in [0, 1]$, based on how close the bat's target is the loudness parameter can vary greatly depending on the volume. (B_o) And a minimum of noise() B_{min}

$$f_i = f_{min} + (f_{max} - f_{min})\beta$$ (2)

$$v_i^t = v_i^{(t-1)} + (x_i^t - x^*)f_j$$ (3)

$$x_i^t = x_i^{(t-1)} + v_i^t \qquad (4)$$

Where;

β: drawn from a uniform distribution, a random vector $\in [0, 1]$.

x^*: After comparing all possible locations among all bats, the current best location (solution) was found. f_i. a frequency that is drawn consistently from $[f_{min}, f_{max}]$ A The local search alters the present best solution according to the equation and uses a stochastic process with direct exploitation.

$$x_{new} = x_{old} + \in A^t \qquad (5)$$

Where;

$\in$: is an arbitrary number [-1,1].

A^t: is the volume level that the best performers have on average at this moment.

r_i: is the pulse emission rate. As soon as the prey is found, each bat becomes less loud and emits pulses at a faster rate. The following equations express loudness and pulse emission:

$$A_i^{t+1} = \alpha A^t \qquad (6)$$

$$r_i^{(t+1)} = r_i^{(0)}[1 - exp(-yt)] \qquad (7)$$

$$A_i^t \to \theta \; and \, r_i^t \to r_i^0 ast \to \infty \qquad (8)$$

Where;

α: *is constant* $0 < \alpha < 1$

γ: *is constant* $\gamma > 0$

5. Result and Discussion

This section evaluates the efficacy of suggested techniques by comparing the accuracy, precision, and recall of various methods, such as K-Nearest Neighbor (KNN), Multinomial Naive Bayesian (MNB), and Support Vector Machines (SVM).

Accuracy

The degree to which a measurement corresponds to its actual value is referred to as its accuracy. The term "Accuracy" refers to the degree to which multiple measurements carried out under the same conditions provide consistent outcomes.

Figure 78.2 shows the similar accuracy measure values. Compared to existing methods like KNN, which has an accuracy rate of 95.48%, MNB, which has an accuracy rate of 58.76%, and SVM, which has an accuracy rate of 93.71%, the proposed method BFBSSO value is 98.57%. The proposed BFBSSO performs well and is more accurate than previous approaches.

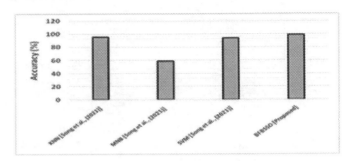

Fig. 78.2   Comparison of acuuracy

Precision

In robotics, precision is the capacity of an industrial robot to repeatedly bring its end effectors to the same position and orientation. This skill is often referred to as repeatability.

Figure 78.3 demonstrates that the suggested strategy might deliver performance outcomes that are better than those attained by the study's present methodologies. The precision of the proposed approach is 97.64%, which performs better than existing outcomes. KNN, MNB, and SVM precision rates are 95.49%, 49.45%, and 93.90%.

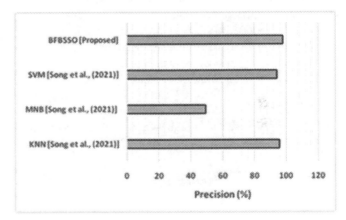

Fig. 78.3   Comparison of precision

Recall

The ability of a model to correctly identify each important data contained inside a data collection is referred to a recall.

Figure 78.4 compares the data for the recall measures. Recall rates for KNN 95.48%, MNB 58.76%, SVM 93.71%, and BFBSSO 99.07%. The proposed method performed better than the current results, with a recall of 99.07%.

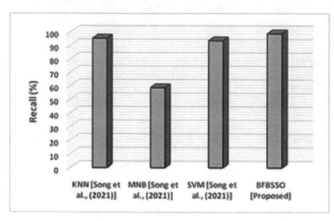

Fig. 78.4 Comparison of recall

6. Conclusions

Currently, new applications like electronic circuit designs, network routing, computer visual effects, biopharmaceutical designs, computational biology, etc. are increasingly inspiring path planning development. The best anticipated performance of the best path planning algorithms is BFBSSO, with a forecast accuracy of about 99.9%. In order to fully utilize the global optimization capabilities of robotic algorithms, we have presented a BFBSSO algorithm is highly effective when compared with KNN, SVM, and MNB. Based on creating a global optimal control path and then discovering the globally optimal path, this algorithm can instantly optimize the local path. The suggested BFBSSO's path planning performance is superior to that of existing relevant path planning algorithms, and it may be used for both worldwide and local path planning.

Reference

1. Campbell, S., O'Mahony, N., Carvalho, A., Krpalkova, L., Riordan, D. and Walsh, J., 2020, February. Path planning techniques for mobile robots a review. In *(2020) 6th International Conference on Mechatronics and Robotics Engineering (ICMRE)* (pp. 12-16). IEEE.
2. Sun, F., Chen, Y., Wu, Y., Li, L. and Ren, X., (2022). Motion Planning and Cooperative Manipulation for Mobile Robots with Dual Arms. IEEE Transactions on Emerging Topics in Computational Intelligence, 6(6), pp.1345-1356.
3. Grover A., Mohan Kumar R., Angurala M., Singh M., Sheetal A., Maheswar R."Rate aware congestion control mechanism for wireless sensor networks." Alexandria Engineering Journal. (2022); 61(6):4765-4777. DOI: 10.1016/j.aej.2021.10.032.
4. Zafar, M.N. and Mohanta, J.C., (2018). Methodology for path planning and optimization of mobile robots: A review. Procedia computer science, 133, pp.141-152.
5. Mavrogiannis, C., Baldini, F., Wang, A., Zhao, D., Trautman, P., Steinfeld, A. and Oh, J., (2021). Core challenges of social robot navigation: A survey. arXiv preprint arXiv:2103.05668.
6. Sharma N., Mangla M., Yadav S., Goyal N., Singh A., Verma S., Saber T. "A sequential ensemble model for photovoltaic power forecasting."Computers and Electrical Engineering. 2021; 96. Article No.: 107484. DOI: 10.1016/j.compeleceng.2021.107484.
7. Ahmad, R. and Sharma, G. (2022). Mobile Robot Navigation in Dynamic Environment using Fuzzy Logic. Research Journal of Engineering Technology and Medical Sciences (ISSN: 2582-6212), 5(01).
8. Ajeil, F.H., Ibraheem, I.K., Azar, A.T. and Humaidi, A.J. (2020). Autonomous navigation and obstacle avoidance of an omnidirectional mobile robot using swarm optimization and sensors deployment. International Journal of Advanced Robotic Systems, 17(3), p.1729881420929498.
9. Sánchez-Ibáñez, J.R., Pérez-del-Pulgar, C.J. and García-Cerezo, A.,(2021). Path planning for autonomous mobile robots: A review. Sensors, 21(23), p.7898.
10. Patle, B.K., Pandey, A., Parhi, D.R.K. and Jagadeesh, A.J.D.T., (2019). A review: On path planning strategies for navigation of mobile robot. Defence Technology, 15(4), pp.582-606.
11. Li, B., Liu, H. and Su, W., (2019). Topology optimization techniques for mobile robot path planning. Applied Soft Computing, 78, pp.528-544.
12. Mittal S., Bansal A., Gupta D., Juneja S., Turabieh H., Elarabawy M.M., Sharma A., Bitsue Z.K. "Using Identity-Based Cryptography as a Foundation for an Effective and Secure Cloud Model for E-HealthComputational Intelligence and Neuroscience. 2022; (2022). Article No.: 7016554. DOI: 10.1155/2022/7016554.
13. Yang, G., Chen, Z., Li, Y. and Su, Z., (2019). Rapid relocation method for mobile robot based on improved ORB-SLAM2 algorithm. Remote Sensing, 11(2), p.149.
14. Song, Q., Li, S., Yang, J., Bai, Q., Hu, J., Zhang, X. and Zhang, A.,(2021. Intelligent optimization algorithm-based path planning for a mobile robot. Computational Intelligence and Neuroscience, 2021.

Note: All the figures in this chapter were made by the authors.

Advancing Sustainable Science and Technology for a Resilient Future – Sai Kiran Oruganti et al. (eds)
© 2024 Taylor & Francis Group, London, ISBN 978-1-032-79020-6

79

Design of a Bio-Inspired Framework for Underwater Robotic Communication System

Ankit Punia*

Chitkara University Institute of Engineering and Technology, Chitkara University, Punjab, India.

Abstract: For robot communication and detection, interference and jamming constitute critical challenges, particularly in underwater sensor networks. This study develops a novel enhanced genetic bee colony optimization (EGBCO) technique and describes its research in a number of studies to help with scheduling within robotic swarms. The chromosome and the ant serve as biological models for the scheduling method. To increase scalability as well as provide independence for the methods, innovative dispersed frameworks for both are also being researched. The modelled findings indicate the model viability and robustness for planning underwater swarm interaction and detection. Additionally, the study demonstrates how the technique to calculate the swarm size in a distributed mode may also be used to obtain greater swarm intelligence. The presentation also includes an actual robot experiment employing platforms for underwater swarm robots.

Keywords: Underwater robot communication, Bio-inspired method, Enhanced genetic bee colony optimization (EGBCO), Detection

1. Introduction

Over the past ten years, underwater robotics have become increasingly common, supporting a range of duties such as inspections of ship hulls and harbour safety, maintenance-inspection-repair (MIR), and intelligence surveillance reconnaissance (ISR). Unfortunately, the sensing capabilities of autonomous underwater vehicles (AUVs) are constrained. First off, using the global positioning system (GPS) is impossible due to GPS signal attenuation in the water (McConnell, et al., 2022).

Hence, multi-robot systems have drawn lots of interest from researchers all around the world. One of the most crucial responsibilities in underwater multi-robot systems is formation control. In order to accomplish specified geometries, it coordinates several underwater robots (Zhou, et al., 2022).

A growing number of applications for underwater robotics are being developed today, from data collecting and mapping to inspection and surveillance. Although totally autonomous underwater navigation is still a challenge, teams of human divers and autonomous robots are frequently needed for underwater missions. Typically, human divers are the mission leaders and control the robots throughout mission execution (Islam, et al., 2019).

The employment of several remotely operated vehicles (ROV) or intelligent aerial vehicles (I-AUVs) in cooperative applications is a recent robotic treatment trend (Surrounded by water electric robots for involvement). As an illustration, some robots carry out a certain duty while others carry out additional tasks, including visual surveying, to offer the operator visual input on how the operation is going. ROVs or autonomous underwater vehicles (AUVs) and their operators typically communicate with each other through umbilical cables or sound transducers (Centelles, et al., 2019).

Since the two most prevalent modalities for human connection are severely constrained, robot communication with humans underwater is rather difficult. Equipment noise can hide, attenuate, and distort sound. Although vision is usually present, it is frequently impaired or changed. The de facto response in such a difficult setting is to embrace the AUV as a quiet partner that listens but does not speak (Fulton, et al., 2019).

*ankit.punia.orp@chitkara.edu.in

DOI: 10.1201/9781003490210-79

2. Related Works

The article provides a tactical decision support framework for a management decision working with an automated guided machine that integrates using the vertically sound-speed profile for communicating and navigating underwater (AUV) (Bhatt. et al., 2022).

(Bali, et al., 2022 and Muthukumaran, et al., 2022) introduce a robot communications framework based on movements in this paper that makes nonverbal communicating in between robotic underwater vehicles and humans divers (AUVs). In contrast to conventional Voice, light, or radio frequency-based AUV communications, they develop a gesture language for AUV-to-AUV communication that is easily understood by divers viewing the dialogue.

(Singh, et al., 2023 and Gupta, et al., 2022) build an emulation framework and a tractable mathematical model for transmitting information via water using acoustic waves. Due to its size, diversity of properties based on the scenario, kind of water body (lakes, rivers, tanks, sea, etc.), and location of the water body, water is one of the most difficult mediums to simulate.

3. Proposed Methodology

For underwater robotic communication systems, a variety of data sets are accessible, depending on the particular study issue and application.

The underwater acoustic signals from a network of hydrophones in the Coral Sea are included in the CORAL data set. It is made to aid in underwater communication research, particularly in the creation of signal processing algorithms and communication protocols.

Artificial bee colony algorithm

The artificial bee colony algorithm (ABC) has three individuals: working bees, observers, and exploring bees. Hunting bees are used when working bees transmit messages. Responsibility for pursuing sources of protein, observer bees are tasked with identifying sources of protein in the performing region, and employed bees are responsible for finding sources of protein. Hunting bees are used when working bees transmit messages. bees working. Employed bees change into scouting bees if a Bees on the job leave their food source behind and observer bees.

A food source's place suggests a potential resolution when utilizing the artificial bee colony method to solve a planning problem. The target function values of the optimization problem are represented by the amounts and level of the food source nectar.

Hired bees begin by searching the immediate region for the appropriate solutions (food sources) and then choose those

that are more fit. Observer bees receive data about answers after all hired bees have finished their searches.

Employed bees and neighbouring bees conducting a search area as

$$X_{i,} = V_i + O_i(V_i - V_l)$$
$$W_{i,} = Z_i + O_i(Z_i - Z_l) \tag{1}$$

Where, (V_i, Z_i), $i = 1, 2 ..., M$, which is random selected, $l \neq i$, $Q_i \in [-1,1]$,

This is a random number and can be reduced properly. The observation bees choose the solution with a given probability in accordance with the information they have learned. The more the fitness, the greater the likelihood of selection. The fitness function is:

$$e_j = \frac{1}{\left(T - T^0 - dm\right)^s \left(T - T^0 - dm\right)} \tag{2}$$

Onlooker bees select solutions (food sources) in accordance with the probability value and is calculated as

$$O_j = \frac{e_j}{\sum_{j=1}^{M} e_j} \tag{3}$$

After that, observer bees check about the area to find a workable option (a food source). A solution is eliminated if it has undergone L finite cycles but has not yet been improved. L is a regulating variable.

Scouting bees produce a new solution to update the initial solution as follows, assuming that the th solution is discarded:

$$V_j = V_j^{min} + rand(0,1)\left(V_j^{max} - V_j^{min}\right) \tag{4}$$
$$Z_j = Z_j^{min} + rand(0,1)$$

Genetic algorithm

To increase the effectiveness of path planning, the Obstacle Avoidance Algorithm (OAA) and Distinguish Algorithm (DA) are used to create the first society. While DA is used to determine whether a path is possible or not, OAA is specially built for creating the first crowd.

Improved genetic algorithm approaches are suggested to make the procedure of coming up with a workable pathways initially in the populace simpler. Two algorithms are created in this work to solve these difficulties. The first population is generated using both a Distinguish Algorithm (DA) and an Obstacle Avoidance Algorithm (OAA), ensuring that every member of the first generation is viable. Certain specific genetic operators (deletion, refinement, mutation, and crossover) are based Preparing routes for moving robots is based on domain knowledge and heuristics. Simulation

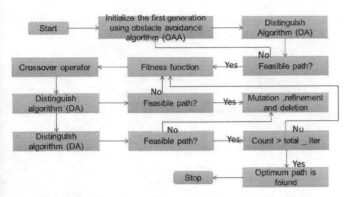

Fig. 79.1 Flowchart of Proposed Genetic Algorithms

simulations show how well the suggested knowledge-based evolutionary path-planning algorithm for robotic systems works. By implementing the simulation studies in real time on a Team AmigoBotTM robot, an actual mobile robot, the results are further tested and verified.

Initialization of the Population by OOA

In the earlier described genetic algorithms, it is presumed that nearly all of the members of the first population were produced at random. This is straightforward, however if employed in target tracking, it will result in a significant number of infeasible paths, and infeasible paths ought to be ignored. In this work, an OAA that is intended to produce a first group of people is suggested. The mobile robot must avoid the obstacle(s), whether they are either vertically horizontal direction, as much as is physically possible.

Distinguish Algorithm (DA)

A DA created expressly to determine whether plan can be implemented or not. This algorithm will position practical paths in the feasible group and impossible paths in the impossible groups. The fundamental goal of this DA is to determine whether points between nodes m and n fall within the obstacle region. Thus can infer that there are no obstacles between node m and cluster n if none of them fall into any of the obstacle areas. Until all of the nodes in a particular path are studied, a same procedure is used again. A practicable path is one that leads to success.

4. Result and Discussion

Positioning error

Underwater robotic systems typically estimate the robot's position using a combination of inertial navigation and acoustic positioning. Some of the typical positioning error algorithms employed by underwater robotic systems include the following:

Range-based positioning: Range-based positioning algorithms calculate the distance between a robot and a group of acoustic

beacons using the time-of-flight of sound waves. The robot's position is then inferred via trilateration using the range measurements as a starting point. Underwater positioning systems like the Long Baseline (LBL) and Ultra Short Baseline (USBL) systems frequently employ range-based positioning. Figure 79.2 depicts the Positioning errors comparison. The proposed (EGBCO) is better than existing method.

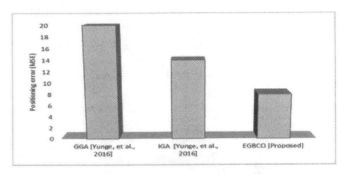

Fig. 79.2 Positioning error

Time interval for detection of underwater Target

The length of time it takes to identify an underwater target relies on a number of variables, including the size and shape of the target, the surrounding environment (such as water quality, noise levels, etc.), and the detecting technology being applied. Long-range underwater targets can be detected using acoustic detection technologies like sonar, although the detection time varies on the type of sonar and the target's range. Because sound travels slowly through water, low-frequency sonar, for instance, may identify huge targets like submarines at distances of several kilometres, but the discovery process can take several minutes.

From Fig. 79.3 it is concluded that the proposed enhanced genetic bee colony optimization (EGBCO) features a 2% increase in efficiency for underwater targets over time interval.

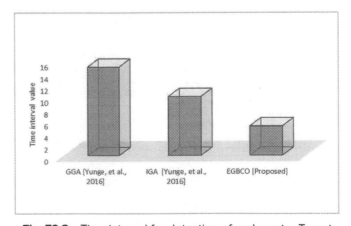

Fig. 79.3 Time interval for detection of underwater Target

5. Conclusion

Enhanced genetic bee colony optimization (EGBCO) is used in this study to build a positioning system with a focus on the underwater robotic communication system and in accordance with the marine positioning idea of underwater acoustic positioning. The setup of the system is simple. The technology is also extremely stable and reliable. The enhanced genetic bee colony optimization (EGBCO) has a consistent performance and can solve positional coordinates precisely. Although underwater acoustic location is appropriate for marine situations, it may be affected by things like water quality, noise, and interference. To guarantee the dependability and precision of the positioning system, the research has to recognize and take into account these environmental limits.

References

1. McConnell, J., Huang, Y., Szenher, P., Collado-Gonzalez, I. and Englot, B., 2022, October. DRACo-SLAM: Distributed Robust Acoustic Communication-efficient SLAM for Imaging Sonar Equipped Underwater Robot Teams. In 2022 IEEE/RSJ International Conference on Intelligent Robots and Systems (IROS) (pp. 8457-8464). IEEE.

2. Zhou, Y., Wang, W., Zhang, H., Zheng, X., Li, L., Wang, C., Xu, G. and Xie, G., 2022. Underwater robot coordination using a bio-inspired electrocommunication system. Bioinspiration & biomimetics, 17(5), p.056005.

3. Islam, M.J., Ho, M. and Sattar, J., 2019. Understanding human motion and gestures for underwater human–robot collaboration. Journal of Field Robotics, 36(5), pp.851-873.

4. Centelles, D., Soriano-Asensi, A., Martí, J.V., Marín, R. and Sanz, P.J., 2019. Underwater wireless communications for cooperative robotics with uwsim-net. Applied Sciences, 9(17), p.3526.

5. Fulton, M., Edge, C. and Sattar, J., 2019, May. Robot communication via motion: Closing the underwater human-robot interaction loop. In 2019 International Conference on Robotics and Automation (ICRA) (pp. 4660-4666). IEEE.

6. Bhatt, E.C., Howard, B. and Schmidt, H., 2022. An embedded tactical decision aid framework for environmentally adaptive autonomous underwater vehicle communication and navigation. IEEE Journal of Oceanic Engineering, 47(4), pp.848-863.

7. Bali M.S., Gupta K., Koundal D., Zaguia A., Mahajan S., Pandit A.K. "Smart architectural framework for symmetrical data offloading in IoT." Symmetry. 2021; 13(10). Article No.: 1889. DOI: 10.3390/sym13101889.

8. Singh J., Singh S., Singh S., Singh H. "Evaluating the performance of map matching algorithms for navigation systems: an empirical study." Spatial Information Research. 2019; 27(1):63-74. DOI: 10.1007/s41324-018-0214-y.

9. Muthukumaran, V., Natarajan, R., Kaladevi, A.C. et al. Traffic flow prediction in inland waterways of Assam region using uncertain spatiotemporal correlative features. Acta Geophys. (2022). https://doi.org/10.1007/s11600-022-00875-8

10. S. K. Gupta, B. Pattnaik, V. Agrawal, R. S. K. Boddu, A. Srivastava and B. Hazela, "Malware Detection Using Genetic Cascaded Support Vector Machine Classifier in Internet of Things," 2022 Second International Conference on Computer Science, Engineering and.2022.9936404.

Note: All the figures in this chapter were made by the authors.

Advancing Sustainable Science and Technology for a Resilient Future – Sai Kiran Oruganti et al. (eds)
© 2024 Taylor & Francis Group, London, ISBN 978-1-032-79020-6

Smart Environmental Monitoring System Using Enhanced Bayes Bayesian Optimized Robotics Platform

80

Aashim Dhawan*

Chitkara University Institute of Engineering and Technology,
Chitkara University, Punjab, India

Abstract: The two types of environments that affect skin wound healing are the internal environment, which is the environment of the neighboring portion beneath the wound site, and the exterior, which is close to the wound surface. Both kinds of environments are important for wound healing and might result in continual or sluggish wound healing. Although other earlier research offered wound care remedies, they concentrated on a specific environmental component, such as the pH value, moisture content of the wound, or healing enzymes. Realistically speaking, evaluating environmental impact based on the identification of one or two criteria is meaningless because both types of environments have a large number of additional aspects that must be taken into account during the inquiry. Moreover, prior research did not categorize general healing as being continuous or impeded based on the influence of the external environment. In this study, we suggested an innovative radial support vector machine with enhanced Bayesian optimization (R-SVM+EBO) approach for outside monitoring the environment to take into account the impact of the external environment on the recovery process of wounds. By giving patients a smart wound management tool to track and supervise wound healing from house, this study helps patients. When compared to other methods for smart environmental monitoring, the suggested method achieves the best performance.

Keywords: Smart environmental monitoring, Wound healing, Radial support vector machine with enhanced Bayesian optimization (R-SVM+EBO)

1. Introduction

Smart environmental monitoring system evaluates leaks or spills of hazardous materials such as biochemical, microbiological, radioactive, nuclear, or highly flammable materials, quick and efficient pollution control is required. Portable terrestrial and overhead robotic systems may be used to locate, separate, track, map, and forecast the dissemination of such compounds to reduce their pollution and waste and the risk they represent to soil health (He, et al., 2020). The increasing lifespans in industrialized nations have made it imperative to discover appropriate innovations to transform how health providers are thought of and run entirely. Intelligent home technology has emerged as a trend to provide socially and economically feasible answers to a few of the difficulties presented by cellular technology and telephony (Tamasi, et al., 2022). Finds direct and encryption method breakthroughs are second to information inventions that enable data manipulation. Another is still an important study topic, whereas the initial argument is growing increasingly specific, specifically for combustible substances that the petroleum and natural gas sector often releases (Bahrami, et al., 2023). It focuses mostly on experimenting with the various options and evaluating the results. This is not feasible for systems in the real world, where operating safely and within environmental limits are essential requirements. For instance, machine learning techniques in computers often deal with significant previous uncertainty, and errors may seriously harm the robot and its surroundings (Venkatesh, et a., 2019 and Kaur, et al., 2022). Robots are becoming an essential component of both daily life and work. Conventional and sophisticated robots integrate biomedical engineering, computer software, extensive sensor fusion, and department of electronics, biomaterials, character recognition, and other disciplines to create a sophisticated, dependable device. Robotics' growth and progression

*aashim.dhawan.orp@chitkara.edu.in

DOI: 10.1201/9781003490210-80

provide conclusive evidence of the scientific community and development's progress (Berkenkamp, et al., 2021). The study considers both internal and extrinsic elements affecting wound healing, providing a more complete picture of the recovery process. The study's conclusions and approach may not be generalizable. The research may not apply to all wound healing situations. The findings must be confirmed in other locations and patient populations. Here, we addressed the revolutionary R-SVM+EBO strategy, which combines improved Bayesian optimization with radial support vector machines.

2. Related Works

Manufacturing operations are gradually using robots. By acquiring fresh duties and adjusting to account for uncertainty, the robots support good skills. To imitate the installation of a transmission, the operation entails putting a sprocket into its rectangular shape spindle in a secure presence and enabling the robots to adapt and improve their behaviors to complete the setup despite assignment uncertainty (Roveda, et al., 2021). In an independent study conducted at several public hospitals, a sophisticated annotation, the multi-access behavioral monitoring program, may identify issues with people's lifestyles that are related to underlying severe illnesses. Employing scientific results, the effectiveness of these portable Internet of Things systems with audio-visual aids is assessed and described in terms of correctness, profitability, maximum transition probability, latency, and lower electricity usage (Manogaran, et al., 2019). Researchers really have to develop technology that can identify and track actions that individuals commonly do as part of everyday routines in the interest of tracking the operational fitness of occupants of smart homes. Such splitting intelligent patch is regarded as one of the phylogenetic biologies in condition having checked of affect the functionality physical measurement techniques using a multi-sensory approach (Sunhare, et al., 2022 and Khadda, et al., 2021). Throughout current history, Bayesian optimization has drawn a lot of interest as an universal tool for finding the most significant number of difficult-to-evaluate optimization methods. The nature and quantity of specimens that may be collected are also determined by the ecosystem's condition and the detecting apparatus. Several investigations on using a robotic system to observe geographical interactions were conducted (Pitera, et al., 2022). Mobile robots are a good foundation for sensitive multi-responsive composites because they can give them a better physiological sense of balance, and associated development is an important aspect. An enhanced morphology adaptability in inter settings reacting to both internal and outside inputs may be achieved by a system that has "educational and recognize" capabilities, as well as fundamental "physical" and "information awareness" (Wen et al., 2018).

Based on a review of existing literature, the suggested method improves the smart monitoring system for wound healing by using a radical support vector machine integrated with an enhanced Bayesian optimization algorithm and structural methods are applicable.

Radial Support Vector Machine (R-SVM): The primary applications of RSVM, a kind of artificial learning exercise, are classifying and simple regression. They evaluate data and identify structures or determination criteria inside this information. The total dimension count is referred to as the classification model of the collection, and it is used by RSVM to create high energy in a Hilbert space that divides various class borders. Several different kinds of data may be handled using RSVM. Both are two types of circular: full spheres and delineated clusters. The RSVM's objective is to classify the various pairs according to their attributes. It has different lengths in total. The peripheral path or border, w.x-b=0, is the first. The nearest training spots for both divisions are shown by the curves w.x-b =1 and w.x-b =-1. The attribute values are the rectangles located on the excessive self. An exception is the completed oval in the opposite subclass. To reduce the likelihood of an approximate solution, the RSVM is trying to optimize the geometrical distance that separates the sides of the support vectors. The extension potential rises with falling RSVM classification numbers because the high energy relies on the number of vertices.

Enhanced Bayesian Optimization (EBO): To optimize the parameters of every mixed-race mechanism, EBO is an efficient solution $f(x)$. EBO was helpful for algorithm setup and is currently a well-liked advanced option. To estimate a complete factorial component, BO is already using prior beliefs. It instead seems to be using these beliefs to choose the following arrangement to test x_n. It then validates x_n by using the actually black component, considers proximal opinion that used the investigated efficiency, and repeats this operation until a deterring criterion is satisfied. BO seems to be using preconceived ideas to estimate a process variables method named $\hat{f}(x)$, implements $\hat{f}(x)$ to decide which subsequent structure to test, assessing $f(x_n)$ through using actually black component, quantifies inferiorly conviction using the investigated effectiveness $f(x_n)$, and continues the procedure until a preventing criterion has been satisfied. Algorithm 1 demonstrates a procedural code for Bayesian Optimization (BO).

Algorithm 1 Bayesian Optimization
Inputs: block-box function f, BO Algorithm $\hat{f}$, Parameter Space
$Y \longleftarrow 0$
for $a = 1, 2, \ldots$ **do**
Select $x_n \in$ args $\max_{a \in X} \hat{f}(a; y)$
Evaluate $b_n \longleftarrow f(x_n)$
Update $\longleftarrow y \cup (a_n, b_n)$
Check for Exit Criteria
end for

The three most commonly used options for ML problems are the Cluster Semi-parametric Estimation, Quadratic Processes, and Stochastic Plantation regression model for constructing the regression line of $\hat{f}(x)$. The algorithms use acquiring functions to allow for an import and export between discovering and exploiting data, and the most common options are Potential of Improving (PI), Expecting Increase in performance (EI), and Upper Confidence bound (UCB). The dark-skinned transfer function, which could in our research depicts the functionality of a Hidden layer with a population of solutions x, is notably non-convex. Moreover, $\hat{f}(x)$ may be assessed at any point x, although determining the criteria might take a long time since evaluating $f(x)$ necessitates running the DNN whole development operation.

3. Result and Discussion

We used Classifier to build the suggested method, which needed pre-labeled data sets for testing. There were 800 cases in the sample we created for teaching our proposed framework (Sattar, et al., 2019). Table 80.1 contains a complete plan for the training data collection and a comprehensive summary of the training dataset examples for each external component. We take into account the local ambient circumstances as well as the provided dataset.

Table 80.1 Data collections

Class	Instances	Location
Constant	180	Cottage → 300
Affected	120	
Constant	120	Garden → 250
Affected	130	
Constant	70	Clinic → 100
Affected	30	
Constant	80	Company → 150
Affected	70	

The proposed methodology R-SVM+EBO (Radial Support Vector Machine + Enhanced Bayesian Optimization) value is compared to existing technologies like VPG (Voronoi-based path generation), GP (Gaussian Process), and Hi-MPC (Hierarchical- Model Predictive Control). In our research, we assess parameters like accuracy and error rate.

Accuracy

A difference between the result and the correct number is caused by inadequate precision. The percentage of actual outcomes reveals how balanced the data is overall. Accuracy is assessed using equation (1).

$$\text{Accuracy} = \frac{TP + TN}{TP + TN + FP + FN} \qquad (1)$$

Figure 80.1 shows the comparable values for the accuracy measures. Compared to existing methods like VPG, which has an accuracy rate of 81%, GP, which has an accuracy rate of 83%, and Hi-MPC, which has an accuracy rate of 89%, the recommended method R-SVM+EBO value is 93%. The suggested R-SVM+EBO performs well in the monitoring environment in a smart way and has higher accuracy than other methods.

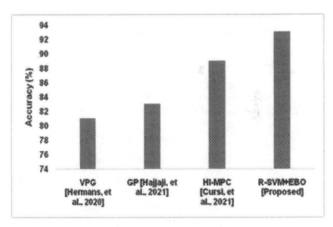

Fig. 80.1 Comparison of accuracy

Error Rate

Inaccurate responses from survey respondents might lead to a measuring error known as a classification error. Nominal categorical data have the potential for false adverse and false positive outcomes. There is a standardized way to evaluate its effectiveness. The True Positives (TP)/False Negatives (TN) ratio is positive, indicating that the instance class has been correctly predicted. Mistake: incorrectly determining the category of an instance (False Positives (FP)/False Negatives (FN)). Classification errors for the proposed and current modalities are shown in Fig. 80.2. This indicates that the

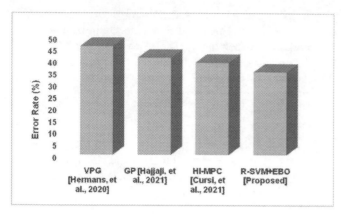

Fig. 80.2 Comparison of error rate

recommended method is more accurate and does not have any errors.

$$Error = Error/total = FP + FN/TP + TN + FP + FN \quad (2)$$

The corresponding Error rate values are shown in Fig. 80.2. The suggested method's R-SVM+EBO value is 35%, which is lower than current methods like VPG, which has a 46% error rate, GP, which has a 41% accuracy rate, and Hi-MPC, which has a 39% error margin. The proposed R-SVM+EBO works well in environmental tracking in a clever manner and is more effective than other approaches.

4. Conclusion

Smart monitoring system is the interest of examining how the environmental factors affect the recovery process of wounds, article offered an original radial support vector machine with enhanced Bayesian optimization (R-SVM+EBO) approach for outdoor heath monitoring. The suggested strategy offers a low error rate (35%) and high accuracy (93%). Robots incorporate the actuators and sensors in their environments and will accomplish more in the future. This makes environmental service robots cheaper and more complex. This discovery raises questions about how to create robot applications and intelligent settings together and evaluate the final environmental systems. Future studies may include long-term wound monitoring and predictive analytics. Long-term data analysis reveals wound healing outcome factors and patterns. Proactive measures and customised treatments may result..

References

1. Tamasi, M.J., Patel, R.A., Borca, C.H., Kosuri, S., Mugnier, H., Upadhya, R., Murthy, N.S., Webb, M.A. and Gormley, A.J., (2022). Machine Learning on a Robotic Platform for the Design of Polymer–Protein Hybrids. Advanced Materials, 34(30), p. 2201809.
2. Venkatesh, V., Raj, P., Kannan, K. and Balakrishnan, P., (2019). Precision centric framework for activity recognition using Dempster Shaffer theory and information fusion algorithm in smart environment. Journal of intelligent & fuzzy systems, 36(3), pp. 2117–2124.
3. Berkenkamp, F., Krause, A. and Schoellig, A.P., (2021). Bayesian optimization with safety constraints: safe and automatic parameter tuning in robotics. Machine Learning, pp. 1–35.
4. Roveda, L., Magni, M., Cantoni, M., Piga, D. and Bucca, G., (2021). Human–robot collaboration in sensorless assembly task learning enhanced by uncertainties adaptation via Bayesian Optimization. Robotics and Autonomous Systems, 136, p. 103711.
5. Manogaran, G., Shakeel, P.M., Fouad, H., Nam, Y., Baskar, S., Chilamkurti, N. and Sundarasekar, R., (2019). Wearable IoT smart-log patch: An edge computing-based Bayesian deep learning network system for multi access physical monitoring system. Sensors, 19(13), p. 3030.
6. Sunhare, P., Chowdhary, R.R. and Chattopadhyay, M.K., (2022). Internet of things and data mining: An application oriented survey. Journal of King Saud University-Computer and Information Sciences, 34(6), pp. 3569–3590.
7. Wen, J., He, L. and Zhu, F., (2018). Swarm robotics control and communications: Imminent challenges for next generation smart logistics. IEEE Communications Magazine, 56(7), pp. 102–107.
8. Sattar, H., Bajwa, I.S., Ul-Amin, R., Mahmood, A., Anwar, W., Kasi, B., Kazmi, R. and Farooq, U., (2019). An intelligent and smart environment monitoring system for healthcare. Applied Sciences, 9(19), p. 4172.
9. Kaur R., Sood A., Lang D.K., Bhatia S., Al-Harrasi A., Aleya L., Behl T. "Potential of flavonoids as anti-Alzheimer's agents: bench to bedside."Environmental Science and Pollution Research. (2022); 29(18):26063-26077. DOI: 10.1007/s11356-021-18165-z.
10. Khadda Z.B., Fagroud M., Karmoudi Y.E., Ezrari S., Berni I., De Broe M., Behl T., Bungau S.G., Houssaini T.S. "Farmers' knowledge, attitudes, and perceptions regarding carcinogenic pesticides in fez meknes region (Morocco)."International Journal of Environmental Research and Public Health. (2021); 18(20). Article No.: 10879. DOI: 10.3390/ijerph182010879.

Note: All the figures and table in this chapter were made by the authors.

Advancing Sustainable Science and Technology for a Resilient Future – Sai Kiran Oruganti et al. (eds)
© 2024 Taylor & Francis Group, London, ISBN 978-1-032-79020-6

81

Automated Robotic Navigation with Obstacle Detection Using a Deep Learning Algorithm

Komal Parashar*

Chitkara University Institute of Engineering and Technology, Chitkara University, Punjab, India

Abstract: The navigation of an autonomous Pioneer P3-DX wheeled robot around obstacles is presented in this research employing a hale-optimized modular neural network (WO-MNN). The deviation between the MNN's actual and predicted values is reduced by the WO method. The range of objects is entered by the suggested un-tuned MNN and WO-MNN from ultrasonic sensors. Also, the steering angle of the differential motor of the autonomous Pioneer P3-DX robotic system is outputted. When we examined the outcomes of WO-MNN with MNN without tuning, we discovered that WO-MNN offered an improved path and traveled closer to the destination. The efficiency of WO-MNN in autonomous Pioneer P3-DX wheel navigation has been confirmed by a comparison study between MNN with and without tuning and WO-MNN. We have also contrasted it to previously established approaches to demonstrate the validity and real-time use of the WO-MNN technique.

Keywords: Navigation, Autonomous robotic system, Obstacles, Whale-optimized modular neural network (WO-MNN)

1. Introduction

Path planning is the most frequent issue with robot navigation since robots must move from their starting point to their destination while avoiding obstacles. Navigating through a network of city streets, a disaster-prone area, or a war-torn environment can be challenging. These drawbacks include finding new and complicated routes around dead ends and obstacles or wrecks possibly (Kumar et al., 2017). Path planning is a crucial component of mobile robots used for cleaning, monitoring, and delivery.

It is now possible to create small-scale automated robots because of the availability of lightweight and affordable sensors like Lidar, cameras, IMUs (Inertial Measurement Units), and depth sensors. Under some circumstances, pathways might not be accessible. This research suggests a method for blocking specific map portions through virtual obstacles, as such cases are simple enough for people to understand. As a result, the mobile robot's global path planner will automatically design a safer path. In the map, a simulated obstacle is not the same as a genuine one (Ravankar et al., 2019). Ideally, current route planners, such as those in the Robotics Operating System (ROS), can locate appropriate routes through these difficult navigational locations., given that a viable path exists.

However, based on our experience with actual deployments, robots frequently encounter significant difficulties when acting securely and moving along the intended route. Numerous factors can contribute to this unpleasant behavior, such as noise, sensor calibration issues, mistakes in robot localization, and inaccuracies in motion implementation (Moreno, et al., 2019). One essential mobile robot technology that sets them apart from other manufactured robots is automated guidance control. An automated navigation program controls the robot's navigation, which uses various navigational methods, including magnetic, optical, and inertial, to steer the robot around obstructions and acquire its location in a given environment (Gupta et al., 2019).

Due to deep convolutional neural networks' achievement in ImageNet Classification, deep learning has made significant progress in several different fields. With just raw observation data, deep understanding has enormous promise for solving robot control issues from beginning to finish. Using camera and LiDAR for robot guidance has been effectively implemented using deep learning (Ruan et al., 2019).

*komal.parashar.orp@chitkara.edu.in

DOI: 10.1201/9781003490210-81

2. Related Works

The automated navigation of agricultural robots is required for many more intelligent agriculture activities. Agricultural robot navigation typically uses machine vision (MV) because of the superior visual information and low hardware costs. (Wang, et al., 2022). In addition to selecting a collision-free section, the multilayer perceptron (MLP) neural network also regulates the robot's speed for each motion. According to simulation findings, the strategy is effective and provides a nearly ideal path for the mobile robot to reach its intended place (Singh and Thongam, 2019). The mobile robot's primary technology is autonomous navigation control. This study examines the mobile robot's navigation method—the invention of a particle filter-based simultaneous localization and mapping (SLAM) method. To create the navigation algorithm and run tests on it, it is integrated with the Vector Field Histogram (VFH) obstacle avoidance algorithm (Wang 2019). A low-complexity, collision-free navigation technique for mobile robots is called collision-aware mobile robot navigation in grid environments. The suggested strategy uses a predetermined decision table for navigation, a hybrid approach for path planning, and for localizing mobile robots using the Radio Frequency-based Identification technique. (Tripathy et al., 2021). DRL-based navigation methods and Deep Reinforcement Learning (DRL) frameworks. Then, a careful comparison and analysis of the connections and contrasts between four common application scenarios: internal navigation, multiple robot navigation, local obstacle avoidance, and social navigation. (Zhu and Zhang., 2021).

3. Proposed Method

Automated robotic navigation with obstacle detection involves designing a robot that can move through an environment and avoid obstacles without human intervention. This requires the integration of several technologies, including sensors, artificial intelligence, and control systems. Artificial neural networks are used in deep learning, a type of machine learning, to address complicated issues. Deep learning algorithms may be utilized for applications like speech recognition, natural language processing, and picture recognition since they are made to learn from massive volumes of data.

4. Whale-Optimized Modular Neural Network (WO-MNN)

The Whale-Optimized Algorithm (WOA) is a swarm intelligence method suggested for ongoing optimization issues. It has been demonstrated that this algorithm performs as well as or even performs better than some existing algorithmic strategies. WOA has drawn inspiration from the humpback whales' hunting techniques. According to WOA, each solution is a whale. In this approach, a whale tries to fill in a new location in the search area that is used as a reference for the best member of the group. In order to find their prey and launch an attack, whales employ two different methods. The prey is enclosed in the first, while bubble nets are made in the second. In terms of optimization, whales hunt for prey by exploring their environment and utilizing it during an attack.

The robot has eight ultrasonic sensors based on the Active-Media Pioneer P3- DX. According to its technical specifications, a Polaroid 6500 Series Sonar Ranging Module, which contains eight ultrasonic sensors, has an average absolute accuracy of 1% throughout the range of 15 cm to 10 m. However, considering the sensor interference brought on by beam dispersion and the size of the inside environment, we employed them up to around 4 m as its maximum range. The variable θg indicates the target's relative orientation The intended orientation and the eight sensor data are inputs to the single-layer perceptron (SLP) NN, which outputs a steering command. The local information from the sensor readings and the global (goal) information may be considered to develop the sequence of steering instructions that simultaneously execute obstacle avoidance and target searching. A sigmoid function transforms the NN's output into a continuous actual number that ranges from -1 to +1. The steering angle, which ranges from -90 to +90 with 1 accuracy, is then linearly transferred to this real-valued output.

A single NN controller can rarely handle the entire work because the environment for navigation is frequently complex, partially unknown, and unexpected. If only one NN is utilized, it must have a complicated internal structure and many parameters to handle the highly nonlinear navigation issue. As a result, many nonlinear problems have been solved with strong generalization and fault-tolerance capabilities using a Modular neural network MNN) utilizing the divide-and-conquer strategy. The MNN can simply and rapidly identify appropriate local solutions since it uses several specific NN modules, each covering a distinct local environment.

The general equation for a modular neural network can be expressed as follows:

$$y = f(g(h(x1)), g(h(x2)), ..., g(h(xn))) \qquad (1)$$

Where y is the network's output, $x1$, $x2$, ..., and xn are its inputs, h is a module that processes those inputs, g is a module that combines those processed inputs, and f is a module that generates the network's final output.

A neural network or any other suitable machine-learning technique can implement each module. Feature extraction, categorization, and decision-making are a few examples of the various subtasks the modules may be built to carry out. These subtasks can be integrated to create a comprehensive system that can address more challenging issues, such as robotic navigation. Several learning strategies, including reinforcement learning, unsupervised learning, and supervised learning, can be used to learn the weights and parameters of the neural network. The modular neural network may be trained to make predictions or control the actions of a robot in real time.

5. Result and Discussion

Analysis of the (WO-MNN) methodology in comparison to other methods, including the Convolutional Neural Network (CNN), Unmanned Surface Vehicle - Micro Aerial Vehicles (UAV-MAV) are discussed in this section. The findings showed that the WO-MNN technique correctly identified every attack. The WO-MNN system was seen to have achieved an average accuracy of 99%, a detection rate of 95% (Table 81.1).

Table 81.1 Accuracy and detection rate

Method	Accuracy (%)	Detection Rate (%)
UAV-MAV	85	60
CNN	81	75
WO-MNN [Proposed]	99	95

Accuracy: Comparing the number of correct predictions and Total number of predictions, accuracy is assessed. Using the formula given.

$$Accuracy = \frac{Number\ of\ correct\ predictions}{Total\ number\ of\ predictions} \quad (2)$$

The accuracy of the suggested system is shown in Fig. 81.1. While the proposed method achieves the desired accuracy of 99%, UAV-MAV has only reached 85%, and CNN has only achieved 81%. It shows that the proposed course of action is more effective than the present one, as

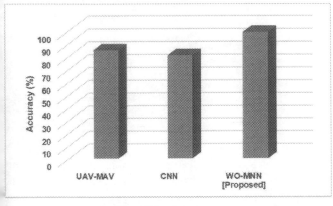

Fig. 81.1 Accuracy

Detection Rate: Detection rate (DR), where TP and FN are the totals of true positive and false negative samples, is the ratio of true positive and the total nonself samples recognized by the detector set.

$$DR = \frac{TP}{TP + FN} \quad (3)$$

The suggested system's detection rate is shown in Fig. 81.2. TP and FN stand for a true positive and a false negative, respectively. The predictions of precision consumption made by the proposed and existing systems are discussed. Whereas UAV-MAV has a 60% detection rate, CNN has a 75% detection rate, and the proposed WO-MNN system has a 95% detection rate. It indicates that the recommended approach is more effective than the present one.

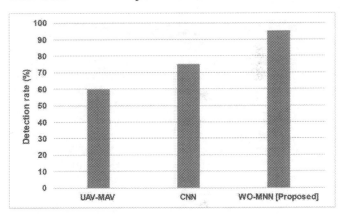

Fig. 81.2 Detection rate

6. Discussion

The UAV-MAV approach fails if both classes share the distribution's mean values. A distinct regression problem axis cannot be found using linear discriminant analysis in this situation. CNN will only succeed if the data is organized into all various classes since it will attempt to locate k's closest neighbor, but all the points are random. Logistic regression is the machine learning technique often employed to solve classification problems. If there are only two classes to predict, it is helpful for binary classification issues. Compared to the current technology, the suggested strategy is more effective.

7. Conclusion

Robotic systems that can move and navigate through surroundings independently without assistance from people are used in automated robotic navigation. This is often achieved by using sensors, algorithms, and computer vision to see and interpret the surroundings, as well as decision-making mechanisms that allow the robot to make the best decisions possible in light of its awareness of the environment. We found that WO-MNN presented a better path and went closer to the target when comparing the results of WO-MNN with MNN without tuning. An analysis of MNN with and without tuning and WO-MNN has validated the effectiveness of WO-MNN in autonomous Pioneer P3-DX wheel navigation. The study shows that, as compared to the untuned MNN, the WO-MNN technique delivers a better path and proximity to the destination. This shows how well the WO-MNN works to improve the Pioneer P3-DX wheeled robot's navigational capabilities. The study focuses on the WO-MNN navigation method used by the wheeled Pioneer P3-DX autonomous robot. The review might not take into account a variety of environmental situations or scenarios, which would limit the applicability of the conclusions.

References

1. Kumar, R., Jitoko, P., Kumar, S., Pillay, K., Prakash, P., Sagar, A., Singh, R. and Mehta, U., (2017). Maze-solving robot with automated obstacle avoidance. *Procedia Computer Science*, *105*, pp. 57–61.
2. Ravankar, A., Ravankar, A., Hoshino, Y. and Kobayashi, Y., (2019, July). Virtual obstacles for safe mobile robot navigation. In *(2019) 8th international congress on advanced applied informatics (IIAI-AAI)* (pp. 552–555). IEEE.
3. Moreno, F.A., Monroy, J., Ruiz-Sarmiento, J.R., Galindo, C. and Gonzalez-Jimenez, J., (2019). Automatic waypoint generation to improve robot navigation through narrow spaces. *Sensors*, *20*(1), p. 240.
4. Grover A., Mohan Kumar R., Angurala M., Singh M., Sheetal A., Maheswar R. "Rate aware congestion control mechanism for wireless sensor networks." *Alexandria Engineering Journal.* (2022); 61(6): 4765–4777. DOI: 10.1016/j.aej.2021.10.032.
5. Ruan, X., Ren, D., Zhu, X., & Huang, J. (2019, June). Mobile robot navigation based on deep reinforcement learning. In *(2019) Chinese control and decision conference (CCDC)* (pp. 6174–6178). IEEE.
6. Wang, T., Chen, B., Zhang, Z., Li, H. and Zhang, M., (2022). Applications of machine vision in agricultural robot navigation: A review. *Computers and Electronics in Agriculture*, *198*, p. 107085.
7. Singh, N.H. and Thongam, K., (2019). Neural network-based approaches for mobile robot navigation in static and moving obstacles environments. *Intelligent Service Robotics*, *12*(1), pp. 55–67.
8. Wang, L. (2019). Automatic control of mobile robots based on autonomous navigation algorithm. *Artificial Life and Robotics*, *24*(4), 494–498.
9. Mittal S., Bansal A., Gupta D., Juneja S., Turabieh H., Elarabawy M.M., Sharma A., Bitsue Z.K. "Using Identity-Based Cryptography as a Foundation for an Effective and Secure Cloud Model for E-Health." *Computational Intelligence and Neuroscience.* (2022); (2022). Article No.: 7016554. DOI: 10.1155/2022/7016554.
10. Zhu, K. & Zhang, T. (2021). Deep reinforcement learning based mobile robot navigation: A review. *Tsinghua Science and Technology*, *26*(5), 674–691.

Note: All the figures and table in this chapter were made by the authors.

Advancing Sustainable Science and Technology for a Resilient Future – Sai Kiran Oruganti et al. (eds)
© 2024 Taylor & Francis Group, London, ISBN 978-1-032-79020-6

A Wall-Climbing Automated Robot Based on a Neural Network for Reinforced Concrete Structural Monitoring

82

Nimesh Raj*

Chitkara University Institute of Engineering and Technology,
Chitkara University, Punjab, India

Abstract: Advanced methods must be used to evaluate man-made concrete structures in order to assure quality and preserve the sustainability of aging infrastructure. As more severe aging issues arise, there is an immediate need for structural health monitoring (SHM) of concrete. Hence, in this study, we offer a wall-climbing robotic system (WCRS) to visually inspect concrete buildings based on a dual objective deep convolution neural network (DODCNN) and 3D semantic reconstruction to generate a 3D map that is overlaid with flaws. Our proposed approach uses pixel-level segmentation to identify cracks and spallings and then marks them on a 3D map for more precise analysis. We train on the dataset with 12,000 iterations and create a semantic dataset with 820 labeled images. To make the defect-marked 3D map, we provide a 3D semantic fusion technique. Our WCRS system can conduct a reliable 3D matric examination, as field tests and experimental findings show.

Keywords: Structural health monitoring (SHM), Wall-climbing robot, Dual objective deep convolution neural network (DODCNN)

1. Introduction

The concrete industry's products are among the most frequently used synthetic materials both in industry and civil projects globally because of their superior mechanical strength and porosities. Stainless rebar reinforcements have enhanced structural components' performances despite altering the microscopic concrete mixture material parameters (Rodrigues, et al., 2021). Thermal degradation, geological assault, endurance cycle, ecological consequences, aging, poor implementation, inadequate maintenance, etc., are only a few of the problems that Reinforced Concrete (RC) structures recognize. The concrete mixture provides sticky binding zones among particles of varying sizes in concrete, in which the compressive stress occurs and where it starts the microscopic cracks (Basu, et al., 2021). The total composites of contemporary civilization are RC structures. Construction of industrial facilities, infrastructures for transportation and energy, and social infrastructure all utilize this composite material. Arrangements of almost every size, shape, and purpose may be made conceivable by the unique qualities of RC. They can be implemented using any technological or architectural approach (Shevtsov, et al., 2022). The foundation of the world economy is the civil infrastructure that includes buildings, roads, and bridges. The bulk of this infrastructure is constructed of RC structures. The unique combination of RC structure durability, the flexibility of form, and the actual product have encouraged its employment in many projects, leading to its extensive use as a structural material. Monitoring pertinent physiochemical characteristics is crucial to respond quickly and avoid significant structural damage, given the range of variables that affect the degradation of RC (Colozza, et al., 2021). Compared to static loads, impact loading causes damage to structures at a greater rate of strain. Due to the diverse response of concrete, the maintenance of RC structures following such hits could be improved. These constructions are often RC structures, and because of their variety, they respond to high stresses in a complicated way. Thus, the effect evaluation of the RC structures has been the focus of the current work (Negi, et al., 2019). The goal is to eliminate the difficulties involved in manually inspecting tall, inaccessible concrete structures so that efficient and precise condition monitoring is possible.

*nimesh.raj.orp@chitkara.edu.in

DOI: 10.1201/9781003490210-82

2. Related Works

Fan, et al., 2021 examined many kinds of fiber optic sensor samples to assess corrosion in RC. Monitoring corrosion in RC is essential for determining the early health status of structures and enabling optimal asset management. Taffese, et al., 2019 provided a conceptual model for RC structure durability monitoring and evaluation during their period of use. Traditional sustainability evaluations include intensive, sophisticated, and expensive analysis of samples that were removed severely from the structures. Kaur, et al., 2019 embedded Lead Zirconate Titanate sensors in the form of concrete accelerometer sensors for monitoring damage, retrofitting, long-term strength, and first-stage fatigue of RC structures. The ready-to-use concrete vibration sensor is packaged uniquely, making it ideal for embedding in RC buildings. Kukreja, et al., 2021 extended the lifespan of concrete structures; RC is a frequently utilized building material. Evaluated for various, non-destructive testing procedures, and other monitoring techniques are often used in RC structural monitoring. Lilhore et al., 2022 assessed the functioning and behavior of the construction, including stresses, strains, vibration, and temperature, a variety of sensors and devices are commonly used throughout the monitoring process. RC structural monitoring checks buildings, bridges, and tunnels for safety, integrity, and longevity. This study presents a wall-climbing robotic system (WCRS) that uses a dual objective deep convolution neural network (DODCNN) and 3D semantic reconstruction to create a 3D map overlaid with faults for the visual assessment of concrete structures.

3. Proposed Method

Wall-climbing Robotic System

A prototype wall-climbing robot comprises a Negative Pressure Adhesive Module (NPAM), a sensor, two drive wheels, the engines and control systems that go with them, a braking system, a DC-DC transformer, an Intel NUC computer for its signal processing. Each motor driver is used to modify the camera angle. It achieves a dynamic equilibrium of adhesive forces and high flexibility by varying the NPAM's propeller speed, which keeps a stable pressure gradient within a suction chamber. It may thus travel both on soft and hard surfaces and pass over shallow grooves since it doesn't need complete encapsulation like ejector methods. In addition to producing standard Red-Green-Blue (RGB) video pictures, the RGB-D camera offers distance readings. A first perspective of the landscape, received from the recording device over Wireless transmission, is shown via the controller graphic user interface (GUI) on a desktop computer. The

GUI also adjusts the cleaning power, bringing light angle and Lighting illumination. The robot is 16.5 x 13 x 8 inches and weighs 12 pounds. It has a carrying capacity of up to 20 lbs.

Dual Objective Deep Convolution Neural Network and 3D Reconstruction for Concrete Inspection

To perform a metric 3D inspection of the concrete structure, propose to use an RGB-D camera-based semantic reconstruction system. For an RGB-D camera, it outputs both RGB image $I = \{(w_j, x_i, d) \in P^3 | 0 \leq j\, n, 0 \leq i \leq n, d = 0, 1\}$, and the corresponding depth image $E = \{e\}$. Then, obtain the point cloud at each pose $M = (Q, s)$ using the pinhole model as

$$\begin{bmatrix} V \\ Z \\ Y \end{bmatrix} = [Q, s]^{-1} L[w, x] \tag{1}$$

Where $[v, z, y]$ denotes the point cloud, $[Q, s]$ indicates the rotation and translation at a location. L is the camera's intrinsic parameter, and $[w, x]$ is a pixel in an RGB image.

Given the current image frame j_o and previous image frame j_v, it first performs feature extraction and matching to obtain the matching feature pairs $E_{JO} \to E_{Jv}$, and then, the pose estimation can be achieved through the following optimization,

$$[Q, s] = \arg\min_{Q,s} \sum_{j \in \{1,\ldots.M\}} K_o \left(E_{Jv}(j) - \pi \left(Q.E_{JO}(j) + s \right)^2_{\Sigma.} \right) \tag{2}$$

Where argmin denotes the linear regression process, $K_o()$ is the Huber Loss based cost function, and $\|\|\Sigma.$ indicates the covariance weighted sum toward a robust convergence. $\pi(.)$ is an inverse projection.

Development of Dual Objective Deep Convolution Neural Network

The most widely used methods of deep learning in image processing are DODCNN. Generally defined, the DODCNN's convolution layer serves as its central processing unit and is responsible for extracting relevant significant features using a variety of convolution kernels (CK). A CK is a group of values, and various weight values provide varied extracting outcomes. Whereas later CK concentrate on semantics, the earlier seeds are more likely to extract position information about objects in the images. In order to continuously improve the weighting of certain kernels throughout development, the error back-propagation approach selects a single element values as from pooling kernel (PK), like the max pooling kernel, and down-samples the output of successive convolution layers. The pooling layer lowers the danger of the DODCNN being over-fitted while reducing the size of the images. To categorize of inputs images as well as forecast its outcomes, the fully connected layer, integrates the qualities learned from the preceding layers and is coupled by a soft-max.

Figure 82.1 displays the quantity and shape of the convolution and PK utilized in the study and the general framework of the proposed DODCNN, created using the deep network builder. Initially, the networks used two convolutional in sequence with and CK, which move over the input image.

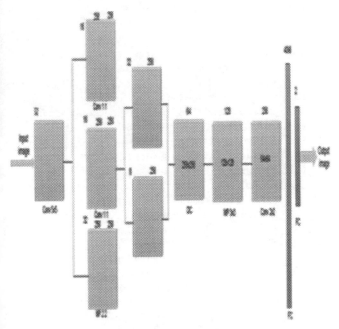

Fig. 82.1 Development of dual objective deep convolution neural network

$$T_{output} = \frac{T_{input} - T_{kernel} + 2T_{padding}}{T_{stride}} + 1 \qquad (3)$$

Where T_{input} and T_{output} stand for the actual size of the input and output matrix, the dimensions of the CK, padding, and sliding stride are indicated by the variables T_{kernel}, $T_{padding}$, and T_{stride}. All of the convolution layers in the research had a stride of 1, and the corresponding 2×2 and 5×5 CK padding was considered to be one and 2. Next, in order to reduce the model's variables and conserve processing capability, the DODCNN method was altered to incorporate a max pooling layer. The only variable that its max pooling operation retains is the maximum value in the PK, which is denoted by the following symbol:

$$z = max(V_{ji}) \qquad (4)$$

Where V_{ji} stands for the PK element of the method, the max pooling layers' sliding stride were set at two during this

whole trial, which indicates that the output matrix of the max pooling layers is just one-fourth as large as the input matrix. Simultaneous initiation modules comprised of 3×3, and 5×5 CK and a 2×2 max pooling layer was included in the network in order to extract several features at once. It should be noted that a 1×1 CK was added to the module to speed up processing while also reducing the number of model parameters. The output of such parallel layers was also combined in a depth concatenation layer. This method incorporated a number of convolution layers and top pooling layers, which equate to 33 kernel and 22 kernels, prior to the models, achieved the completely connected layer. Each input pixel is divided into classes representing bugholes or normal pixels in the final layer, in which a softmax function follows a fully linked layer. When i^{th} class is chosen from m^{th} class, the probability is given by:

$$o(z = m \mid v) = \frac{f^{uv+a}}{\sum_{i=1}^{m} f^{uv+a}} \qquad (5)$$

Where uv + a is the softmax method's input and u and a are the weighting coefficient and biases, correspondingly. The activation function, frequently referred to as ReLU, were chosen to provide non - linearities after the convolution layer. This is because, in contrast to the prior sigmoid ($e(v)$ = $(1 + d^{-1})^{-1}$) and tanh($e(v)$ = tanh (v)) functions, which are more volatile and have gradient descent issues, the ReLU method might allow for effective and rapid development of the DODCNN models with massive data sets. The ReLU function is described.

$$e(v) = \begin{cases} 0 & v < 0 \\ v & v \geq 0 \end{cases}; \qquad (6)$$

Semantic 3D Defects Map

To construct the semantic 3D map, suggest using a 3D conditional random filter-based fusion because of the pose [and the inspection results across the images. Equations (1) and (2) are used to conduct the registration while continuously updating the color data from the inspection network. Hence, the map may continue to be updated by changing the RGB-D camera.

Wall-climbing Robot Test

Test the robot's adhesive force on flat surfaces and smooth vertical surfaces; the findings are displayed in Table 82.1. Measure the pull-up adhesive force when the robot is put on the ground surface by choosing different adhesive suction motor speeds in the percentages about the maximum speed.

According to Table 82.1, the pull-up force varies from 4.63 to 32.63 Kg. Measure the load capacity, which ranges from

Table 82.1 Adhesive force testing for wall-climbing robot

Adhesive motor speed (percentage %)	15	20	25	30	35	40	45	50
Vertical pull-down adhesive force Kg	N/A	1.5	5.6	6.7	8.1	9.8	11.3	9.4
Ground pull-up adhesive force Kg	4.63	5.64	8.35	11.04	11.03	26.4	25.5	32.63

1.5 to 9.4 Kg. When the adhesive suction motor speed is too low, the robot cannot stick to vertical walls and is listed as not available for load carrying.

4. Conclusion

Modern inspection tools must be used to check human-made concrete structures to ensure the construction's quality and maintain the sustainability of the deteriorating infrastructure. Using an RGB-D sensor and a 3D reconstruction technique based on neural networks, introduce a wall-climbing robot for SHM. Hence, in this work, we provide WCRS for visually inspecting concrete structures based on a DODCNN and 3D semantic reconstruction to produce a 3D map overlaid with faults. A cutting-edge dataset with pixel-level labeling and an Inspection network was created for semantic segmentation. To provide metrics for gauging the state of concrete structures, the visual inspection results were recorded in the 3D model. The outcomes of the field tests demonstrate the value of our robotic inspection system's vertical mobility and visual inspection capabilities.

References

1. Rodrigues, R., Gaboreau, S., Gance, J., Ignatiadis, I. and Betelu, S. (2021). Reinforced concrete structures: A review of corrosion mechanisms and advances in electrical methods for corrosion monitoring. Construction and Building Materials, 269, p.121240.
2. Basu, S., Thirumalaiselvi, A., Sasmal, S. and Kundu, T., (2021). Nonlinear ultrasonics-based technique for monitoring damage progression in reinforced concrete structures. Ultrasonics, 115, p.106472.
3. Shevtsov, D., Cao, N.L., Nguyen, V.C., Nong, Q.Q., Le, H.Q., Nguyen, D.A., Zartsyn, I. and Kozaderov, O., (2022). Progress in Sensors for Monitoring Reinforcement Corrosion in Reinforced Concrete Structures—A Review. Sensors, 22(9), p.3421.
4. Colozza, N., Tazzioli, S., Sassolini, A., Agosta, L., di Monte, M.G., Hermansson, K. and Arduini, F., (2021). Multiparametric analysis by paper-assisted potentiometric sensors to diagnose and monitor reinforced concrete structures. Sensors and Actuators B: Chemical, 345, p.130352.
5. Negi, P., Chhabra, R., Kaur, N. and Bhalla, S., (2019). Health monitoring of reinforced concrete structures under impact using multiple piezo-based configurations. Construction and Building Materials, 222, pp.371-389.
6. Fan, L. and Bao, Y., (2021). Review of fiber optic sensors for corrosion monitoring in reinforced concrete. Cement and Concrete Composites, 120, p. 104029.
7. Taffese, W.Z., Nigussie, E. and Isoaho, J., (2019). Internet of things-based durability monitoring and assessment of reinforced concrete structures. Procedia Computer Science, 155, pp. 672–679.
8. Kaur, N., Bhalla, S. and Maddu, S.C., (2019). Damage and retrofitting monitoring in reinforced concrete structures and long-term strength and fatigue monitoring using embedded Lead Zirconate Titanate patches. Journal of Intelligent Material Systems and Structures, 30(1), pp. 100–115
9. Kukreja, V. and Kumar, D., (2021), September. Automatic classification of wheat rust diseases using deep convolutional neural networks. In (2021) 9th International Conference on Reliability, Infocom Technologies and Optimization (Trends and Future Directions)(ICRITO) (pp. 1–6). IEEE.
10. Lilhore U.K., Imoize A.L., Lee C.-C., Simaiya S., Pani S.K., Goyal N., Kumar A., Li C.-T. "Enhanced Convolutional Neural Network Model for Cassava Leaf Disease Identification and Classification."Mathematics. (2022); 10(4). Article No.: 580. DOI: 10.3390/math10040580.

Note: All the figure and table in this chapter were made by the authors.

Advancing Sustainable Science and Technology for a Resilient Future – Sai Kiran Oruganti et al. (eds)
© 2024 Taylor & Francis Group, London, ISBN 978-1-032-79020-6

Automated Marine Robot Navigation System Using Machine Learning Algorithm

83

Nittin Sharma*

Chitkara University Institute of Engineering and Technology,
Chitkara University, Punjab, India

Abstract: Autonomous marine robots must be able to move in challenging settings. The development of advanced navigation systems to transfer marine robots through one location to another has taken many years of engineering and academics. Despite their general effectiveness, there is currently a study focus on creating machine learning (ML) methods to deal with the same issue. Yet, there haven't been many clear comparisons of this issue's conventional and contemporary viewpoints. In the environment of conventional navigation, we suggest a new binary marine predator optimized artificial neural network (BMPO-ANN) in position planning as well as control in marine robot navigation throughout this study. Here, BMPO is utilized in the suggested system to improve ANN functionality. The experimental findings of this work demonstrate that the suggested strategy outperforms current methods in marine robot navigation.

Keywords: Autonomous marine robots, navigation, Machine learning (ML), Binary marine predator optimized artificial neural network (BMPO-ANN)

1. Introduction

Information technology and intelligent development have altered the way numerous organisation operates and are headed. In the main traffic marine industry, information have largely replaced digitalization and application of information technology. The machine learning methodology is quickly emerging as a highly effective method for smart transportation systems as a result of significant advancements in machine learning over the past few decades. The marine industry uses techniques involving machine learning in a number of areas, including ship categorization, feature detection, collision avoidance, risk awareness, and object tracking. The two main implementation fields are independent marine navigating and coastal monitoring (Kollmitz, et al., 2020).

This submarine fleet is advancing into industrialization and intelligence as a result of the quick growth of technological advances the use of algorithms, machine learning, and big data. Autonomous navigating is based on the idea of smart ship visual perception. It can, without a doubt, lower the number of interrupt marine traffic crashes, increase the safety of ship management, and streamline shipping routes, energy usage, and operating costs. Intelligent ship visual perception is essential for marine monitoring and identification. It can recognize harmful ship kinds, keep an eye on the surroundings, and decide how to prevent collisions with other spaceships, all of which are essential to the growth of autonomous vessels (Liu, et al., 2019 and Malik, et al., 2022).

Maritime navigating has historically been carried out solely by human effort over the ages. Yet, today's marine system enables the submarine's navigating crew reduce navigating mistakes. Such technology will soon combine to create a smart tracking system. Such a system will be able to direct its user in figuring out the almost ideal course for avoiding ship collisions. In the future, people will create reliable "smart" robots that can steer warships through terminals and rivers on their own (Statheros, et al., 2019, Navaneetha, et al., 2022).

Robotic devices, aircraft, and warships are examples of robotic systems that have problems with navigating. In a two-dimensional (2D) or three-dimensional (3D) system, the primary objective of navigating is to determine an efficient

*nittinsharma1943@gmail.com

DOI: 10.1201/9781003490210-83

and inefficient beginning level routes to a destination address while navigating. Robust robot navigation systems are needed for internal robotic devices, autonomous guided cars for warehouses, and transportation robots in their dynamic surroundings. Its guidance problem became the subject of academic inquiry across the globe for the past 20 years. Using a number of various strategies is one popular method (Zhu and Zhang, 2021, Baliyan, et al., 2021).

2. Related Works

Manderson and Dudek, 2019 was recommended this device makes it possible to perform numerous machine vision jobs concurrently while maneuvering close to barriers, which makes it possible to perform observational activities like assessing the condition of coral reefs. While a general-purpose CPU analyses data to conduct visual Simultaneous Localization and Mapping, the incorporation of a GPU enables us to use deep neural networks for collision avoidance, automatic item detection, and classification (SLAM)

The landscape of traditional navigation systems was suggested by Xiao et al. in 2022. They reviewed recent research that uses robotics guidance employing techniques for organizing and regulating movements. Each analyzed material is divided into different groups that illustrate how the instructional tactics relate to conventional teaching techniques.

Wang, et al., 2022 and Chaudhry and Sachdeva, 2021 proposed a hybrid deep learning approach for physically classifying the landscapes that autonomous robots meet. A recommended implementation strategy's purpose and a convolutional neural network are combined in a methodology. Machine to maintain excellent categorisation performance similarly increasing effectiveness of activity, taking into account the restricted computational power on robotic systems and the necessity for higher prediction performance. The key concept is that a multi-class categorization is completed using a convolutional neural network, and a two-class classification is created concurrently using a supported vector machine.

3. Proposed Methodology

The primary goal of this research is to create a model that is capable of correctly assessing the automated marine robot navigation in the future using (BMPO-ANN) from machine learning algorithm.

Dataset

A key aspect for marine object recognition and a necessary for training deep learning frameworks is a huge and trainable dataset. Here, we first provide a collection of common high resolution datasets that contain a variety of objects and can be used as source domain datasets during transfer learning's pre-training stage. We have gathered the underwater and surface object datasets for marine object recognition, as well as the widely used datasets, to help with that follows.

Binary Marine Predator Optimization

Both Brownian and Levy movements used by marine predators as part of their food-finding techniques served as the primary inspiration for the current optimization algorithm known as BMPO. The prey itself is a predator when it hunts for food, hence the BMPO views both predator and prey as search agents.

Initialization is done at randomized for each search agent in the starting group. R is a random number between 0 and 1, while A_{max}, A_{min}, and are the upper and lower bounds for parameters.

The elite predators are represented numerically by BMPO in a distinct matrix called E. The preys are expressed quantitatively by some other matrices named P.

Algorithm. 1 BMPO:

1. Create a random population at first in the solution space (p).
2. Evaluate objective space for the created populations (E).
3. Estimate the resultant population's (R) non-dominated rank (NDR).
4. The crowding distance (CDI) for each front that has been obtained.
5. Improve the remedy utilizing BMPO.
6. Combine the equivalent expression.
7. Estimate the created population's objective space.
8. Sort the people using their NDR and CD levels.
9. Alter the initial members.

Artificial Neural Network

The ANN links different kinds of data together. Instead of the usual logical thinking carried out by machines, the ANN achieves understanding as intuition. Figure 83.1 shows the artificial neural network flow on marine robot navigation. The many studies have been conducted to see whether ANNs can be used to forecast marine traffic because their key advantage is their capacity to accommodate uncertainties and information that is continuous.

$$y = F(\sum wj * xj - \theta mj = 1 \qquad (1)$$

The back-propagation (BP) network is one ANN scheme. The back-propagation neural network architecture is composed of rows or layers of different nodes that are completely interlinked. Loads are the interconnections that give the

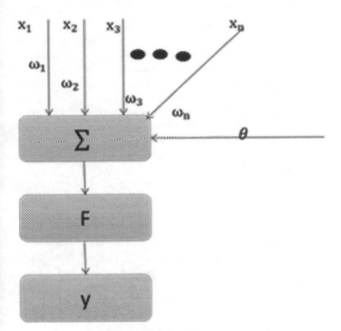

Fig. 83.1 Artificial neural network

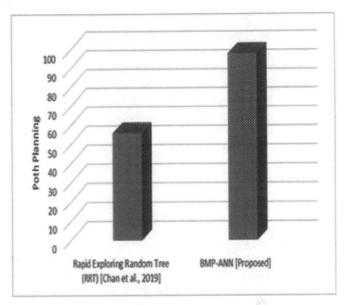

Fig. 83.2 Path planning

ANN the ability to store information gained throughout the process of A learning B. Moreover, mistakes are computed throughout this procedure. Such mistakes are utilized to change all of the loads by back-propagating from the output neurons to all of the hidden neurons. This process of learning proceeds until the mistake is brought down to the predefined minimum values.

4. Result and Discussion

This step follows the theoretical completion of research on automated marine robot navigation system scheduling processes by the suggested machine learning algorithm based on path planning and control efficiency.

Path Planning

Navigation for marine robots is comparable to path planning for other autonomous robot varieties, but there are a few extra features and difficulties because of the underwater environment. When operating in deep, dark waters with poor sight, marine robots like autonomous underwater vehicles (AUVs) and remotely operated vehicles (ROVs) may need to escape underwater impediments including reefs, boulders, and other marine life. The suggested approach's path planning is superior to the current strategy, as seen in Fig. 83.2.

Control Efficiency

As marine robots to complete their missions and navigate successfully, control effectiveness is essential. Marine robots, such as AUVs and ROVs, often travel through the water

and engage with their surroundings using a combination of propulsion, maneuvering, and control systems. The control effectiveness of the suggested strategy is better than the current approach, as shown in Fig. 83.3.

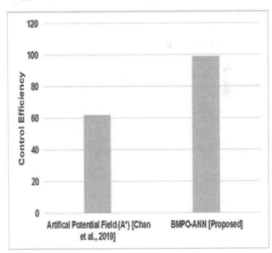

Fig. 83.3 Control efficiency

5. Conclusion

The problem of normatively, private, and ethical robot navigation is resolved in this paper. Initially, using (BMPO-ANN) machine learning, the robot is taught to mimic unexpected human behavior that occurs in real time. To do this, based on the human's current condition, danger zones are created around him or her, and the robot is educated to avoid them. The suggested formulation was tested using cutting-edge techniques, and it performed well by studying real-time human behavior to create safe and secure robot navigation that complies with societal expectations. The robot's vision and sensing skills are vital to its navigation because of their precision and dependability. The robot's ability to effectively detect and avoid hazardous zones may be hampered by constraints or inaccuracies in sensory inputs. To further evaluate its efficacy and generalizability, the proposed formulation should be tested in real-world situations including a wide variety of settings, human behaviors, and robot platforms.

References

1. Kollmitz, M., Koller, T., Boedecker, J. and Burgard, W., 2020. Learning human-aware robot navigation from physical interaction via inverse reinforcement learning. In 2020 IEEE/RSJ International Conference on Intelligent Robots and Systems (IROS) (pp. 11025–11031). IEEE.

2. Liu, B., Wang, S.Z., Xie, Z.X., Zhao, J. and Li, M., 2019. Ship recognition and tracking system for intelligent ship based on deep learning framework. TransNav: International Journal on Marine Navigation and Safety of Sea Transportation, 13.a

3. Statheros, T., Howells, G. and Maier, K.M., 2019. Autonomous ship collision avoidance navigation concepts, technologies and techniques. The journal of Navigation, 61(1), pp. 129–142.

4. Zhu, K. and Zhang, T., 2021. Deep reinforcement learning based mobile robot navigation: A review. Tsinghua Science and Technology, 26(5), pp. 674–691.

5. Chaudhry, A.K. and Sachdeva, P., 2021. Microplastics' origin, distribution, and rising hazard to aquatic organisms and human health: socio-economic insinuations and management solutions. Regional Studies in Marine Science, 48, p.102018.

6. Manderson, T. and Dudek, G., 2019, October. Gpu-assisted learning on an autonomous marine robot for vision-based navigation and image understanding. In OCEANS 2018 MTS/IEEE Charleston (pp. 1–6). IEEE.

7. Baliyan A., Kukreja V., Salonki V., Kaswan K.S. "Detection of Corn Gray Leaf Spot Severity Levels using Deep Learning Approach." 2021 9th International Conference on Reliability, Infocom Technologies and Optimization (Trends and Future Directions), ICRITO 2021. 2021;

8. Xiao, X., Liu, B., Warnell, G. and Stone, P., 2022. Motion planning and control for mobile robot navigation using machine learning: a survey. Autonomous Robots, 46(5), pp. 569–597.

9. Navaneetha Krishnan Rajagopal, Naila Iqbal Qureshi, S. Durga, Edwin Hernan Ramirez Asis, Rosario Mercedes Huerta Soto, Shashi Kant Gupta, S. Deepak, "Future of Business Culture: An Artificial Intelligence-Driven Digital Framework for Organization Decision-Making Process", Complexity, vol. 2022, Article ID 7796507, 14 pages, 2022.

10. Wang, W., Zhang, B., Wu, K., Chepinskiy, S.A., Zhilenkov, A.A., Chernyi, S. and Krasnov, A.Y., 2022. A visual terrain classification method for mobile robots' navigation based on convolutional neural network and support vector machine. Transactions of the Institute of Measurement and Control, 44(4), pp. 744–753.

11. Malik A., Saggi M.K., Rehman S., Sajjad H., Inyurt S., Bhatia A.S., Farooque A.A., Oudah A.Y., Yaseen Z.M. "Deep learning versus gradient boosting machine for pan evaporation prediction." Engineering Applications of Computational Fluid Mechanics. 2022.

Note: All the figures in this chapter were made by the authors.

Advancing Sustainable Science and Technology for a Resilient Future – Sai Kiran Oruganti et al. (eds)
© 2024 Taylor & Francis Group, London, ISBN 978-1-032-79020-6

84

Nabi: A Social Platform where Mental Health Matters

Cerin Gabriel B.*, **Cui Catherine Joyce P.**[1],
Domingo Leo Enrico A Nidoy y Roy Allen A.[2], **Estrella Noel E.**[3]

College of Information and Computing Sciences University of Santo Tomas, Manila, Philippines

Abstract: The stigma surrounding mental illness in the Philippines has deterred many Filipinos from seeking help, as they fear negative judgement from others. Additionally, family members often struggle to understand and support their child's emotions if they haven't experienced similar challenges themselves. To address these barriers, developers created Nabi, a progressive web application for the Youth For Mental Health Coalition Inc. One of the key features of Nabi is the ability to maintain anonymity, allowing users to vent freely without the fear of being identified. By utilizing the Hybrid Development Methodology, the developers ensured that all features were thoroughly tested and aligned with the identified requirements before conducting User Acceptance Tests. The implementation of Nabi is expected to significantly enhance the mental health response capability of the Youth For Mental Health Coalition Inc., while providing a valuable resource for individuals seeking various forms of support.

Keywords: Mental health, Social platform, Hybrid software development cycle.

1. Introduction

In a study by Rivera and Antonio (2017) focusing on the Philippine context, it is stated that stigma affects the willingness of a person to seek help. This stigma contributes to the minimal support available for individuals with mental health problems in the country. Inaccessibility is another key issue, as highlighted by Martinez et al. (2020), who found that mental illness is the third most common disability in the Philippines. Filipinos face challenges in seeking formal help from mental healthcare professionals due to factors such as the fear of losing face, a sense of shame, adherence to societal norms that stigmatize mental illness, and financial constraints. Limited human resources and facilities further hinder access to mental health services, as Estrada et al. (2020) discovered in their study, which revealed a shortage of child psychiatrists and psychologists in the country.

With this, the developers saw an opportunity to create an accessible social platform to establish a safe space for individuals suffering from mental health issues. The prevalence of mobile devices and the increased reliance on virtual platforms during the pandemic facilitated the integration of health consultations and activities into online spaces. Leveraging this opportunity, the developers created Nabi, a progressive web application that allows users to connect anonymously with others who share similar experiences, providing a supportive environment. The platform aims to help users understand and overcome their mental struggles by encouraging engagement with mental healthcare professionals.

The developers believe that Nabi will be a valuable tool in assisting individuals facing mental health difficulties. By addressing the stigma and providing a platform for anonymous support, Nabi seeks to minimize the number of cases where individuals struggle alone. Through this application, users can find a safe space, connect with others who understand what they're going through, and access the necessary professional help. The development of Nabi reflects the developers' commitment to improving mental health outcomes and providing much-needed support to individuals in need.

*Corresponding author: gabrielbcerin@gmail.com
[1]catherinejpcui@gmail.com, [2]domingoleo0611@gmail.com, [3]neestrella@ust.edu.ph

DOI: 10.1201/9781003490210-84

2. Methodology

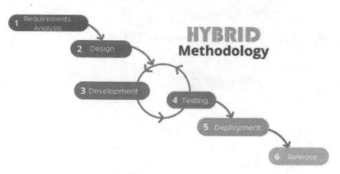

Fig. 84.1 Hybrid software development life cycle

The Hybrid Software Development Life Cycle was chosen by the developers of Nabi, a mental health project, for its advantages. It combines Iterative and Waterfall methodologies. The hybrid approach allowed the developers' validators to actively participate and make changes to features, thanks to its flexibility. Additionally, the iterative approach facilitated early detection of flaws during testing.

The development process of Nabi followed a structured approach. It began with requirement analysis to identify functional and non-functional requirements, as well as project scope and limitations. Consultations with validators, guidance counselors, and psychologists ensured Nabi would provide a safe space for addressing mental health. Next, interfaces, modules, and components were designed to meet the project's objectives, followed by development and coding. The iterative nature of the methodology helped resolve errors efficiently during testing.

After meeting all requirements, Nabi underwent User Acceptance Testing to assess its installability as a progressive web application. Finally, Nabi was released to the Youth for Mental Health Coalition, marking its official launch.

3. Research Method

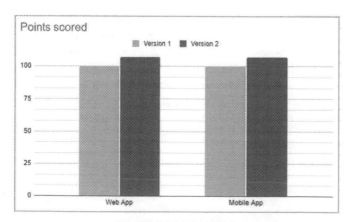

Fig. 84.2 Unit testing

In the unit tests conducted for the web application and mobile installed, 100 out of 107 test cases passed in Version 1, with 7 minor errors. In Version 2, all 107 test cases passed for both applications.

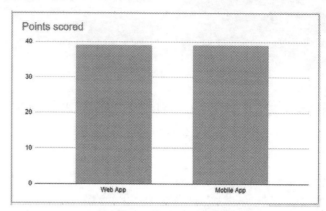

Fig. 84.3 Integration testing

Figure 84.3 shows the result of the conducted integration testing after the unit tests for both mobile and web applications of Nabi. The application attained a 100% passing rate for the test cases.

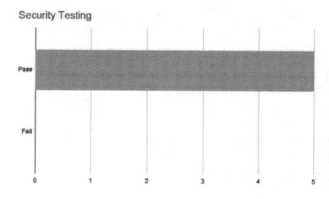

Fig. 84.4 Security testing

In the security test, Nabi achieved a 100% passing rate for all test cases. The quality assurance analyst confirmed the

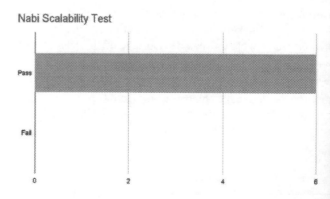

Fig. 84.5 Scalability testing

encryption of sensitive information and the use of HTTPS for accessing Nabi.

In the scalability test, Nabi achieved a 100% passing rate with all 6 test cases passing. In the performance test using Google Lighthouse, the initial version failed with a score of 34 in performance but scored well in accessibility, best practices, and SEO. After improvements, Nabi passed the performance test with a score of 63.

Table 84.1 Installability testing

Brand	Version	Install	Uninstall	Status
iPhone XS	iOS 14.7.1	Yes	Yes	PASS
OPPO A3s	Android 8.1 Oreo	Yes	Yes	PASS
Huawei Y9	Android 9 Pie	Yes	Yes	PASS
iPhone 6s	iOs 13.3	Yes	Yes	PASS

The table displays the compatibility of Nabi with different Android and iOS versions, starting from the optimal requirement.

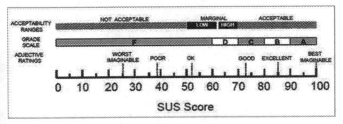

Fig. 84.6 System usability scale

In the Usability test, 55 responses were collected, including 49 user registrations, 5 professional registrations, and 1 administrator. The UAT questionnaire comprised 10 questions on system usability, assessed using the System Usability Scale Score.

Table 84.2 Calculation of Nabi's score

Questions	Average score of 55 respondents	Odd Numbers (x-1)	Even Numbers (5-y)
Question 1	3.890909091	2.890909091	
Question 2	2.345454545		2.654545455
Question 3	4.181818182	3.181818182	
Question 4	2.072727273		2.927272727
Question 5	4.236363636	3.236363636	
Question 6	2.054545455		2.945454545
Question 7	4.327272727	3.327272727	
Question 8	2.2		2.8
Question 9	4.185185185	3.185185185	
Question 10	2.109090909		2.890909091
Total:		15.82154882	14.21818182
Final Score:		**75.0993266**	

Table 2 presents the final System Usability Scale (SUS) score for Nabi as 75.099 out of 100. Nabi is deemed acceptable and provides a good user experience.

4. Conclusion

The developers successfully created Nabi, a social platform for individuals struggling with their mental health, in collaboration with the Youth For Mental Health Coalition, Inc. They conducted alpha and beta testing following a test plan to assess the achievement of specific objectives and functional and non-functional requirements. In Version 1, Nabi passed 93.458% of the unit testing cases, with minor user interface issues as the only failures. In Version 2, Nabi passed all 100% of unit testing and integration testing cases.

Nabi fulfils specific objectives, including, providing a discussion forum with helpful tags, a comment section, and a personal feed. All 23 test cases for the forum on the user and mental healthcare professional side, as well as the 2 test cases on the admin side, passed successfully. The developers also implemented a mood tracker allowing users to record and view their moods. The tracker suggests consultations if users record negative moods for five consecutive days. The personal journal feature was also successfully implemented. The mood tracker and personal journal passed all relevant test cases.

Furthermore, the developers created a blog page where mental health-related articles and announcements are featured. Four test cases for the user/mental healthcare professional side and eight test cases for the admin side passed in unit testing. Nabi includes side and top navigation bars for easy access to its main features and account settings, respectively. All 16 test cases involving the navigation bars passed successfully.

The developers incorporated an evaluation form for users to provide feedback on their sessions with mental healthcare professionals. All six test cases for the user/mental healthcare professional side and both test cases for the admin side passed.

The Usability test collected 55 responses for User Acceptance Testing, with the majority of participants being users. The feedback received indicated that Nabi was highly regarded, as many respondents expressed a strong desire to use it frequently. The platform was found to be user-friendly, well-integrated, and not unnecessarily complex, according to most participants. Additionally, the majority of respondents felt confident in their ability to quickly learn how to use the system and disagreed with the idea that it was burdensome or required extensive learning. With a System Usability Scale Score of 75.099, Nabi has been deemed a good and acceptable system.

In summary, the developers successfully created Nabi, a social platform designed to assist individuals facing mental health challenges. They conducted comprehensive testing to ensure that all functional and non-functional requirements were met. Nabi successfully achieved its specific objectives, such as providing a discussion forum, mood tracker, personal journal, blog page, navigation bars, and an evaluation form. User feedback from the Usability test was overwhelmingly positive, confirming the system's usability and widespread acceptance. Overall, Nabi stands as a well-designed and effective solution for supporting mental health.

References

1. Rivera, A. B., & Antonio, C. T. (2017). Mental Health Stigma Among Filipinos: Time For A Paradigm Shift. Philippine Journal of Health Research and Development. Retrieved from https://www.researchgate.net/publication/326412761_Mental_Hea   lth_Stigma_Among_Filipinos_Time_For_A_Paradigm_Shift

2. Martinez, A.B., Co, M., Lau, J. et al. Filipino help-seeking for mental health problems and associated barriers and facilitators: a systematic review. Soc Psychiatry Psychiatr Epidemiol 55, 1397–1413 (2020). Retrieved from https://doi.org/10.1007/s00127-020-01937-2

3. Estrada, C.A., Usami, M., Satake, N. et al. Current situation and challenges for mental health focused on treatment and care in Japan and the Philippines - highlights of the training program by the National Center for Global Health and Medicine. BMC Proc 14, 11 (2020). Retrieved from https://doi.org/10.1186/s12919-020-00194-0

4. Smith-Merry J., Goggin G., Campbell A., McKenzie K., Ridout B., Baylosis C. (2019). Social connection and online engagement: Insights from interviews with users of a mental health online forum. JMIR Mental Health.

5. Parekh, R., & Givon, L. (2019). What is psychotherapy? psychiatry.org. Retrieved from: https://www.psychiatry.org/patients-families/psychotherapy

6. Kummervold, Per & Gammon, Deede & Bergvik, Svein & Johnsen, Jan-Are & Hasvold, Toralf & Rosenvinge, Jan. (2002). Social support in a wired world: Use of online mental health forums in Norway. Nordic journal of psychiatry. 56. 59-65. 10.1080/08039480252803945.

7. Gordon, C. C., Bowman, N., Goodboy, A. K., & Wright, A. (2019). "Anonymity and Online Self-Disclosure: A Meta-Analysis". Communication Reports. Retrieved from: https://doi.org/10.1080/08934215.2019.1607516

8. Tandel, S. S., & Jamadar, A. (2018, September). Impact of Progressive Web Apps on Web App Development. ijirset.com. Retrieved from http://www.ijirset.com/upload/2018/september/21_Impact.pdf

Note: All the figures and tables in this chapter were made by the authors.

Advancing Sustainable Science and Technology for a Resilient Future – Sai Kiran Oruganti et al. (eds)
© 2024 Taylor & Francis Group, London, ISBN 978-1-032-79020-6

85

PLANTSYS: A SMS Based Automatic Plant Irrigation and Monitoring System

Engr. Jose Marie B. Dipay*

Research Management Cluster Coordinator, Research Management Office,
Assistant Professor IV, Institute of Technology, Polytechnic University of the Philippines

Lester B. Sarmiento

Polytechnic University of the Philippines

Abstract: This study aimed to create Plantsys: A SMS based automatic plant irrigation and monitoring system. It identified challenges in implementing standard irrigation systems for plants, suggested substitute treatments, and evaluated user approval of a prototype observation system based on those characteristics. A questionnaire and in-person interviews conducted by the researchers were part of the mixed-methods technique employed in the study. Participating were five (5) agriculturists, thirty (30) gardeners, and eighty(80) farmers from various locations. A few issues with the conventional method of plant irrigation include inadequate plant monitoring, ambiguity on who should be doing what, and difficulties keeping track of plants' irrigation histories. The building of a sms based automatic plant irrigation and monitoring system, as well as the creation of a system that is more user-friendly than the traditional way for agriculturists, gardeners, and farmers, should be included in the system in order to solve these difficulties. The functionality (4.62), dependability (4.52), usability (4.68), and performance (4.59) of Plantsys: A SMS based automatic plant irrigation and monitoring system were rated as excellent or good by respondents (4.61). Governmental organizations will probably suggest using the current method to keep track of plant irrigation.

Keywords: Plant, Networked sensors, Smart processing algorithms, Irrigation and monitoring system, SMS

1. Introduction

In a relatively short amount of time and in a variety of ways, science has now influenced every aspect of people's lives, simplifying and enhancing them in numerous ways. Because databases are essential to setting up a trustworthy plant irrigation and monitoring system, many developers of irrigation and monitoring systems rely on them. The study's findings were used to develop a new method for controlling and overseeing the irrigation and monitoring system.

Farmers have relied on irrigation systems for more than ten years. The irrigation and monitoring system has come a long way since it first began, according to historical statistics. Early on, it became clear that setting up a sms based automatic plant irrigation and monitoring system would be essential for monitoring plants. Numerous parts of the world have improperly monitored numerous plants for many years.

Proactive plant irrigation is made possible by sms-based plant irrigation and monitoring systems, which are a crucial part of the agriculture sector.

The discipline of plant monitoring has benefited significantly over time from irrigation and monitoring systems. In order to lessen the burden of unanticipated excess irrigation on plants, the current concept of irrigation and monitoring prioritizes minimizing the incidence of plant damage. Let's discuss how computers, permanent surveillance systems, and automated monitoring equipment can make it easier to keep track on plants. Learn how to use availability theory knowledge to observation management. Talk about the language necessary for irrigation and monitoring system, the availability of appropriate solutions, usage of both automated and conventional irrigation technologies, and any other factors that must be taken into account while monitoring plants.

*jmdipay@gmail.com

DOI: 10.1201/9781003490210-85

In addition to processing irrigation and monitoring data with the utmost integrity and avoiding often occurring agriculture problems. The author of the study "Plantsys: A SMS based automatic plant irrigation and monitoring system" offers a method since it could be challenging for farmers to regulate, irrigate, or monitor plants.

2. Methodology

In the present study, a comprehensive methodology was implemented to collect both qualitative and quantitative data. The qualitative component of the investigation encompassed conducting extensive interviews with diverse stakeholders, including agriculturists, gardeners, and farmers. These interviews were carried out confidentially, facilitating frank exchanges regarding the difficulties encountered with conventional plant monitoring systems and the prospective characteristics that could alleviate these concerns.

For the quantitative aspect, a structured questionnaire based on the International Organization for Standardization (ISO 9126) guidelines was used to assess user satisfaction with Plantsys: A SMS based automatic plant irrigation and monitoring system. The sample size for the survey was determined through a careful consideration of the target audience. We aimed to gather responses from a diverse group of users, including individuals with varying levels of expertise in agriculture and plant monitoring. The sample size was calculated to ensure statistical significance and a representative understanding of user satisfaction levels.

The selection of participants for the in-depth interviews and the survey was purposefully done, taking into account the knowledge and proficiency of the respondents. Through involving specialists in the respective field and individuals who utilize plant monitoring systems, our objective was to obtain a comprehensive outlook on the difficulties encountered and the suggested remedies.

3. Results and Discussion

The rapid advancements in science have significantly transformed various aspects of human life, offering simplified and enhanced solutions. In the context of plant irrigation and monitoring, the reliance on robust databases is paramount. Many developers in the field acknowledge the pivotal role databases play in establishing trustworthy plant irrigation and monitoring systems. This study's findings have been instrumental in innovating a novel method for efficient control and oversight of plant irrigation and monitoring systems.

The Difficulties Experienced while doing Conventional Observation on Plants

During the interviews conducted, three major challenges with traditional plant irrigation and monitoring methods emerged:

Improper Plant Monitoring

Conventional irrigation and monitoring systems often lead to inadequate plant monitoring, risking plant health and agricultural yields.

Duplication and Confusion

The use of traditional observation methods often results in duplication and confusion, hindering streamlined data collection and analysis processes.

Difficulty in Maintaining Records

Managing plant monitoring records manually proves to be arduous, leading to inaccuracies and inefficiencies in the monitoring process.

The Opinions of Respondents on How to Build Features that Would Help Solve the Issues that have been Experienced

In response to these challenges, respondents provided valuable insights into the features that could effectively address these issues:

Efficient SMS-Based System

The implementation of an PLANTSYS emerged as a solution that not only expedites the process of creating irrigation and monitoring setups but also simplifies complex tasks. Through instant communication, this technology ensures timely and accurate data exchange, enhancing the overall efficiency of plant monitoring.

User-Friendly Interface

PLANTSYS offers a user-friendly interface tailored to the needs of agriculturists, gardeners, and farmers. Compared to conventional methods, this approach significantly reduces complexity, making it more accessible and manageable for users from diverse backgrounds.

Respondents' Level of Acceptance Toward the Developed System

Table 85.1 On functionality

Functionality	Score
Send/Receive SMS	4.67
Water Dispense	4.57
Sensor Accuracy	4.59
Data Synchronization	4.65
Overall Score	**4.62**

The degree to which respondents found PLANTSYS to be functionally acceptable. Excellent levels of agreement are present. The respondents have demonstrated that they have a high level of approval for the system when it comes to the degree of acceptability in terms of the functionality of PLANTSYS. (Dhanaraju et al., 2022) have emphasized the importance of seamless SMS communication in agricultural automation systems, highlighting its role in enhancing user experience and system efficiency. The mean overall score of 4.62 was used to calculate this data.

Table 85.2 On reliability

Reliability	Score
System Uptime	4.55
Error Handling	4.50
Response Time	4.48
Fault Tolerance	4.57
Overall Score	**4.52**

The reliability findings, which in Table 85.2 show the respondents' level of satisfaction with the system's dependability, relate to Plantsys: A SMS based automatic plant irrigation and monitoring system. It is deemed to be a Very Acceptable level of agreement. The outcome reveals that Plantsys: A SMS based automatic plant irrigation and monitoring system, has a high level of approval among the respondents in terms of reliability. (Dhanaraju et al., 2022) also emphasized the reliability in automated systems is crucial for gaining user trust. The Reliable systems ensure consistent performance, minimizing errors and downtime, leading to increased user satisfaction and system credibility. The total average score, which was 4.52 points, was used to determine this data.

Table 85.3 On usability

Usability	Score
System Uptime	4.65
Error Handling	4.68
Response Time	4.71
Fault Tolerance	4.70
Overall Score	**4.68**

Usability data, which table 3 show the respondents' level of acceptance about the system's usability, refer to Plantsys: A SMS based automatic plant irrigation and monitoring system. It is thought that the level of agreement is Very Acceptable. The responses have shown that they have a high level of approval for the system when it comes to how PLANTSYS work. (Saiz-Rubio, 2020) came that intuitive interfaces and straightforward usability significantly enhance the adoption

of technology in agricultural settings. User-friendly systems lead to enhanced productivity and reduced learning curves for users. The mean overall score of 4.68 was used to calculate this data.

Table 85.4 On performance

Performance	Score
Plant Health Analysis	4.62
Water Efficiency	4.56
Crop Yield	4.58
Resource Optimization	4.61
Overall Score	**4.59**

Performance findings shows in table 4 that the respondents' level of satisfaction with the system's performance, refer to Plantsys: A SMS based automatic plant irrigation and monitoring system.

It is considered that the level of agreement is Very Acceptable. The following outcomes demonstrate the survey's findings: When it comes to the level of acceptability in terms of PLANTSYS's performance, the respondents had a high degree of approval for the system. (Levidow, Les, et al., 2014) has highlighted the importance of efficient irrigation systems in maximizing crop yield and minimizing water wastage. Performance-driven systems are crucial for ensuring optimal plant health and resource utilization. This data was taken from the 4.59 mean score obtained across all categories.

4. Summary

Table 85.5 Overall meaning score

Overall	Score
Functionality	4.62
Reliability	4.52
Usability	4.68
Performance	4.59
Overall Score	**4.61**

Table 85.5 show the total mean score of 4.61 is considered extremely satisfactory.

PLANTSYS has significantly contributed to the body of knowledge in the field of agricultural technology and plant monitoring. The research explored the challenges faced by traditional plant irrigation and monitoring methods and proposed innovative solutions through the development of PLANTSYS.

The study identified common issues in conventional plant observation, such as improper monitoring, duplication, confusion, and difficulties in record-keeping. Through in-depth interviews and user feedback, the research introduced two key solutions: the implementation of an efficient SMS-based system and the design of a user-friendly interface tailored to agriculturists, gardeners, and farmers.

The outcome of this research, revolutionizes the way plant irrigation and monitoring are conducted. By leveraging SMS technology, it enables real-time communication and data exchange, addressing the challenges of delayed information and inadequate monitoring. The system's user-friendly interface ensures accessibility for users with varying levels of expertise, promoting widespread adoption in agricultural practices.

This study's significance lies in its practical contributions to the agricultural sector. Plantsys enhances the efficiency of plant monitoring, minimizes errors associated with traditional methods, and maximizes agricultural productivity. By streamlining data collection and analysis, it empowers users to make informed decisions, optimizing water usage, and ensuring the overall health of plants.

References

1. K. Sirohi, A. Tanwar, Himanshu, and P. Jindal, "Automated irrigation and fire alert system based on hargreaves equation using weather forecast and ZigBee protocol," in 2016 2nd International Conference on Communication Control and Intelligent Systems (CCIS), 2016, pp. 13–17.

2. K. M. Vanthi, R. Kavipriya, D. Divyapriya, and M. Ambika, "Arduino Based Smart Irrigation System," Int. J. Adv. Res. Comput. Commun. Eng., vol. 7, no. 3, pp. 225–228, 2018.

3. A. Agrawal, V. Kamboj, R. Gupta, M. Pandey, V. Kumar Tayal, and H. P. Singh, "Microcontroller Based Irrigation System Solar Powered Using Moisture Sensing Technology," in 2018 8th International Conference on Cloud Computing, Data Science & Engineering (Confluence), 2018, pp. 324–327.

4. P. Srivastava, M. Bajaj, and A. S. Rana, "Overview of ESP8266 Wi- Fi module based Smart Irrigation System using IOT," in 2018 Fourth International Conference on Advances in Electrical, Electronics, Information, Communication and Bio-Informatics (AEEICB), 2018, pp. 1–5.

5. U. Mehta et al., "Designing of a mobile irrigation system," in 2015 2nd Asia-Pacific World Congress on Computer Science and Engineering (APWC on CSE), 2015, pp. 1–6.

6. K. K. Kishore, M. H. S. Kumar, and M. B. S. Murthy, "Automatic plant monitoring system," in 2017 International Conference on Trends in Electronics and Informatics (ICEI), 2017, pp. 744–748.

7. V. V. Kumar, R.Ramasamy, S.Janarthanan, and Communication, "Implementation of Iot in Smart Irrigation System Using Arduino Processor," Int. J. Civ. Eng. Technol., vol. 8, no. 10, pp. 1304–1314, 2017.

8. G. Shruthi, B. S. Kumari, R. P. Rani, and R. Preyadharan, "A-real time smart sprinkler irrigation control system," in 2017 IEEE International Conference on Electrical, Instrumentation and Communication Engineering (ICEICE), 2017, pp. 1–5.

9. B. Doraswamy, "Automatic Irrigation System using Arduino UNO," Ieee, vol. 08, no. 04, pp. 635–642, 2017.

10. S. K. Abdullah and R. F. Chisab, "Programing and Implementation of Wireless Monitoring Automatic Control System for Irrigation Greenhouse using ATMEGA328P-PU-AVR Microcontroller," J. Univ. Kerbala, vol. 15, no. 4, pp. 302–311, 2017.

Note: All the tables in this chapter were made by the authors.

Advancing Sustainable Science and Technology for a Resilient Future – Sai Kiran Oruganti et al. (eds)
© 2024 Taylor & Francis Group, London, ISBN 978-1-032-79020-6

Design, Fabrication, and Performance Evaluation of Ayungin (Leiopotherapon Plumbeus) Smoke Dryer

86

Marjun E. Caguay*

Mindoro State University, Oriental Mindoro, Philippines

Abstract: Ayungin is a highly perishable product that requires preservation in order to be stored for a longer time. Oriental Mindoro is considered as one of the largest producers of dried Ayungin in the Philippines however, the drying process still uses the traditional method of sun drying and smoke drying which affects the quality of the dried product. Thus, the objective of the study was to design, fabricate, and evaluate the performance of an Ayungin dryer as support to local processors. Results indicated that the 2.68 m³/min airflow rate was the most effective with 88% drying efficiency and moisture reduction rate of 4.49 kg/hr. with a significant 69.4% decrease in the fuel consumption compared to traditional smoke dryer. Cost analysis revealed that the dryer could generate an additional income for fisherfolk, amounting to 158,839.46 Php/year with 0.1 yr. payback period and 42.14% rate of return making the dryer a cost-efficient device.

Keywords: Ayungin, Airflow rate, Dryer, Biomass fuel

1. Introduction

The food sector plays an essential role in the countries' economy which continuously grows due to the ever-changing needs of consumers. In the modern era, the consumer's awareness of the benefits of consuming nutritious food, such as fish, has made the demand and making it as one of the highest consumable foodstuffs. (FAO,2020). The Philippines is rich in inland aquatic resources with 646,336 hectares of swamplands, existing fishponds, and other inland resources. It is one of the top fish producing countries in the world with its total production of 4.4 million metric tons of fish, crustaceans, mollusks, and aquatic plants in 2019 (Tancuco, 2021).

In 2020, Oriental Mindoro recorded a total of 251 metric tons of fish, contributing to about 62.7 percent of the MIMAROPA region total volume of production in the Inland Municipal subsector (PSA, 2020). According to the Bureau of Fisheries and Aquatic Resource (2021), one of the high valued fish produced in this province is the Ayungin or silver perch (Leiopotherapon plumbeus) with 186 metric tons volume of production in 2021 alone.

Ayungin is a highly perishable product that requires preservation in order to be stored for a longer time. One way of preserving Ayungin is drying, and is also being used by the fisherfolks of Naujan, Oriental Mindoro particularly the dried Ayungin producers in the area. The fisherfolks tend to sun-dry the Ayungin fish for about 2 -3 days in a good weather condition while in rainy season, the fisherfolks uses an improvised smoke dryer made from woods and scrap materials. These methods are effective way of drying an Ayungin fish however, there is a major concern of the Good Agricultural Practice (GAP) in the current process. In sun drying, due to the product is exposed to the environment, it is prone to food contamination and domestic animal attacks. In using the improvised dryer, due to the materials used in the dryer is not made from food grade materials results also to the product being prone to contamination. Additionally, due to varying heat source of the improvised dryer results in a change in the dried product's taste which affects the market quality and cost of the product. Hence, this study aims to design, fabricate and evaluate an Ayungin (Leiopotherapon plumbeus) smoke dryer to help the local processors/fisherfolks of dried Ayungin producers in the area through applying engineering knowledge in bringing safe

*abeminscat@gmail.com

DOI: 10.1201/9781003490210-86

and acceptable product to the consumers, while benefiting also the fisherfolks by increasing their income by adopting the developed device.

Generally, the study aimed to design, fabricate, and evaluate an Ayungin (Leiopotherapon plumbeus) smoke dryer. Specifically, the study aimed to:

1. Design and fabricate an Ayungin smoke dryer;
2. Evaluate the performance of the fish dryer in terms of: drying efficiency; drying capacity; drying time; moisture reduction rate; fuel consumption; electric consumption; and
3. Perform simple cost analysis of the dryer.

2. Materials and Methods

Conceptualization of the Study

Figure 86.1 shows the conceptual framework of the study following the input – process – output method. This presents the major activities done in the conduct of the study.

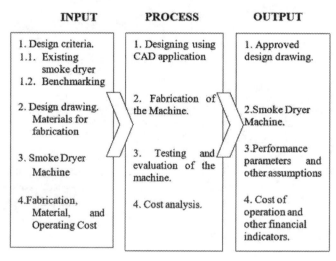

Fig. 86.1 Conceptual framework of the study

Design and Fabrication of the Dryer

The dryer was designed using Solidworks™ software application. An electricity blower was used for air circulation, and a coconut shell charcoal was used as heat source. Figure 2 shows the design of the dryer aims to remove a 67% moisture content of fresh Ayungin. The design is composed of six (6) major components namely: drying chamber, furnace, exhaust vent, frame, blower, and rubberized wheel.

Performance Test and Evaluation

Fifteen (15) kilograms of fresh Ayungin ranging from 5-10 cm in length was used in each trial for the evaluation of the dryer. The performance of the dryer was evaluated on its

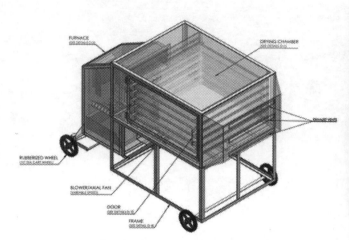

Fig. 86.2 Perspective view of the dryer

drying efficiency, drying capacity, drying time, moisture reduction rate, fuel consumption and electric consumption. Performance was calculated using the following equations:

1. Drying Efficiency

$$D_E = \frac{\text{heat utilized, kj/hr}}{\text{Heat supplied to the burner, kj/hr}} \quad (1)$$

2. Drying Capacity

$$D_c = \frac{W_i}{T_D} \quad (2)$$

Where: D_C = drying capacity, kg/hr
W_i = initial weight of test material, kg
T_D = actual drying time, hr

3. Moisture Content Reduction Rate

$$MCr = \frac{W_i - W_f}{T_D} \quad (3)$$

Where: MCr = moisture reduction rate, kg/hr
W_f = final weight content of the fish, kg
W_i = initial weight of test material, kg
T_D = actual drying time, hr

3. Results and Discussion

Description of the Dryer

The dryer aims to remove a 67% of the moisture content of Ayungin fish, and reduce the timeliness of operation while conforming with the drying standard particularly the Good Agricultural practice in drying products. Figure 86.3 shows the fabricated Ayungin while the Fig. 86.4 and Fig. 86.5 shows the fresh and dried Ayungin product. All parts of the dryer were fabricated using locally available materials and technology.

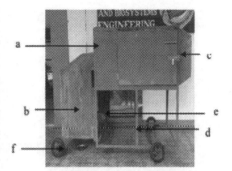

Fig. 86.3 Fabricated Ayungin smoke dryer

	PARTS
a	Drying Chamber
b	Furnace
c	Exhaust Vent
d	Frame
e	Blower
f	Wheel

Fig. 86.4 Fresh Ayungin drying arrangement

Fig. 86.5 Ayungin after 1 hour of continuous drying

Dynamic Simulation of Heated Air in the Dryer

Dynamic simulation was used in the study to predict the possible distribution of heated air in the drying chamber. Figure 6 shows the temperature present in all the parts of the drying chamber in a particular energy input. A five (5) kilogram of charcoal was fueled in the furnace in order to produce heat. This can produce 60 °C of heated air that pushes on the drying chamber using a 3 m³/min airflow rate. The figure illustrates the temperature, red being the highest and blue being the lowest.

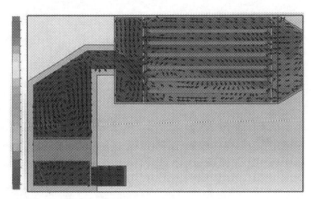

Fig. 86.6 Temperature profile and airflow pattern inside the dryer

Performance Evaluation of the Ayungin Fish Dryer

The dryer was evaluated in terms of the drying efficiency, drying capacity, drying time, moisture reduction rate, fuel consumption, and electric consumption as affected by different airflow rate of the fan/blower (1.34 m³/min, 2.01 m³/min, and 2.68 m³/min).

1. Drying Capacity

Refers on how many kilograms of Ayungin that the device can be dried in an hour. It was measured in kilogram per hour (kg/h). It was found out that the 2.01 m3/min air flow rate has the highest drying capacity mean with 6.50 kg/h followed by 2.68 m³/min and 1.34 m³/min with 6.44 kg/h and 6.16 kg/h, respectively.

2. Drying Efficiency

Refers on how efficient was the dryer in drying the Ayungin. Treatment 3 (T3) with airflow rate of 2.68 m3/min was found to be the most efficient with 88 % drying efficiency, followed by T2 (2.01 m3/min) and T1 (1.34 m3/min) with 84 % and 82 %, respectively.

3. Moisture Reduction Rate

The moisture reduction rate refers to the moisture content that has been removed in the Ayungin for every hour expressed in kg/h. T2 has the higher moisture reduction rate mean with 4.49 kg/h than T1 and T3 with 4.18 kg/hr and 4.48 kg/h, respectively. This means that T2 removed larger and faster amount of moisture content present in the initial weight of the Ayungin than those on T1 and T3.

4. Drying Time

The drying time refers to the total time spent in drying the Ayungin including the rearrangement, and collection of dried products. Treatment 3 recorded the fastest average drying time at 2.75 hrs while the T1 and T2 has an average operating time of 3.88 hours and 3.20 hours, respectively. The results of the analysis of variance indicate a significant distinction between treatments 1 and 2, as well as treatments 1 and 3. However, no significant difference was observed between treatments 2 and 3.

5. *Fuel Consumption*

The fuel consumption of the dryer refers to the heat source consumed of the dryer during the drying operation expressed in kg/hr. T2 (1.89 kg/h) has the highest fuel consumption followed by T3 (1.82 kg/h) and T1 (1.81 kg/h) respectively.

6. *Electric Consumption*

The electric consumption of the dryer refers to the electrical power consumed of the dryer during the drying operation expressed in kWh. Treatment 1 (T1) has the highest electric consumption with 0.03157 kWh followed by Treatment 2 (T2) and Treatment 3 (T3) with 0.02994 and 0.02905 kWh respectively.

Comparison of Conventional and Developed Dryer

Corresponding mean values on the comparison of the locally improvised dryer and developed Ayungin dryer are presented in Table 86.1. Results revealed that the conventional dryer and the developed dryer has a capacity of 1.5 kg/hr and 2.75 kg/hr respectively. Use of the developed dryer almost doubled the capacity of the locally improvised dryer. The increase in capacity would also increase or expedite the Ayungin drying production while reducing manual labor requirement. The table also shows the superiority of the developed dryer as compared to the improvised dryer.

Table 86.1 Comparison of locally improvised and newly developed dryer

Treatment	Capacity (kg/hr)	Drying Efficiency (%)
Ayungin dryer	2.75	88
Local dryer	1.5	34

Cost Analysis of Ayungin Dryer

Table 86.2 shows the summary of cost analysis in using the dryer and cost curve of using the machine was drawn in Fig. 86.7. The initial cost of the dryer was 31, 295 Php/year with an assumed life span of five years. A dryer capacity of 2.75 hours/batch and 5 kilograms of dried Ayungin was used in the calculation. The dryer salvage value was assumed to be a 10% of the initial investment of the dryer.

Table 86.2 Cost analysis of using the dryer

Particular		
Annual Fixed Cost	4,162.11	Php/yr
Depreciation Cost	3,293.1	Php/yr
Interest on Investment	503.1125	Php/yr
Tax and Insurance	365.9	Php/yr
Annual Variable Cost	372,806.426	Php/yr
Power Cost	288,691.68	Php/yr
Repair & Maintenance	914.75	Php/yr
Operator's Wage	83,200	Php/yr
Break-even Point	342.04	Kg/yr
Net Income	158,839.46	Php/yr
Rate of Return	42.14	%
Payback Period	0.10	yr

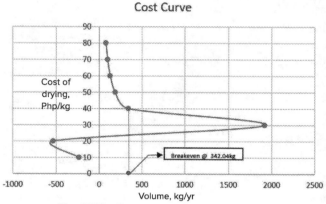

Fig. 86.7 Cost curve of using the dryer

4. Conclusion

Based on the test, observation and data gathered, the following conclusion are drawn:

1. The effectiveness of the developed dryer in reducing drying time, labor, and expenses for Ayungin has been demonstrated. Furthermore, the implementation of this machine offers the potential for additional income to processors and fisherfolks. It is worth noting that the dryer is constructed using locally available materials and local manufacturing technologies.

2. Based on various performance parameters such as drying efficiency, drying capacity, drying time, moisture reduction rate, fuel consumption, and electric consumption, the machine has demonstrated satisfactory performance. In fact, when comparing the developed Ayungin dryer to the locally improvised dryer, the T-test analysis revealed that the developed dryer exhibited twice the capacity and efficiency.

3. The machine is economical to use at a cost of 376,968.54 Php/yr operating performance and with an initial investment cost of 18,295.00 Php/yr. It would benefit the fisherfolks with a net income of 158,839.46 Php/yr. The dryer needs to dry a total of 342.04 Kg/yr

in order to breakeven and would have a rate of return of 42.14% and a payback period of 0.1 years or 27 working days.

5. Acknowledgement

First and foremost, the author wishes to express profound gratitude to the Almighty God, the ultimate provider of wisdom and strength.

The author extends heartfelt appreciation to Engr. Rosemarie Kate Mutya and Engr. Jermaine A. De Castro for their invaluable support in the successful completion of this research study.

The author also acknowledges and expresses gratitude to all individuals who, although not specifically mentioned, offered prayers, assistance, and support. The study would not have been possible without their presence and contributions.

A heartfelt thank you to everyone involved, and may all glory be attributed to God.

References

1. Bureau of Fisheries and Aquatic Resources (2021). Philippine Fisheries Profile 2015. Quezon: Bureau of fisheries and aquatic resources.
2. Food and Agriculture Organization of the United Nations. (2022). Calculation of Machine Rates. Retrieved from Fao. Org. https://www .fao.org/3/T0579E/t0579e05.html
3. Hunt, D. (2001). Farm Power and Machinery Management 10th Edition. Iowa State University Press, 2121 South State Avenue, Ames, Iowa 50014.
4. Philippine Agricultural Engineering Standard. 2010. Agricultural machinery – Fruit Dryer – Specifications. Department of Trade and Industry. PNS/PAES 248:2010.
5. Philippine Agricultural Engineering Standard. 2010. Agricultural machinery – Fruit Dryer – Method of Test. Department of Trade and Industry. PNS/PAES 249:2010.
6. Philippine Statistics Authority (2020). Fisheries Situation Report, 2017 January-December. Quezon: PSA.
7. Tancuco, V. (2021). In Numbers: The Philippines' fisheries sector. Rappler. https://www.rappler.com/environment/philippines-fisheries-sector-statistic

Note: All the figures and tables in this chapter were made by the authors.

Advancing Sustainable Science and Technology for a Resilient Future – Sai Kiran Oruganti et al. (eds)
© 2024 Taylor & Francis Group, London, ISBN 978-1-032-79020-6

87

Lived Experiences of Parents with Child Diagnosed with Retinoblastoma

John Carlo A. Abaniel*
President Ramon Magsaysay State University, Iba, Zambales, Philippines

Abstract: This study identified the stressors and coping mechanisms of parents having a child with Retinoblastoma using phenomenological approach. Participants were three parents with child diagnosed with retinoblastoma, two mothers and one father aged 30-45 years old from National Capital region. Findings of the study revealed that common stressors faced by parents of an Rb child medical, economical and psychological stressors. These include child's treatment or medical procedures, financial strain, acceptance that a child has cancer, and child's health conditions. Parents were also found to be resilient of the situation they went through because of the following: a) social, b) spiritual and c) emotional coping mechanism. Based on the findings, the following recommendations were made: family and friends of parents of Rb children may provide emotional, psychological and even financial support to the affected family. Social groups such as parents who face the same situation, charitable institutions, medical staff may organize groups which can support the affected families.

Keywords: Cancer, Rb children, Retinoblastoma

1. Introduction

Retinoblastoma (Rb) is the most prevailing malignancy of the eye in infancy and childhood accounting to 3 percent of all pediatric cancers. According to the International Journal of Ophthalmology research done in 2018, it is estimated that 1 per 15 – 20,000 live births which corresponds to around 9,000 new cases every year worldwide. According to American Cancer Society, a staging is used to give the doctors an outlook for saving he child's vision. To determine the best option, doctors divide retinoblastoma into 2 main groups; Intraocular and extraocular retinoblastoma. Intraocular Rb means that the cancer is still within the eye. Extraocular Rb on the other hand, means that the cancer cells are found outside the eye. It may be found in the eye socket, or spread to other parts of the body such as brain or bone marrow. Rb may affect one or both eyes, or even affect other organs at the same time, such as the brain. One of the most common tumor of the eye in the Philippines is known as retinoblastoma (Domingo, 2015). According to the records of Philippine General Hospital (PGH), most of the Rb affected eyes were enucleated at the advanced stage of the disease. Retinoblastoma (Rb) is considered as a rare genetic

eye cancer which develops in late-stage fetal development or early infancy. Most patients are diagnosed before 3 years of age (Cross, 2010). Since retinoblastoma is a rapidly growing tumor, immediate treatment at the onset of symptoms is very important. (Domingo, 2018) If untreated, mortality rate is 99% (Espiritu,et al., 2004). A three series study was done by the University of the Philippines-Philippines General Hospital (UP-PGH) and found an increasing number of new cases from 40 per 100, 000 to 237 per 100,000 from years 1967-2001. Because Rb is highly malignant and invasive, if left untreated the mortality rate is 99 percent. Thus, the main goal of physicians is patient survival. Keeping the eye globe and preserving the vision comes later. Families having children diagnosed with cancer consider the journey as sequence of rough times of distress and suffering (Woodgate and Degner, 2003). It is expected of parents to take care of their child Parents agreed that they can identify their child's pain better than medical professional or staff. Holm, et al. (2003) found out that parents participated in their children's medical care through advocacy. They spoke for their children to let medical persons know what they needed. These efforts of parents were found in diagnosis phase and treatment phase. Parents of children diagnosed with cancer faces challenges

*abanieljohncarlo@gmail.com

DOI: 10.1201/9781003490210-87

at the diagnostic stage, treatment and even after treatment stage. Acccording to McGrath (2001), shock, disbelief, anxiety and sadness were the common reactions during the diagnosis of cancer. Although there are literatures discussing the challenges faced by parents of children having diagnosed with cancer, few literatures were found on analyzing the challenges faced by parents of children having diagnosed with Retinoblastoma. This prompted the researcher to conduct the study to explore the lived experience of parents with child diagnosed with retinoblastoma.

The study aimed to explore the lived experiences of parents of children diagnosed with Retinoblastoma. The study specifically seeks to answer the following questions:

1. What are the common stressors experienced by the parents of an Rb child?
2. What are the coping strategies of parents to deal with experience of having a child diagnosed with Retinoblastoma?

2. Research Methodology

The study employed phenomenological research design. This research design was employed to profoundly describe how parents with Rb child found their experiences. Through phenomenology, their stressors and coping mechanisms were identified. Qualitative approaches used in the study were open-ended questionnaire and individual interviews. The participants of the study were three parents with child diagnosed with retinoblastoma, two mothers and one father aged 30 to 45 years old from National Capital Region (NCR). Children were diagnosed with retinoblastoma before the age of two and had undergone enucleation surgery . The participants were parents who have been taking care of their child since diagnosed up to present. The guide questions were formulated in congruence with Patterson, Holm, Patterson and Gurney's (2003) research entitled " Parental Involvement and Family-centered care in the diagnostic and treatment phases of childhood cancer: Results from a Qualitative study. The researcher asked permission from the respondents to conduct the interview and they were assured that their personal information will be held confidential. Interviews were recorded and transcribed. For the qualitative data analysis, the interview transcript was analyzed by using content analysis. Anonymity of the participants was also ensured in this study. Participants were also aware of the implications of their participation in this research.

3. Results and Discussions

Answers to the open ended questionnaires and interview results were encoded. Manual coding was applied in the qualitative data analyses. There are different stressors faced

by parents of an Rb child, namely Medical, economic and psychological stressors. Parents had mixed emotions when they learned about their child's condition.They were stressed out with so many concerns like what treatment should the child undergo, how to finance such treatment, will their child be saved, how will the child cope with the situation as he or she undergoes treatment or as he or she grows up. Not only the parents were bothered by this situation, it is also extended to other family members such as grandparents and siblings. As according to Woodgate and Degner (2003), families having children diagnosed with cancer consider the journey as sequence of rough times of distress and suffering.

Theme 1: Medical stressor

The medical stressors are the mode of treatment or medical procedure and child's health. The child's treatment is the emerging concern of an Rb parent since diagnosis. Parents are concerned that a more invasive treatment will be required because a more invasive treatment such as chemotherapy could lead to side effects such as weak immune system, hair loss, loss of appetite and could seriously harm their child's health and recovery. One participant said:

Worries that my child has to undergo a more invasive treatment.

Parents are also concerned about their child's health: nutritional requirement, worried that the child will have infection after the treatment. Participants mentioned when asked about their concerns:

Health of our child, because of the treatment, she is often sick because her immune system is getting weaker.

We are always worried that the would might get infected.

Parents are concerned of the side effects of the surgery and the treatment. For parents who have their children undergo chemotherapy, they were deeply concerned of the nutrition and immune system of their children as they were adversely affected by the treatment. As according to Yiu and Twinn (2001), during the diagnosis parents face threats if the side effects of chemotherapy, invasive medical procedures and fear of relapse.

Theme 2 : Economic stressor

Economic stressor is one of the major considerations of parents. The process of treating the child exhausts not only physically or mentally but also financially. Parents would seek for the best treatment for their child survival but they had to face the burden of paying a large amount of money. The following are some of the participants' responses:

We had experienced so many problems, first is our finances, we do not know where to get money for medication.

We were thinking, if we are financially-abled for the treatment.

Theme 3: Psychological stressor

Acceptance would be the psychological stressor of parents, but upon knowing the child's condition, they had to immediately accept it and do a follow through to the next stage which is the treatment of the child. According to Bisht et al. (2019) psychological stress symptoms are evident in family caregivers of cancer patients. Upon the initial diagnosis, they are stressed due to having to cope with burdens such as physical, social, emotional and financial. One participant said:

Firstly is acceptance. It was very difficult for us to accept that my son is having cancer.

Four themes emerged under the coping mechanism. The coping mechanisms identified are social, emotional and spiritual coping mechanisms.

Theme 1: Social coping mechanism

Social coping mechanism emerged from sub-themes: family and friends and social groups. Because of immense stress faced by parents during diagnosis, treatment and recovery, they needed shoulders to lean on. They have found the much needed support from parents, families, friends and even social groups. Family and friends provided emotional support through words of encouragement and prayers. The following is a statement of one of the participants:

Knowing that these people are behind us in trying times, family, relatives and friends helped cope from the problems they face during diagnosis and treatment.

Social groups such as parents of Rb survivors, charity institutions and healthcare providers helped parents coped with the situation.

Hearing from other parents somehow relieves their anxieties and gives them assurance that their child will survive and thrive. Meeting other families with a child with cancer can make families feel better (Zomerlei,2015). Parents also found connection with other parents with Rb child across the world through the internet. Parents recognized that experienced families might offer information and emotional support to new Rb families of similar disease profile and treatment (Downie, 2017). Beddard, et al. (2019) also found out that support from stranger and members of support group who also had the same experience, were seen as valuable in supporting families. One of the participants' remarks on how they coped with the situation were as follows:

Parents of children facing the same situation and institutions which give support to children experiencing the same situation.

Parents also found social support from charity institutions. There were charity institutions which offer help for families having child diagnosed with cancer. The programs conducted by charity institutions cheer their children up despite of having to undergo surgery and treatment. Trust in healthcare providers also help families cope with the situation. One participant emphasized:

Trust in ophthalmologist- oncologist of my child.

Having a complete trust on their doctors help parents cope with their apprehensions about maintaining child's health after surgery and treatment. According to Downie (2017), Rb parents benefited from structured support provided to them from the time of diagnosis and integrated with medical treatment and follow up.

Theme 2: Emotional coping mechanism

Another coping mechanism identified is emotional coping strategy. A parental instinct of being strong for the sick child has made parents continue fighting, no matter what the hurdles may be for the health of their children. Seeing their child play, do daily normal activities as if nothing happens after surgery eases parents' pain. This serves as an assurance that their child will be alright. One participant said when asked about how he she coped:

Knowing that my son has coped faster than me, and they will as long as they know that they always got your back.

Theme 3: Spiritual coping mechanism

Spiritual coping mechanism was also identified in this research derived from sub category Faith. This theme emerged from axial codes trust in God and prayers. Seeking for any form of comfort and security from God, helps to ease their sufferings and anxieties. Prayers helped parents know that somebody else is in control over the situation. In the study of Pishkuni, et al. (2018) parents feel better and less disappointed when they pray and feel spiritual support. Strong religious beliefs were one of the most influential factors in dealing with pediatric cancer. One of the parents' remarks is as follows:

Prayers really helped in making us trust God more and to be strong and optimistic,

Another parent added:

Just do everything you can and leave the rest to Him. When I pray I tell Him this so you have given me challenges, please give me strength to overcome it.

4. Conclusion and Recommendations

Parents indeed face adversities during the diagnosis, surgery and treatment of retinoblastoma. Common stressors identified in this study are (a) medical procedures or treatment, (b) financial strain, (c) acceptance that a child has cancer, and (d) child's health conditions. These stressors affect parents psychologically, physically, emotionally and economically. Resiliency of parents who face this situation came from

(a) family and friends, (b) social groups, (c) resilient child and (d) faith. Family and friends have played vital roles in coping from psychological and financial stress parents of Rb children have suffered from Social groups such as parents with children who had retinoblastoma, charity institutions and medical staffs also have helped parents cope with the situation. Emotional scaffold may also be provided by a resilient child. Faith and prayers helped parents know that somebody else is in control over the situation. Based on the findings, the following recommendations were made: family and friends of parents of Rb children may provide emotional, psychological and even financial support to the affected family. Social groups such as parents who face the same situation, charitable institutions, medical staff may organize groups which can support the affected families. A study aligned to this may be conducted involving more participants to capture other concerns of parents of Rb children. Roles of medical staffs in helping parents cope with the situation may also be investigated. More participants may be engaged in another phenomenological study on the lived experience of parents having children diagnosed with Retinoblastoma.

References:

1. Bisht. S., Chawla B., Tolahunase, M., Mishra, R., Dada, R. (2019). Impact of yoga based lifestyle intervention on psychological stress and quality of life in the parents of children with retinoblastoma. Annals of Neurosciences, 26(2).

2. Beddard, N., McGeechan, G.J., Taylor, J., Swainston, K. (2019). Childhood eye cancer from a parental perspective: lived experience of parents with children who have retinoblastoma. European Journal of Cancer Care. Doi: 10.1111/ecc.13209

3. Domingo, R.E., Manganip, L.E., Castro, R.M. (2015). Tumors of the eye and ocular adnexa at the Philippine Eye Research Insitute: A 10-year review. Clin Ophthalmol, 9, 1239–1247.

4. Downie, R. (2017). Dads and Dyads: stress and coping when a child has retinoblastoma . (unpublished thesis). University of Western Ontario, Ontario, Canada.

5. Holm, K.E., Patterson, J.M., and Gurney, J.G. (2003). Parental involvement and family-centered care in the diagnostic and treatment phases of childhood cancer: results from a qualitative study. J Pediatr Oncol Nurs, 20 (6), 301–13. Doi: 10.1177/1043454203254984

6. McGrath, P., Paton, M.A.,and Huff, N. (2004). Beginning treatment for pediatric acute myeloid leukaemia; diagnosis and the early hospital experience. Scand J Caring Sci, 18 (4), 358–367.

7. Woodgate, R.L. and Degner, L.F. (2003a) A substantive theory of keeping the spirit alive: the Spirit within Children with cancer and their families. J Pediatric Oncol Nurs, 20 (3), 103–119.

8. Yiu, J.M. and Twinn, S. (2001). Determining the needs of Chinese parents during the hospitalization of their child diagnosed with cancer: an exploratory study, Cancer Nurs, 24 (6), 483–489.

Printed in the United States
by Baker & Taylor Publisher Services